MXenes for Energy Storage Applications

MXenes for Energy Storage Applications focuses on two-dimensional transition metal carbides and nitrides (MXenes), covering the structure, precursors, types of compositions, and unique properties of MXenes. It highlights various promising applications of MXenes-based materials across different energy fields, such as catalysis and energy conversion, including hydrogen evolution reaction, oxygen evolution reaction, oxygen reduction reaction, supercapacitors, batteries, and so forth.

This book provides advanced solutions for all the issues and challenges related to energy storage and conversion:

- It covers a large range of MXenes from their basics to the advancements and their energy applications.
- It provides a detailed description of the structures, precursors, types of compositions, and unique properties of MXenes.
- It discusses synthesis techniques of MXenes along with their pros and cons.
- It summarizes the promising energy applications, including catalysis and electrochemical reactions for hydrogen production to batteries.
- It briefly explains recent progress, challenges, and practical limitations with proposed solutions.

This book is aimed at researchers and graduate students in materials science, chemical engineering, and nanomaterials.

Muhammad Rafique received his Ph.D from the University of Engineering and Technology, Lahore, Pakistan. Currently, he is working as Assistant Professor in Department of Physics, University of Sahiwal, Sahiwal.

Muhammad Bilal Tahir received his Ph.D from the University of Gujrat, Gujrat, Pakistan. Currently, he is working as Head of Department and Associate Professor in the Department of Physics, Khwaja Fareed University of Engineering and Information Technology, Rahim Yar Khan.

Saira Anwar received her Master of Philosophy (M. Phil) in Physics from the University of Sahiwal, Sahiwal, and works as Research Scholar in the Department of Physics, University of Sahiwal.

Emerging Materials and Technologies
Series Editor: Boris I. Kharissov

The *Emerging Materials and Technologies* series is devoted to highlighting publications centered on emerging advanced materials and novel technologies. Attention is paid to those newly discovered or applied materials with potential to solve pressing societal problems and improve quality of life, corresponding to environmental protection, medicine, communications, energy, transportation, advanced manufacturing, and related areas.

The series takes into account that, under present strong demands for energy, material, and cost savings, as well as heavy contamination problems and worldwide pandemic conditions, the area of emerging materials and related scalable technologies is a highly interdisciplinary field, with the need for researchers, professionals, and academics across the spectrum of engineering and technological disciplines. The main objective of this book series is to attract more attention to these materials and technologies and invite conversation among the international R&D community.

Advancements in Nanomaterials for Energy Conversion and Storage
Edited by Piyush Kumar Sonkar and Vellaichamy Ganesan

2D Materials-Based Sensors
Technology and Applications
Vinod Kumar Khanna

Metal Organic Framework Derived Materials
Design Strategies and Applications
Gomathi Nageswaran, Varsha M V, Arun Kumar Rajasekaran and M Shashank Rao

Hydrogen Production, Storage, and Utilization
Technologies and Applications
Abbas Tcharkhtchi, Hamidreza Vanaei, Albert Lucas and Sedigheh Farzaneh

MXenes for Energy Storage Applications
Emerging Characteristics, Compositions, and Synthesis Methods
Muhammad Rafique, M. Bilal Tahir, and Saira Anwar

For more information about this series, please visit: *www.routledge.com/Emerging-Materials-and-Technologies/book-series/CRCEMT*

MXenes for Energy Storage Applications

Emerging Characteristics, Compositions, and Synthesis Methods

Muhammad Rafique, M. Bilal Tahir, and Saira Anwar

CRC Press
Taylor & Francis Group
Boca Raton London New York

CRC Press is an imprint of the
Taylor & Francis Group, an **informa** business

Designed cover image: Getty Images

First edition published 2025
by CRC Press
2385 NW Executive Center Drive, Suite 320, Boca Raton FL 33431

and by CRC Press
4 Park Square, Milton Park, Abingdon, Oxon, OX14 4RN

CRC Press is an imprint of Taylor & Francis Group, LLC

© 2025 Muhammad Rafique, M. Bilal Tahir, and Saira Anwar

Reasonable efforts have been made to publish reliable data and information, but the author and publisher cannot assume responsibility for the validity of all materials or the consequences of their use. The authors and publishers have attempted to trace the copyright holders of all material reproduced in this publication and apologize to copyright holders if permission to publish in this form has not been obtained. If any copyright material has not been acknowledged, please write and let us know so we may rectify in any future reprint.

Except as permitted under U.S. Copyright Law, no part of this book may be reprinted, reproduced, transmitted, or utilized in any form by any electronic, mechanical, or other means, now known or hereafter invented, including photocopying, microfilming, and recording, or in any information storage or retrieval system, without written permission from the publishers.

For permission to photocopy or use material electronically from this work, access www.copyright.com or contact the Copyright Clearance Center, Inc. (CCC), 222 Rosewood Drive, Danvers, MA 01923, 978-750-8400. For works that are not available on CCC, please contact mpkbookspermissions@tandf.co.uk

Trademark notice: Product or corporate names may be trademarks or registered trademarks and are used only for identification and explanation without intent to infringe.

ISBN: 978-1-032-73756-0 (hbk)
ISBN: 978-1-032-73759-1 (pbk)
ISBN: 978-1-003-46576-8 (ebk)

DOI: 10.1201/9781003465768

Typeset in Times
by Apex CoVantage, LLC

"In the honour of my loving, late father Abdul Sattar, whose wisdom continues to inspire and guide me every day"

Muhammad Rafique

Contents

Authors

Muhammad Rafique received his Ph.D. from the University of Engineering and Technology, Lahore, Pakistan. Currently, he is working as an assistant professor in Department of Physics, University of Sahiwal, Sahiwal. His research interests include condensed matter physics, laser matter interactions, energy storage materials, energy storage and conversion devices (supercapacitors, fuel cells, solar cells, etc.) photocatalysis, water splitting and water purification, etc. On the basis of his contributions, he has been declared a 2% scientist of the world.

M. Bilal Tahir received his Ph. D from University of Gujrat, Gujrat, Pakistan. Currently, he is working as Head of Department and Associate Professor in Department of Physics, Khwaja Fareed University of Engineering and Information Technology, Rahim Yar Khan. He has published many articles in international reputed and impact factor journals. He also has contributed many chapters in various books and authored international books. His research interests include photo-catalysis, water splitting and hydrogen production, dye degradation, industrial wastewater treatments, energy storage and conversion devices, nano-composite as energy storage, photocatalytic materials, etc.

Saira Anwar received her Master of Philosophy (M. Phil) in Physics from University of Sahiwal, Sahiwal and working as Research Scholar in Department of Physics, University of Sahiwal. Her research interests are 2D materials, energy storage and conversion devices, supercapacitors, nano-composite as energy storage materials, etc.

Preface

The idea of writing a book on two-dimensional (2D) transition metal carbides and nitrides (MXenes) came about due to the significant impact and rapid advancement produced by this family of 2D nanomaterials in the material sciences. This book provides detailed knowledge of the structure, precursors, compositions, and unique properties of MXenes. It also describes numerous top-down and bottom-up synthesis techniques with their pros and cons. Besides, this book highlights various promising applications of MXenes-based materials across different energy fields such as catalysis and energy conversion including hydrogen evolution reaction (HER); oxygen evolution reaction (OER) and oxygen reduction reaction (ORR); supercapacitor electrode materials for symmetric, asymmetric, and hybrid supercapacitors; and battery electrodes for metal-ion, metal–air, and hybrid-ion batteries. Furthermore, the recent progress and practical limitations are also briefed. This book provides advanced solutions for all the issues and challenges related to energy storage and conversion on a global level and aims to accomplish sustainable development goals (SDGs) for developing countries. Therefore, this book would be a valuable reference for students, researchers, professionals, and policy-makers in order to provide reliable and sustainable materials for energy applications. There are 11 chapters in total. Chapter 1 briefly summarizes the energy storage and conversion applications of 2D nanomaterials and presents a detailed discussion of discovery, structural design, and recent developments in the field of MXenes. Chapter 2 deals with various factors that influence the characteristics of MXenes, such as geometry and energies, variable valency and tunable interlayer spacing of MXenes, effects of the mass of metals, dopants and intercalants, as well as their stability, and surface chemistries. Chapter 3 focuses on a detailed discussion of both theoretically predicted and experimentally synthesized MAX and non-MAX phase precursors of MXenes and surface terminations, with recent researches and advancements. Furthermore, a number of theoretical calculations of carbide and nitride MXenes are also summarized to study the behavior and stability of various structures. Chapter 4 is devoted to the classification of different types of structures and compositions of MXenes. Chapter 5 outlines the variety of approaches, both theoretical and experimental, that can strategically and successfully introduce dopants to modify the structure of MXenes. Examples of the preparation of doped MXenes (i.e., metals, halides, nitrogen, and sulfur) into M, X, and T positions of MXenes, their accomplishments, and their importance in developing new materials are also presented. The experimental investigations, both in situ and ex situ techniques, are discussed to produce the elemental doping or substitution of MXenes. Furthermore, the energy storage and conversion applications of hetero-MXenes are also presented, along with the latest research and advancements in this field. Chapter 6 discusses the two primary approaches (i.e., top-down and bottom-up) that are being used by researchers to synthesize 2D MXenes. The top-down strategies include non-hydrofluoric acid/fluoride-free etching, electrochemical etching, molten salt etching, alkali etching, and topochemical synthesis. Additionally, a wide range of bottom-up approaches include chemical/physical vapor deposition, hydrothermal/solvothermal synthesis, spray coating, dip coating, drop casting, plasma-enhanced pulsed-laser deposition, hot press techniques, vacuum-assisted filtration, and solution processing techniques. Furthermore, strategies to synthesize nitride MXenes are also summarized along with the recent research and advancements. Chapter 7 focuses on the characteristics and unique features of MXenes—that is, structural and magnetic properties; flexibility and mechanical strength; electrical and electronic conductivities; chemical and thermal stabilities; optical, optoelectronic, and plasmonic properties; hydrophilicity and hydrogen storage feature; nonlinear optics and optical detection; and photothermal effects and ultrafast dynamics. Chapter 8 deals with the catalytic activity of a wide range of pristine and doped MXenes for energy conversion applications. These applications include carbon dioxide (CO_2) hydrogenation, synthetic reactions, CO_2 reduction, nitrogen reduction reaction (NRR), hydrogen evolution reaction (HER), oxygen reduction reaction (ORR), and oxygen evolution reaction (OER). Chapter 9 presents a detailed discussion of the electrochemical properties and energy storage mechanisms of MXenes for different types of supercapacitors, that is, symmetric, asymmetric, hybrid, and micro supercapacitors. MXenes combined with other electro-active elements provide better electrochemical performance. Therefore, the recent advancements in developing the composites of MXenes with carbon, metal oxides, and conductive polymer, and their corresponding applications as supercapacitor electrodes, are also summarized. Non-lithium energy storage techniques are gaining popularity owing to their high energy densities and low cost. Therefore, Chapter 10 outlines the lithium and non-lithium rechargeable energy storage devices made of MXene-based composites, including sodium-ion, aluminum-ion, zinc-ion, potassium-ion, magnesium-ion, calcium-ion, and metal–air batteries. Despite the latest researches and advancements in this field, there are many important issues that the MXene community has not addressed yet. Therefore, Chapter 11 highlights the challenges faced by researchers and the future perspectives that should be taken into account in order to deal with these issues. Individuals who require further in-depth coverage should refer the research papers, reviews, or books cited in the list of references.

Acknowledgments

The authors would like to express their deep and sincere gratitude to Prof. Dr. Muhammad Shahid Rafique, Dean, Faculty of Natural Sciences, Humanities and Islamic Studies, University of Engineering and Technology, Lahore, for his immense knowledge, invaluable guidance, and important suggestions. We would like to extend our thanks to Dr. Ghulam Nabi, Associate Professor, Department of Physics, University of Gujrat, and Dr. Muhammad Shakil, Associate Professor, Institute of Physics, the Islamia University of Bahawalpur, for their continued support and encouragement. We also thank all the team members of CRC Press for their guidelines, feedbacks, and proofreading, which have helped shape the current edition of this book. Last but not least, we are extremely grateful to our families and friends for their unconditional love and never-ending prayers.

Dr. Muhammad Rafique
Assistant Professor
Department of Physics
University of Sahiwal, Sahiwal, Pakistan

Dr. M. Bilal Tahir
Associate Professor
Director, Institute of Physics
KFUEIT, Rahim Yar Khan, Pakistan

Saira Anwar
Research Scholar
Department of Physics
University of Sahiwal, Sahiwal, Pakistan

1 Introduction and Fundamentals

MXenes (pronounced as "max-eens") belong to the family of two-dimensional (2D) nanomaterials. These materials have a layered structure made from transition metal carbides, nitrides, and carbonitrides. MXenes can be synthesized by exfoliating their bulk crystal, called MAX or non-MAX phases, that first came to light in 2011. They exhibit high volumetric capacitance and excellent conductivity, unlike most 2D materials. MXenes can potentially be composed of millions of combinations of transition metals like titanium or molybdenum and nonmetallic elements like carbon or nitrogen. The main challenge is to identify the stable ones among them. This is what makes this class of materials so fascinating. Researchers have estimated that there are still over one million stable MXene compounds that can be discovered through a high-throughput computational platform, which scans through the formation and adsorption energies of alloying configurations.

The unique composition and structure of MXenes make them versatile materials. Some MXenes can split water with low overpotentials, which makes them ideal for electrocatalytic water splitting. On the other hand, some MXenes have a wide range of electrochemical stability, making them perfect for supercapacitor electrodes. This wide range of properties makes MXenes an attractive choice for various applications. To sum up, MXenes are useful materials suitable for various applications, including energy storage, transparent electronics, sensors, electromagnetic interference (EMI) shielding, and catalysis. These applications involve the use of MXenes as thin and flexible films, embedded in matrix materials, anchored in hybrid materials, and even as wearable fabrics. A thorough understanding of the structural stabilities and tribological properties of MXenes is essential for all of these applications [1–7].

MXenes and MXene-based composites have primarily been researched for their use in electrochemical energy storage. Their exceptional electrical and electronic conductivity makes them an attractive option for batteries and supercapacitors. As per theoretical predictions, MXenes are more efficient than other 2D materials in altering the desired chemical and structural properties. This makes them promising anode materials for secondary batteries. In particular, lighter MXenes (i.e., n = 1) such as Ti_2C, V_2C, Nb_2C, and Sc_2C are expected to have higher theoretical gravimetric capacities than heavier ones (i.e., n >1) like M_3X_2 or M_4X_3. In comparison to conventional electrode materials, bare MXenes exhibit substantially reduced specific gravimetric capacity due to their large molecular mass. These materials also exhibit certain drawbacks such as restacking, simple oxidation, and poor electrical conductivity. To address these issues, researchers have explored the possibility of creating composites by integrating MXenes with other materials such as stable carbon compounds. The goal is to enhance the performance of MXenes. So far, successful synthesis of MXene-based composites has been achieved by using metals, metal

DOI: 10.1201/9781003465768-1

sulfides, metal oxides, mesoporous carbon, graphene, high-quality graphene, carbon nanotubes, carbon fibers, and poly-pyrrole, among others. This enhances their capability to be used as an electrode material in batteries [8–10].

Besides their use in batteries, MXenes are also efficient electrode materials for supercapacitors. There are certain factors that influence the energy storage mechanisms of MXene-based supercapacitors, such as morphology of the electrode material and choice of an electrolyte. For an aqueous electrolyte, metal ions interact with the water molecules, thereby creating an electric double layer on the surface of the electrode. However, in non-aqueous electrolytes, large potential differences on the surface can gradually disrupt the double-layer structure. The capacitance on the MXene surface is mostly generated by the chemical reactions. Although MXenes have a layered structure, they experience ion redeposition during charging/discharging cycles, leading to constrained energy density that limits their potential. To overcome this problem, another electrochemically active material is often combined with the MXene to create a composite material [9]. MXenes are also ideal co-catalysts as they possess fast charge transfer, hydrophilicity, strong metal conductivity, surface functionalization, large surface area, exceptional aqueous solution stability, etc. MXenes are often considered to be superior catalysts for various energy conversion processes, including the oxygen evolution reaction (OER), hydrogen evolution reaction (HER), oxidation of carbon (CO), and reduction of carbon dioxide (CO_2R), among others. They can play the role of the primary catalyst, co-catalyst, or supporting substrate in different processes [11, 12].

1.1 2D NANOMATERIALS AND THEIR ENERGY APPLICATIONS

"Two-dimensional (2D) nanomaterials" refers to a novel family that exhibits sheet-like structures with lateral sizes exceeding 100 nm, or sometimes up to a few micrometers. Such materials are often only a few atoms thick and are composed of either a single or a few atomic layers. The discovery of graphene by exfoliation of the graphite in 2004, using Scotch tape, marked a revival of ultrathin 2D nanomaterials [13]. 2D nanomaterials research is advancing rapidly due to the abundance of their distinctive features. Confining reaction electrons in a 2D structure, particularly on single-layer nanosheets, allows for precise electron movement without interlayer interactions. They possess a large surface area because of their horizontal dimension, making them ideal for supercapacitors, rechargeable batteries, and catalysis. With numerous surface atoms, it is easy to create more active sites on the surface, thereby improving a material's inherent characteristics. Being only one atom thick, they stand out as the top choice for developing a new generation of electrical devices. These materials exhibit high carrier mobility and strong thermal conductivity due to their 2D structure [14–18].

Due to their numerous benefits, 2D materials have become a popular area of research for various applications. This group includes hexagonal boron nitride (h-BN), transition metal dichalcogenides (TMDs), layered metal oxides (MOs), covalent-organic frameworks (COFs), black phosphorus (BP), layered double hydroxides (LDHs), graphitic carbon nitride (g-C_3N_4), and metal–organic framework nanosheets (MOFs) (as illustrated in Fig. 1.1 [15]). Recently, many novel members like hybrid

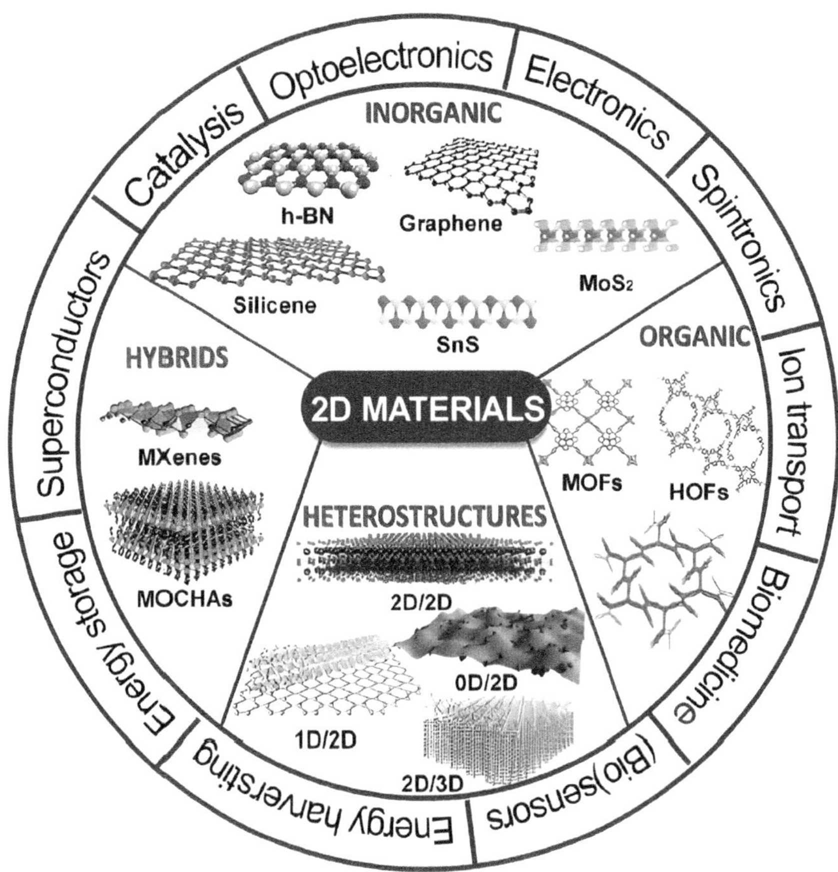

FIGURE 1.1 Classification of 2D materials and their applications.

Source: Reprinted from [15] Scarano D. and Cesano F.: Graphene and Other 2D Layered Nanomaterials and Hybrid Structures: Synthesis, Properties and Applications. *Materials* 2021, 14(23), 7108. Open access.

perovskites, silicates, and MXenes that combine organic and inorganic elements have also been synthesized [13, 16–18]. For a complete understanding of their properties and functionalities, it is essential to measure their size, composition, defects, thickness, vacancies, crystal phase, doping, strains, and electronic states accurately. All of these exceptional structural stabilities and characteristics make ultrathin 2D nanomaterials one of the most popular research areas in chemistry, material science, condensed matter physics, and nanotechnology [13–15]. The electrochemical energy storage and conversion properties and applications of some of these 2D nanomaterials are summarized in this section.

Graphene, an allotrope of carbon, is obtained by the exfoliation of graphite. Each atom in the network forms covalent bonds with three neighbors through pi-bonds [13]. A hexagonal structure is the building block of monolayer h-BN, composed of an equal number of nitrogen and boron atoms. h-BNs are known as white graphene due to their similarity with traditional graphene. But in contrast to graphite, this material is thermally conducting and electrically insulating. Therefore, at high temperatures, in a vacuum or an oxidizing atmosphere, it displays better lubricating properties [13, 19, 20]. Another analog of graphite is g-C_3N_4, which has a layered structure with van der Waals (vdW) interactions. Its crystal structure can be viewed as a N-doped graphite structure, as it is formed through sp^2 hybridization of carbon and nitrogen. The two diverse structural compositions for this 2D nanomaterial are s-triazine and condensed tri-s-triazine [13, 16]. 2D TMDs (with the crystal structure formula of MX_2) are chemically adaptable with a tunable bandgap. Each TMD structure comprises three atomic layers, where one M is placed between two X-layers. In this X–M–X layered structure, M belongs to groups IV–VII and X is group 16 of periodic table [13, 19, 20].

Bridgman created BP for the first time in 1914, and it caught researchers' attention once more as a promising 2D nanomaterial after a century. The direct bandgap of BP ranges from around 0.3 eV to 2.0 eV, in contrast to TMDs having indirect bandgaps. The large interlayer distance causes the rapid ion diffusion within the BP layers, which results in an enhanced electrochemical double-layer capacitance. LDHs share structural similarities with brucite, which consists of two layers of hydroxide sandwiching a layer of cationic metal atoms. The hydrated layers in between the sheets contain intercalated anions, which balance the positive charges. LDHs have unique characteristics, such as the memory effect, whereby materials lose their layered structure during the annealing process but regain it during the rehydrating phase [13, 19, 20]. Transition metal oxyhalides are a class of inorganic compounds, having the generic formula MOX. Each layer of the multilayer crystal structure of MOX is made up of two halide layers sandwiching a corrugated metal–oxygen plane. The metal halides are represented as MXn (where X represents halogen) [13].

MOFs are hybrid substances made of both organic and inorganic elements joined together by covalent coordination bonds. 2D MOFs have a porous, crystalline, and tunable structure with a large surface area. In addition to changing their crystalline structure and other chemical characteristics, 2D MOFs can also change their geometry. For instance, depending on the application, the form, size, and volume of the pores can be tuned. Despite the fact that MOFs have excellent electrochemical properties, their high crystallinity still has a number of negative effects, such as short cycle stability, reduced electrical conductivity, and insufficient electrolyte diffusion into the porous structure [13, 16, 19, 20]. Since 2005, COFs have developed into a brand-new class of significant crystalline polymers with ordered porosity structures and programmable topological patterns that adhere to challenging chemistry procedures similar to MOF nanosheets. Strong covalent bonds between organic units made up of elements such as H, C, N, O, B, and Si result in the formation of COFs. Van der Waals stacking could cause individual layers to combine with the bulk material. As a result, thinner COFs often have better activity compared to their stacked bulk counterparts [13, 16].

These materials have been widely explored in several electrocatalytic applications, such as OER, HER, ORR, and CO_2R. 2D TMDs are superior HER electrocatalysts due to their high surface-to-volume ratio and distinctive electronic structures (thickness-dependent). Graphene sheets doped with different materials have also been investigated as potential HER catalysts. According to density functional theory (DFT) calculations, uniform combination of N-graphene with g-C_3N_4 produces a mediated adsorption–desorption behavior that aids in the overall kinetics of HER. Several materials are being researched to enhance the catalysis of OER. Some of these include 2D chalcogenides such as sulfides and selenides, transition metal nitrides, and MOFs. In recent years, doped g-C_3N_4 and transition metal (oxy) hydroxides (TMOs) have also gained attention due to their cost effectiveness and easy scale-up preparation. A link has been established between the intrinsic activity of these materials and the regional structures and compositions. LDHs emerge as standard substitutes for OER. In fact, in alkaline medium, Ni-, Co-, or Fe-based LDHs have demonstrated promising OER activities [16, 17, 21, 22].

Despite all the catalytic applications, there are certain challenges with 2D nanomaterials that need to be addressed. Black phosphorene's instability hinders large-scale use, with challenges in elemental doping, surface functionalization, and 2D/2D heterojunction formation. In comparison to traditional catalysis, research on COF-based photocatalysis is in its beginning and has various challenges such as low product selectivity and catalytic efficiency, along with a limited range of organic reactions. There is still a long way to go for 2D chalcogenides to be practically applied due to various challenges such as poor photocatalytic stability and efficiency. Although still in their early stages, solutions like making nanostructures and altering surfaces have the potential to alleviate problems caused by the small specific surface area of materials such as g-C_3N_4. Despite the significant advancements in LDH photocatalysis, there are still several issues that act as a major barrier to their further application. One of the critical problems is the synthesis of monolayer LDHs on a large scale. Therefore, there is a strong demand for simple and more effective solutions. Identifying the active sites at the atomic level is still a challenging task due to the rapid active intermediates and complexity of the actual structure. The presence of heteroatoms and defects make graphene a complex catalyst for a given reaction [23, 24].

Supercapacitor performance is assessed based on four main criteria: chemical stability, specific surface area, electrical conductivity, and surface termination. To create highly efficient materials, various techniques have been used to adjust these variables, including heteroatom doping, defect engineering, formation of 2D heterostructure assembly, and thickness/size controlling [16]. Graphene can enhance the performance of supercapacitor electrodes due to its charge mobility and large surface area. Due to its unique structure and exceptional characteristics, graphene has a greater theoretical capacitance than many other carbon materials, up to 550 F g^{-1}. Conductive polymers such as 2D polypyrrole (PPy), poly(3,4-ethylenedioxythiophene) (PEDOT), and polyaniline (PANI) are desirable electrode materials for supercapacitors as they store charge through the doping/de-doping process. TMDs are materials with a large surface area and multiple oxidation states, which enable them to form an electrical double layer. This property, in combination with their strong electrochemical activity at the edges, allows TMDs to store a significant amount of energy in supercapacitors through quick and reversible redox charge storage techniques [13]. Due to their excellent structural stabilities, high values of capacitance, and eco-friendliness, 2D transition metal oxides and hydroxides are promising materials for supercapacitor electrodes [25, 26].

Despite advancements in design, procedures, and new materials, barriers remain for flexible supercapacitors. Creating 2D nanomaterials that function efficiently as pseudocapacitors while maintaining good electric conductivity remains a challenging task. This is because most pseudocapacitive electrode materials are insulators that do not facilitate ion and electron transport during electrochemical operations. As a result, there is a need to develop materials that can meet these requirements. Hybridizing insulating pseudocapacitive materials with conducting graphene still leads to problems at the hetero-structural interface. Weak forces of contact in the hybrid structure cause these problems, rather than the crystalline bonding hetero-structure [27]. Supercapacitors require novel electrodes such as phosphorene, but they have stability issues in ambient environments. To overcome this limitation, it is necessary to conduct a thorough investigation and find ways to properly encapsulate or functionalize phosphorene while preserving its electrochemical properties. The use of an inert environment could be a potential solution to address the stability problems [25].

There is extensive exploration of 2D nanomaterials for various storage devices, such as Li-ion batteries (LIBs), metal–air batteries, and Na-ion batteries. Numerous 2D nanomaterials that resemble graphene have been synthesized and studied for their potential applications in rechargeable batteries. MoS_2, which has been extensively researched as a possible anode material for LIBs, exhibit a larger theoretical capacity, that is, 670 mAh g^{-1}, than a conventional graphite anode. BP is commonly used as an anode material due to its high Li-ion absorption and its capability to expand up to 300% during electrochemical cycling. Similarly to cathodes, 2D materials are also utilized to control the electrode–electrolyte interfaces. Carbon-based electrodes play an important role in the performance of batteries. Graphene provides a reversible capacity greater than that of commonly used graphite as an anode. Nanostructuring TMDs provides a promising alternative for cathodes in LIBs [13, 17, 28]. Expanded graphite has been investigated along with ultrathin 2D nanomaterials such as TMDs, MoO_{3-x} nanosheets, and BP as electrode materials for sodium-ion batteries (SIBs) [13, 17]. These materials with high lateral diameters could reduce shuttle effects in lithium-sulfur (Li-S) batteries by trapping polysulfides and preventing their dissolution in the electrolyte. The affinity of carbon for sulfur makes carbon-based 2D nanomaterials a suitable host for Li-S batteries [29].

Although 2D nanomaterials have shown promising battery applications, they are still facing a number of challenges in this field. The mechanism for adsorption and catalysis in the cathode of Li-S batteries requires further investigation despite widespread reports of usage. Current understanding of the atomic-level chemical interaction between the polysulfides and sulfur host is insufficient to evaluate the mechanism of electrode materials. Although the 2D sheets have a large surface area, they cannot contain lithium polysulfides (LiPSs), and restacking of the nanosheets causes the collapse of material characteristics. Finding cheaper electrode materials is crucial as Li-S batteries manufacturing costs are high compared to that of commercial LIBs. It is worth noting that using 2D materials can significantly reduce issues such as shuttle effect, sulfur conductivity, and volume changes during charging/discharging cycles. However, this approach also limits the benefit of sulfur energy density as it decreases the loading of the active material [30–32].

1.2 THE RISE OF MXENES

Although many of these materials have only been studied for academic purposes, some have caught the attention of researchers due to their appealing characteristics that have led to practical applications. However, researchers have faced several challenges when working with these 2D materials, such as their hydrophobicity and instability in air. The discovery of MXenes had a significant impact on the field of material science. These materials possess metallic conductivity and hydrophilicity, and consist of large flakes, along with being easy to process and having reasonably high yields [5]. In addition to creating intercalation compounds, MXenes can also intercalate with organic molecules and inorganic salts, expanding their potential uses. MXene is a 2D material that has many advantages over other materials. Studies have shown that the bandgap of these materials can be tuned using their surface terminations. For example, pristine MXenes are metallic conductors, while OH- or F-terminated MXenes are commonly semiconductors with narrow bandgaps. In addition, the conductivity of electronically conductive multilayer MXenes is comparable to or better than that of multilayered graphene. MXenes are also hydrophilic, which means they are easily dispersed in aqueous solutions. This is in contrast to graphene, which does not exhibit hydrophilic nature [11].

What are MXenes exactly and how were they discovered? Have there been any recent developments in this field? Additionally, what is the structural design of MXenes? Lastly, what specific combination of characteristics makes MXenes unique when compared to other materials? In the following sections, each of these questions will be answered in a concise manner.

1.2.1 DISCOVERY

MXenes are significant materials due to their structural and chemical stabilities, hardness, and strong electrical conductivity. These materials have been extensively studied as bulk ceramic materials for many years, primarily for high-temperature applications. However, reducing the dimensionality of transition metals from bulk three-dimensional (3D) solids to 2D sheets and 1D nanoribbons has been challenging due to their strong C–N interactions in all directions, mainly covalent/metallic bonds. Unconventional methods are needed to lessen their dimensionality. After years of research on MAX phases, numerous layered carbides, nitrides, and carbonitrides with a wide range of characteristics have been synthesized. In 1996, Barsoum and El-Raghy conducted an important study by creating the first single-phase samples of pure Ti_3SiC_2. They found that this material was easily tunable and a superior electrical conductor. In 1997, they discovered that Ti_3SiC_2 was part of a group of 50 phases, most of which were originally identified in the 1960s by Nowotny and colleagues. The discovery of Ti_4AlN_3 in 1999 led to the classification of these phases as the $M_{n+1}AX_n$ or MAX phases [33].

A team of researchers from Drexel University discovered a new type of 2D inorganic substances in 2011, which they named MXenes. The discovery created a lot of interest among the research community worldwide [6]. The researchers transformed the traditional 3D representation of MAX phases, known as titanium aluminum carbide (Ti_3AlC_2), into a 2D structure with distinct characteristics. They used a chemical process called exfoliation to remove the aluminum from Ti_3AlC_2 particles by immersing them in hydrofluoric acid at room temperature. This produced 2D Ti_3C_2 nanosheets, also known as MXene, which are analogous to graphene [34]. They showcased the applicability of the selective etching method on various MAX phases using the aluminum A-layer, which is one of the frequently found A-elements in MAX phases. As a result, they named this novel family as MXenes to signify its relation to MAX phases and their dimensionality [33]. MXenes are structurally two-dimensional with a thickness of several atoms, along with a size on the order of several micrometers [35].

Generally, MXenes are a family of 2D transition metal carbides, nitrides, and carbonitrides with various surface functional groups [36]. These materials can be fabricated from their parent compounds called MAX phases. These 3D precursors for MXenes have the formula $M_{n+1}AX_n$. Structurally, the MAX phases consist of octahedral blocks made of transition metal carbide/nitride sheets, with X at the centers of these octahedrons, held together by A-layers. These materials are stacked hexagonally ($P6_3/mmc$ space group). The etching of A-layers (using an acid) produces 2D $M_{n+1}X_n$ layers of MXenes. The eventual chemical formula of MXenes is $M_{n+1}X_nT_x$, where T_x represents the surface terminations (-OH, -O, -F, etc.). In all of these formulas, M stands for transition metals (i.e., Sc, Ti, V, Cr, Mn, Y, Zr, Nb, Mo, Hf, Ta, and W), element A belongs to the A group (i.e., group IIIA or IVA), X is carbon or nitrogen, and n = 1–4. MXene thickness can be changed by altering the value of n, resulting in different multilayered structure formations (i.e., M_2XT_x to $M_3X_2T_x$, and $M_4X_3T_x$) [1–3, 8, 33, 37].

1.2.2 DEVELOPMENT

MXenes are a rapidly growing family of 2D nanomaterials with a wide range of compositions, structures, synthesis techniques, and applications [4]. The number of publications on MXenes has significantly increased in the past few years. In May 2019, the 2nd International Conference on MXenes was held. The conference covered a broad range of topics, including their structure, synthesis, and properties. The researchers also highlighted the practical applications of MXenes in areas such as separation membranes, energy storage and conversion, optics, biomedicine, and electronics. Unlike the first conference, which focused mainly on MXenes for energy, this conference showcased the diverse range of potential uses of MXenes. The field is experiencing rapid growth due to various factors, and there are several evidences to this expansion. One way to measure the growth of MXenes is by counting the research organizations that have studied and published their findings on this material in peer-reviewed journals. This extensive research has revealed a visual increase in the studies and synthesized compositions [5].

All of the four components of the formula (i.e., M, X, T, and n) have expanded their range of compositions in $M_{n+1}X_nT_x$. A publication in *ACS Nano* from 2012 describes the synthesis of M_2X, M_3X_2, and M_4X_3 (i.e., n = 1–3), which led to the discovery of this family of sizable 2D materials. In 2020, the variety of 2D carbides and carbonitrides was further expanded with the publication of M_5C_4 (n = 4). Solid solutions such as $(Ti,V)_2CT_x$, $(Ti,Nb)_2CT_x$, and $(V,Nb)_2CT_x$ are being synthesized. These solid solutions allow for precise control over the desired properties of the material. In 2021, researchers discovered high-entropy MXenes that contain multiple M elements. Over the span of a year, several research organizations worldwide published five different compositions of these high-entropy MXenes. It has been reported that there are additional carbonitride-based solid solutions at the X site. Solid solutions can be formed on M, X, and T sites, which results in the existence of an almost infinite amount of 2D materials with various properties. Currently, there are ongoing efforts to create 2D borides, which could increase the X element of the structure [5, 7].

Although the first ordered MXene was produced in 2014 and reported in 2015, this subfamily of MXenes has given rise to a plethora of novel compounds [5]. In the 2020s, MXenes with uniform surface functionalities, such as -O, -S, Cl, Br, and those without surface terminations (bare MXenes), were successfully created. However, it was only in the 2010s that computational studies began exploring the uniform surface groups [6]. The community of researchers working on MAX phase materials was thrilled with the rapid synthesis of new 2D carbide phases with an ordered structure. After 20 years of research on MAX phases and MXenes, scientists have started creating new MAX phases along with other layered carbides and nitrides. Since 2017, many new double transition metal MAX precursors have been developed, and their properties, particularly their magnetic properties, have been investigated [5]. Recently, scientists discovered a new class of materials called transition metal carbo-chalcogenides. These materials are composed of M_2C layers with chalcogen (such as S) as a surface termination. For instance, Nb_2CS_2 and Ta_2CS_2 are some of the compositions of these phases. These compounds are formed by exfoliating layered transition metal carbo-chalcogenides. They offer a new avenue for connecting the gap between MXenes and other 2D materials [6].

The proportion of various MXene publications also changed with the growth of MXene compositions between the end of 2018 and the present. In different studies, researchers analyzed publications on MXenes through 2018 and found only around 1,300 [6]. Till October 2023, there were more than 17,000 publications on MXenes. These publications include review articles (3,573), research articles (12,047), encyclopedia (98), book chapters (552), conference abstracts (51), editorials (43), mini reviews (22), short communications (590), and so on. The number of publications has increased by an order of magnitude in under 5 years. The rapid progress of MXenes is intriguing, especially the way these materials are spreading into new fields of study like biology, mechanics, electronics, and electromagnetics. In 2019, energy and catalytic uses made up 47% of all MXene articles, compared to today, where these topics only account for around 30% of all MXene articles. Despite the pandemic, research in energy storage, conversion, and catalytic applications continues to thrive. There has been an over eightfold increase in publications in this field, indicating that it is not slowing down. The rapid expansion of these new fields is the primary reason for the rise in the share of new study fields in all MXene articles [6].

In the past years, another important advancement has been made in exploring the new methods for synthesizing MXenes without using fluoride salts. Most of the early MXene production methods involved using aqueous or molten salts containing fluoride. However, a recent discovery suggests that an electrochemical process that does not involve fluoride could be used, such as diluted hydrochloric acid. However, it may be difficult to use selective electrochemical etching techniques at a large scale. In 2019, scientists developed new MAX phases and fluorine-free MXenes by utilizing molten zinc chloride salt. This method allows individuals who are interested in MXenes to conduct their experiments without the use of chemicals that contain or produce hydrofluoric acid (HF). This advancement has the potential to significantly broaden the scope of experimental research on MXenes. Currently, top-down and bottom-up techniques are reported as the two primary techniques for synthesizing 2D MXenes. In top-down techniques, bulk materials are exfoliated by selective MAX phase etching to produce monolayer or multilayer films. Bottom-up approaches start with atoms and build up to a layered structure (e.g., chemical vapor deposition (CVD), hydrothermal, and solvothermal synthesis techniques) [5].

Figure 1.2 [8] shows the schematic illustration of the rise in the discovery of different structures and compositions of MXenes in the last decade. Due to the large, under-researched family of MXenes and emerging synthesis techniques, along with their unique combination of properties, there is immense potential for various applications. Despite all these developments, MXene research is still in its early stages, and we believe that there will be numerous captivating discoveries in the future [5].

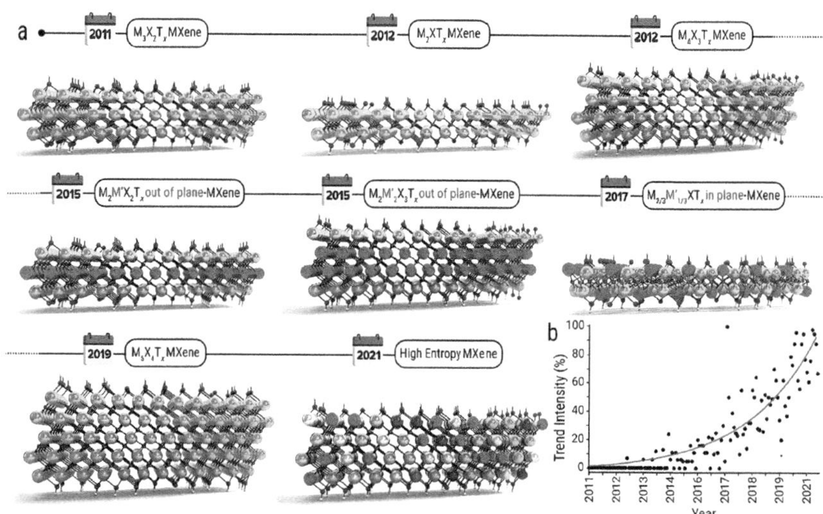

FIGURE 1.2 (a) Schematic illustration of different MXene structures. $M_3X_2T_x$ MXene (top left) was the first discovered composition in 2011, followed by the discovery of M_2XT_x and $M_4X_3T_x$ in 2012. The family of MXenes was expanded by out-of-plane ordering (in 2015) followed by in-plane ordering (in 2017). $M_5X_4T_x$ solid solution MXene and high-entropy $M_4X_3T_x$ MXenes are among the latest additions to the MXene family. (b) Monthly Google Trends data, collected over the decade from users' engagement with the word "MXene." Since 2015, there has been a considerable increase in trend intensity.

Source: Reprinted from [8] *Materials Today Advances*, 13, M. Dadashi Firouzjaei, M. Karimiziarani, H. Moradkhani et al., MXenes: The two-dimensional influencers, 100202, Copyright 2022, with permission from Elsevier.

1.2.3 LAYERED STRUCTURE

MXene has a clearly defined micro-sized hard layered structure due to its ceramic nature. The wide diversity of MAX phases helps adjust the structure and properties of MXenes, which directly affect the MX layers. MXenes are a type of material that have a hexagonal close-packed structure. Specifically, the metal atoms (M) of the M_2X type are arranged in an ABABAB pattern. However, in M_3X_2 and M_4X_3 structures, where M atoms are organized in a face-centered cubic structure, the ordering is often different and follows an ABCABC pattern. The surface chemistry and stability of MXenes are significantly influenced by this configuration of transition metals. It is difficult to determine the precise structure of MXenes because of the unavoidable terminations connected to their etched surface. Theoretical computing is frequently used in research on the structure of MXenes. MXenes differ from other 2D nanomaterials in a number of ways due to all of these capabilities. All of these innovative aspects of MXene may help researchers become more enlightened and imaginative [35, 38–40].

According to the recent investigations, MXenes can be found in six different types of structures. These structures include one M element (i.e., M_2XT_x, $M_3X_2T_x$, and $M_4X_3T_x$); out-of-plane ordered double M elements (represented as $M'_2M''X_2$ or $M'_2M''_2X_3$), also called o-MXenes; in-plane ordered double M elements or i-MXenes (represented as $(M'_{2/3}M''_{1/3})_2X$); in-plane MXenes with ordered divacancies (represented as $M_{1.33}X$); solid solutions on M sites (represented as $(M'_yM''_{1-y})_2X$, $(M'_yM''_{1-y})_3X_2$, etc.); and solid solutions on X sites (represented as $M_2(C_yN_{1-y})T_x$ and $M_3(C_yN_{1-y})_2T_x$) [33, 41, 42]. Two M elements used to be the most prevalent number in solid solution MXenes. However, researchers have shown that with the appropriate 2D carbides, alloy MAX phases with high entropy can be created. They synthesized MXenes with four M elements represented as $(M1M2M3M4)C_3T_x$, that is, $TiVNbMoC_3T_x$ and $TiVCrMoC_3T_x$ by removing A-layers of $TiVNbMoAlC_3$ and $TiVCrMoAlC_3$, respectively. This indicates that these brand-new MXene compositions would make good candidates for use in energy conversion, catalysis, energy storage, etc. [35]. The layer spacing of the (002) crystal plane in 2D MXenes is extremely large (1 nm), larger than the layer spacing of the graphite layer (0.337 nm). On the surface or interface of MXenes, there are several groups like the O, OH, or F group; more significantly, layer spacing could be somewhat increased by a number of ways [43].

The performance of MXenes are also greatly influenced by the surface functional groups. They are typically created through selective etching using HF solution. Therefore, typically, -O, -F, and -OH-functional species can terminate the surface of MXenes. The latter can have an impact on a variety of MXene properties, including conductivity, biocompatibility, conductivity, and electron mobility [35]. For MXenes, a variety of uniform surface termination options need to be taken into account to identify the configurations with lowest energy. These decisions are based on where the termination groups are placed, and they are expected to be favorably adsorbed at the hollow locations on the surface of MXenes [37]. Single-layered or multilayered MXenes can be obtained by an essential process, that is, delamination. Altering the interlayer spacing is another method for modifying the characteristics of MXenes. Intercalation is a method for achieving it. For instance, the conductivity of MXenes changes when intercalants are added. MXenes have metallic conduction by nature, but when the c-lattice spacing increases, they behave more like semiconductors. Therefore, intercalation can be employed to efficiently change the characteristics of MXenes [35].

By using intercalators, the structure of multilayered MXenes can be changed since the van der Waals (vdW) forces and hydrogen bonds replace the stronger M–A bonding. MXenes still have interlaminar interactions that are 2–6 times stronger than those of graphite. As a result, the transparent tape method of mechanical peeling cannot produce a single layer of MXene. Typically, mechanical stirring followed by intercalation can be used to delaminate MXenes. The intercalator is inserted into the 2D sheet in two processes: ultrasonic treatment and centrifugation. The concentration of MXene in the solution is influenced by both the synthesis technique and the type of intercalator [11]. Researchers have confirmed the potential use of various chemicals as intercalating agents, including urea, dimethyl sulfoxide (DMSO), and isopropylamine. Other intercalating compounds were also examined, achieving interlayer distance increases of up to three times, including NH_4HF_2, tetramethylammonium hydroxide (TMAOH), tetrabutylammonium hydroxide (TBAOH), and aryl diazonium salts in water [42].

1.2.4 UNIQUE COMBINATION OF PROPERTIES

MXenes have a unique set of characteristics, and although we have learned a lot about them over the past 10 years, much more research is still necessary to completely comprehend their features [44]. MXenes are hydrophilic and capable of bonding to various materials due to electromagnetic wave absorption, negative zeta potential, and functionalized surfaces, which have all led to a wide range of uses [5]. Energy storage (batteries and supercapacitors) was the first known use of MXenes, and majority of the studies till date are based on these applications. With investigations on theragnostic, biosensor, dialysis, and neurological electrodes, the biomedical application of MXenes has emerged as one of the most popular study subjects. Another area where MXenes are replacing other nanomaterials is their electromagnetic applications, such as printed antennas and interference shielding (EMI). When combined with other 2D materials, these materials exhibit prospective building blocks for vdW heterostructures, which are still extensively studied. MXenes may function as conductive layers in these heterostructures while also altering the electrical characteristics of other 2D materials [3, 5]. The properties and applications of MXenes are illustrated in Fig. 1.3 [5].

The capacity of MXenes to withstand oxidation is influenced by a variety of factors, including their composition, morphology (such as single- or multilayered flakes, films, or powder), flake size, structural defects, and storage conditions (such as colloidal solution, in vacuum or air). It is typical to keep the 2D flakes in organic solvents or water as a colloidal solution because MXenes are produced through solution processing. Practically any method used to generate inks can be utilized to create MXene films and patterns due to the significant negative zeta potential of MXene colloidal solutions. Recent researches have demonstrated that the key

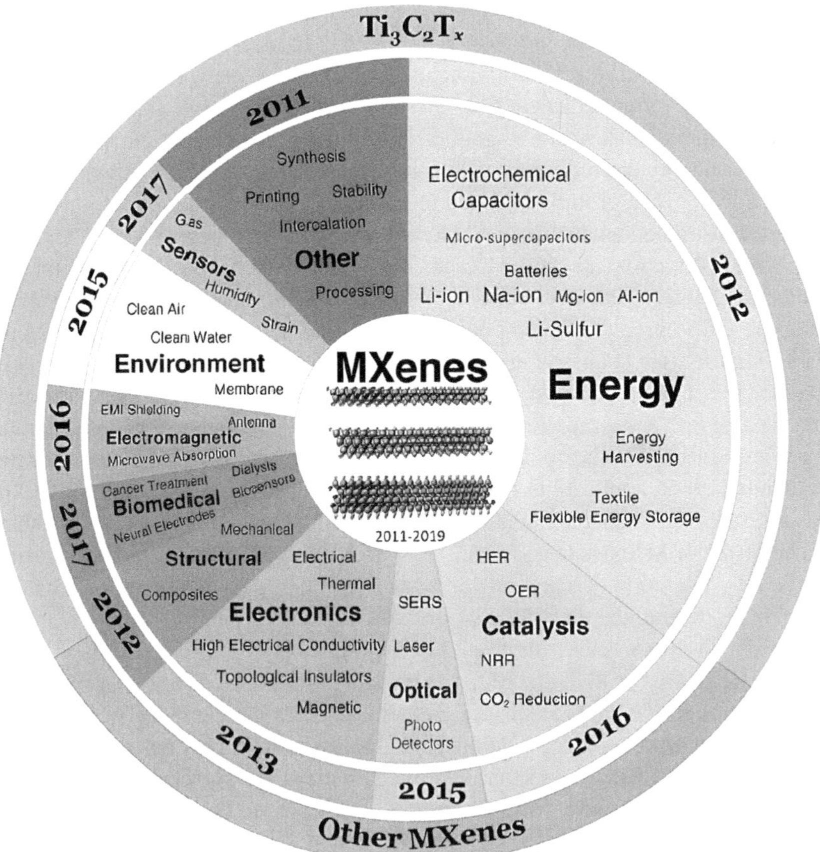

FIGURE 1.3 Explored applications and properties of MXenes to date. The center pie chart shows the ratio of publications in each explored application/property of MXenes with respect to the total number of publications on MXenes. The middle pie chart ring, with the same colors, shows the starting year for exploration of each application/property of MXenes.

Source: Reprinted with permission from [5] Y. Gogotsi, B. Anasori. 2019. The Rise of MXenes. *ACS Nano.*, 13, 8491–8494. Copyright 2019, American Chemical Society.

factors in MXene degradation are water and MXene hydrolysis. When MXene is stored in aqueous solutions, its decomposition can be postponed by storing it under argon at lower temperatures and with less dissolved oxygen in water. When MXene is removed from water by deposition techniques including spraying, spin coating, or filtering, it becomes substantially more stable in air for up to months [44].

The mechanical behavior of MXene is quite interesting because of the stronger M–C and M–N bonding. Previous researches suggested that their elastic parameters should be two times greater than that of MAX phases. Although their bending characteristic (1050 GPa) is maximum despite having elastic capabilities, it is 2–3 times less than that of graphene. Due to the presence of surface functionalities, MXenes can form better composites with polymeric matrixes as compared to the graphene.

Numerous papers have emphasized how Young's modulus of both carbide and nitride MXenes decreases with the increase in the number of layers [36, 44]. Surface strains are crucial for determining the magnetic characteristics that MXene materials display. Due to the strong covalent bonding between M, X, and T, the majority of the MXene components are nonmagnetic. Additionally, the magnetic properties of these materials are caused by an increase in the density of electrons, which can be described by d group orbitals assigned to M elements [36, 37, 45].

The electronic conductivities of MXenes are of primary significance. Similar to their MAX precursor, all bare MXenes are metallic in nature; however, they can become semiconductors or metals upon surface terminations. Some experimental studies have also predicted the topological insulating behavior of a few MXenes [44]. By altering the solid solution formation, functional groups, or material balance, all of these features can be changed. The high electronic conductivity of MXenes makes them capable for various applications [36, 37, 45]. MXenes also exhibit notable optical properties. Since 2016, the optical properties of these materials have been researched, and a number of applications, including mode-locked lasers, photothermal therapy, and surface-enhanced Raman spectroscopy, have come to light. By switching the chemistry of MXenes (e.g., from Ti_2CT_x to V_2CT_x), the optical properties can be significantly altered [44]. The transmittance percentage may also be somewhat optimized by modifying the ion intercalation and thickness of material [36].

Due to their distinct characteristics, MXenes can improve the electrocatalytic activity of the HER, OER, ORR, etc. MXenes are suited for use as catalyst supports due to characteristics such as a wide surface area and superior conductivity, which result in fast electron transport and great stability [46]. MXenes are highly desirable for energy storage applications due to a special mix of characteristics. The electrochemical characteristics of MXenes have been thoroughly investigated. The majority of MXene's publications to date have dealt with energy-related subjects. Surface terminations, more specifically oxygen, produce surfaces that resemble transition metal oxides and are redox active. High electrical conductivity is a property of MXenes. The electrochemically active sites can quickly transfer electrons due to high conductivity. Fast ion transport resulted from MXene's 2D structure and the simplicity of cation intercalation. Depending on the electrolyte, charge storage takes on several forms. In alkaline and aqueous electrolytes, MXenes exhibit primary double-layer capacitive behavior; however, in acidic electrolytes, they exhibit pseudocapacitive behavior and ultrahigh capacitance. Because MXenes may accommodate different metal cations, their applications in lithium-, sodium-, potassium-, magnesium-, and aluminum-ion batteries have widely been investigated. All of these electrochemical features make MXenes efficient electrode materials in batteries and supercapacitors [44].

1.3　A COMPARATIVE STUDY OF 2D MATERIALS AND MXENES

Despite the fact that there were many 2D materials before MXenes were found, the majority of them were semimetals, insulators, or semiconductors with poor carrier concentrations and electronic conductivities. Importantly, few 2D materials could be produced in quantities large enough to be used in industries other than microelectronics. With the exception of h-BN and graphene, solution-processed materials have relatively small sizes due to low mechanical strength that can cause 2D sheets to break during delamination. Many of these substances were also unstable in air and hydrophobic in nature [33]. Although TMDs are the subject of extensive research, there are always other materials that can perform better in practical applications, with the exception of electronic applications, which may still be a long way off. Because strong and conductive multilayer sheets can be made in vast quantities by mechanically shearing natural graphite, graphene has found widespread use in composites. Applications where graphene derivatives perform better than other materials include conductive additives, heat spreaders for cell phones, and corrosion-resistant paint additives. Simple and cost-effective large-scale graphene synthesis is still difficult, and layer stacking hinders the development of stable graphene suspensions [47].

Researches have shown that the already available 2D nanomaterials lack certain features that restrict their uses in respective applications. The discovery of 2D MXenes with ease of processing, metallic conductivity, tunable compositions, hydrophilicity, biocompatibility, and reasonably high yields, thus, had a significant impact on the entire field of materials science [33, 47]. The usage of MXenes is generally gaining popularity across the majority of applications examined because they exhibit high electrical conductivity along with other structural advantages. They share many benefits of graphene, such as good adherence to substrates, but they do not require the heat or chemical reduction procedure required for graphene oxide. These materials quickly disperse in aqueous substrates without any additive, and this is in contrast to graphene nanoparticles. This results in Joule heaters with lower temperatures than graphene, sometimes even competing with metallic nanomaterials. Their enhanced conductivity allows for better signal attenuation in EMI shielding at the same thickness as graphene [48].

MXenes have unique features that make them superior to other 2D materials. They exhibit the highest electrical conductivity than all other reported 2D materials. Figure 1.4 [48] illustrates the year-on-year percentage change in the published articles with graphene and MXenes. Compared with graphene, the conductivity of MXene films can vary by orders of magnitude. Because of their highly organized structure, MXenes can have their nature predicted by theoretical calculations. The plasmon resonance of these films can be found throughout the visible and far-infrared spectrum. The emission of MXenes is predominantly caused by size effects, such as defects and quantum confinements, in terms of a luminescence phenomenon. These materials have the potential to be utilized in optoelectronics, plasmonics, and photonics. One such application that has gained a lot of interest is photodynamic cancer therapy, where MXenes display high efficiency in converting light to heat. Additionally, the chemical characteristics of the surface layer are significantly influenced by the particular transition metal used in their production. MXenes can be compared to hydrophilic 2D metals that can dissolve in water or electrically conducting clay. MXenes can provide stable dispersions in a variety of solvents without the use of surfactants or additives. Such MXene dispersions are excellent for creating solar, plasmon, and polymer composite applications since they can be sustained throughout time [11, 36, 43, 49, 50, 51].

The 2D layered geometric structure of MXenes enables quicker ion transport, and reduction/oxidation reactions create a pseudocapacitive mechanism of storing

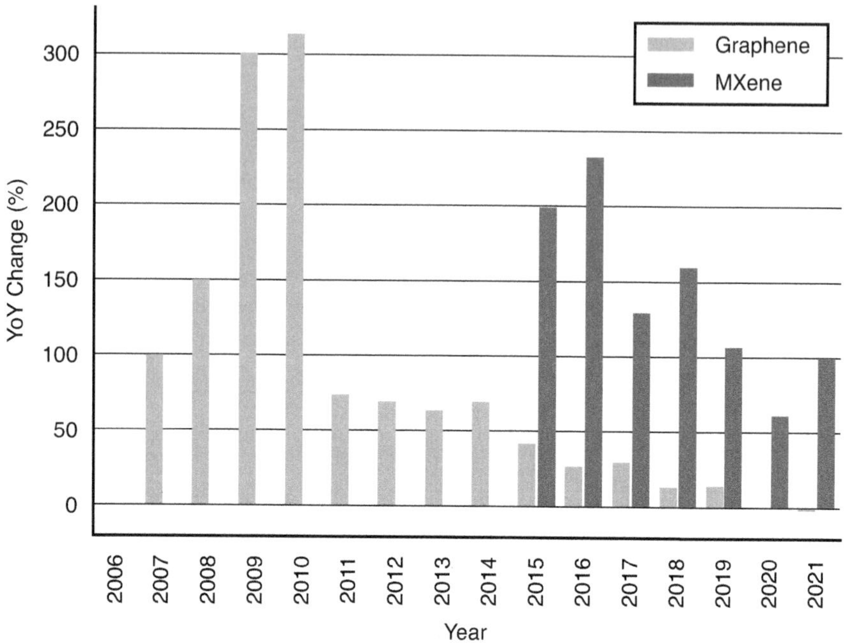

FIGURE 1.4 Year-on-year percentage change of published articles with graphene and MXenes over the last 15 years.

Source: [48] V. Orts Mercadillo, K. Chio Chan, M. Caironi, et al.: Electrically Conductive 2D Material Coatings for Flexible and Stretchable Electronics: A Comparative Review of Graphenes and MXenes. *Advanced Functional Materials*. 2022, 32, 2204772. Copyright Wiley-VCH GmbH. Reproduced with permission.

charge. These adaptable qualities confirm MXene's status as an active component for developing supercapacitors and batteries. The appeal of having a variety in content and structure is found in this family of 2D nanomaterials. A recent study has reported the synthesis of a MXene film with sizable area possessing remarkable EMI shielding effects (SE) and excellent tensile strength, which are not present in other 2D materials. It has also been suggested to employ MXenes as a negative electrode to create concentration gradients of higher strength in acidic electrolytic solutions, thereby increasing the energy density. Due to their smaller capacitance and narrow potential gradient, carbon-based components have a substantially lower level of this property. As a result, MXenes are chosen as catalysts for these components in order to maximize energy conservation through the hybridization process [11, 36, 43, 49, 50].

REFERENCES

[1] Wyatt, B.C., A. Rosenkranz, and B. Anasori, *2D MXenes: tunable mechanical and tribological properties*. Advanced Materials, 2021. **33**(17): p. 2007973.

[2] Naguib, M., et al., *25th anniversary article: MXenes: a new family of two-dimensional materials*. Advanced Materials, 2014. **26**(7): p. 992–1005.

[3] Ng, L., et al., *Printing of Graphene and Related 2D Materials*. 2018, Springer.

[4] Johnson, D., et al., *Holdups in nitride MXene's development and limitations in advancing the field of MXene*. Small, 2022. **18**(17): p. 2106129.

[5] Gogotsi, Y. and B. Anasori, *The Rise of MXenes*. 2019, ACS Publications, p. 8491–8494.

[6] Anasori, B. and Y. Gogotsi, *MXenes: trends, growth, and future directions*. Graphene and 2D Materials, 2022. **7**(3–4): p. 75–79.

[7] Gogotsi, Y. and Q. Huang, *MXenes: Two-Dimensional Building Blocks for Future Materials and Devices*. 2021, ACS Publications. p. 5775–5780.

[8] Firouzjaei, M.D., et al., *MXenes: The Two-Dimensional Influencers*. 2022, Elsevier. p. 100202.

[9] Yang, J., et al., *MXene-based composites: synthesis and applications in rechargeable batteries and supercapacitors*. Advanced Materials Interfaces, 2019. **6**(8): p. 1802004.

[10] Pang, J., et al., *Applications of 2D MXenes in energy conversion and storage systems*. Chemical Society Reviews, 2019. **48**(1): p. 72–133.

[11] Meng, W., et al., *Advances and challenges in 2D MXenes: from structures to energy storage and conversions*. Nano Today, 2021. **40**: p. 101273.

[12] Zhu, J., et al., *Recent advance in MXenes: a promising 2D material for catalysis, sensor and chemical adsorption*. Coordination Chemistry Reviews, 2017. **352**: p. 306–327.

[13] Tan, C., et al., *Recent advances in ultrathin two-dimensional nanomaterials*. Chemical Reviews, 2017. **117**(9): p. 6225–6331.

[14] Gupta, A., T. Sakthivel, and S. Seal, *Recent development in 2D materials beyond graphene*. Progress in Materials Science, 2015. **73**: p. 44–126.

[15] Scarano, D. and F. Cesano, *Graphene and other 2D layered nanomaterials and hybrid structures: synthesis, properties and applications*. 2021, MDPI. p. 7108.

[16] Yang, F., et al., *Recent progress in two-dimensional nanomaterials: synthesis, engineering, and applications*. FlatChem, 2019. **18**: p. 100133.

[17] Khan, A.H., et al., *Two-dimensional (2D) nanomaterials towards electrochemical nanoarchitectonics in energy-related applications*. Bulletin of the Chemical Society of Japan, 2017. **90**(6): p. 627–648.

[18] Chen, Y., et al., *Two-dimensional metal nanomaterials: synthesis, properties, and applications*. Chemical Reviews, 2018. **118**(13): p. 6409–6455.

[19] Ng, L.P.W., et al., *Structures, properties and applications of 2D materials*, in *Printing of Graphene and Related 2D Materials: Technology, Formulation and Applications*. 2019. Springer. p. 19–51.

[20] Vargas-Bernal, R., *Graphene against other two-dimensional materials: a comparative study on the basis of electronic applications*, in *Two-Dimensional Materials: Synthesis, Characterization and Potential Applications*; Nayak, P.K., Ed. 2016. Intech. p. 103–121.

[21] Jin, H., et al., *Emerging two-dimensional nanomaterials for electrocatalysis*. Chemical Reviews, 2018. **118**(13): p. 6337–6408.

[22] Chen, Y., et al., *Two-dimensional nanomaterials for photocatalytic CO_2 reduction to solar fuels*. Sustainable Energy & Fuels, 2017. **1**(9): p. 1875–1898.

[23] Yang, Y., et al., *2020 roadmap on two-dimensional nanomaterials for environmental catalysis*. Chinese Chemical Letters, 2019. **30**(12): p. 2065–2088.

[24] Qin, D., et al., *Recent advances in two-dimensional nanomaterials for photocatalytic reduction of CO_2: insights into performance, theories and perspective*. Journal of Materials Chemistry A, 2020. **8**(37): p. 19156–19195.

[25] Kumar, K.S., et al., *Recent advances in two-dimensional nanomaterials for supercapacitor electrode applications*. ACS Energy Letters, 2018. **3**(2): p. 482–495.

[26] Mendoza-Sánchez, B. and Y. Gogotsi, *Synthesis of two-dimensional materials for capacitive energy storage*. Advanced Materials, 2016. **28**(29): p. 6104–6135.

[27] Peng, X., et al., *Two dimensional nanomaterials for flexible supercapacitors*. Chemical Society Reviews, 2014. **43**(10): p. 3303–3323.

[28] Mohanty, R., A. Mishra, and J. Khatei, *Two-dimensional nanostructures for advanced applications*, in *Adapting 2D Nanomaterials for Advanced Applications*. 2020, ACS Publications. p. 1–31.

[29] Jana, M., et al., *Rational design of two-dimensional nanomaterials for lithium–sulfur batteries*. Energy & Environmental Science, 2020. **13**(4): p. 1049–1075.

[30] Xu, H., et al., *Review on recent advances in two-dimensional nanomaterials-based cathodes for lithium–sulfur batteries*. EcoMat, 2023. **5**(2): p. e12286.

[31] Liu, B., J.-G. Zhang, and G. Shen, *Pursuing two-dimensional nanomaterials for flexible lithium-ion batteries*. Nano Today, 2016. **11**(1): p. 82–97.

[32] Mao, J., et al., *Two-dimensional nanostructures for sodium-ion battery anodes*. Journal of Materials Chemistry A, 2018. **6**(8): p. 3284–3303.

[33] Naguib, M., M.W. Barsoum, and Y. Gogotsi, *Ten years of progress in the synthesis and development of MXenes*. Advanced Materials, 2021. **33**(39): p. 2103393.

[34] Naguib, M., et al., *Two-dimensional nanocrystals produced by exfoliation of Ti_3AlC_2*. Advanced Materials, 2011. **23**(37): p. 4248–4253.

[35] Pogorielov, M., et al., *MXenes—A new class of two-dimensional materials: structure, properties and potential applications*. Nanomaterials, 2021. **11**(12): p. 3412.

[36] Kumar, J.A., et al., *Methods of synthesis, characteristics, and environmental applications of MXene: a comprehensive review*. Chemosphere, 2022. **286**: p. 131607.

[37] Khazaei, M., et al., *Recent advances in MXenes: from fundamentals to applications*. Current Opinion in Solid State and Materials Science, 2019. **23**(3): p. 164–178.

[38] Peng, J., et al., *Surface and heterointerface engineering of 2D MXenes and their nanocomposites: insights into electro- and photocatalysis*. Chem, 2019. **5**(1): p. 18–50.

[39] Razack, A. and O. Salim, *MXene based materials for energy storage and separation*. 2020, UNSW Sydney.

[40] Xu, X., et al., *MXenes with applications in supercapacitors and secondary batteries: a comprehensive review*. Materials Reports: Energy, 2022. **2**(1): p. 100080.

[41] Champagne, A. and J.-C. Charlier, *Physical properties of 2D MXenes: from a theoretical perspective*. Journal of Physics: Materials, 2020. **3**(3): p. 032006.

[42] Ronchi, R.M., J.T. Arantes, and S.F. Santos, *Synthesis, structure, properties and applications of MXenes: current status and perspectives*. Ceramics International, 2019. **45**(15): p. 18167–18188.

[43] Sun, S., et al., *Two-dimensional MXenes for energy storage*. Chemical Engineering Journal, 2018. **338**: p. 27–45.

[44] Anasori, B. and Û.G. Gogotsi, *2D Metal Carbides and Nitrides (MXenes)*. Vol. 416. 2019, Springer.

[45] Abbasi, N.M., et al., *Recent advancement for the synthesis of MXene derivatives and their sensing protocol*. Advanced Materials Technologies, 2021. **6**(10): p. 2001197.

[46] Ahmad Junaidi, N.H., et al., *A comprehensive review of MXenes as catalyst supports for the oxygen reduction reaction in fuel cells*. International Journal of Energy Research, 2021. **45**(11): p. 15760–15782.

[47] Gautam, R., N. Marriwala, and R. Devi, *A review: study of MXene and graphene together*. Measurement: Sensors, 2023. **25**: p. 100592.

[48] Orts Mercadillo, V., et al., *Electrically conductive 2D material coatings for flexible and stretchable electronics: a comparative review of graphenes and MXenes*. Advanced Functional Materials, 2022. **32**(38): p. 2204772.

[49] Wang, Y., et al., *MXenes: focus on optical and electronic properties and corresponding applications*. Nanophotonics, 2020. **9**(7): p. 1601–1620.

[50] Bhat, A., et al., *Prospects challenges and stability of 2D MXenes for clean energy conversion and storage applications*. npj 2D Materials and Applications, 2021. **5**(1): p. 61. 51. Ashraf, N., et al., *A review of the interfacial properties of 2-D materials for energy storage and sensor applications*. Chinese Journal of Physics, 2020. **66**: p. 246–257.

2 Metal Nitrides and Carbides

The materials known as transition metal carbides, nitrides, and carbonitrides exhibit unique extraordinary characteristics, such as hardness, metallic luster paired with vibrant colors, high melting point, and metallic structures. They also exhibit strong thermal and electrical conductivities. This set of characteristics has attracted a lot of interest and led to theoretical research and technical implementations. Carbonitrides, which likewise exhibit ratio-dependent changes in their solid-state characteristics, are frequently formed via the solid-solution formation of carbides and nitrides. There is a lot of room for modification with this behavior in carbonitrides. Liquid metals can readily saturate carbides and, to a lesser extent, carbonitrides and nitrides. This property is useful in hard metals and ceramics. MXenes have a high degree of chemical stability; alkaline solutions and dilute acids—aside from oxidizing acids and hydrofluoric acid—do not easily attack them. Their free energy of production is correlated with thermal stability, which decreases as the number of groups increases [1]. Various characteristics of MXenes, including geometry and energies, variable valency, tunable interlayer spacing, the effect of the mass of metals, doping effects, and effects of intercalations, stability, and surface chemistries, are discussed below in detail.

2.1 GEOMETRY AND ENERGIES

The transition metals in the MAX phases are organized in a close-packed structure with X atoms located in the octahedral interstitial spaces. These precursors are responsible for the $P6_3/mmc$ space group and hexagonal symmetry of MXenes (as illustrated in Fig. 2.1 [3]). Since the M-A bond is metallic in nature and M-X bond has a mixed covalent/metallic/ionic character, the A-layer of the MAX phases can selectively be etched to turn them into MXenes [2, 3]. The as-produced metal surfaces will undergo surface terminations (T_x) following the removal of the "A" element. It has been observed experimentally that MXenes exfoliated by hydrofluoric acid (HF) or acid solutions will result in outer layers terminated with −F or −OH groups. Additionally, O-terminated surfaces could be produced chemically or by post-processing −OH terminations, for instance by means of an ultraviolet-ozone cleaning method or thermal treatment [4]. Furthermore, a recent study revealed that when MXenes are etched in an environment containing Cl, Cl atoms can also functionalize the surface of the compound. The strong M-T bonding is indicated by negative formation energies, which suggests that MXenes are most likely to be fully terminated [2].

The combined adsorption of terminations on the surfaces of model structures makes it challenging to predict the electrical characteristics of MXenes. As a result,

DOI: 10.1201/9781003465768-2

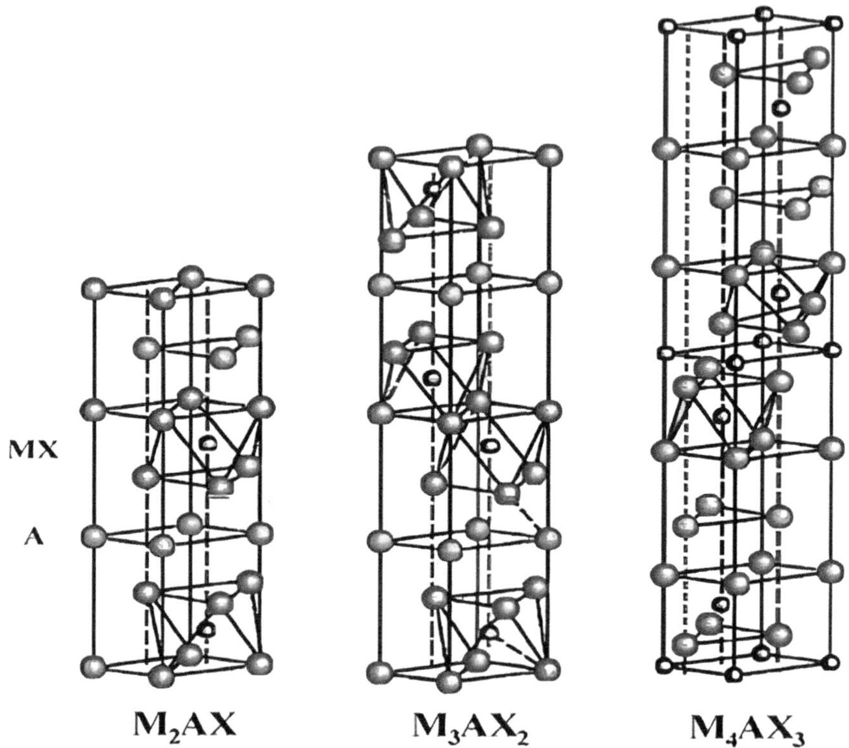

FIGURE 2.1 Structural representation of MAX phase (P6$_3$/mmc) symmetry.

Source: Reprinted from [3] Biswas S. and Alegaonkar P. S.: MXene: Evolutions in Chemical Synthesis and Recent Advances in Applications. *Surfaces* 2022, 5(1), 1–34. Open access. The source of material *Thin Solid Films*, 518, P. Eklund, M. Beckers, et al., The M$_{n+1}$AX$_n$ phases: Materials science and thin-film processing, 1851–1878, Copyright 2010, permission from Elsevier is acknowledged.

very few theoretical investigations have used mixtures of these groups, and the majority have used the models with homogeneous adsorption of only one group. Although these ideal structures are probably unattainable right now, they aid in our understanding of the chemistry of these materials and direct for future research [5]. In order to investigate the geometrical structures and characteristics of a number of possible MXenes, such as those created via the formation of M$_2$X(OH)$_2$, M$_2$XO$_2$, and M$_2$XF$_2$, first-principal calculations are being carried out. Two different sorts of hollow sites can be found on the surfaces of MXenes: those where an X atom is available under them and those where none is available. Therefore, there are various options for the uniform surface terminations adsorbed at the hollow sites, which should be considered for obtaining the lowest energy configurations [5].

On the surface of MXenes, there are three potential locations for these terminations, that is, between three nearby X atoms below M, on top of the X atoms, and on top of M elements. According to density functional theory (DFT) studies, surface groups on both sides of MXene sheets (site I) have been demonstrated to be the most stable configuration due to their minimal steric hindrance. However, when the transition metals are unable to supply enough electrons to both surface terminations and X, site II becomes more favorable. Specifically, –OH and –F groups only need one electron from the metal surface to stabilize their adsorption location, but O-terminations need two electrons. O-terminations containing low valency metals typically have a site II or a combined site I and II configuration [2–4, 6, 7].

The surface groups are adsorbed at hollow sites, where the M-X bonding states reach their maximum occupancy. It was also discovered that elastic stiffness in O-terminated MXenes is very low in comparison with flakes with –OH and –F terminations. On the other hand, the MXenes with terminations were rather flexible compared to those without surface terminations [3]. Additional thermodynamic simulations reveal that MXenes become more thermodynamically stable as their surfaces reach full functionalization. This makes sense because a functionalized M_2XT_2 (T = F, OH, or O) MXene is formed when a metal is surrounded by multiple ions in a variety of complexes and crystals [5, 8]. In order to determine the thermodynamic stability and its dependence on their chemical composition, researchers evaluated the binding energies of 2D MXene surfaces. The findings indicated that all MXene surfaces, with the exception of M = Sc, are totally saturated with oxygen, confirming the more thermodynamically stable nature of MXene surfaces with full functionalization [8].

As for multilayer crystal formations of MXenes, they can develop in various stacking orders, just like graphite does. Therefore, before looking at the electrical structures of multilayer MXenes, it is crucial to determine the proper stacking sequence. However, the effects of van der Waals interactions on structural stability need further research. Examining the impact of octahedral crystal field on the d orbitals of M can help explain the adsorption energies of surface groups on MXenes. For instance, a study on $Ti_3C_2T_2$ (T= O, F, H, OH) revealed that the created crystal field caused the 3d orbitals of surface Ti atoms to split, forming pseudo gaps whose values indicate the stability of $Ti_3C_2O_2$, $Ti_3C_2H_2$, $Ti_3C_2(OH)_2$, $Ti_3C_2F_2$, and Ti_3C_2, in that order. Additionally, they also observed how different functionalization combinations affected the stabilities of different materials (i.e., $Ti_2CO_xF_y(OH)_z$, $Ti_3C_2O_xF_y(OH)_z$, and $Nb_4C_3O_xF_y(OH)_z$). Therefore, the least energetically favorable materials are fully O-terminated (OH-terminated) Ti_2C, Ti_3C_2, and Nb_4C_3 [5].

When formation energies of $M_{n+1}X_nO_2$ were calculated, it was found that all of the formation energies are positive. Among the synthesized MXenes, V_2CO_2 has the highest positive value (+0.285 eV/atom) and may be regarded as a threshold. Generally, only those MXenes that have formation energies lower than those of V_2CO_2 are likely to be synthesized. The effect of temperature- and time-dependent degradation of MXenes, which could cause MXenes to change into bulk transition metal carbides or oxides, is another significant concern. For instance, it has been shown in experiments that heating 2D Ti_3C_2 results in the transformation of TiO_2 particles into thin graphitic carbon sheets. Thermodynamic studies indicate that Ti_3C_2 may become TiF_3 and TiF_4 due to the high HF concentration [5].

The computation study revealed that every MXene without surface functionalization was metallic. While the majority of the other MXenes were metallic, Ti_2CO_2, Hf_2CO_2, and Zr_2CO_2 were semiconducting as a result of influence of terminations on the electronic structure of pristine MXenes. $Cr_2C(OH)_2$, Cr_2CF_2, Cr_2NO_2, Cr_2NF_2, and $Cr_2N(OH)_2$ have ferromagnetic ground states. Generally, good thermoelectric materials are semiconductors with narrow gaps [4, 7]. The final properties of a surface termination group are mutually influenced by the structural composition of MXenes and their respective species. Most of the time, MXene structures are over-simplified as homogeneous terminating species, which does not correspond to the actual circumstances. As a result, more accurate models must be developed to capture this intricate system, since multiple surface groups may coexist and random absorption frequently occurs simultaneously. Furthermore, reports of multilayer stacking in practice highlight how crucial it is to examine interlayer interactions [9].

2.2   VARIABLE VALENCY

The difference between the energy levels and the electron work function (WF) indicates that least amount of energy is required to remove electrons from the surface of MXenes. The induced surface dipole created by the surface terminations and the material's shift at the Fermi level due to electron redistribution can both be related to the change in the WF of bare/pristine MXenes. The temperature-sensitive WF of metallic MXenes ranges from 1.8 to 8 eV. O-termination raises the WF of MXenes in comparison to the bare surface, while the OH group lowers it. Depending on the specific material, the effect of –F group on the WF of MXenes varies. Remarkably, WF is comparatively low for all hydroxyl-terminated MXenes—it is even lower than for Sc metal. Compared to Pt, which has the highest WF of all metals, several O-terminated MXenes have comparatively high WF (W_O). In comparison to W_O and W_{OH}, F-terminated MXenes exhibit an intermediate WF (W_F). Additionally, there is a positive association between W_F and W_O and a negative correlation between W_{OH} and W_O. The change in dipole moment brought about by the terminations is responsible for the change in their corresponding WF [10].

Fluorine can cause a positive or negative dipole moment on the surface depending on the choice of material, and the hydroxyl termination frequently causes a negative dipole moment, thereby lowering the WF. Therefore, in order to generate an ultralow (or ultrahigh) WF, it is important to modulate the –OH group of single photonic crystals [10]. The difference between energy per atom and the most stable configuration is also used to determine the formation energies of nanosheets (NSs). The formation energy of MXenes is the difference in energy between the 2D material and the corresponding 3D phases. Interestingly, despite the fact that relatively few nitride MXenes have been synthesized yet, a number of them are projected to be stable. There are a few possible explanations for this, but the most convincing one would be that in the formation energy calculations, water molecules have not been taken into account as competing species. Additional explanations could include the effect of entropy at high temperatures or the inadequate DFT representation of the triple bond in N_2 [11, 12].

The common oxidation states of M are represented by three groups, which are about 3+ for Sc and Cr; 4+ for Ti, Zr, and Hf; and 5+ for Nb and Ta. This arrangement makes sense because carbon is the favored X element for high oxidation metals because it receives more valence electrons than nitrogen, and vice versa. The

$M_2XT_2 > M_3X_2T_2 > M_4X_3T_2$ trend is typically followed by the MXene formation energies. This is what makes it reasonable because thicker MXenes have a greater surface-to-volume area ratio [11, 12]. The WF of MXene is not only a property of fundamental interest in surface research, but it is also a metric with uses in energy storage, catalysis, electronics, etc. MXenes with low WF values are used in electronics, but those with high values are preferred in sensing and optoelectronics. The highly adjustable surface chemistry of MXenes has led to an increase in their application as substrates for noble metal catalysts. The dipole moment at the metal–support interface and the degree of charge transfer are both indicated by the WF [13].

T. Schultz et al. [14] demonstrated that the WF of $Ti_3C_2T_x$ can be changed by using metal–organic thin layers of donor or acceptor molecules. This allows one to align the energy level to an organic semiconductor that is deposited later on, from Fermi energy at the lowest unoccupied to the highest occupied molecular level. Moreover, it was demonstrated that, contrary to theoretical assumptions, an O-terminated surface does not result in an exceptionally high WF value. Additionally, the WF of MXenes was computed at various annealing temperatures. The results showed that, depending on the surface composition of the material, the value ranged from 3.9 to 4.8 eV. After vacuum heating, the WF increased from 3.9 to 4.8 eV, most likely as a result of carbon-dominated pollution, the presence of OH species, and water desorption. At higher temperatures, fluorine desorption caused the WF to drop to 4.1 eV [15]. Different adsorption sites for surface groups of $Ti_3C_2T_x$ are illustrated in Fig. 2.2 [15].

Large-area MXenes (Ti_3C_2) were shown to be effective as electrical contacts in another work, demonstrating the production of n-type and p-type oxide thin-film transistors (TFTs). The WF of the MXene (i.e., contact material), the SnO valence band, and the n-type ZnO conduction band were determined using X-ray photoelectron spectroscopy. The measured values of the MXene, E_C (electron affinity), and E_V

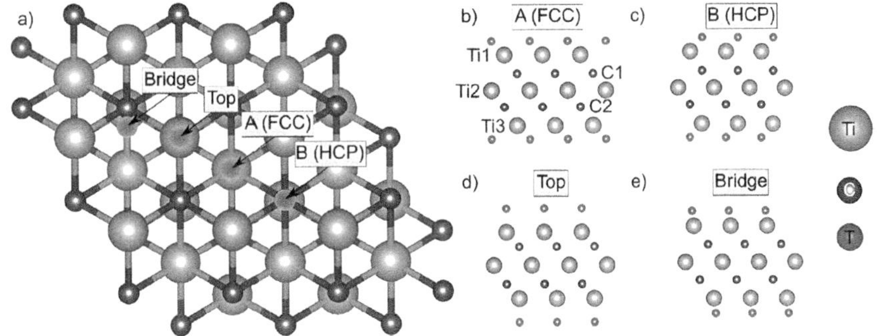

FIGURE 2.2 Schematic illustration of the different adsorption sites of $Ti_3C_2T_x$. (a) Top and side views. (b) A (FCC) adsorption site (on top of Ti atom in the third atomic layer), (c) B (HCP) adsorption site (on top of C atom in the second atomic layer), (d) top adsorption site (on top of a Ti atom in the first atomic layer), and (e) bridge adsorption site (between topmost Ti and C atoms).

Source: Reprinted with permission from [15] T. Schultz, Nathan C. Frey, K. Hantanasirisakul, et al. 2019. Surface Termination Dependent Work Function and Electronic Properties of $Ti_3C_2T_x$ MXene. *Chem. Mater.*, 31, 6590–6597. Copyright 2019, American Chemical Society.

(ionization energy) were in the order of 4.60, 4.58, and 4.65 eV, respectively. These numbers show that there is little potential barrier into the transistor channel for carrier injection [10].

2.3 TUNABLE INTERLAYER SPACING

During the selective etching and exfoliation of MAX phases, an accessible interlayer spacing is created in the MXene structures. This space provides an abundance of active sites for ion storage. The electrochemical performance of layered MXenes is significantly influenced by the interlayer distance. In general, the energy-storage capability of HF-etched MXenes will be hindered by their limited interlayer area. These materials will offer a reduced diffusion barrier and greater accessibility by expanding the interlayer gap. Thus, the interlayer channel influences the carrier ion diffusion in MXene-based electrodes. Therefore, by intercalating molecules or ions, it is possible to decrease or increase the interlayer distance, thereby ultimately improving their electrochemical performance. Intercalants can also help ensure long-term cycling stability. Remarkably, compared to conventional bare MXenes, the cycling stability and metal-ion storage capacity of modified MXene-based materials are substantially higher [12, 16].

MXenes are capable of accepting a variety of intercalants, that is, ionic, inorganic, and organic species, due to the weak bonding between M-X layers. A variety of multivalent cations (Li^+, Na^+, K^+, Mg^{2+}, etc.) can be incorporated into M-X layers, suggesting their use in metal-ion batteries. These devices can provide high capacitance values (i.e., 300 F cm^{-3}) better than all carbon-based electrochemical double-layer capacitors (EDLCs). To extend the interlayer gap of MXenes, polar organic molecules of various sizes, including urea, hydrazine, isopropanol, tetrabutylammonium hydroxide (TBAOH), dimethyl sulfoxide (DMSO), and hydrazine, can also be intercalated. Similar to $Ti_3C_2T_x$, when hydrazine was intercalated, the lattice parameter increased from 19.5 Å to 25 Å. Producing "MXene paper," which entails sonication in water and filtration of the supernatant colloidal solution, also requires intercalation with such molecules [7, 17, 18].

A technique for considerably improving the electrochemical behavior of $Ti_3C_2T_x$ MXenes through surface modification and cation intercalation was described by J. Li et al. [19]. They observed that after K^+ intercalation and –OH/–F removal, pseudocapacitance was observed to be three times higher than that of the pure material. Additionally, MXene sheets showed a notable enhancement in the capacitance (i.e., 517 F g^{-1} at 1 A g^{-1}). The lowest terminating surface group concentration and extensive interlayer voids of Ti_3C_2 are responsible for this enhanced electrochemical performance. Alkalized Ti_3C_2 nanoribbons were created by P. Lian et al. [20] by continuously shaking pristine Ti_3C_2 MXene in an aqueous KOH solution. The resulting a-Ti_3C_2 anodes demonstrated excellent Na/K storage due to the extended interlayer spacing, narrow widths, and 3D interconnected porous frameworks. Similar to this, A.V. Mohammadi et al. [21] showed how to assemble unstable 2D V_2CT_x MXene flakes into very stable pseudocapacitive electrodes with improved electrochemical characteristics. At these scan rates, the specific capacitances delivered by the

C-V_2CT_x electrodes are higher than those of the d-$Ti_3C_2T_x$ film electrodes, demonstrating remarkable high-rate capabilities.

In another study, V_2C MXene with altered interlayer spacing for desired storage capacity was demonstrated by C. Wang et al. [22]. Through strong V–O–Co interaction, the cobalt ions were securely intercalated into the V_2C MXene interlayer to generate a novel interlayer expanded structure. Intercalated V_2C MXene electrodes provide considerably ultralong cycling stability of over 15,000 cycles in addition to demonstrating better capacity of up to 1117.3 mAh g^{-1} at 0.1 A g^{-1}. By tuning the interlayer spacing and V–O–Sn bonding, C. Wang et al. [23] demonstrated Sn^{4+}-decorated V_2C material that not only exhibited better lithium-ion battery performance but also possessed good cycling stability. Sn^{4+}-decorated Ti_3C_2 nanocomposites were synthesized by J. Luo et al. [24] using a simple liquid-phase immersion method assisted by polyvinylpyrrolidone (PVP). Owing to the potential "pillar effect" of Sn between alk-Ti_3C_2 layers, the nanocomposites demonstrated excellent capacitance retention after 50 cycles and a superior reversible volumetric capacity (much higher than that of a graphite electrode). Based on these findings, PVP-Sn(IV)@Ti_3C_2 nanocomposites present an intriguing possibility for high-efficiency lithium-ion batteries.

Fe^{2+} ions were added to MXene dispersion by Y. Deng et al. [25], and they function as linkers to join the NSs together to create a three-dimensional (3D) MXene network. When employed as a supercapacitor electrode, the resultant hydrogel efficiently prevents the MXene NSs from restacking and significantly increases their surface utilization, leading to a high-rate performance. Guo et al. [26] achieved $Ti_3C_2T_x$ by removing gallery Al atoms in Ti_3AlC_2 using a straightforward hydrothermal etching process, therefore reserving suitable Al interlayers ($Ti_3C_2T_x$@Al). Instead of solitary $Ti_3C_2T_x$ flakes, $Ti_3C_2T_x$@Al maintains a stable layered structure, preventing flake restacking. Consequently, the $Ti_3C_2T_x$@Al-based all-solid-state supercapacitor (ASSS) shows stable performance at various bending states and produces a high capacitance of 242.3 mF cm^{-2} at 1 mV s^{-1}. A Lewis-basic halides treatment was proposed by T. Zhang et al. [27], and simultaneously surface termination and interlayer spacing of different MXenes can be tailored. Benefiting from the plentiful desolated halogen anions and cations in Lewis-basic halides in a molten state, a nucleophilic reaction replaced the –F termination, increasing the interlayer gap. In order to improve ionic conductivity, M. Lu et al. [28] used Al^{3+}-based pre-intercalation to develop an enlarged ion transfer channel that permits further enlargement during the charging process.

The 2D layered $Ti_3C_2T_x$ were tailored by Y. Li et al. [29] using different cations, such as K^+, Ca^{2+}, Mg^{2+}, Al^{3+}, and NH^{4+}, transported into the MXene interlayers by ion exchange and pre-intercalated with Li^+. The Al^{3+}-$Ti_3C_2T_x$ sheets provided the maximum specific capacity (i.e., 175 F g^{-1} at 0.3 A g^{-1}) by increasing the interlayer by approximately 17.02%. An easy alkali metal ion pillaring technique was described by J. Luo et al. [30] to improve the kinetics and capacity of Ti_3C_2 for sodium-ion storage. Following Na^+ pillaring, the Na-Ti_3C_2 NSs demonstrated an excellent cycling stability for 2,000 cycles (at 2.0 A g^{-1}) along with a high reversible capacity. The formation of pillared Ti_3C_2 with controlled interlayer spacings was reported by P. Simon et al. [31]. By exchanging ions with Sn(+IV) ions, these materials underwent further intercalation. The confinement effect, which restricts volume growth, and enhanced ion accessibility to the structure are both responsible for the improved electrochemical performance seen in the data.

Because of the limitation of small volume with most cations, it is highly anticipated to investigate intercalation of MXenes with polymers and organic compounds to further increase interlayer spacing [16]. Li^+-intercalated $Ti_3C_2T_x$ and ion-exchanged trimethyl alkylammonium (AA) cations were synthesized by M. Ghidiu et al. [32]. The AA cations were studied as interlayer pillars and were observed to increase the resistivity of conductive MXenes. J. L. Hart et al. [33] demonstrated how intercalation can cause inter-flake effects to bring about changes from metallic to semiconductor-like transport. The understanding of magnetic, semiconducting, and topologically insulating MXenes may be facilitated by these discoveries, which establish the foundation for intercalated MXenes, thereby offering enhanced electronic conductivity. By etching pristine Ti_3AlC_2 MAX in NH_4F along with choline chloride and oxalic acid, J. Wu et al. [34] developed a water-free ionothermal synthesis method. The DES-Ti_3C_2 anodes consequently showed improved lithium storing capability. O. Mashtalir et al. [35] observed that intercalation of hydrazine and N,N-dimethylformamide increases the c parameters of surface-functionalized f-Ti_3C_2.

S. Zheng et al. [36] fabricated micro supercapacitors (MSCs) by pre-intercalation of ionic liquid and observed high areal and volumetric energy density when operated at 3 V in 1-butyl-3-methylimidazolium tetrafluoroborate (EMIMBF$_4$). According to research by M. Peng et al. [37], p-phenylenediamine (PDA)-MXene exhibits pore structure for better charge-transport characteristics, improved electrolyte-accessible surface area, and promoted Zn^{2+} ion storage. A straightforward method for achieving high magnesium storage capacity for Ti_3C_2 MXene was presented by M. Xu et al. [38] through the pre-intercalation of cetyltrimethylammonium bromide (CTAB). DFT studies confirmed that intercalated CTA^+ cations lower the Mg^{2+} diffusion barrier, thereby improving the reversible insertion/desertion of ions between the layers significantly. J. Luo et al. [39] also used a simple liquid-phase CTAB and an Sn^{4+}-based approach to create pillared Ti_3C_2. In comparison to traditional MXene materials, Lithium-ion capacitors (LIC) displays increased energy and power density with the CTAB–Sn(IV)@Ti_3C_2 anode and an AC cathode.

2.4 EFFECT OF THE MASS OF METALS

Using DFT calculations, researchers investigated the mechanical properties of pristine and terminated MXenes with M_2XT_2 composition. Elastic constants showed that, in contrast to pristine nitrides, there is a positive association between the mass of the transition metal and the stiffness of pristine carbides. Furthermore, the Young Modulus of the nitrides was marginally greater than that of carbides. In a recent study, the mechanical properties of monolayered Janus MXenes (i.e., M_2X) were examined theoretically, to determine the impact of −F and −OH functional groups. It was discovered that surface functionalization and the bulk of the transition metal influence mechanical characteristics. According to the results, the elastic characteristics of MXenes are impacted by asymmetric functionalization [40]. Since the N atom offers an extra valence electron above the C atoms, resulting in stiffer M-X bonds in all pristine M_2X structures, the in-plane stiffness of M_2C is somewhat lower than that of M_2N. Nonetheless, the in-plane stiffness of Mo_2N is marginally less than that of Mo_2C because of the H structure of Mo_2X. Furthermore, monolayer M_2X has a lower in-plane stiffness than single-layer h-BN and graphene, which can be improved upon asymmetrical surface functionalization [40].

Using molecular dynamics (MD) computations, the impact of thickness of layers on the elastic and structural characteristics of 2D $Ti_{n+1}C_n$ was also investigated. It was observed that by reducing the layer thickness, the Young's modulus of MXenes may be considerably increased. The Young modulus was determined to be 597 GPa for Ti_2C, 502 GPa for Ti_3C_2, and 534 GPa for Ti_4C_3, with an interpolation error of 10% and a strain less than 0.01. Ti_2C (with 3 atomic layers) was found to have the highest Young's modulus. These outcomes line up with other theoretical predictions. Another study used DFT calculations to report similar results regarding the impact of mono-layer thickness on the elastic characteristics of the $Ti_{n+1}C_n$ and $Ti_{n+1}N_n$ MXenes. It was reported that the Young's moduli of MXenes decrease with increasing monolayer thickness. For Ti_2C, Ti_3C_2, and Ti_4C_3, the $Ti_{n+1}C_n$ Young's moduli were determined to be 601, 473, and 459 GPa, respectively. These values are in good agreement with previous theoretical calculations [3, 40].

According to cohesive energy estimations, MXenes with thicker monolayers have superior structural stability. According to estimations of adsorption energy, $Ti_{n+1}N_n$ has a greater inclination to stick to the terminal groups, suggesting that nitride-based MXenes have more active surfaces. More notably, beyond the OH-terminated carbide/nitride-based MXenes, essentially free electron states are discovered, particularly in $Ti_{n+1}N_n(OH)_2$. These materials offer the transmission channels for electron transport without nuclear scattering. It is observed that nitride MXenes exhibit higher electrical conductivity than carbides. We cannot assert that additional parameters, such as yield stress and associated strain, are accurately replicated theoretically in the absence of experimental data, even though the estimated Young's moduli are nearly identical. In conclusion, the mechanical behavior of the two-dimensional (2D) titanium carbides is accurately described by the traditional MD technique that was used. This approach can be extended to investigate the structural and mechanical features of additional bare and surface-functionalized MXenes [41, 42].

2.5 DOPING EFFECTS AND EFFECTS OF INTERCALATIONS

MXenes hold outstanding potential; however, they are quite prone to performance degradation. Research on MXenes has several significant obstacles, including layer restacking, oxidation susceptibility, limited flexibility, high contact resistance, biocompatibility, and cytotoxicity. "Doping," or adding external elements to 2D materials, is useful for adjusting the materials' chemical and physical characteristics. Doped MXenes have been shown to be an effective technique to address these issues, similar to previous layered and 2D systems. In addition to enhancing their characteristics, doping the MXenes has allowed them to be used in previously unexplored areas such as sensors and catalysis [43]. In the last few years, there has been a tremendous increase in the production of hetero-MXenes through extensive experimental work and theoretical studies. In general, hetero-MXenes can retain the superior qualities

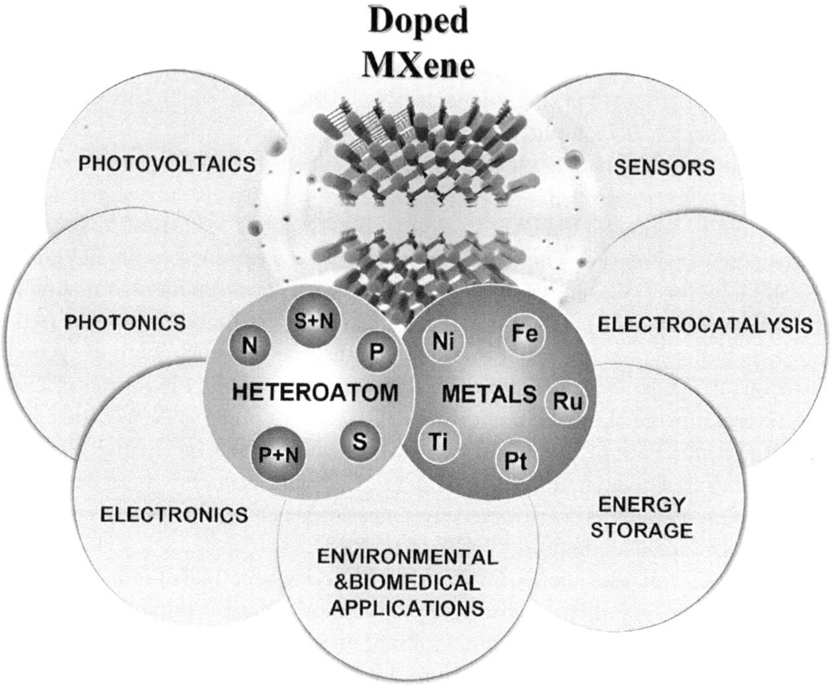

FIGURE 2.3 Doping of MXenes by metals and heteroatoms and their corresponding applications.

Source: Reprinted from [43] *Progress in Materials Science*, 139, A. Dey et al., Doped MXenes—A new paradigm in 2D systems: Synthesis, properties and applications, 101166, Copyright 2023, with permission from Elsevier.

of their parent compounds while potentially altering the surface terminations and lattice structure to produce more active sites. The hetero-MXenes outperform pristine MXenes in a number of applications, including as energy storage (rechargeable batteries and SCs), electrocatalysts [hydrogen evolution reaction (HER), oxygen evolution reaction (OER), oxygen reduction reaction (ORR), and nitrogen reduction reaction (NRR)], and sensors, because of their unique characteristics (as illustrated in Fig. 2.3 [43]) [44].

For the purpose of designing, engineering, and functionalizing doped MXenes, both theoretical and experimental methods are being investigated. Doping for the M, X, or T positions has therefore been studied, and the results have been categorized as M-, X-, or T-doped MXenes. All three components can be occupied by heteroatoms like O, N, S, or P. Halides can take the place of T elements, and transition metals can be added to the doped MXenes to take the place of any or both M and X positions [43, 44]. After pretreating Ti_3C_2 with CTAB, J. Luo et al. [45] successfully intercalated the S atoms to expand the interlayer spacing via Ti-S bonding. This was achieved at an annealing temperature of 450 °C. They found that the resultant CT-S@ Ti_3C_2_450 electrode provides an improved Na-ion capacity of 550 mAh g^{-1} (at 0.1 A g^{-1}) and better capacitive retention (after 5,000 cycles at 10 A g^{-1}). Similarly, X. Guo et al. [46] reported that P-V$_2$C/NiCo-LDH retains a specific capacity of 1,077 mAh g^{-1} after 700 cycles (at 500 mA g^{-1}). In a similar manner, Yoon et al. [47] presented an approach to enhance the capacitance of SC electrodes through the synthesis of N-Ti$_2$CT$_x$ using p-C$_3$N$_4$, which serves as both an intercalant and a nitrogen source.

For symmetric electrochemical capacitors (ECs), C. Yang et al. [48] produced N-Ti$_3$C$_2$ film electrodes that were flexible, freestanding, and compact. They reported that in a 3M H$_2$SO$_4$ electrolyte, the urea-assisted film displays an ultrahigh volumetric capacitance. The high-capacity SC material described by C. Yang et al. [49] is based on N-d-Ti$_3$C$_2$, which are made by carbonizing delaminated Ti$_3$C$_2$ combined with urea. These NSs in two dimensions show a high specific capacitance together with a strong capacity retention ability. In order to synthesize a novel kind of N-Ti$_3$C$_2$T$_x$, Y. Wen et al. [50] post-etched annealed Ti$_3$C$_2$T$_x$ in ammonia. This material showed promising use as electrode. When optimized, the resulting doped MXene materials showed much greater electrochemical capacitances than the un-doped Ti$_3$C$_2$T$_x$ materials, measuring 192 F g^{-1} and 82 F g^{-1} in 1 M H$_2$SO$_4$ and 1 M MgSO$_4$ electrolyte, respectively. Another simple method for improving the HER catalysis of N-Ti$_3$C$_2$T$_x$ was given by T. A. Le et al. [51] utilizing ammonia as a heat treatment. In comparison to pure Ti$_3$C$_2$T$_x$, the experimental results demonstrated that doped material annealed at 600 °C exhibits superior HER activity.

Y. Tang et al. [52] used the inexpensive industrial substance urea as a nitrogen source to create new N-Ti$_3$C$_2$ NSs through a simple one-step hydrothermal synthesis. The N-Ti$_3$C$_2$ as received demonstrates a high cycling stability and a maximum specific capacitance of 156 F g^{-1} (at 5 mV s^{-1}). Even with the aforementioned advancements, there are still many scientific questions concerning synthetic strategies, structure–property relationships, and intrinsic mechanisms for performance enhancement. Our knowledge of the specifics surrounding hetero-MXenes is still in its initial stages. First of all, the concepts of substituted and doped MXenes are still not quite clear. MXenes with low dopant concentrations (ppm or lower) should be referred to as doped MXenes based on their heteroatom content, particularly if they are MXene-based catalysts and have single atoms at the MXene defect sites. Higher heteroatom content MXenes, on the other hand, are referred to as substituted MXenes and would have distinct characteristics from doped MXenes. More research into the production mechanisms and possible uses of doped and substituted MXenes will benefit from a much clearer differentiation between the two [44].

The functions of metal ions in applications and processing of MXenes were compiled by Y. Long et al. [53]. Three categories are established based on how they alter the MXene structure: etching, cross-linking, and intercalation. The electrostatic attraction between the surface groups (negatively charged) and the metal cations is the driving force behind intercalation and cross-linking, and this ion exchange behavior allows further metal ions to be inserted. Because of the transition metal's redox activities in MXenes, charge transfer appears to be the main cause of in-plane defect creation in the etching process. According to N. C. Osti et al. [54], MXene with K$^+$ intercalation has a more homogeneous, uniform structure and can retain water stability between its individual layers better than with pure MXene. Water in this MXene becomes significantly less mobile, improving the composite's stability against environmental changes and possibly improving its operational stability over a wide variety of applications. f-$Mo_2Ti_2C_3$ was utilized as a SC electrode for the first time ever by D. Gandla et al. [55] using 1 M EMIMTFSI (1-Ethyl-3-methylimidazolium bis(trifluoromethylsulfonyl)imide) in acetonitrile electrolyte.

In conclusion, various organic molecules, cations, and polymers can intercalate and enlarge the interlayer of MXenes. This will have the advantage of facilitating charge transport, providing additional electrochemical active site. Even with all of these efforts, a great deal of important subjects needs to be explored. For instance, more research is still needed to determine how heteroatom doping may affect electrochemical performance. Furthermore, it remains difficult to precisely and selectively manipulate the surface groups. Enlarged ionic resistance in an electrode–electrolyte interface may result from surface modification, which can also reduce the electrode material's capacity to donate electrons and alter other structural and chemical characteristics. It should be mentioned that theoretical computations can be a potent tool for demonstrating rules for comprehending the mechanics behind heteroatom doping's effects on the electrochemical performance of MXenes [16].

Furthermore, greater cooperation in combining the benefits of sophisticated characterization methods with theoretical computations is needed in order to fully comprehend the intercalation mechanism and advance the energy storage capabilities of MXene-based materials. Because of the physical and chemical instability, these materials are currently difficult to employ on a broad scale for industrial applications. Regarding future research, a greater emphasis should be placed on the safe and moderate synthesis of high-quality MXenes. In order to achieve the controlled synthesis of MXene-based electrode materials, it is also crucial to investigate more effective species that might be inserted into MXene sheets. Finally, apart from the intriguing energy storage abilities, there are still a plethora of alternative applications for MXene-based materials. For instance, the intercalation process may have an impact on 3D printing, light heat, artificial intelligence, machine learning, and so forth [16].

2.6 STABILITY AND SURFACE CHEMISTRIES

The development of MXenes for various applications has drawn increasing attention due to their unique morphologies and electrical, mechanical, and magnetic properties. However, it has been recognized that the processing and commercial use of MXene-based devices need to be improved because of environmental instability. The combined impacts of air, moisture, and light can easily cause MXenes to disintegrate into distorted structures and be adorned with oxides. Degradation mechanisms have the potential to compromise structural integrity and impact MXene activity seriously. High temperatures and oxidizing environments (air, O_2, CO_2, etc.) greatly compromise the stability of these materials. The basic chemical composition, along with the choice of surface group and M-layers, plays a major role in their stability. MXene structural changes can occur at high temperatures, even without any external source of oxidation, and are frequently caused by a mixture of molecular hydrogen between the NSs and surface terminations [56, 57].

It is expected that oxidation reactions will initiate at the functional sites that are more likely prone to oxidation. The effective way to improve the intrinsic thermal stability of MXenes is to remove surface groups by thermal annealing them in an inert environment or vacuum. Heat treatment may cause multiple changes in the crystal structure at the same time, including coupling, defect repair, and a decrease in interlayer spacing. Furthermore, synthesized MXenes terminated with –Cl exhibit greater stability compared to those terminated with –F and –O, and additional termination systems may yet be discovered that offer the possibility of producing MXenes with enhanced performance. Lewis-acid and Lewis-base molten salts are two examples of molten inorganic salt compositions that can be used to undertake substitution or elimination reactions in order to tune or customize surface terminations (e.g., T = Br, Cl, S, Se, or Te) [56]. From a materials perspective, oxidation has also been examined, and the edges, defects, and vacancies of MXenes have been found to be oxidation-prone regions. However, a thorough comprehension of the ways in which water molecules engage with these sensitive locations is still unclear [2, 57, 58].

Primitive MXene NS oxidation takes place close to the defected areas. The mechanism by which structural defects affect the oxidation of methylene is intricate and involves the production of internal electric fields that facilitate the migration of electrons and Ti-cations, resulting in the formation of larger vacancies and TiO_2 crystals. Therefore, oxidation can be reduced by removing the internal electric fields that are naturally generated close to structural flaws. Nevertheless, structural flaws will inevitably arise from the chemical and mechanical exfoliation procedures. The desired MXene NSs with fewer defects can be achieved by environmentally friendly and safe synthesis methods, such as by using a LiF–HCl or difluoride (e.g., KHF_2, $NaHF_2$, and NH_4HF_2) etching solutions instead [56, 57]. Because many MXene-based devices are often made using these etchants, their endurance is extremely important. Not only can MXenes dissolved in aqueous solutions experience oxidation, but their rate of oxidation is also considerably higher than that of MXenes that are freestanding [56].

Thus, more research is needed to completely understand the mechanism of MXene degradation in an aquatic environment. The environmental stability of MXenes can be increased by implementing a well-thought-out and uniform surface encapsulation method that not only increases surface smoothness but also serves as an additional barrier to surroundings [56]. Antioxidation techniques that provide effective means of maintaining these materials in their aqueous dispersion state are currently restricted to expecting an extension of MXenes' shelf life. Determining the total impact of these antioxidants on the performance of MXenes in diverse applications requires additional research. Further research is needed to understand the interactions among molecules and specific orientation of the antioxidants across the surface, as antioxidation techniques mostly rely on the surface capping of MXenes. It is also important to consider the concentration and total elimination of these antioxidants. Therefore, the focus of research should shift to investigating more powerful antioxidant compounds that may be able to efficiently stabilize MXenes at lower concentrations and either eliminate the requirement for removal entirely or make removal easier [2, 57, 58].

In addition to extending the storage period, dividing MXene NSs with bigger lateral sizes from those with smaller ones will increase MXene activity for various uses.

Enhancing storage conditions in a refrigerated and deoxygenated atmosphere is a crucial measure to safeguard MXenes against oxidation. Storing in organic solvents devoid of moisture can also significantly improve stability, as the presence of water facilitates the diffusion of ions (e.g., Ti^{4+} and O^{2-}) involved in the oxidation process. The stability of MXenes in organic solvents has been discovered, which concurrently opens up new processing avenues for MXenes. For example, inks and MXene-based polymer composites can be used for 3D additive manufacturing. Additionally, MXene particles can also be totally spared from the undesirable degradation brought on by the aqueous environment by being isolated from the colloidal solution. MXenes undergo severe oxidation when exposed to oxidizing environments such as air and oxygen, aqueous solutions containing oxygen, corrosive solutions containing H_2O_2 and NaOH, and anodic, hydrothermal, and UV light. If MXenes are shielded from extreme heat by inert atmospheres such as Ar, N_2, and He, they can also retain their original multi-layer structure throughout lengthy temperature ranges [56, 59].

A. Mishra et al. [60] conducted first-principle calculations on the stability of M_2CO_2 functionalized with –O group. There are two possible structural phases for M_2CO_2, depending on where the O atoms are located (as illustrated in Fig. 2.4 [60]).

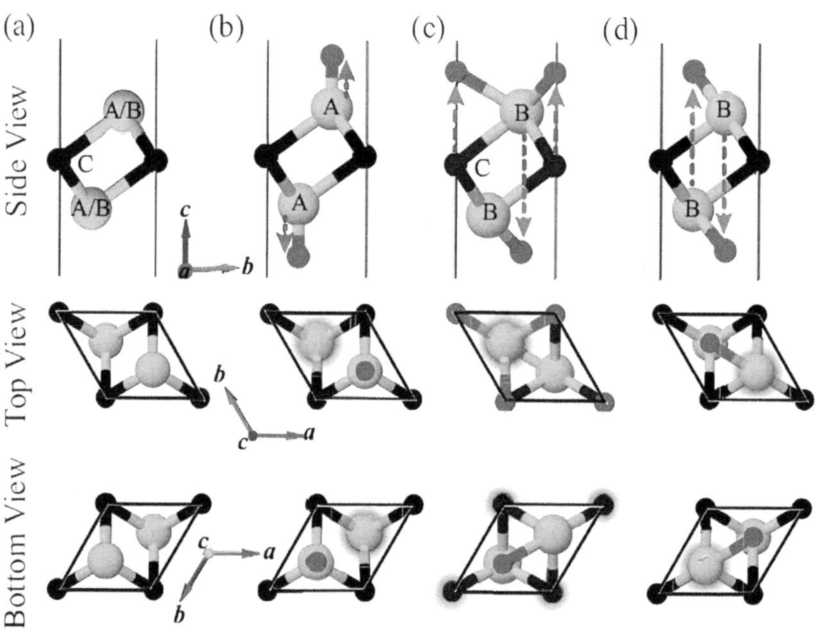

FIGURE 2.4 Structures of the different possible phases of O-functionalized MXenes together with the structure of pristine MXene. (a) Structure of pristine MXene with adsorption sites, labeled as A, B, and C. (b), (c), and (d) Structures of O-functionalized AA, CB, and BB′ phases, respectively.

Source: Reprinted with permission from [60] A. Mishra, P. Srivastava, A. Carreras, et al. 2017. Atomistic Origin of Phase Stability in Oxygen-Functionalized MXene: A Comparative Study. *J. Phys. Chem. C*, 121, 34, 18947–18953. Copyright 2017, American Chemical Society.

The O atom can be located exactly on top of the M on one side of the MXene. Conversely, it can be found on top of the C atom (CB phase) or on top of the M on the opposing side (BB′ phase). They discovered that the CB phase is stable for metals like Sc and Y, while the BB′ phase is stable for other metals. Using a molten salt ($FeCl_2$) etching method, L. Liu et al. [61] synthesized Cl-terminated material (designated as $Ti_3C_2Cl_x$) for the first time. Next, by post-heating in Li_3N-containing electrolytes, substitution from Cl$^-$ to N$^-$ was carried out. The findings demonstrated that the creation of high-performance electrodes in a hitherto unreachable limit of energy storage is possible through control of the MXene surface chemistry.

X. Zha et al. [62] examined the structural, electrical, and mechanical characteristics of the M_2CT_2 using first-principle DFT computations. The results demonstrate that W_2CO_2 has the highest mechanical strength, and Sc_2CO_2 has the shortest interlayer thickness. The M_2CO_2 structures, Ss_2CF_2, Mo_2CF_2, and $Sc_2C(OH)_2$, all displayed semiconducting qualities in terms of electrical properties. The stability and optimization of the $Hf_3C_2F_2$ monolayer utilizing ab initio molecular dynamics (AIMD) by DFT study were presented by R. P. Jadav et al. [63]. The steady structure, metallic character, and low diffusion energy barrier of the $Hf_3C_2F_2$ monolayer, they observed, indicate that a metal anode material can be used for the rechargeable storage device. Similarly, Z. Li et al. [64] studied the thermal stability of 2D carbide MXene, i.e., Ti_3C_2.

The thermal stability of MXene–organic hybrid polystyrene (PS) nanocomposites was studied by Z. Zhang et al. [65]. First, bulk MAX (Ti_3AlC_2) material was exfoliated for the MXene NSs. To increase their dispersibility in dimethylformamide, these NSs were subsequently terminated with a cationic surfactant. The resulting PS nanocomposite exhibited good thermal stability, even with a negligible loading of $O-Ti_3C_2$ (e.g., 2wt%). A number of MXenes have been synthesized so far; however, the majority are carbides, with the exception of a few nitrides and carbonitrides. Only a few MXene types—such as $Ti_3C_2T_x$ and Nb_2CT_x—have drawn experimental studies on their ability to oxidize. Despite expectations regarding binding energy derived from theoretical calculations, the differences in stability among different MXenes remain poorly understood. It is imperative to investigate more favorable and promising MXenes (such as $Zr_3C_2T_x$ and $Nb_4C_3T_x$) for use in harsh settings, particularly while operating at a high temperature [56], as the inherent qualities of MXenes are weakened by oxidative breakdown. By regulating various synthesis parameters, such as the quality of MAX phase, the type of acid etchants, ultrasonication, storage media, and pH, it is possible to significantly postpone the oxidative decomposition of MXenes [66].

One of the most difficult tasks in using high-quality MXene for extended periods of time while maintaining optimum repeatability of outcomes is storing it after careful synthesis. For this reason, if kept in an inert atmosphere and at a lower temperature, an aqueous dispersion of MXene with a neutral pH can be kept for a few months. Eliminating dissolved oxygen and water is advantageous because they are the primary causes of oxidation. By using their dispersions in organic solvents instead of aqueous ones, the oxidation stability of MXenes can be increased by reducing their exposure to water and oxygen molecules. The oxidative breakdown response of aqueous $Ti_3C_2T_x$ MXene dispersions remains poorly understood, despite encouraging experimental advances. It is also difficult to completely stop $Ti_3C_2T_x$ and other MXenes from oxidizing, particularly when they are transferred from labs to businesses to be used in the creation of commercial products. As a result, MXenes have motivated researchers to look into the kinetics and mechanisms of their oxidation as well as create fresh plans of action to increase their oxidation stability [66].

REFERENCES

[1] Lengauer, W., *Transition metal carbides, nitrides, and carbonitrides*, in *Handbook of Ceramic Hard Materials*. 2000, Wiley-VCH. p. 202–252.

[2] Hantanasirisakul, K. and Y. Gogotsi, *Electronic and optical properties of 2D transition metal carbides and nitrides (MXenes)*. Advanced Materials, 2018. **30**(52): p. 1804779.

[3] Biswas, S. and P.S. Alegaonkar, *MXene: evolutions in chemical synthesis and recent advances in applications*. Surfaces, 2021. **5**(1): p. 1–34.

[4] Khazaei, M., et al., *Novel electronic and magnetic properties of two-dimensional transition metal carbides and nitrides*. Advanced Functional Materials, 2013. **23**(17): p. 2185–2192.

[5] Khazaei, M., et al., *Recent advances in MXenes: from fundamentals to applications*. Current Opinion in Solid State and Materials Science, 2019. **23**(3): p. 164–178.

[6] Halim, J., *Synthesis and Transport Properties of 2D Transition Metal Carbides (MXenes)*. Vol. 1953. 2018, Linköping University Electronic Press.

[7] Wang, C., S. Chen, and L. Song, *Tuning 2D MXenes by surface controlling and interlayer engineering: methods, properties, and synchrotron radiation characterizations*. Advanced Functional Materials, 2020. **30**(47): p. 2000869.

[8] Sun, S., et al., *Two-dimensional MXenes for energy storage*. Chemical Engineering Journal, 2018. **338**: p. 27–45.

[9] Wang, Y., et al., *MXenes: focus on optical and electronic properties and corresponding applications*. Nanophotonics, 2020. **9**(7): p. 1601–1620.

[10] Gong, Y., et al., *Emerging MXenes for functional memories*. Small Science, 2021. **1**(9): p. 2100006.

[11] Ashton, M., et al., *Predicted surface composition and thermodynamic stability of MXenes in solution*. The Journal of Physical Chemistry C, 2016. **120**(6): p. 3550–3556.

[12] Lei, Y.-J., et al., *Tailoring MXene-based materials for sodium-ion storage: synthesis, mechanisms, and applications*. Electrochemical Energy Reviews, 2020. **3**: p. 766–792.

[13] Roy, P., et al., *Predicting the work function of 2D MXenes using machine-learning methods*. Journal of Physics: Energy, 2023. **5**(3): p. 034005.

[14] Schultz, T., et al., *Work function and energy level alignment tuning at Ti3C2Tx MXene surfaces and interfaces using (metal-) organic donor/acceptor molecules*. Physical Review Materials, 2023. **7**(4): p. 045002.

[15] Schultz, T., et al., *Surface termination dependent work function and electronic properties of $Ti_3C_2T_x$ MXene*. Chemistry of Materials, 2019. **31**(17): p. 6590–6597.

[16] Tang, J., et al., *Interlayer space engineering of MXenes for electrochemical energy storage applications*. Chemistry—A European Journal, 2021. **27**(6): p. 1921–1940.

[17] Meng, W., et al., *Advances and challenges in 2D MXenes: from structures to energy storage and conversions*. Nano Today, 2021. **40**: p. 101273.

[18] Razack, A. and O. Salim, *MXene-Based Materials for Energy Storage and Separation*. 2020, UNSW Sydney.

[19] Li, J., et al., *Achieving high pseudocapacitance of 2D titanium carbide (MXene) by cation intercalation and surface modification*. Advanced Energy Materials, 2017. **7**(15): p. 1602725.

[20] Lian, P., et al., *Alkalized Ti_3C_2 MXene nanoribbons with expanded interlayer spacing for high-capacity sodium and potassium ion batteries.* Nano Energy, 2017. **40**: p. 1–8.

[21] VahidMohammadi, A., et al., *Assembling 2D MXenes into highly stable pseudocapacitive electrodes with high power and energy densities.* Advanced Materials, 2019. **31**(8): p. 1806931.

[22] Wang, C., et al., *Atomic cobalt covalently engineered interlayers for superior lithium-ion storage.* Advanced Materials, 2018. **30**(32): p. 1802525.

[23] Wang, C., et al., *Atomic Sn^{4+} decorated into vanadium carbide MXene interlayers for superior lithium storage.* Advanced Energy Materials, 2019. **9**(4): p. 1802977.

[24] Luo, J., et al., *Sn^{4+} ion decorated highly conductive Ti_3C_2 MXene: promising lithium-ion anodes with enhanced volumetric capacity and cyclic performance.* ACS Nano, 2016. **10**(2): p. 2491–2499.

[25] Deng, Y., et al., *Fast gelation of $Ti_3C_2T_x$ MXene initiated by metal ions.* Advanced Materials, 2019. **31**(43): p. 1902432.

[26] Guo, M., et al., *Flexible $Ti_3C_2T_x$ @ Al electrodes with ultrahigh areal capacitance: in situ regulation of interlayer conductivity and spacing.* Advanced Functional Materials, 2018. **28**(37): p. 1803196.

[27] Zhang, T., et al., *Simultaneously tuning interlayer spacing and termination of MXenes by Lewis-basic halides.* Nature Communications, 2022. **13**(1): p. 6731.

[28] Lu, M., et al., *Tent-pitching-inspired high-valence period 3-cation pre-intercalation excels for anode of 2D titanium carbide (MXene) with high Li storage capacity.* Energy Storage Materials, 2019. **16**: p. 163–168.

[29] Li, Y., et al., *Tunable energy storage capacity of two-dimensional Ti3C2Tx modified by a facile two-step pillaring strategy for high performance supercapacitor electrodes.* Nanoscale, 2019. **11**(45): p. 21981–21989.

[30] Luo, J., et al., *Tunable pseudocapacitance storage of MXene by cation pillaring for high performance sodium-ion capacitors.* Journal of Materials Chemistry A, 2018. **6**(17): p. 7794–7806.

[31] Simon, P., *Two-dimensional MXene with controlled interlayer spacing for electrochemical energy storage.* ACS Nano, 2017. **11**(3): p. 2393–2396.

[32] Ghidiu, M., et al., *Alkylammonium cation intercalation into Ti_3C_2 (MXene): effects on properties and ion-exchange capacity estimation.* Chemistry of Materials, 2017. **29**(3): p. 1099–1106.

[33] Hart, J.L., et al., *Control of MXenes' electronic properties through termination and intercalation.* Nature Communications, 2019. **10**(1): p. 522.

[34] Wu, J., et al., *Highly safe and ionothermal synthesis of Ti_3C_2 MXene with expanded interlayer spacing for enhanced lithium storage.* Journal of Energy Chemistry, 2020. **47**: p. 203–209.

[35] Mashtalir, O., et al., *Intercalation and delamination of layered carbides and carbonitrides.* Nature Communications, 2013. **4**(1): p. 1716.

[36] Zheng, S., et al., *Ionic liquid pre-intercalated MXene films for ionogel-based flexible micro-supercapacitors with high volumetric energy density.* Journal of Materials Chemistry A, 2019. **7**(16): p. 9478–9485.

[37] Peng, M., et al., *Manipulating the interlayer spacing of 3D MXenes with improved stability and zinc-ion storage capability.* Advanced Functional Materials, 2022. **32**(7): p. 2109524.

[38] Xu, M., et al., *Opening magnesium storage capability of two-dimensional MXene by intercalation of cationic surfactant.* ACS Nano, 2018. **12**(4): p. 3733–3740.

[39] Luo, J., et al., *Pillared structure design of MXene with ultralarge interlayer spacing for high-performance lithium-ion capacitors.* ACS Nano, 2017. **11**(3): p. 2459–2469.

[40] Ibrahim, Y., et al., *The recent advances in the mechanical properties of self-standing two-dimensional MXene-based nanostructures: deep insights into the supercapacitor.* Nanomaterials, 2020. **10**(10): p. 1916.

[41] Borysiuk, V.N., V.N. Mochalin, and Y. Gogotsi, *Molecular dynamic study of the mechanical properties of two-dimensional titanium carbides Tin+1Cn (MXenes)*. Nanotechnology, 2015. **26**(26): p. 265705.

[42] Zhang, N., et al., *Superior structural, elastic and electronic properties of 2D titanium nitride MXenes over carbide MXenes: a comprehensive first principles study*. 2D Materials, 2018. **5**(4): p. 045004.

[43] Dey, A., et al., *Doped MXenes—a new paradigm in 2D systems: synthesis, properties and applications*. Progress in Materials Science, 2023: p. 101166.

[44] Gao, L., et al., *Hetero-MXenes: theory, synthesis, and emerging applications*. Advanced Materials, 2021. **33**(10): p. 2004129.

[45] Luo, J., et al., *Atomic sulfur covalently engineered interlayers of Ti_3C_2 MXene for ultrafast sodium-ion storage by enhanced pseudocapacitance*. Advanced Functional Materials, 2019. **29**(10): p. 1808107.

[46] Guo, X., et al., *Constructing P-doped self-assembled V 2 C MXene/NiCo-layered double hydroxide hybrids toward advanced lithium storage*. Materials Advances, 2023. **4**(6): p. 1523–1533.

[47] Yoon, Y., et al., *A strategy for synthesis of carbon nitride induced chemically doped 2D MXene for high-performance supercapacitor electrodes*. Advanced Energy Materials, 2018. **8**(15): p. 1703173.

[48] Yang, C., et al., *Flexible nitrogen-doped 2D titanium carbides (MXene) films constructed by an ex situ solvothermal method with extraordinary volumetric capacitance*. Advanced Energy Materials, 2018. **8**(31): p. 1802087.

[49] Yang, C., et al., *Improved capacitance of nitrogen-doped delaminated two-dimensional titanium carbide by urea-assisted synthesis*. Electrochimica Acta, 2017. **225**: p. 416–424.

[50] Wen, Y., et al., *Nitrogen-doped $Ti_3C_2T_x$ MXene electrodes for high-performance supercapacitors*. Nano Energy, 2017. **38**: p. 368–376.

[51] Le, T.A., et al., *Synergistic effects of nitrogen doping on MXene for enhancement of hydrogen evolution reaction*. ACS Sustainable Chemistry & Engineering, 2019. **7**(19): p. 16879–16888.

[52] Tang, Y., et al., *Synthesis of nitrogen-doped two-dimensional Ti_3C_2 with enhanced electrochemical performance*. Journal of the Electrochemical Society, 2017. **164**(4): p. A923.

[53] Long, Y., et al., *Roles of metal ions in MXene synthesis, processing and applications: a perspective*. Advanced Science, 2022. **9**(12): p. 2200296.

[54] Osti, N.C., et al., *Effect of metal ion intercalation on the structure of MXene and water dynamics on its internal surfaces*. ACS Applied Materials & Interfaces, 2016. **8**(14): p. 8859–8863.

[55] Gandla, D., F. Zhang, and D.Q. Tan, *Advantage of larger interlayer spacing of a $Mo_2Ti_2C_3$ MXene free-standing film electrode toward an excellent performance supercapacitor in a binary ionic liquid–organic electrolyte*. ACS Omega, 2022. **7**(8): p. 7190–7198.

[56] Cao, F., et al., *Recent advances in oxidation stable chemistry of 2D MXenes*. Advanced Materials, 2022. **34**(13): p. 2107554.

[57] Razium, A.S. and F. Baomin, *Progression in the Oxidation Stability of MXenes*. 2023, Springer.

[58] Anasori, B. and Û.G. Gogotsi, *2D Metal Carbides and Nitrides (MXenes)*. Vol. 416. 2019, Springer.

[59] Zheng, Z., et al., *The oxidation and thermal stability of two-dimensional transition metal carbides and/or carbonitrides (MXenes) and the improvement based on their surface state*. Inorganic Chemistry Frontiers, 2021. **8**(9): p. 2164–2182.

[60] Mishra, A., et al., *Atomistic origin of phase stability in oxygen-functionalized MXene: a comparative study*. The Journal of Physical Chemistry C, 2017. **121**(34): p. 18947–18953.

[61] Liu, L., et al., *Tuning the surface chemistry of MXene to improve energy storage: example of nitrification by salt melt*. Advanced Energy Materials, 2023. **13**(2): p. 2202709.

[62] Zha, X.-H., et al., *Role of the surface effect on the structural, electronic and mechanical properties of the carbide MXenes*. Europhysics Letters, 2015. **111**(2): p. 26007.

[63] Jadav, R.P., et al., *Structural stability and electronic properties of 2D MXene $Hf_3C_2F_2$ monolayer by density functional theory approach*. Biointerface Research in Applied Chemistry, 2022. **13**(2): p. 152.

[64] Li, Z., et al., *Synthesis and thermal stability of two-dimensional carbide MXene Ti_3C_2*. Materials Science and Engineering: B, 2015. **191**: p. 33–40.

[65] Zhang, Z., et al., *Thermal stability and flammability studies of MXene–organic hybrid polystyrene nanocomposites*. Polymers, 2022. **14**(6): p. 1213.

[66] Iqbal, A., et al., *Improving oxidation stability of 2D MXenes: synthesis, storage media, and conditions*. Nano Convergence, 2021. **8**(1): p. 1–22.

3 Composition and Structure of MXenes

3.1 BRIEF DESCRIPTION OF MAX PHASES AND MXENES

MAX phases are the primary materials that are selectively etched to yield two-dimensional (2D) MXenes [1]. The discussion started in the 1960s when researchers led by Hans Nowotny in Vienna found over 100 novel carbides and nitrides, including the Ti_3SiC_2 and Ti_3GeC_2 families, known as the H phases. During the mid-1990s, Barsoum and El-Raghy synthesized samples of pure Ti_3SiC_2, which showed that this material exhibits an extraordinary combination of properties, including good thermal and electrical conductivity, tunable structure, and oxidation resistance [2]. Afterward, Barsum and his coworkers were successful in developing numerous other, related compounds from the same family, such as Ti_2AlC, Ti_3GeC_2, and Ti_4AlN_3, with different molecular compositions (i.e., 211, 312, and 413, respectively). These materials were called MAX phases, and eventually, this term took the place of the earlier one. These phases were observed to exhibit hexagonal $P6_3/mmc$ space group symmetry [1].

The MAX phases are represented by a general formula, that is, $M_{n+1}AX_n$ (where n = 1, 2, or 3). It is common to refer to the various MAX stoichiometries as 211, 312, and 413, for n = 1, 2, and 3, respectively. Transition metals (M) are usually Sc, Ti, Zr, Hf, V, Nb, Ta, Cr, and Mo, which primarily belong to groups 3–6 of the periodic table. The X element is either C or N, while the A element is mainly found in groups 13–16 [2–8]. The distinctive crystal structure possessed by MAX phases is formed by alternate stacking of the nearly close-packed M- and A-layers, with the octahedral sites of layers of M elements intercalated by X atoms (forming M_6X octahedral). Changing the value of n simply modifies the number of M_6X layers [9]. For instance, the most basic crystal structure of Ti_2AlC consists of a MAX phase with four atom layers, Ti–C–Ti–Al, stacked periodically along its cross-section. The two layers of closely spaced atoms make up the Ti bilayers. The Al monolayer combines two periodic units of the Ti–C–Ti tri-layer, whereas the C monolayer fills the octahedral positions between two Ti atomic layers. The Al atoms are located at the central positions of trigonal prisms, whereas the Ti_6C octahedral structure is comparable to that of rock salt [3].

It was discovered that the MAX phases, which would ordinarily be thermodynamically unstable, can stabilize when a new element is added to their composition at the M or A positions. In 2008, new members of this family emerged, with the general formula $(M_xM'_{1-x})_{n+1}AX_n$, where the M locations are represented by a solid solution of two distinct transition metals. Since the elements in the periodic table near one other have comparable atomic sizes and electrical structures, combining them will result in a solid solution. The first solid-state MAX phases were

DOI: 10.1201/9781003465768-3

discovered to be $(V_{0.5}Cr_{0.5})_2AlC$, $(Cr_{2/3}Ti_{1/3})_3AlC_2$, $(V_{0.5}Cr_{0.5})_4AlC_3$, $(Cr_{5/8}Ti_{3/8})_4AlC_3$, and $(V_{0.5}Cr_{0.5})_5Al_2C_4$. All of these compounds enable the combination of benefits, including resistance to oxidation and corrosion at high temperatures [1]. The layered structure along with the metallic-covalent character of the relatively weak M-A bonds and strong M-X bonds gives rise to the extraordinary features of the MAX phases [2, 9].

While MAX phases with ceramic nature exhibit hardness, low density, and exceptional resistance to corrosion, their high electrical and thermal conductivities and outstanding machinability make them comparable to metallic materials. Due to their special qualities, MAX phases are a promising material for rotating electrical contacts and bearings, high-temperature structural applications (such as heat exchangers, nozzles, and heating components), and materials that prevent wearing and corrosion. These novel characteristics originate from the metallic M-A bonds and M-X bonds, which combine covalent, metallic, and ionic properties. As a result, these materials have strong connections that prevent breakage by any means, in contrast to graphite and transition metal dichalcogenides (TMDs), which exhibit weak van der Waals interactions. Consequently, the synthesis of 2D MXenes from these MAX phases was achieved for the first time in 2011 by chemically exfoliating the A-layers [5, 6].

The first ordered MAX phase alloy was found in 2014. It was an M_3AX_2 structure, in which alternate M-layers of one metal formed a phase without any planar ordering. In 2017, an in-plane ordering with a similar 2:1 ratio between M elements was found in an M_2AX phase. This discovery led to the development of out-of-plane and in-plane notations, which distinguish between the two material groupings [1, 2, 9]. MXenes currently have four distinct compositions (i.e., M_2XT_z, $M_3X_2T_z$, $M_4X_3T_z$, and $M_5X_4T_z$), depending on the precursor that was employed (as shown in Fig. 3.1 [10]). The quaternary MAX phases have two different types of chemical order: in-plane order (i-MAX) and out-of-plane order (o-MAX). Recently, the A and M′ components from the $(M_{2/3}M'_{1/3})_2AlC$ i-MAX phase were etched to provide $M_{1.33}XT_z$ MXenes. After their discovery, several in-plane orders have been reported, such as $(Mo_{2/3}Sc_{1/3})_2AlC$, $(Mo_{2/3}Sc_{1/3})_2GaC$, $(Mo_{2/3}Y_{1/3})_2AlC$, $(Mo_{2/3}Y_{1/3})_2GaC$, $(W_{2/3}Sc_{1/3})_2AlC$, $(W_{2/3}Y_{1/3})_2AlC$, $(V_{2/3}Zr_{1/3})_2AlC$, $(Cr_{2/3}Sc_{1/3})_2AlC$, $(Cr_{2/3}Y_{1/3})_2AlC$, and $(Cr_{2/3}Zr_{1/3})AlC_2$ [10, 11].

A large number of the recently identified o-MAX phases are based on aluminum, which permits their conversion to MXenes. For instance, Mo_2Ga_2C produces Mo_2CT_z, whereas Mo_2GaC is resistant to etching. In other instances, the 2D flakes that are produced are unstable in the etching solution. For instance, Cr_2AlC disappears upon etching. Nevertheless, $Cr_2TiC_2T_z$ has been effectively etched from the ordered MAX phase Cr_2TiAlC_2. Al is the A element and Cr is the outermost layer in both scenarios [10]. Researchers showed in 2021 that the corresponding 2D carbides can also be employed to synthesize high-entropy alloy MAX-phases. Following the effective removal of $TiVNbMoAlC_3$ and $TiVCrMoAlC_3$, the A-layers were selectively etched to provide $TiVNbMoC_3T_x$ and $TiVCrMoC_3T_x$, respectively. The synthesized materials' stoichiometric ratios for the transition metals M1, M2, M3, and M4 were 1:1:1:1. These two represent the first successful syntheses of high-entropy MXenes from MAX phases to date. Future studies are expected to concentrate on their attributes because these $(M1M2M3M4)C_3T_x$ MXenes should have improved

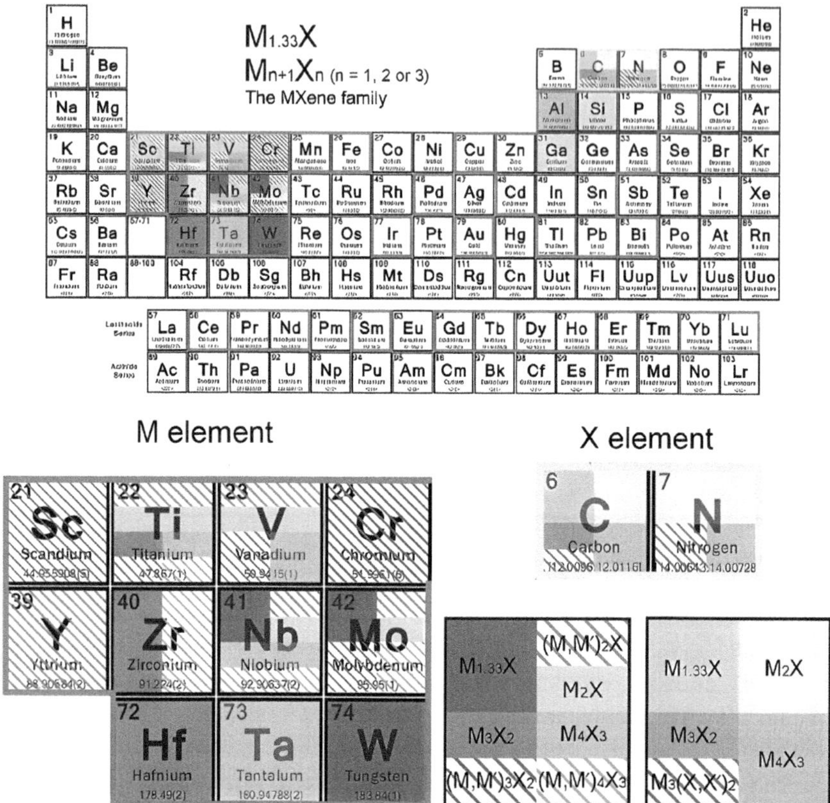

FIGURE 3.1 Periodic table with elements used to synthesize different compositions of MXenes. The elements are framed in green (M), orange (A), and yellow (X), respectively.

Source: Reprinted from [10] *Trends in Chemistry*, 1, L. Verger, V. Natu, M. Carey, M. W. Barsoum, MXenes: An Introduction of Their Synthesis, Select Properties, and Applications, 656–669, Copyright 2019, with permission from Elsevier.

qualities. This indicates that high-entropy alloys make many novel MXene compositions possible. They might make good candidates for use in energy harvesting and storage, catalysis, and other related fields [1].

It is challenging to model or simulate the chemical exfoliation of MAX phases in detail due to the complicated nature of the process, which involves several reaction kinetics and dynamics. However, by theoretically analyzing the exfoliation energies, we may still uncover some interesting candidates that can probably be exfoliated in the future. Exfoliation of MAX phases require the use of acid treatment, such as an HF solution, to remove the A element. The exfoliation of Ti_3AlC_2 into 2D Ti_3C_2 by using the HF acid has been theoretically investigated by researchers. It was demonstrated that HF molecules are adsorbed at the edge of metal atoms after dissociating into H and F radicals, thereby weakening the Al-Ti bonds and consequently creating an interlayer gap. In turn, the interlayer gap makes it easier for HF molecules to penetrate further. Following these procedures, AlF_3 and H_2 are finally removed, and the fluorinated MXene remains [4].

Static computations, on the other hand, can also be useful for quickly screening the bond characteristics of several MAX phase configurations. Mechanical exfoliation of M_2AlC has been studied theoretically based on tensile and shear module studies. It was demonstrated that M-Al bonds are disrupted by high tensile stress, causing the M_2C and Al layers to separate. On the other hand, it would be a viable technique to produce MXenes on a huge scale for a very low price [4]. A recent study examined the use of electron microscopy to investigate the MAX phases and their corresponding MXenes. Furthermore, they described how structural imaging may be used to monitor the dynamic process in real time, making it a growing tool for structural customization of MXenes [12].

3.2 MAX AND NON-MAX PRECURSORS FOR MXENES

The 2D $M_{n+1}X_n$ layers, which were first inherited from the MAX phases, are termed MXenes because they exhibit a hexagonal symmetry similar to that of graphene. The discussion of MAX and non-MAX precursors is presented here in detail. A number of MAX phases exist experimentally. Among them, Ti_2AlC, Ti_2AlN, Ti_3AlC_2, Ti_3SiC_2, Ti_4AlN_3, V_2AlC, V_4AlC_3, Nb_2AlC, Nb_4AlC_3, Zr_3AlC_2, and Ta_4AlC_3 have already been exfoliated into Ti_2C, Ti_2N, Ti_3C_2, Ti_4N_3, V_2C, V_4C_3, Nb_2C, Nb_4C_3, Zr_3C_2, and Ta_4C_3 MXenes [4]. Each MAX phase has distinct parameters and circumstances toward different etchants, which should not generalize it for all the MXenes. For example, M_4AX_3 requires comparatively flexible etching conditions than M_2AX and M_3AX_2. It has been noted that only a few elements, including Al and Si, have been successfully etched to produce MXenes from the minimum of 10 A elements present in these phases. A recent study reported the use of powder metallurgy to synthesize $(Mo_{0.8}V_{0.2})_5AlC_4$. The formation of the corresponding MXene through the effective etching of Al makes it possible to synthesize new compositions [13].

It is crucial to comprehend how chemistry controls and fine-tunes physical qualities because a multitude of synthetic isostructural solid solutions of MAX phases exist [14]. Apart from the above-mentioned varieties, solid solutions can also exhibit MAX phases, as demonstrated in $Ti_2Al_{(1-x)}Sn_xC$, $Ti_2Al(C_{0.5},N_{0.5})$, and $Ti_3Al(C_{0.5},N_{0.5})_2$. This makes it possible to prepare MXenes with various compositions, resulting in various chemical and physical properties that could be applied as needed in particular situations [8]. The only solid solutions in the MAX phase known to exist on the M sites up until the end of 2014 were random ones, including $(Ti_xCr_{1-x})_2AlC$, $(Ti_xNb_{1-x})_2AlC$, $(Ti_{1/2}V_{1/2})_3AlC_2$, and $(Cr_{5/8}Ti_{3/8})_4AlC_3$ [15]. A solid solution of distinct M′ and M″, A′ and A″, or C and N has been observed in numerous experiments to produce alloy MAX phases, which in turn lead to diverse alloy MXenes. Different alloys have been experimentally manufactured in this regard, including TiNbC, $(Ti_{0.5}Nb_{0.5})_2C$, Ti_3CN, $(V_{0.5}Cr_{0.5})_3C_2$, $(Nb_{0.8}Zr_{0.2})_4C_3$, and $(Nb_{0.8}Ti_{0.2})_4C_3$ [4]. To generalize, we say that $(Ti, V)_2AlC$, $(Ti, Nb)_2AlC$, $(Nb, V)_2AlC$, $(Mo, V)_4AlC_3$, and $(Nb, Zr)_4AlC_3$ have been etched and corresponding solid-solution MXenes are being synthesized [16].

The discovery of ordered quaternary material, that is, $(M', M'')_{n+1}AlC_n$, is a significant recent advancement. In contrast to the solid solutions, both metal atoms on the M site are different in these ordered phases. One M' (such as Ti) would be the ideal site in the composition, whereas another transition metal M'' (such as Cr) would be the nonequivalent site. Cr_2TiAlC_2 and V_2CrAlC_2 were the early phases in which this was described. This should therefore ideally have a set stoichiometry. These phases seem to show a high degree of order in practice; however, they may not be completely occupied [13–15]. For example, in Mo_2TiAlC_2 and $Mo_2Ti_2AlC_3$, the elemental planes have a stacking sequence, that is, Mo-Ti-Mo-Al-Mo-Ti-Mo and Mo-Ti-Mo-Al-Mo-Ti-Mo, respectively. The octahedral positions between the metals are occupied by the C atoms in both scenarios. For ordering, Mo requires high energy to occupy the positions where C is already arranged in an fcc pattern. This kind of arrangement is limited to some MAX phases (with n = 2 or higher), and it cannot be achieved for all M elements [2, 13, 16, 17].

Rules for the generation of i-MAX can be developed based on theoretical and empirical evaluations. This will improve our understanding of the underlying processes leading to their formation and enable the prediction of future phases to be synthesized. A 2:1 ratio for M1:M2, difference in sizes of both metals (with M2 being larger), and a favorable small-sized A are all necessary for the creation of i-MAX phases. These criteria are thought to be the most crucial. Intermixing of M is also expected, but further research is needed to determine how a departure from the optimal 2:1 ratio affects stability and what effects it has on the conversion of i-MAX into i-MXene [2, 13, 15, 17]. Therefore, the 2D structures produced from i-MAX phases can be referred to as i-MXenes, in analogy to 2D MXenes. Most of these phases, such as $(M'_{2/3}Zr_{1/3})_2AlC$ (for $M' = Cr$, V) and $(M'_{2/3}M''_{1/3})_2AlC$ (where $M' = Mo$, Cr, W, and $M'' = Y$, Sc) have already undergone exfoliation and have been experimentally produced. On the other hand, Sc/Y and Al dissolve during the exfoliation process, forming 2D $M'_{1.33}C$ MXenes [4]. On the other hand, efforts have been undertaken to substitute Sc with alternative metals due to its high cost and scarcity. These methods were used to synthesize $(V_{2/3}Zr_{1/3})_2AlC$, $(W_{2/3}Y_{1/3})_2AlC$, $(Mo_{2/3}Y_{1/3})_2AlC$, and $(Mo_{2/3}RE_{1/3})_2GaC$ (where RE is rare earth metal) [1].

There are fewer studies on nitride and carbonitride MXenes than on C-based MXenes. This is partly due to the challenges associated with their synthesis, as acids can dissolve the nitride layers. It continues to be expected that, with further hard work, other novel MAX phases containing nitrogen will eventually be created and that 2D nitrides will be created by etching these phases. Other novel strategies, such as nitriding MXenes to substitute nitrogen for part or all of the carbon, have been effectively applied to obtain Mo_2NT_x, V_2NT_x, and nitrogen-doped Ti_2CT_x [13]. Gogotsi and associates synthesized titanium nitride $(Ti_4N_3T_x)$ for the first time in 2016. By heating in a molten fluoride salt, the resulting MXenes product is formed from their respective MAX phases. In a different investigation, the multilayered Ti_2NT_x product was obtained by immersing Ti_2AlN in HCl-KF solution. Centrifugation is used to transform the multidimensional Ti_2NT_x (where T is the termination group) into a few-layered material following sonication in dimethyl sulfoxide (DMSO) [6]. According to the density functional theory (DFT) calculations, $Ti_{n+1}N_n$ exhibits higher cohesive and formation energies than corresponding $Ti_{n+1}C_n$ [18, 19].

Later, it was discovered that other layered compounds can also be utilized for MXene synthesis, in addition to the MAX phases. Sc, Zr, and Hf are among the metals that are more likely to generate quaternary $(MC)_n[Al(A)]_4C_3$ (n = 1, 2, 3) or layered ternary $(MC)_nAl_3C_2$ carbides. These compounds have alternate MC layers that exhibit greater chemical stability compared to Al_4C_3, as the latter is more susceptible to hydrolysis in acidic conditions. Consequently, a weaker connection and greater reactivity are shown in the Al–C sublayers, exfoliated by etching. For instance, the HF etching method of separating $Zr_3Al_3C_5$ and $Hf_3[Al(Si)]_4C_6$ was reported in order to get $Zr_3C_2T_x$ and $Hf_3C_2T_x$ MXenes. $Hf_3C_2T_x$ is produced by further etching the Hf–C and Al–C sublayers in the unit cell. The addition of silicon to the Al–C sublayer was observed to speed up the etching of layered ternary carbide due to the weak interfacial adhesion between the Hf–C and Al(Si)–C. The fact that silicon has a greater atomic charge (2.36) than aluminum (2.19) explains the weakening of this bond. Because of this, the common carbon layer and the Al-containing layer are bound considerably more strongly [1, 2, 6, 8, 19, 20].

The above considerations suggest that the $(MC)_n[Al(A)]_mC_{m-1}$ compounds might serve as possible building blocks for MXenes. A number of these members have been found through experimental investigations. Specifically, $ThAl_4C_4$, $Zr_2Al_3C_4$, $Zr_3Al_3C_5$, $Hf_2Al_3C_4$, $Hf_3Al_3C_5$, $ScAl_3C_3$, $U_2Al_3C_4$, UAl_3C_3, and $YbAl_3C_3$ were reported years ago, while $LuAl_3C_3$, YAl_3C_3, $CeAl_3C_3$, $DyAl_3C_3$, $ErAl_3C_3$, $TmAl_3C_3$, $ZrAl_4C_4$, $ZrAl_8C_7$, $Zr_2Al_4C_5$, $Zr_3Al_4C_6$, $Zr[Al(Si)]_4C_4$, $Zr_2[Al(Si)]_4C_5$, $Zr[Al(Si)]_8C_7$, $[(ZrY)]_2Al_4C_5$, $Zr_3[Al(Ge)]_4C_6$, $Hf_2Al_4C_5$, $Hf_3Al_4C_6$, and $HfAl_4C_4$ were developed after 2006 [2, 16]. In 2015, layered Mo_2Ga_2C was successfully employed to create MXenes in addition to carbides containing Al. Based on this material, Mo_2C MXene was first obtained in the same year. Thin films along with the bulk material in the form of Mo_2Ga_2C were synthesized. This multilayer carbide resembles the Mo_2GaC 211 MAX phase, with the addition of one extra Ga layer. In this instance, the two gallium layers are stacked so that they rest symmetrically on top of one another. Consequently, it has been demonstrated that molecules other than the MAX phases can be used to synthesize MXenes [1, 2, 20].

3.3 SURFACE TERMINATIONS FOR MXENES

The experimental implementations revealed that the stability of MXenes increases with the surface functionalization (T = OH, O, F, Cl, Br, etc.); thus, fully terminated MXenes are observed to be more stable [20, 21]. Ti_3C_2 MXene, for example, can have at least three distinct notations: $Ti_3C_2O_2$, $Ti_3C_2(OH)_2$, and $Ti_3C_2F_2$. The properties of these materials are heavily influenced by surface functional groups, which in turn affect the uses of materials [22]. Thus, a complete analysis of these functional groups is crucial in order to understand their influence on the characteristics of MXenes. As the temperature rises, the −OH groups can be converted to generate the O-terminating groups. In addition, transition metals (like Pb) and alkaline earth metals (like Ca) often replace H atoms in −OH functional groups, making them unstable. Because of this, the MXene surface can create some mixture of O, OH, and F groups [23–25]. When MXenes are used for adsorptions, −F groups are typically not favored; therefore, −O and −OH groups are considered to be significantly more stable. This demonstrates that, despite their larger atomic weight, MXenes have the ability to show improved adsorptive efficacies when compared to other nanomaterials based on carbon, such as graphene. It is known that the metallic constituents that do not terminate are highly reactive and have a higher chemical reactivity than other constituents. Future research in this area must focus on MXenes' adsorptive properties without the functional group associations [22].

Due to increased thermodynamic stability, all experimentally synthesized MXenes exhibit these terminations. However, the most stable functionalization is still up for debate, and it may vary depending on the specific materials and synthesis techniques. According to theoretical research, the termination stability of a number of MXenes, including $Ti_4N_3T_x$, $Ti_3C_2T_x$, $Nb_4C_3T_x$, and M_2CT_x (M = Ti, Zr, and Hf) compounds, increases from –OH to –F to –O. The substitution of alkali or transition metals for hydrogen atoms and the conversion of hydrogen to –O at high temperatures are responsible for the decreased stability of OH terminations. Further, it was discovered that M_2CT_x compounds had a higher likelihood of –F termination. The HF technique's order of termination stability, –F > –O > –OH, was confirmed experimentally. In contrast, the HCl–LiF procedure and thermal treatments produce –O > –F > –OH. However, most simulation studies ignore the etching solution, and hence it is unclear why they differ in terms of the functionalization sequence [2, 5].

Numerous novel terminations, such as N- and Cl-functionalized MXenes, have been studied from a theoretical perspective. Through first-principle calculations, researchers examined N-terminated Nb_2C and Ta_2C and discovered that these MXenes are not only stable but also display ultrahigh carrier mobility that is comparable to or greater than graphene's, as well as direct bandgap semiconducting characteristics. Another work examined the electrochemical process of Cl^- ion storage onto MXenes using a first-principle method. It was shown that the kinetic barrier needed for Cl ion diffusion on a 2D surface is significantly lower than the barrier needed for diffusion in bulk materials. Large reversible Cl capacity and high-rate capabilities were anticipated as a result [2]. Ti_3C_2 MXene with different surface groups (i.e., O, S, Cl, Br, and I) was formed with the Lewis acid molten salt etching method. Moreover, solvothermal and hydrothermal techniques can be used to replace, remove, and modify MXene surface termination. Similarly, surface terminations of –NH, –Se, –Te, and –P were created for Ti_2C, Nb_2C, and Ti_3C_2 MXenes. The advancement of MXene termination preparation technology offers a material science foundation for investigating MXene surface terminations, and further technological avenues for controlling performance of MXenes as functional materials [26, 27].

The comparison of adsorption free energies of various species in the vacancy of fully terminated MXenes is illustrated in Fig. 3.2 [27]. A series of first-principle calculations demonstrated that after accepting the surface groups, the low dimensionality of MXenes results in significant differences between their magnetic and electronic properties. According to the studies, Ti_2CO_2, Zr_2CO_2, Hf_2CO_2, Sc_2CO_2, $Sc_2C(OH)_2$, and Sc_2CF are semiconducting with bandgaps ranging from 0.24 to 1.8 eV. Furthermore, the findings showed that at low temperatures, $Cr_2C(OH)_2$, $Cr_2N(OH)_2$, Cr_2CF_2, Cr_2NF_2, and Cr_2NO_2 can all become magnetic [28, 29]. Using the DFT study, researchers examined the relative stabilities of CB and BB′, two structural phases of M_2CO_2 (M = groups III–V). For any M_2C MXene, the negative formation energies

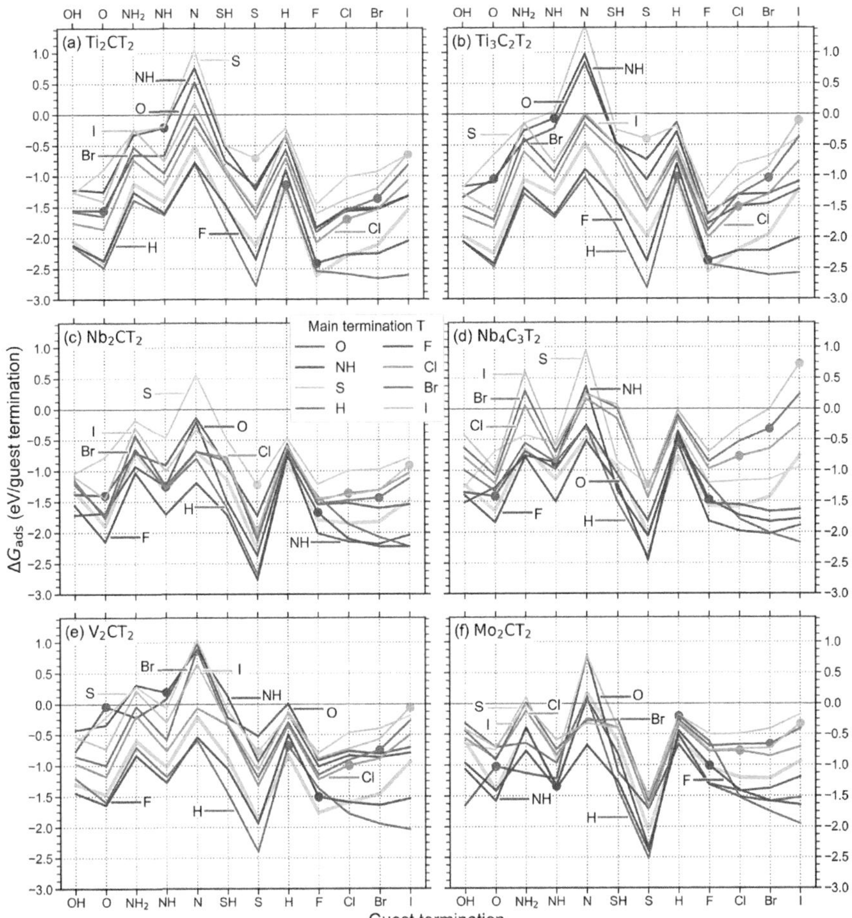

FIGURE 3.2 Comparison of the adsorption free energies of different chemical species in the vacancy of fully terminated MXenes in a hydrogen atmosphere. The different colors refer to the surface terminations, while the x-axis of the respective graph refers to the guest terminations. The dots indicate the data points for which the guest adsorbate is the same as the main termination and indicate whether a fully single-terminated MXene is stable.

Source: Reprinted with permission from [27] J. Björk and J. Rosen. 2021. Functionalizing MXenes by Tailoring Surface Terminations in Different Chemical Environments. *Chem. Mater.*, 33, 23, 9108–9118. Copyright 2021, American Chemical Society.

of the BB′ and CB phases show a strong likelihood of their synthesis. These results suggest that if M belongs to groups III–V, then these phases are energetically more favorable. Similarly, phonon computations presented the dynamical stability of BB′-Sc_2CO_2 and CB-Ti_2CO_2 MXenes [30].

In another DFT calculation, the electrical and mechanical properties of M_2CT_2 MXene were examined. Generally speaking, MXenes functionalized with oxygen have larger mechanical strengths and smaller lattice parameters than those functionalized with hydroxyl and fluorine groups. W_2CO_2 has the highest mechanical strength, while Sc_2CO_2 has the shortest interlayer thickness. Different materials display semiconducting properties, that is, M_2CO_2 (where M = Sc, Ti, Zr, Hf, or W), M_2CF_2 (M = Sc or Mo), and $Sc_2C(OH)_2$ [31]. In conclusion, MXene surface terminations are critical to both the emerging characteristics and the stability of the MXene structure. The research on surface terminations of MXenes is limited due to the preparation method's fairly simple surface termination types. Therefore, creating novel MXene preparation techniques and regulating MXene surface termination are essential to tune the MXene surface termination. To customize features for particular uses, like termination by non-inherent species, the current research must advance toward total control of the surface terminations [2].

3.4 THEORETICAL STUDIES OF CARBIDE AND NITRIDE MXENES

Density functional calculations were used by G. Gao et al. [32] to show that, under standard conditions, Nb_2C and $Nb_4C_3O_2$ are O-terminated, while V_2C, Ti_2C, and Ti_3C_2 are terminated by a mixture of –O and –OH groups. The structural stability of different compositions and predictions for their corresponding energy applications were summarized by C. Zhan et al. [33]. They also described the ways in which big data, machine learning, and high-throughput computing can expand the capabilities of the MXene family. In order to estimate their uses in supercapacitors and electrical devices, Y. Xin et al. [34] used ab initio DFT calculations to study the work functions and quantum capacitances of $Nb_{n+1}C_nZ_2$ (where Z = O, OH, OCH_3, or F). They showed that the capacitance limitation of graphene-based electrodes can be overcome by these materials. With the exception of the Nb_2C sheet, the functional groups have very little effect at negative electrodes. The work function of the MXene sheets decreases when F and O atoms are added, whereas it increases when OH and OCH_3 groups are present. The formation energies for MXenes of the same thickness are as follows: $Nb_{n+1}C_nO_2 < Nb_{n+1}C_n(OH)_2 < Nb_{n+1}C_nF_2 < Nb_{n+1}C_n(OCH_3)_2$.

DFT calculations were carried out by Q. Meng et al. [35] in order to investigate the electronic characteristics and possible uses of Zr_2C, Zr_2CO_2, Zr_3C_2, and $Zr_3C_2O_2$ in sodium-ion batteries. While Zr_2CO_2 and $Zr_3C_2O_2$ can be semiconductors and metals, respectively, the bare Zr_2C and Zr_3C_2 act as a magnetic metal. In the meantime, density of states (DOS) investigation revealed that following Na adsorption, all of these MXenes exhibit metallic nature. The possible applications of sodium-ion batteries (NIBs) are favored by these metallic properties. Threshold photoemission electron microscopy (PEEM) was used by J. Chen et al. [36] to describe the interlayer electron-transfer mechanism in Mo-based MXene flakes. According to the observations, interlayer electron transport is suppressed when the gap between stacked nanosheets expands. Lithium and tetrabutylammonium hydroxide (TBAOH)-intercalated MXene flakes were observed to exhibit a less sensitive interlayer electron-transport process to temperature fluctuations compared to intercalation-free flakes. Time-resolved ultrafast observations demonstrate that intercalation leads to an extended photoinduced carrier lifetime. These findings suggested that the most significant factor influencing the interlayer electrical dynamics is the spacing distance between stacked nanosheets.

Through DFT study, K. D. Fredrickson et al. [37] examined the mechanical and electrical effects of water intercalation and surface groups on layered Ti_2C and Mo_2C MXenes. It was discovered that c (lattice parameter) varied considerably depending on the functional group and increased dramatically when water was intercalated. One monolayer of O was discovered to functionalize both Ti_2C and Mo_2C under zero applied potential. The adsorbate coverage was altered by applying a potential, which resulted in the systems being H-covered at negative potentials instead of O-covered and, in certain situations, in a metal–insulator transition. Based on the first-principle DFT calculation, X. Zha et al. [38] observed that Mo_2C exhibits various applications other than supercapacitors. This is because they exhibit super-high electrical conductivity, small molar volume, high mechanical strength, and favorable thermal conductivity. In an effort to finding promising anode materials for lithium-ion batteries, X. Zhao et al. [39] conducted a theoretical investigation on M_2C, M_2N, MC_2, and MN_2 MXenes. V_2N and Ti_2N have theoretical capacities of 924 and 975 mAh g^{-1}, respectively. In comparison to graphite electrodes, Ti_2N and V_2N exhibit greater stability due to their lower deformation rate and less energy change.

DFT was employed by J. Zhu et al. [40] to study the energy storage and structural characteristics of lithium-decorated bare and functionalized Zr_2C. They reported low diffusion barriers and high Li-specific capacities for Zr_2C and Zr_2CS_2. The most advanced positive and unlabeled (PU) machine learning method was modified by N. C. Frey et al. [41] to theoretically predict the stability of 2D MXenes for synthesis. This model predicted that 18 MXenes and 111 MAX phases are stable to be synthesized, including structures like Hf_4N_3, Sc_3C_2, and W_4C_3. The 20 most promising MAX phases that have a high possibility of being synthesized and etched to produce new MXenes were also determined. In order to estimate how thermodynamic stability of these compounds would rely on their chemical composition, M. Ashton et al. [42] compared the binding energies of different surface groups on MXene surfaces using first-principle simulations. The results indicated that oxygen is saturated on the surfaces of all MXenes during synthesis, with the exception of M = Sc. With lower formation energies, materials like $Ti_3N_2O_2$, $V_3N_2O_2$, $Ti_4N_3O_2$, $Cr_4N_3O_2$, and $V_4N_3O_2$ would be easier to synthesize.

Y. Xie et al. [43] theoretically investigated the structural stability and electronic characteristics of $Ti_{n+1}C_n$ and $Ti_{n+1}N_n$ with different surface groups. They noticed that in the absence of terminations, MXenes exhibit magnetically ordered ground states. With the exception of Ti_2CO_2, which is expected to be semiconducting, all other materials were observed to be metallic. Thicker MXenes (with greater n value) have a computed DOS at the Fermi level that is significantly larger than that of thin MXenes. This suggests that surface chemistry and electronic conductivity will be different. With the same functional groups, carbides and nitrides generally behave differently. Another DFT study was used to methodically take advantage of the anchoring effects of $Ti_{n+1}C_n$ for lithium–sulfur batteries [44]. The findings showed that Ti_2CO_2 and $Ti_3C_2O_2$ may interact strongly with Li_2S_n species with notable binding strength to successfully immobilize it. This is because lithium ions in polysulfides are attracted to O atoms in MXene monolayers. In particular, even while the Li-S bonds got weakened, the Li_2S_n species can be preserved quite intact.

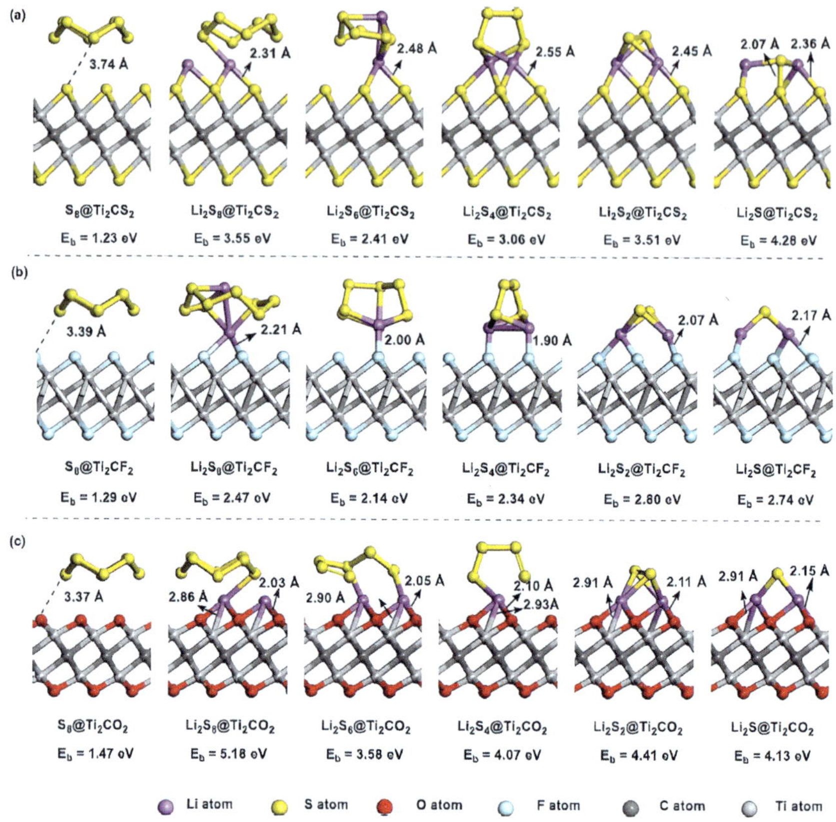

FIGURE 3.3 Stable adsorption configuration of polysulfides on (a) Ti_2CS_2, (b) Ti_2CF_2, and (c) Ti_2CO_2. Corresponding binding energies are also given.

Source: Reprinted with permission from [45] Q. Zhang, X. Zhang, Y. Xiao, et al. 2020. Theoretical Insights into the Favorable Functionalized Ti_2C-Based MXenes for Lithium–Sulfur Batteries. *ACS Omega*, 5, 45, 29272–29283. Copyright 2020, American Chemical Society.

Q. Zhang et al. [45] compared the electronic conductivity, kinetic conversion ability, and adsorption ability of pristine and functionalized Ti_2C surfaces (i.e., O, OH, F, and S). The study was based on DFT. It was discovered that Ti_2CO_2, Ti_2CF_2, and Ti_2CS_2 exhibit effective polysulfide adsorption (as illustrated in Fig. 3.3 [45]) when compared to pure and $Ti_2C(OH)_2$ surfaces. The greatest surface adsorption energy is exhibited by Ti_2CO_2, followed by Ti_2CS_2 and Ti_2CF_2. MXenes can operate as superior terahertz (THz) sensing materials, as demonstrated by Y. I. Jhon et al. [46] using systematic DFT calculations. Regardless of stacking degree, extinction coefficients and optical absorptions were seen at the THz range in Ti_3C_2. In order to model the pseudocapacitive performance of MXenes in the aqueous H_2SO_4 electrolyte, C. Zhan et al. [47] computationally screened the electrodes. They reported that nitrides perform better than carbides in terms of pseudocapacitive efficiency. In particular, $Zr_{n+1}N_nT_x$ was expected to have the best areal capacitance, whereas Ti_2NT_x was expected to exhibit a high gravimetric capacitance.

Ball and stick models of molecules (a) $S_8@Ti_2CS_2$ with E_b equals 1.23 eV, Li_2S_8 $@Ti_2CS_2$ with E_b equals 3.55 eV, $Li_2S_6@Ti_2CS_2$ with E_b equals 2.41 eV, $Li_2S_4@Ti_2CS_2$ with E_b equals 3.06 eV, $Li_2S_2@Ti_2CS_2$ with E_b equals 3.51 eV and $Li_2S@Ti_2CS_2$ with E_b equals 4.28 eV, (b) $S_8@Ti_2CF_2$ with E_b equals 1.29 eV, $Li_2S_8@Ti_2CF_2$ with E_b equals 2.47 eV, $Li_2S_6@Ti_2CF_2$ with E_b equals 2.14 eV, $Li_2S_4@Ti_2CF_2$ with E_b equals 2.34 eV, $Li_2S_2@Ti_2CF_2$ with E_b equals 2.80 eV and $Li_2S@Ti_2CF_2$ with E_b equals 2.74 eV, and (c) $S_8@Ti_2CO_2$ with E_b equals 1.47 eV, $Li_2S_8@Ti_2CO_2$ with E_b equals 5.18 eV, $Li_2S_6@Ti_2CO_2$ with E_b equals 3.58 eV, $Li_2S_4@Ti_2CO_2$ with E_b equals 4.07 eV, Li_2S_2 $@Ti_2CO_2$ with E_b equals 4.41 eV and $Li_2S@Ti_2CO_2$ with E_b equals 4.13 eV.

3.5　RECENT RESEARCH AND ADVANCEMENTS

Recently, various compositions of MAX phase precursors (other than Ti-based) are being explored, that is, solid solutions and ordered double-transition metals. As discussed above, researchers are also discovering non-MAX phase precursors to synthesize MXenes. A number of studies are predicting the behavior and stability of these materials using theoretical calculations. The number of studies on nitride MXenes is also increasing with the discovery of structural properties, compositions, theoretical predictions, and different synthesis methods. Currently, at least 17 nitride MXenes and 28 nitride MAX phases have been identified as existing and being thermodynamically stable; however, because MAX phase synthesis and etching are challenging, the majority of these phases are just theoretical [48]. Numerous new materials have been produced as a result of surface chemistry modifications and compositional diversity. These materials are being investigated for various applications, such as in situ spectroelectrochemistry, optoelectronics, catalysis, energy storage, wireless communication, medicine, and sensing [48].

Using DFT calculations, N. Zhang et al. [49] examined the structural, mechanical, and electrical properties of bare $(Ti_{n+1}C_n/Ti_{n+1}N_n)$ and functionalized $(Ti_{n+1}C_nT_2/Ti_{n+1}N_nT_2)$ materials. They found that the values of in-plane Young's modulus in nitrides are greater than those in carbides, and the value decreases in both structures as monolayer thickness increases. According to cohesive energy estimates, MXenes with thicker monolayers have superior structural stability. According to adsorption energy calculations, $Ti_{n+1}N_n$ has a greater tendency to stick to the terminal groups, which implies that nitride-based MXenes have more active surfaces. A. Jurado et al. [50] reported the interaction of carbon dioxide with nitride MXenes of different thicknesses to find their potential use in carbon capture and storage (CCS) devices. They examined the basal (0001) surface plane of $M_{n+1}N_n$ nitrides and compared it to similar outcomes for the extended (001) and (111) surfaces of the bulk rock salt compounds. The outcomes demonstrated that while the thickness effect is generally minor but not insignificant, the composition of nitrides has a noticeable impact on these substrates.

Based on a thorough investigation of a crystal field theory model, H. Kumar et al. [51] predicted a number of M_2NT_x structures that exhibit magnetic properties.

They reported inherent ferromagnetism in Ti_2NO_2, Mn_2NT_x, and Cr_2NO_2, with different surface groups. The intrinsic half-metallic transport behavior, strong ferromagnetism, and high magnetic moments contribute to these MXenes being excellent prospects for spintronic applications. S. The et al. [52] theoretically studied the lattice structures, electrical and magnetic characteristics, and molecular oxygen content of nitrides, such as V_2NO_2, V_2NF_2, Mo_2NO_2, Mo_2NF_2, Mn_2NO_2, and Cr_2NF_2, with intrinsic bandgaps. These nitride MXenes show orbital ordering, which can lead to magnetoelectric or magnetoelastic coupling depending on the situation. Specifically, Cr_2NF_2 is a ferro-elastic material and the direction of the ferro-elastic strain is associated with the spiral magnetization propagation vector. In V_2NO_2, Cr_2NF_2, and Mo_2NF_2, magnetic order is associated with polar displacements, and the ferroelectric phase exists as an excited state. An ab initio evolutionary approach with changing composition was used by S. Yu et al. [53] to study the stability of different Zr–N compounds. In addition to Zr_2N, the recently found Zr_xN_y compounds have zirconium or organized nitrogen vacancies in rock salt frameworks.

Using first-principle calculations, V. Mehta et al. [54] investigated the electrochemical characteristics of Mo_2N to be utilized as an electrode in lithium-ion batteries. Similarly, the strong electronic conductivity and metallic nature of the Na-adsorbed Mo_2N monolayer implied that it could find usage in sodium-ion batteries [55]. Additionally, the charge transfers from the monolayer to the Na-adsorbed atom showed that the adsorbed metal atom is in a cationic state. Furthermore, S-terminated nitride as anodes for Li/Na ion batteries was investigated by V. Shukla et al. [56] using DFT calculations. It has been determined that these two 2D materials (V_2NS_2 and Ti_2NS_2) in particular are better suited for Li-ion batteries, which have respective estimated theoretical capacities of 308.28 mAh g^{-1} and 299.52 mAh g^{-1}. Mono-layered VN_2 was studied by Y. Dong et al. [57] as a potential anode material for batteries. First, phonon spectra were calculated to show the great stability. Furthermore, the strong electrical conductivity and metallic characteristics suggested a good capacity and superior rate performances. Based on first-principle calculations, H. Liu et al. [58] also predicted that the V_2N monolayer would be a desirable anode material for lithium, sodium, and magnesium ion batteries.

A thorough understanding of how different chemical species interact with different MXenes is necessary, given the importance of surface terminations for properties and recent advancements in controlling these groups. Numerous theoretical investigations have been conducted with an emphasis on various characteristics of 2D carbide and nitride MXenes [59]. Although theoretical efforts have been made to comprehend the termination of MXenes, general conclusions regarding surface chemistry and its regulation by the selection of transition metals and adsorbates remain unclear. Moreover, a comprehensive investigation that goes beyond traditional terminations is encouraged by the novel opportunities for surface termination substitution. Recently, twelve distinct surface terminations (O, OH, N, NH, NH_2, S, SH, H, F, Cl, Br, and I) and six distinct MXenes that have been experimentally discovered (Ti_2C, Nb_2C, V_2C, Mo_2C, Ti_3C_2, and Nb_4C_3) were focused by researchers using electronic structure theory [26, 27].

The preference of an element to adsorb in its hydrogenated form depends on the environment that is chosen. For instance, in a hydrogen atmosphere, oxygen is predicted to exist as a mixture of O and OH, but in an oxygen atmosphere, oxygen is expected to exist entirely as O. Conversely, in an environment of hydrogen, sulfur adsorbs as atomic S; however, the terminations disintegrate in the presence of oxygen. A couple of things in particular need to be brought to light. Efforts to synthesize fully fluorine-terminated MXenes are promoted since they may be stable in ambient settings and are suggested to be highly resistive against oxidation. In an oxygen atmosphere, it is thermodynamically advantageous to replace all ending species— apart from fluorine—with oxygen. For certain terminations, though, this substitution might not be kinetically possible. To test MXenes' resistance to oxidation, we specifically suggest evaluating their stability in an oxygenated environment after they have been terminated by Cl and Br [26, 27].

In a hydrogen atmosphere, MXenes completely terminated by sulfur are expected to remain stable. In addition, it is anticipated that exposure to H_2S gas will cause atomic S to replace the majority of other terminations. This is especially true for Mo_2C, where a Mo_2CS_2 MXene is predicted as a result of a complete termination substitution. In a hydrogen environment, nitrogen favors the NH state; nonetheless, the presence of NH_2 groups cannot be ruled out, with the exception of Mo_2C, for which pure $Mo_2C(NH)_2$ is expected (assuming the atmosphere contains just nitrogen and hydrogen). The possibilities of modifying the surface terminations of atomic nitrogen for Ti_3C_2, Nb_2C, and Nb_4C_3 are also predicted through a mild oxidation of the NH and NH_2 terminations. The research only addressed the surface of over-terminated MXenes, demonstrating that fluorine's adsorption power decreases significantly on a surface completely covered with oxygen in contrast to adsorption on a termination site that is freely open. However, more research on this might be beneficial. Gaining an understanding of the mechanics underlying the various substitution reactions of termination groups would also be quite interesting [26, 27].

REFERENCES

[1] Pogorielov, M., et al., *MXenes—A new class of two-dimensional materials: structure, properties and potential applications.* Nanomaterials, 2021. **11**(12): p. 3412.

[2] Anasori, B. and Û.G. Gogotsi, *2D Metal Carbides and Nitrides (MXenes).* Vol. 416. 2019, Springer.

[3] Pang, J., et al., *Applications of 2D MXenes in energy conversion and storage systems.* Chemical Society Reviews, 2019. **48**(1): p. 72–133.

[4] Khazaei, M., et al., *Recent advances in MXenes: from fundamentals to applications.* Current Opinion in Solid State and Materials Science, 2019. **23**(3): p. 164–178.

[5] Ronchi, R.M., J.T. Arantes, and S.F. Santos, *Synthesis, structure, properties and applications of MXenes: current status and perspectives.* Ceramics International, 2019. **45**(15): p. 18167–18188.

[6] Abbasi, N.M., et al., *Recent advancement for the synthesis of MXene derivatives and their sensing protocol.* Advanced Materials Technologies, 2021. **6**(10): p. 2001197.

[7] Garg, R., A. Agarwal, and M. Agarwal, *A review on MXene for energy storage application: effect of interlayer distance.* Materials Research Express, 2020. **7**(2): p. 022001.

[8] Ming, F., et al., *MXenes for rechargeable batteries beyond the lithium-ion.* Advanced Materials, 2021. **33**(1): p. 2004039.

[9] Xu, X., et al., *MXenes with applications in supercapacitors and secondary batteries: a comprehensive review.* Materials Reports: Energy, 2022. **2**(1): p. 100080.

[10] Verger, L., et al., *MXenes: an introduction of their synthesis, select properties, and applications.* Trends in Chemistry, 2019. **1**(7): p. 656–669.

[11] Zhang, C., et al., *Two-dimensional transition metal carbides and nitrides (MXenes): synthesis, properties, and electrochemical energy storage applications.* Energy & Environmental Materials, 2020. **3**(1): p. 29–55.

[12] Alnoor, H., et al., *Exploring MXenes and their MAX phase precursors by electron microscopy.* Materials Today Advances, 2021. **9**: p. 100123.

[13] Naguib, M., M.W. Barsoum, and Y. Gogotsi, *Ten years of progress in the synthesis and development of MXenes.* Advanced Materials, 2021. **33**(39): p. 2103393.

[14] Eklund, P., J. Rosen, and P.O.Å. Persson, *Layered ternary $M_{n+1}AX_n$ phases and their 2D derivative MXene: an overview from a thin-film perspective.* Journal of Physics D: Applied Physics, 2017. **50**(11): p. 113001.

[15] Verger, L., et al., *Overview of the synthesis of MXenes and other ultrathin 2D transition metal carbides and nitrides.* Current Opinion in Solid State and Materials Science, 2019. **23**(3): p. 149–163.

[16] Chen, N., W. Yang, and C. Zhang, *Perspectives on preparation of two-dimensional MXenes.* Science and Technology of Advanced Materials, 2021. **22**(1): p. 917–930.

[17] Halim, J., *Synthesis and Transport Properties of 2D Transition Metal Carbides (MXenes).* Vol. 1953. 2018, Linköping University Electronic Press.

[18] Huang, K., et al., *Two-dimensional transition metal carbides and nitrides (MXenes) for biomedical applications.* Chemical Society Reviews, 2018. **47**(14): p. 5109–5124.

[19] Chakraborty, P., T. Das, and T. Saha-Dasgupta, *1.15 MXene: a new trend in 2D materials science.* Comprehensive Nanoscience and Nanotechnology, 2019. **319**.

[20] Champagne, A. and J.-C. Charlier, *Physical properties of 2D MXenes: from a theoretical perspective.* Journal of Physics: Materials, 2020. **3**(3): p. 032006.

[21] Wang, Y., et al., *MXenes: focus on optical and electronic properties and corresponding applications.* Nanophotonics, 2020. **9**(7): p. 1601–1620.

[22] Kumar, J.A., et al., *Methods of synthesis, characteristics, and environmental applications of MXene: a comprehensive review.* Chemosphere, 2022. **286**: p. 131607.

[23] Peng, J., et al., *Surface and heterointerface engineering of 2D MXenes and their nanocomposites: insights into electro- and photocatalysis.* Chem, 2019. **5**(1): p. 18–50.

[24] Tang, X., et al., *2D metal carbides and nitrides (MXenes) as high-performance electrode materials for Lithium-based batteries.* Advanced Energy Materials, 2018. **8**(33): p. 1801897.

[25] Anasori, B., M.R. Lukatskaya, and Y. Gogotsi, *2D metal carbides and nitrides (MXenes) for energy storage.* Nature Reviews Materials, 2017. **2**(2): p. 1–17.

[26] Tang, M., et al., *Surface terminations of MXene: synthesis, characterization, and properties.* Symmetry, 2022. **14**(11): p. 2232.

[27] Bjork, J. and J. Rosen, *Functionalizing MXenes by tailoring surface terminations in different chemical environments.* Chemistry of Materials, 2021. **33**(23): p. 9108–9118.

[28] Bhat, A., et al., *Prospects challenges and stability of 2D MXenes for clean energy conversion and storage applications.* npj 2D Materials and Applications, 2021. **5**(1): p. 61.

[29] Khazaei, M., et al., *Novel electronic and magnetic properties of two-dimensional transition metal carbides and nitrides.* Advanced Functional Materials, 2013. **23**(17): p. 2185–2192.

[30] Mishra, A., et al., *Atomistic origin of phase stability in oxygen-functionalized MXene: a comparative study.* The Journal of Physical Chemistry C, 2017. **121**(34): p. 18947–18953.

[31] Zha, X.-H., et al., *Role of the surface effect on the structural, electronic and mechanical properties of the carbide MXenes.* Europhysics Letters, 2015. **111**(2): p. 26007.

[32] Gao, G., A.P. O'Mullane, and A. Du, *2D MXenes: a new family of promising catalysts for the hydrogen evolution reaction.* ACS Catalysis, 2017. **7**(1): p. 494–500.

[33] Zhan, C., et al., *Computational discovery and design of MXenes for energy applications: status, successes, and opportunities.* ACS Applied Materials & Interfaces, 2019. **11**(28): p. 24885–24905.

[34] Xin, Y. and Y.-X. Yu, *Possibility of bare and functionalized niobium carbide MXenes for electrode materials of supercapacitors and field emitters.* Materials & Design, 2017. **130**: p. 512–520.

[35] Meng, Q., et al., *Theoretical investigation of zirconium carbide MXenes as prospective high capacity anode materials for Na-ion batteries.* Journal of Materials Chemistry A, 2018. **6**(28): p. 13652–13660.

[36] Chen, J., et al., *Effects of intercalation on the interlayer electron-transfer process in Mo-based multilayered MXene flakes.* The Journal of Physical Chemistry C, 2021. **125**(31): p. 17232–17240.

[37] Fredrickson, K.D., et al., *Effects of applied potential and water intercalation on the surface chemistry of Ti_2C and Mo_2C MXenes.* The Journal of Physical Chemistry C, 2016. **120**(50): p. 28432–28440.

[38] Zha, X.-H., et al., *Intrinsic structural, electrical, thermal, and mechanical properties of the promising conductor Mo_2C MXene.* The Journal of Physical Chemistry C, 2016. **120**(28): p. 15082–15088.

[39] Zhao, X., et al., *Screening MXenes for novel anode material of lithium-ion batteries with high capacity and stability: a DFT calculation.* Applied Surface Science, 2021. **569**: p. 151050.

[40] Zhu, J., et al., *S-functionalized MXenes as electrode materials for Li-ion batteries.* Applied Materials Today, 2016. **5**: p. 19–24.

[41] Frey, N.C., et al., *Prediction of synthesis of 2D metal carbides and nitrides (MXenes) and their precursors with positive and unlabeled machine learning.* ACS Nano, 2019. **13**(3): p. 3031–3041.

[42] Ashton, M., et al., *Predicted surface composition and thermodynamic stability of MXenes in solution.* The Journal of Physical Chemistry C, 2016. **120**(6): p. 3550–3556.

[43] Xie, Y. and P. Kent, *Hybrid density functional study of structural and electronic properties of functionalized $Ti_{n+1}X_n$ (X= C, N) monolayers.* Physical Review B, 2013. **87**(23): p. 235441.

[44] Zhao, Y. and J. Zhao, *Functional group-dependent anchoring effect of titanium carbide-based MXenes for lithium-sulfur batteries: a computational study.* Applied Surface Science, 2017. **412**: p. 591–598.

[45] Zhang, Q., et al., *Theoretical insights into the favorable functionalized Ti_2C-based MXenes for lithium–sulfur batteries.* ACS Omega, 2020. **5**(45): p. 29272–29283.

[46] Jhon, Y., M. Seo, and Y. Jhon, *First-principles study of a MXene terahertz detector.* Nanoscale, 2018. **10**(1): p. 69–75.

[47] Zhan, C., et al., *Computational screening of MXene electrodes for pseudocapacitive energy storage.* The Journal of Physical Chemistry C, 2018. **123**(1): p. 315–321.

[48] Johnson, D., et al., *Holdups in nitride MXene's development and limitations in advancing the field of MXene.* Small, 2022. **18**(17): p. 2106129.

[49] Zhang, N., et al., *Superior structural, elastic and electronic properties of 2D titanium nitride MXenes over carbide MXenes: a comprehensive first principles study.* 2D Materials, 2018. **5**(4): p. 045004.

[50] Jurado, A., et al., *Adsorption and activation of CO_2 on nitride MXenes: composition, temperature, and pressure effects.* ChemPhysChem, 2021. **22**(23): p. 2456–2463.

[51] Kumar, H., et al., *Tunable magnetism and transport properties in nitride MXenes.* ACS Nano, 2017. **11**(8): p. 7648–7655.

[52] Teh, S. and H.-T. Jeng, *Magnetoelastic and magnetoelectric coupling in two-dimensional nitride MXenes: a density functional theory study.* Nanomaterials, 2023. **13**(19): p. 2644.

[53] Yu, S., et al., *First-principles study of Zr–N crystalline phases: phase stability, electronic and mechanical properties*. RSC Advances, 2017. **7**(8): p. 4697–4703.

[54] Mehta, V., et al., *Assessment of Mo$_2$N monolayer as Li-ion battery anodes with high cycling stability*. Materials Today Communications, 2021. **26**: p. 102100.

[55] Mehta, V., et al. *First principles study of Mo$_2$N monolayer as potential anode material for Na-ion batteries*, in *AIP Conference Proceedings*. 2020, AIP Publishing.

[56] Shukla, V., et al., *Modelling high-performing batteries with MXenes: the case of S-functionalized two-dimensional nitride MXene electrode*. Nano Energy, 2019. **58**: p. 877–885.

[57] Dong, Y., et al., *2D-VN$_2$ MXene as a novel anode material for Li, Na and K ion batteries: insights from the first-principles calculations*. Journal of Colloid and Interface Science, 2021. **593**: p. 51–58.

[58] Liu, H., et al., *Two-dimensional V$_2$N MXene monolayer as a high-capacity anode material for lithium-ion batteries and beyond: first-principles calculations*. ACS Omega, 2022. **7**(21): p. 17756–17764.

[59] Fatima, J., et al., *Structural, optical, electronic, elastic properties and population inversion of novel 2D carbides and nitrides MXene: a DFT study*. Materials Science and Engineering: B, 2023. **289**: p. 116230.

4 Classification of MXenes

MXenes can be found in six different types of structures and compositions. These structures include single transition metal elements (i.e., M_2XT_x, $M_3X_2T_x$, $M_4X_3T_x$, and $M_5X_4T_x$); out-of-plane ordered double-transition metal (DTM) elements (represented as $M'_2M''X_2$ or $M'_2M''_2X_3$), also called o-MXenes; in-plane ordered DTM elements or i-MXenes (represented as $(M'_{2/3}M''_{1/3})_2X$); in-plane MXenes with ordered divacancies (represented as $M_{1.33}X$); solid solutions on M sites (represented as $(M'_yM''_{1-y})_2X$, $(M'_yM''_{1-y})_3X_2$, $(M'_yM''_{1-y})_4X_3$, and $(M'_a M''_b M'''_c M^{IV}_{1-a-b-c})_4C_3$); and solid solutions on X sites (represented as $M_2(C_yN_{1-y})T_x$ and $M_3(C_yN_{1-y})_2T_x$). Two M elements used to be the most prevalent number in solid solution MXenes. However, researchers have shown that alloy MAX phases with high entropy can be also be synthesized; for example, MXenes with composition $(M1M2M3M4)C_3T_x$ are being synthesized from their corresponding MAX phase precursors [1–6]. Some of these structural representations and compositions of MXenes are presented in Fig. 4.1 [7], and elements of known MAX phases and MXenes are represented in Fig. 4.2 [8].

4.1 SINGLE TRANSITION METAL MXENES

MXenes are represented by the chemical formula $M_{n+1}X_nT_x$. The thickness of MXenes can be altered by varying the value of n, which results in different multilayered structures. Depending on the value of n, MXenes are classified as M_2XT_x, $M_3X_2T_x$, $M_4X_3T_x$, and $M_5X_4T_x$ [1, 9, 10]. To date, a wide range of theoretical and experimental studies can be found on single transition metal MXenes. The first one M MXene, that is, $Ti_3C_2T_x$, was synthesized by M. Naguib et al. [11] by selective hydrofluoric acid (HF) etching of its MAX precursor (i.e., Ti_3AlC_2). A. Champagne et al. [2] reviewed the structural, electronic, and magnetic properties of bare and functionalized M_2XT_2 MXenes. According to the studies, the MAX phases and the MXenes, which initially exhibited metallic characteristics, can be changed into semiconductors due to the production of bandgap after the addition of surface terminations. However, introducing surface termination does not affect the metallic characteristics of thicker MXenes (with a larger n value). For instance, regardless of the surface termination, Zr_3C_2, Ti_3C_2, Nb_4C_3, Ti_4C_3, etc., display metallic properties [12].

M. Ashton et al. [13] computed the formation energies of freestanding $M_{n+1}X_nT_2$ nanosheets and observed that these energies follow a trend where $M_2XT_2 > M_3X_2T_2 > M_4X_3T_2$. This is because thicker MXenes have a larger volume-to-surface area ratio. However, the bulk MAX phases have an opposite trend, and there are only a few uncommon M and X combinations in M_4AX_3 and M_3AX_2 bulk phases that can be used to produce these MXenes. Notably, five of the six MXene compounds, namely, Nb_2CT_2, Ti_2CT_2, V_2CT_2, $Ti_3C_2T_2$, $Nb_4C_3T_2$, and $Ta_4C_3T_2$, have formation energies that fall within the typical 0.2 eV/atom threshold (found for freestanding 2D materials). V_2CO_2 exhibit a formation energy of 0.285 eV/atom, which is slightly

DOI: 10.1201/9781003465768-4

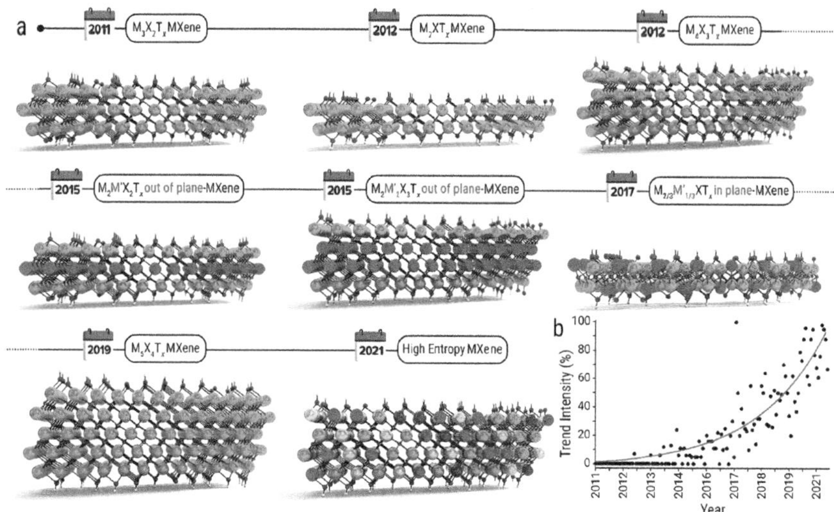

FIGURE 4.1 (a) Possible structures and compositions of MXenes. (b) Monthly Google Trends data, collected over the decade from users' engagement with the word "MXene." Since 2015, there has been a considerable increase in trend intensity.

Source: Reprinted from [7] *Materials Today Advances*, 13, M. Dadashi Firouzjaei, M. Karimiziarani, H. Moradkhani et al., MXenes: The two-dimensional influencers, 100202, Copyright 2022, with permission from Elsevier.

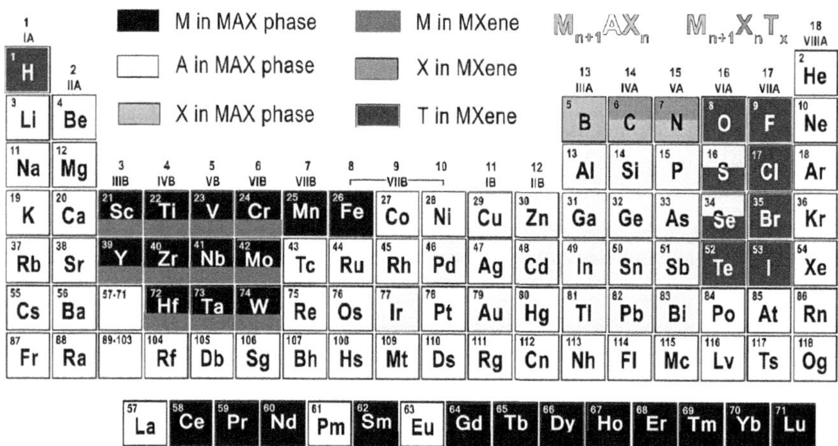

FIGURE 4.2 Elements of MAX phases and MXenes.

Source: Reprinted from [8] M. Pogorielov, K. Smyrnova, et al.: MXenes—A New Class of Two-Dimensional Materials: Structure, Properties and Potential Applications. *Nanomaterials* 2021, 11(12), 3412. Open access.

higher than the threshold and is considered an outlier. The increased metastability of MXenes, in comparison to other 2D materials, is suggested to be due to their complex decomposition paths to their corresponding 3D phases.

Researchers used ab initio density functional theory (DFT) to investigate the work functions and quantum capacitances of $Nb_{n+1}C_nZ_2$ (where n = 1–4) and their potential to be used in supercapacitors and electrical devices. They found that the formation energies of MXenes of the same thickness follow the order: $Nb_{n+1}C_nO_2$ < $Nb_{n+1}C_n(OH)_2$ < $Nb_{n+1}C_nF_2$ < $Nb_{n+1}C_n(OCH_3)_2$. These energies were observed to be weakly dependent on the sheet thickness and strongly dependent on the functional groups. Similarly, DFT calculations were used by Q. Meng et al. [14] to explore the Na adsorption on Zr-based MXenes, that is, Zr_2C, Zr_2CO_2, Zr_3C_2, and $Zr_3C_2O_2$. Y. Bai et al. [15] conducted a DFT analysis on 2D Ti_2CT_2 and $Ti_3C_2T_2$ MXene monolayers to study their crystal and electronic structures, as well as elastic and optical characteristics. The study revealed a significant correlation between the surface-terminated groups produced during the MXene etching process and the crystal structure, optical response, elastic stiffness, and electrical structure.

In their study, H. Zhang et al. [16] used the DFT approach to examine the theoretical properties of three geometries of functionalized materials, that is, M_2CT_2-I, M_2CT_2-II, and M_2CT_2-III (where M = Ti, Zr, and Hf). The studied properties included the structural, electrical, and optical characteristics. The researchers found that the phonon spectra of all materials (except for M_2CO_2-II and M_2CO_2-III) were free from negative frequencies, indicating their kinetic stability. In another study by X. Zha et al. [17], the thermal and electrical characteristics of semiconducting Sc_2CT_2 were theoretically examined. The research found that both –F- and –OH-functionalized MXenes exhibit excellent carrier mobilities. $Sc_2C(OH)_2$ has a nearly isotropic electron mobility, which is different from Sc_2CF_2. They both also have high thermal conductivities that enable them to be utilized for electronic devices. In another study, the MXenes and their corresponding MAX phases were examined using the positive and unlabeled (PU) learning method [18]. The result was the identification of 18 compositions (i.e., Zr_2C, Ta_2C, Hf_2C, Sc_2C, Ti_2N, Nb_2N, Nb_3C_2, Ta_3C_2, Sc_3C_2, W_3C_2, Hf_4C_3, Ta_4N_3, Ti_4C_3, W_4C_3, Zr_4C_3, Sc_4C_3, Mo_4C_3, and Hf_4N_3) having structural stability to be synthesized.

In a study conducted by R. P. Jadav et al. [19], the electronic and structural properties of $Hf_3C_2F_2$ monolayer were investigated using first-principle computations. The team found that the $Hf_3C_2F_2$ MXene monolayer displays remarkable performance characteristics, such as strong conductivity and a stable structure, which indicates potential applications in ab initio molecular dynamics (AIMD) for energy storage. Another study by S. Zhao et al. [20] described a technique to create nanopores in $Nb_4C_3T_x$ flakes by varying the etching duration. When tested in 1M Na_2SO_4, $(NH_4)_2SO_4$, and Li_2SO_4 electrolytes, as-synthesized material exhibited a 50% enhanced rate capability. The long cycling stability and rate performance of $V_4C_3T_x$ MXenes are negatively impacted by the use of binders and compactly stacked layers. To overcome this issue and develop a new member for flexible electrodes, X. Bin et al. [21] delaminated the multilayered $V_4C_3T_x$ with tetrabutylammonium hydroxide (TBAOH) and organized it into a flexible film (without binders). A. Mishra et al. [22] conducted an extensive investigation on the conversion of MXene sheets by introducing LiF, HF, and alkali atoms into the Nb_4AlC_3 MAX phase using first-principle calculations. The bond dissociation energy (BDE) calculations revealed that HF leads to functionalized MXene because H binds strongly to MXene, while F binds weakly to it.

In a research by G. Deysher et al. [23], the successful synthesis and characterization of $Mo_4VC_4T_x$ was reported. The authors achieved this by fabricating its MAX precursor, which was free of any other MAX phase impurities. The study suggested that additional Al-containing MAX phases could allow for the synthesis of more $M_5C_4T_x$ MXenes. A theoretical insight into M_2C, M_2N, MC_2, and MN_2 (where M = V, Cr, Ti, and Sc) suggested that Ti_2N and V_2N are excellent anode materials for lithium-ion batteries due to their good conductivity, thermal stability, low diffusion barrier, and high theoretical capacity [24]. C. Zhan et al. [25] used computational screening to model the pseudocapacitive performance of a group of MXene electrodes in aqueous H_2SO_4 electrolyte. The study found that nitrides exhibit superior pseudocapacitive efficiency compared to carbides. Specifically, $Zr_{n+1}N_nT_x$ MXenes were expected to have the best areal capacitance, while Ti_2NT_x was predicted to have a high gravimetric capacitance. In their study, S. Teh et al. [26] theoretically analyzed the lattice structures, electrical and magnetic properties, and molecular orbital structures of nitride MXenes including V_2NO_2, Mo_2NO_2, Mn_2NO_2, V_2NF_2, Mo_2NF_2, and Cr_2NF_2, which have intrinsic bandgaps.

The structural stability of pure and Li-adsorbed Mo_2N monolayers was theoretically analyzed [27]. Similarly, by using DFT calculations, the electrochemical behavior of Mo_2N was investigated [28]. B. Ranjan et al. [29] demonstrated a method of creating supercapacitive electrodes by synthesizing Mo_2N thin films (on stainless steel substrate) using magnetron sputtering. The electrode showed excellent mechanical and electrochemical performance due to the mechanical strength of the substrate and the large specific surface area of Mo_2N. According to H. Liu et al. [30], the 2D V_2N MXene monolayer would be an excellent choice for anode material in rechargeable batteries. They projected this using first-principle calculations. It is found that the V_2N monolayer is a metallic substance. When it comes to theoretical capacities, the V_2N MXene monolayer has a capacity of 925 mAh g^{-1} for Li ions, 463 mAh g^{-1} for Na ions, and 1850 mAh g^{-1} for Mg ions. In a study by S. Venkateshalu and colleagues, they introduced a new method for synthesizing V_2NT_x using powdered V_2AlN. The process involved sonicating the mixture and selectively removing the "Al" from it by immersing it in a LiF–HCl mixture. The V_2NT_x electrode demonstrated a specific capacitance of 112.8 F g^{-1} in 3.5 M KOH electrolyte [31]. The compositions of all of the theoretically predicted and experimentally synthesized single transition metal carbides and nitrides are given in Table 4.1.

4.2 DOUBLE TRANSITION METAL MXENES

Double transition metal (DTM) MXenes are a second class of MXenes composed of M′ and M″, that is, two separate transition metals. By structure, DTM MXenes are further divided into solid solution and ordered compositions. In ordered MXenes, M′ and M″ occupy specific sites, which are referred to as either in-plane order or out-of-plane order. The formula $M'_{4/3}M''_{2/3}XT_x$ is used to describe a type of ordered MXenes

TABLE 4.1

Experimentally Synthesized and Theoretically Predicted Compositions of Single Transition Metal MXenes

Single Transition Metal MXenes						
M_2X $n = 1$		M_3X_2 $n = 2$		M_4X_3 $n = 3$		M_5X_4 $n = 4$
M_2C	M_2N	M_3C_2	M_3N_2	M_4C_3	M_4N_3	$(Mo_4V) C_4,$
$Sc_2C,$	$Sc_2N,$	$Sc_3C_2,$	$Sc_3N_2,$	$Sc_4C_3,$	$Sc_4N_3,$	$(Ti\ Ta)_5C_4,$
$Y_2C,$	$Y_2N,$	$Y_3C_2,$	$Y_3N_2,$	$Y_4C_3,$	$Y_4N_3,$	$(Ti\ Nb)_5C_4,$
$Ti_2C,$	$Ti_2N,$	$Ti_3C_2,$	$Ti_3N_2,$	$Ti_4C_3,$	$Ti_4N_3,$	$(Mo_{4/5}V_{1/5})_5C_4$ or
$Zr_2C,$	$Zr_2N,$	$Zr_3C_2,$	$Zr_3N_2,$	$Zr_4C_3,$	$Zr_4N_3,$	$(Mo_{0.8}V_{0.2})_5C_4$
$Hf_2C,$	$Hf_2N,$	$Hf_3C_2,$	$Hf_3N_2,$	$Hf_4C_3,$	$Hf_4N_3,$	
$V_2C,$	$V_2N,$	$V_3C_2,$	$V_3N_2,$	$V_4C_3,$	$V_4N_3,$	
$Nb_2C,$	$Nb_2N,$	$Nb_3C_2,$	$Nb_3N_2,$	$Nb_4C_3,$	$Nb_4N_3,$	
$Ta_2C,$	$Ta_2N,$	$Ta_3C_2,$	$Ta_3N_2,$	$Ta_4C_3,$	$Ta_4N_3,$	
$Cr_2C,$	$Cr_2N,$	$Cr_3C_2,$	$Cr_3N_2,$	$Cr_4C_3,$	$Cr_4N_3,$	
$Mo_2C,$	$Mo_2N,$	$Mo_3C_2,$	$Mo_3N_2,$	$Mo_4C_3,$	$Mo_4N_3,$	
$W_2C,$	$W_2N,$	$W_3C_2,$	$W_3N_2,$	$W_4C_3,$	$W_4N_3,$	
Mn_2C	Mn_2N	Mn_3C_2	Mn_3N_2	Mn_4C_3	Mn_4N_3	

M: Sc, Ti, V, Cr, Mn, Y, Zr, Nb, Mo, Hf, Ta, W; **X**: C, N

T_x: O, S, Se, Te, F, Cl, Br, I, H

Source: Experimentally synthesized MXenes are represented in red, and all theoretically predicted MXenes are represented in black.

called in-plane MXenes. Examples of in-plane MXenes include $Mo_{4/3}Y_{2/3}CT_x$. These materials consist of two different types of transition metals arranged in alternating positions within each atomic plane of the M-layer. When certain M″ and A elements are selectively removed, it results in the formation ordered divacancy i-MXene. This material has the general formula $M'_{1.33}XT_x$ and has ordered divacancies left behind by the removal of M″. On the other hand, the term "out-of-plane MXenes" is denoted by the formula $M'_2M''X_2T_x$ or $M'_2M''_2X_3T_x$. This refers to a type of ordered structures, where M′ and M″ occupy distinct atomic planes. In this case, the exterior layers of M′ are enclosed in inner layers of M″, such as $Cr_2TiC_2T_x$, $Mo_2TiC_2T_x$, and $Cr_2VC_2T_x$. Unlike ordered MXenes, solid-solution MXenes consist of two different metals randomly distributed all over the M-layers. Examples include $(Ti,V)_2CT_x$, $(Ti,Nb)_2CT_x$, $(Mo,V)_4C_3T_x$, $(Nb,Zr)_4C_3T_x$, and $(Nb,V)_2CT_x$ [9, 32, 33].

It is worth noting that so far, no DTM nitrides or carbonitrides have been explored. All the DTM MXenes that have been synthesized or studied theoretically are carbides. The composition of DTM MXenes is determined by their parent MAX phases. Although many DTM MXenes have been fabricated, several MAX phases of this type have not yet undergone selective etching to yield their corresponding MXenes. Similarly, while some DTM MXenes are theoretically predicted, they have not been empirically comprehended in their MAX precursor yet. This expands the MXene family, offering numerous novel compositions. DTM MXenes allow the modification of the transition metal atom arrangement to create solid solutions that can be either in-plane or out-of-plane ordered. This exclusive ability in the 2D material field enables control over the electrical, optical, electrochemical, magnetic, and mechanical properties. Currently, there is no direct transition from mono-M to DTM MXenes, which can be studied in the future [32, 33].

4.2.1 IN-PLANE ORDERED DTM MXENES

In-plane order has been observed only in M_2XT_x MXene structure. These materials consist of two different types of transition metals (M′ and M″) arranged in alternating positions within each atomic plane of the M-layer. M′ is usually V, Cr, Mn, Nb, Mo, or W, while M″ is typically Sc, Y, or Zr. For example, an ordered $Mo_{4/3}Y_{2/3}AlC$ phase was produced by reactively sintering a combination of Mo, Y, Al, and C for 20 hours at 1500°C (in an argon environment) [32, 33]. Researchers have conducted a study using $(Mo_{2/3}Y_{1/3})_2AC$ and $(Mo_{2/3}Sc_{1/3})_2AC$ as model systems to investigate the production of i-MAX. Through first-principle computations and experimental studies, they discovered seven new thermodynamically stable i-MAX phases. To achieve this, they utilized group 13 and 14 elements as the A-layer. Some of these phases that have been verified experimentally include $(Mo_{2/3}Sc_{1/3})_2AlC$, $(Mo_{2/3}Y_{1/3})_2AlC$, $(Cr_{2/3}Y_{1/3})_2AlC$, $(Cr_{2/3}Zr_{1/3})_2AlC$, $(Cr_{2/3}Sc_{1/3})_2AlC$, $(V_{2/3}Sc_{1/3})_2AlC$, $(V_{2/3}Zr_{1/3})_2AlC$, $(W_{2/3}Y_{1/3})_2AlC$, $(W_{2/3}Sc_{1/3})_2AlC$, $(Mo_{2/3}Y_{1/3})_2GaC$, and $(Mo_{2/3}Sc_{1/3})_2GaC$ [34]. There are four main factors that contribute to these structures, including a significant electronegativity difference between M″ and A elements, M′ and M″ with a 2:1 stoichiometric ratio, and a considerable size difference between both metals (where M″ is larger) [32–34].

These i-MAX phases are also responsible for the production of i-MXenes that contain ordered divacancies. It has been observed experimentally that the electrochemical performance of Mo can be tuned by substituting $Mo_{4/3}Y_{2/3}CT_x$ with $Mo_{4/3}CT_x$ (i.e., i-MXene with ordered divacancy). The results showed that $Mo_{4/3}CT_x$ exhibits higher volumetric capacitance in acidic electrolytes. However, $Mo_{4/3}Y_{2/3}CT_x$ outperforms its performance in basic electrolytes, suggesting that changes in electrochemical performance are also dependent on the electrolyte used. These changes may be due to the surface terminations, which are affected by the introduction of vacancies and the minority M″, or variations in electrochemically active sites. The reduced Mo concentration in $Mo_{4/3}CT_x$ results in a deficiency of electrons for surface terminations, leading to a greater number of –F terminations than –O terminations. Similarly, $W_{4/3}CT_x$ is a promising HER catalyst and outperforms the activity of Mo-based i-MXenes [32].

By etching A and M2 elements from a quaternary $(Nb_{2/3}Sc_{1/3})_2AlC$ phase, an analogous composition of Nb-based MXene ($Nb_{1.33}C$) is produced. $Nb_{1.33}C$ features small vacancy clusters and disordered vacancies, which is significant for the material's application. Moreover, it has a M:C ratio comparable to an i-MXene with vacancies. Therefore, removing the alloying element and A-layer by etching certain quaternary MAX phase alloys can yield both typical MXenes and i-MXenes with ordered vacancies [34]. In order to fully unlock the potential of DTM MXenes with in-plane ordering for various applications, further research is necessary to gain a complete understanding of the principles of charge transport kinetics in these materials [32]. A study conducted by A. Thore et al. [35] computed the electrical, vibrational, and elastic characteristics of two nano-laminated materials from the i-MAX phase family: $(Mo_{2/3}Y_{1/3})_2AlC$ and $(Mo_{2/3}Sc_{1/3})_2AlC$. The study also compared the characteristics of some hypothetical MAX phases (i.e., Mo_2AlC, Y_2AlC, and Sc_2AlC). The analysis indicated that although the i-MAX phases are in different space groups, they have very similar bonding properties.

In a recent study, I. Persson et al. [36] introduced a new method to modify the composition and structure of MXenes. They reported that $(Mo_{2/3}Y_{1/3})_2AlC$ could be selectively etched using different protocols, resulting in two different products, that is, $Mo_{1.33}C$, by removing both Al and Y atoms, and $(Mo_{2/3}Y_{1/3})_2C$. In KOH electrolyte, $(Mo_{2/3}Y_{1/3})_2C$ displayed a high volumetric capacitance (>1,500 F cm^{-3}), which is 40% greater than that of $Mo_{1.33}C$. However, $Mo_{1.33}C$ showed a higher capacitance (about 1,200 F cm^{-3}) in H_2SO_4 electrolyte. Q. Tao et al. [37] presented a simple and efficient method for inserting ordered divacancies in MXenes in a controlled manner. First, they fabricated an in-plane ordered $(Mo_{2/3}Sc_{1/3})_2AlC$, and then selectively etched the Al and Sc layers to create 2D $Mo_{1.33}C$ sheets with high electrical conductivities. A flexible solid-state supercapacitor was developed by L. Qin et al. [38] using a $Mo_{1.33}C$/PEDOT:PSS (poly(3,4-ethylenedioxythiophene): poly(styrene sulfonic acid)) composite film. The supercapacitor exhibited high volumetric capacitance (568 F cm^{-3}), with an ultrahigh energy and power density. Similarly, another study utilized $Mo_{1.33}C$/PEDOT:PSS to develop high-efficiency polymer solar cells [39]. The research confirmed that this combination is a viable option for use in organic photovoltaics. In a similar way, J. Halim et al. [40] produced $Nb_{1.33}C$ by simultaneous etching of Sc and Al atoms from the $(Nb_{2/3}Sc_{1/3})_2AlC$ phase.

Meanwhile, R. Meshkian et al. [41] used DFT to predict the phase stability of quaternary $(W_{2/3}M^2_{1/3})_2AC$ (where $M^2 = Sc$ and Y). Out of the 18 studied compositions, only two were found to be stable: $(W_{2/3}Y_{1/3})_2AlC$ and $(W_{2/3}Sc_{1/3})_2AlC$. These 3D laminates were selectively etched to yield ordered divacancy-based MXene sheets having a $W_{1.33}C$ basis. In another study, $(Cr_{2/3}Zr_{1/3})_2AlC$ was developed through the hot pressing of a combination of Cr, ZrH_2, Al, and C at a temperature of 1300 °C [42]. The primary phase exhibited an in-plane stoichiometry, accompanied by secondary phases of Cr_2AlC, $ZrAlC_2$, and ZrC, as well as a minor quantity of Al-Cr intermetallic phases based on quantitative chemical composition investigations. The magnetic structure of $(Cr_{2/3}Zr_{1/3})_2AlC$ was theoretically evaluated and was shown to have an antiferromagnetic ground state. Q. Tao et al. [43] reported the discovery of magnetic $(Mo_{2/3}RE_{1/3})_2AlC$ phases, in which the Al Kagome lattice and Mo honeycomb arrangement are covered by bilayers of a quasi-2D triangular lattice. The rare earth (RE) elements that belong to this family are Nd, Ce, Ho, Gd, Tm, Pr, Dy, Sm, Tb, Er, and Lu [44]. The compositions of all of the theoretically predicted and experimentally synthesized in-plane ordered DTM MXenes are given in Table 4.2.

4.2.2 Out-Of-Plane Ordered DTM MXenes

Out-of-plane order is observed in $M_3C_2T_x$ and $M_4C_3T_x$ MXenes. While in-plane order arranges transition metals in a single atomic plane, out-of-plane order refers to a type

TABLE 4.2

Compositions of In-Plane Ordered DTM MXenes

In-Plane Ordered DTM MXenes		
M_2X		
$n = 1$		
$M'_{4/3} M''_{2/3}C$	$(M'_{2/3} M''_{1/3})_2C$	Ordered Divacancy MXenes $M_{1.33}C$
$V_{4/3}Zr_{2/3}C$, $Cr_{4/3}Sc_{2/3}C$, $Cr_{4/3}Y_{2/3}C$, $Cr_{4/3}Zr_{2/3}C$, $Mn_{4/3}Sc_{2/3}C$, $Nb_{4/3}Sc_{2/3}C$, $Mo_{4/3}Sc_{2/3}C$, $Mo_{4/3}Y_{2/3}C$, $W_{4/3}Sc_{2/3}C$, $W_{4/3}Y_{2/3}C$ (Using M': V, Cr, Mn, Nb, Mo, W, and M'': Sc, Y, Zr; total 18 combinations can be made)	$(Mo_{2/3}Sc_{1/3})_2C$, $(Mo_{2/3}Y_{1/3})_2C$, $(W_{2/3}Sc_{1/3})_2C$, $(W_{2/3}Y_{1/3})_2C$ (Total 18 combinations can be made using M' and M'')	$Nb_{1.33}C$, $Mo_{1.33}C$, $W_{1.33}C$, $Cr_{1.33}C$ The combinations of ordered divacancy MXenes can also be made for other M elements

M: Sc, Ti, V, Cr, Mn, Y, Zr, Nb, Mo, Hf, Ta, W; **X**: C, N; T_x: O, S, Se, Te, F, Cl, Br, I, H

Source: The experimentally synthesized MXenes are represented in red, and all theoretically predicted MXenes are represented in black.

of ordered structures, where M' and M'' occupy distinct atomic planes. Structurally, the outer layers of M' are enclosed in inner layers of M''. The structural arrangement of these MXenes is determined during the synthesis of precursor MAX phases. Currently, only a few o-MAX phases are experimentally synthesized, including Mo_2TiAlC_2, Mo_2ScAlC_2, Ti_2ZrAlC_2, Cr_2TiAlC_2, Cr_2VAlC_2, $Mo_2Ti_2AlC_3$, and $Cr_2V_2AlC_3$. Following etching and exfoliation of o-MAX phases, the corresponding structural ordering remains intact in their derivative MXenes. Thus, $Mo_2ScC_2T_x$, $Mo_2TiC_2T_x$, Cr_2TiC_2Tx, and $Mo_2Ti_2C_3T_x$ are among the synthesized DTM o-MXenes [32, 33]. For stable structures of o-MXenes, M' is usually Mo, W, and Cr, while M'' is typically V, Ti, Nb, Sc, Ta, Hf, and Zr. However, researchers can expand the existing pairing selection of M' and M'' by preventing solid-solution formation of these metals during the fabrication of o-MAX phases. The choice of these metal pairs in o-MAX precursors can greatly influence the electrical, optical, electrochemical, and magnetic properties of the resulting o-MXenes. Both experimental and theoretical studies are being carried out to analyze the electronic characteristics of these compositions. Although mono-M $M_3X_2T_x$ and $M_4X_3T_x$ exhibit metallic conductivity, only some of the o-MXenes have the potential to become semiconductors or semimetals [32, 33].

Spin-orbit coupling (SOC) between electrons in the d orbitals of M' and M'' and magnetic ordering of electrons in the d orbital of M' are some of the main reasons for the observed semiconductor properties. A higher electronegativity difference between surface terminations and transition metals improves the ability to generate semiconductor MXenes. Moreover, interactions of surface terminations with metals also affect these properties. Therefore, to create semiconductor MXenes, it is crucial to manage the M' and surface terminations in MXenes [32, 33]. Researchers studied 24 different types of 2D DTM carbide MXenes, with $M'_2M''C_2T_x$ and $M'_2M''_2C_3T_x$

compositions (for M′: V, Cr, Nb, or Ti, and M″: V, Nb, Ti, or Ta). They used DFT calculations to screen these carbides and determine which ones were suitable for use as hydrogen evolution reaction (HER) catalysts. Out of the 18 carbides that were projected to be viable candidates for HER electrocatalysts, it was found that $Mo_2NbC_2O_2$ had the lowest overpotential [44]. Using first-principle computations, Z. Q. Huang et al. [45] carried out a thorough search on $M'_2M''C_2$ (using Ta, Nb, or V for M′ and Hf, Ti, or Zr for M″) with different surface groups, that is, H, O, OH, F, Cl, Br, or I. Most of the structures demonstrated semimetallic band structures and the topological phase. Above all, $M'_2M C_2F_2$, fluorine-terminated MXenes, were topological insulators.

Similarly, X. Feng et al. [46] theoretically determined the surface group preference of 33 distinct o-MXenes, synthesized via HF etching of their corresponding o-MAX precursors. According to the enthalpy and Gibbs free energy calculations, the resultant material will be O-terminated for 4d- or 5d-orbital outermost M, and it will be F-terminated if the outermost M is 3d orbital. Additionally, B. Anasori et al. [47] predicted the presence of two o-MXenes, $M'_2M C_2T_x$ and $M'_2M''C_3T_x$, using the DFT calculations. They also synthesized $Mo_2TiC_2T_x$, $Mo_2Ti_2C_3T_x$, and $Cr_2TiC_2T_x$ to validate the DFT predictions. Based on their calculations, some materials are preferred to be fully ordered at 0 K, including Mo_2TiC_2, Mo_2NbC_2, Mo_2VC_2, Mo_2TaC_2, Cr_2VC_2, Cr_2TiC_2, Cr_2NbC_2, Cr_2TaC_2, Ti_2TaC_2, Ti_2NbC_2, V_2TiC_2, and V_2TaC_2. Partial ordering of a few other MXenes, such as Nb_2VC_2, Ta_2TiC_2, and Nb_2TiC_2, is considered to more stable than the fully ordered one. The Mo and Cr atoms govern the structural and electrochemical characteristics of the 2D flakes since they are at external positions. Using first-principle calculations, C. Si et al. [48] reported the quantum spin Hall (QSH) effect in $Mo_2M_2C_3O_2$ (M = Ti, Zr, or Hf), or $Mo_2Ti_2C_3O_2$, $Mo_2Zr_2C_3O_2$, and $Mo_2Hf_2C_3O_2$.

In another study, the crystalline structure and stabilities of Mo_2TiAlC_2 and $Mo_2Ti_2AlC_3$ were presented [49]. The general ordered stacking sequence was observed to be Mo-Ti-Mo-Al-Mo-Ti-Mo and Mo-Ti-Ti-Mo-Al-Mo-Ti-Ti-Mo for Mo_2TiAlC_2 and $Mo_2Ti_2AlC_3$, respectively. These sequences show that carbon atoms generally occupy the octahedral sites between M-layers. Researchers also investigated the impact of various defects on structural stabilities, electrical conductivity, and electrochemical characteristics of ordered $Mo_2TiC_2T_x$ [50]. The results showed that the surface groups play an important role in the development of defects, where O-terminated materials require more energy for defect formations than F- and OH-terminated MXenes. Another research by R. Meshkian et al. [51] confirmed an out-of-plane order in the MAX phase alloy Mo_2ScAlC_2. The study evaluated phase stability using ab initio calculations based on DFT. The findings showed that, in contrast to the unstable Sc_3AlC_2 and Mo_3AlC_2, the o-MAX phase exhibits stability and high formation enthalpy. Due to extraordinary properties and structural stabilities, these materials have the potential to be used in various applications.

For example, the electrochemical behavior of Mo_2MC_2 (where M = V, Sc, Nb, Zr, Ti, Ta, or Hf) as an anode material of aluminum-ion battery was studied theoretically [52]. In a multilayer adsorption configuration, researchers calculated the average adsorption energies, and the results showed that these structures could adsorb three layers of Al atoms (on both upper and lower surfaces). This resulted in high theoretical capacities, which ranged from 888.98 mAh g^{-1} for Mo_2TaC_2 to

1170.33 mAh g^{-1} for Mo_2ScC_2. Likewise, S. A. Zahra et al. [53] synthesized two different DTM carbides, $Mo_2TiC_2T_x$ and $Mo_2Ti_2C_3T_x$, as bifunctional catalysts for water splitting in alkaline environments. These carbides are non-precious metal (NMP)-based and have the potential to replace precious metals in water splitting processes. Using vacuum-assisted filtration, Mo_2CT_x, $Mo_2TiC_2T_x$, and $Mo_2Ti_2C_3T_x$ can also be processed into binder-free sheets, which exhibit better electric and thermoelectric properties [54]. DFT simulations were used by H. Vovusha et al. [55] to investigate the sensing of S-containing gases, such as sulfur dioxide (SO_2) and hydrogen sulfide (H_2S) on o-MXenes. It was discovered that both H_2S and SO_2 physiosorbed on $M_2TiC_2T_x$ monolayers (where M is either Mo or Cr and T = O, OH, or S). According to the results, SO_2 can strongly adsorb on S-terminated materials, while H_2S can strongly adsorb on O-terminated ones.

It was also studied experimentally that Mo atoms prefer to reside on the outer layers of $[MC]_nM$ structures. When the electrical characteristics of these Mo-based materials were compared to their Ti_3C_2 counterparts, it was discovered that the resistance of the former was increasing with decreasing temperature, unlike metallic-like conductors [56]. N.M. Caffrey and colleagues analyzed the necessary conditions for increasing the yield of o-MXenes (i.e., $(Mo, Ti)_2CT_x$) by utilizing Pourbaix diagrams. They discovered that the high yields can be achieved because o-MXene maintains stability in aqueous solutions over a broad pH and applied potential range, compared to other ionic and molecular species. The experiment results demonstrated that o-MXene is a promising material for various applications [57]. The effect of extensile strain on the magnetic properties of $Cr_2M_2C_3T_2$ (where M = V, Ti, or Nb) was also studied by using DFT calculations [58]. The results of the study showed that for $Cr_2Ti_2C_3O_2$ and $Cr_2V_2C_3O_2$, the ferromagnetic configuration is energetically favorable in the strain-free state. However, $Cr_2Nb_2C_3T_2$ structure exhibits antiferromagnetic arrangement at extensile stresses ranging from −5% to 5%. Furthermore, $Cr_2Ti_2C_3T_2$ and $Cr_2V_2C_3T_2$ undergo ferromagnetic to antiferromagnetic—or vice versa—phase transitions, but $Cr_2Ti_2C_3O_2$ does not exhibit such a transition.

A recent study by Yang et al. [59] explored the electronic and magnetic properties of $Cr_2M'C_2T_2$ (using V or Ti for M′) through first-principle calculations. The study found that the configurations of the systems can be influenced by the choice of M′ and T, along with the bond-coupling interactions. The resulting configurations can be antiferromagnetic ($Cr_2TiC_2(OH)_2$ and $Cr_2TiC_2F_2$), ferromagnetic ($Cr_2VC_2O_2$, $Cr_2VC_2F_2$, and $Cr_2VC_2(OH)_2$), or nonmagnetic ($Cr_2TiC_2O_2$), thereby exhibiting metallic or semiconductor behavior. Z. Jing et al. [60] conducted a study where they showed that pure and functionalized Cr_2TiC_2 MXenes are semiconductors (with moderate bandgap), using SCAN-rVV10+U. In all the observed structures, a high value of the Seebeck coefficient (>400 μV/K) was observed. Among them, Cr_2TiC_2 and $Cr_2TiC_2F_2$ showed the highest values (>800 μV/K and >700 μV/K, respectively). Additionally, the study revealed that $Cr_2TiC_2(OH)_2$ exhibited the strongest surface group-dependent electron thermal conduction and the least lattice thermal conductivity.

W. Sun et al. [61] used the Heyd-Scuseria-Ernzerhof hybrid (HSE06) density functional to show that $TiMn_2N_2F_2$ has a remarkably large bandgap above 1 eV, whereas the pure material retains metallic properties. It is worth noting that not every combination of terminations and metals is expected to be stable, unlike the

TABLE 4.3

All Experimentally Synthesized and Theoretically Predicted Compositions of Out-of-Plane Ordered DTM MXenes

Out-of-Plane Ordered DTM MXenes		
M_3X_2 $n = 2$		M_4X_3 $n = 3$
$(M'_2 M'') C_2$	$(M'_{2/3} M''_{1/3})_3 C_2$	$(M'_2 M''_2) C_3$
$(Sc_2Nb)C_2$, $(Sc_2Ta)C_2$, $(Sc_2W)C_2$, $(Ti_2V)C_2$, $(Ti_2Mn)C_2$, $(Ti_2Nb)C_2$, $(Ti_2Zr)C_2$, $(Ti_2Ta)C_2$, $(V_2Ti)C_2$, $(V_2Ta)C_2$, $(Cr_2Ti)C_2$, $(Cr_2V)C_2$, $(Cr_2Nb)C_2$, $(Cr_2Ta)C_2$, $(Cr_2Ti)N_2$, $(Mn_2Ti)N_2$, $(Mo_2Sc)C_2$, $(Mo_2Ti)C_2$, $(Mo_2V)C_2$, $(Mo_2Zr)C_2$, $(Mo_2Ta)C_2$, $(Mo_2Nb)C_2$, $(Mo_2Hf)C_2$, $(Ta_2Hf)C_2$, $(Hf_2V)C_2$, $(Hf_2Mn)C_2$, $(W_2Ti)C_2$, $(W_2Zr)C_2$, $(W_2Hf)C_2$, $(Nb_2Zr)C_2$, $(Nb_2Hf)C_2$, $(Nb_2Ta)C_2$ (Total 132 combinations can be made using M′ and M″)	$(Cr_{2/3}Ti_{1/3})_3C_2$, $(Mo_{2/3}Ti_{1/3})_3C_2$, $(Mo_{2/3}Sc_{1/3})_3C_2$, $(Ti_{0.62}Ta_{0.38})_3C_2$	$(Ti_2Ta_2)C_3$, $(Ti_2Nb_2)C_3$, $(V_2Nb_2)C_3$, $(V_2Ti_2)C_3$, $(V_2Ta_2)C_3$, $(Cr_2Ti_2)C_3$, $(Cr_2V_2)C_3$, $(Cr_2Nb_2)C_3$, $(Cr_2Ta_2)C_3$, $(Nb_2Hf_2)C_3$, $(Nb_2Ta_2)C_3$, $(Mo_2Ti_2)C_3$, $(Mo_2V_2)C_3$, $(Mo_2Zr_2)C_3$, $(Mo_2Nb_2)C_3$, $(Mo_2Hf_2)C_3$, $(Mo_2Ta_2)C_3$, $(W_2Ti_2)C_3$, $(W_2Zr_2)C_3$, $(W_2Hf_2)C_3$, $(Ta_2Hf_2)C_3$ (Total 66 combinations can be made using M′ and M″)

M: Sc, Ti, V, Cr, Mn, Y, Zr, Nb, Mo, Hf, Ta, W; **X**: C, N T_x: O, S, Se, Te, F, Cl, Br, I, H

Source: Experimentally synthesized MXenes are represented in red, and all theoretically predicted MXenes are represented in black.

more popular Ti-C MXenes. In addition, strong ferromagnetism in $Hf_2MnC_2O_2$ and $Hf_2VC_2O_2$ monolayers was observed [62]. In another study, researchers employed a method that synergistically used DFT and Monte Carlo methods to investigate the electronic properties of magnetic $Hf_2VC_2O_2$ and $Hf_2MnC_2O_2$ [63]. This method allows for realistic surface terminations by creating surface defects such as oxygen vacancies and H adatoms. It was discovered that by introducing hydrogen atoms or surface oxygen vacancies, the magnetic anisotropies and electronic structures can be modified. The compositions of all of the theoretically predicted and experimentally synthesized out-of-plane ordered DTM MXenes are given in Table 4.3.

4.2.3 SOLID-SOLUTION MXENES

Solid-solution MXenes, represented as $(M'_\alpha M''_{1-\alpha})_{n+1}C_n T_x$ ($0 < \alpha < 1$), are unique materials that have two randomly distributed M elements, unlike ordered MXenes. Although other 2D materials, such as TMDs, TMCs, or TMNs, also exhibit solid solutions, ordered structures are exhibited by MXenes only. The structural properties of solid-solution MXenes can be modified by changing the M′ and M″ ratio. These types of compositions are reported for M_2CT_x, $M_3C_2T_x$, $M_4C_3T_x$, and even $M_5C_4T_x$, which only exist as a solid solution till date. Solid-solution MXenes and their corresponding MAX phases cover a wide range of M′:M″, in contrast to ordered MXenes, which exhibit only a few M′:M″ ratios. Based on the M′:M″ atomic ratio, solid-solution MXenes, such as $(Nb_\alpha V_{1-\alpha})_2CT_x$ and $(Ti_\alpha Nb_{1-\alpha})_2CT_x$, can be adjusted to vary their electronic conductivity and electromagnetic interference (EMI) shielding efficiency. This means that their properties can be adjusted to be similar to Nb_2CT_x and V_2CT_x, or Ti_2CT_x and Nb_2CT_x, depending on the ratio. For example, by changing the value of α, the electrical conductivity of $(Mo_\alpha V_{1-\alpha})_4C_3T_x$ increases by more than an order of magnitude, that is, from 15 S cm^{-1} for $(MoV_3)C_3T_x$ to 830 S cm^{-1} for $(Mo_{2.7}V_{1.3})C_3T_x$ [32].

Despite having over thousands of potential solid-solution compositions, there has been limited theoretical and experimental research on solid-solution MXenes since their first publication in 2012. Additionally, not all MAX phases have been transformed into MXenes, such as $(Cr_\alpha Mn_{1-\alpha})_2AlC$ [32]. J. Yang et al. [64] synthesized two novel 2D Nb_4C_3-based solid solutions, that is, $(Nb_{0.8},Zr_{0.2})_4C_3T_x$ and $(Nb_{0.8},Ti_{0.2})_4C_3T_x$, from their respective MAX phases and studied their electrochemical behavior by intercalating lithium ions. In another study, $V_{2-y}Nb_yT_x$ and $Ti_{2-y}Nb_yT_x$ ($0 \leq y \leq 2L$) were synthesized [65]. Moreover, the relationship between the chemistry and charge storage capacity, such as cycling stability and capacitive characteristics in an aqueous electrolyte, was also discovered. The discovery and synthesis of the first magnetic MAX phases were made possible by using DFT to evaluate their phase stability, followed by the production of heteroepitaxial thin films. A paper by A. S. Ingason et al. [66] assessed the properties of these phases. All of the magnetic MAX phases discovered so far, including Cr_2AlC, $(Cr, Mn)_2AlC$, $(Cr, Mn)_2GaC$, Cr_2GeC, $(Cr, Mn)_2GeC$, $(Mo, Mn)_2GaC$, Mn_2GaC, and $(V, Mn)_3GaC_2$, contain either Cr or Mn, or both.

D. Pinto et al. [67] synthesized 2D $Mo_xV_{4-x}C_3$ by selectively etching aluminum from its MAX precursors. Unlike the ordered DTM material, $Mo_xV_{4-x}C_3$ presented a Mo-V solid solution within the M-layers. The $Mo_{2.7}V_{1.3}C_3$ composition exhibited a high electrical conductivity (830 S cm^{-1}) and an impressive volumetric capacitance (about 860 F cm^{-3}) at room temperature. Solid-solution $(Nb_yV_{2-y})AlC$ MAX phases were synthesized and then chemically etched to produce $(Nb_yV_{2-y})CT_x$ MXenes. These MXenes were further oxidized to create corresponding oxides. M. A. Andrade et al. [68] demonstrated that the production of solid solutions speeds up the formation of MXenes and facilitates the etching kinetics of powdered MAX phase. In contrast to pure materials, the synthesis of solid solutions leads to quick formation of MXenes. Electrochemical cycling of oxides formed from $(Nb_yV_{2-y})CT_x$ in lithium-ion cells showed variable intercalation-like behavior. Electrodes derived from V_2CT_x exhibited pseudocapacitive response and redox processes. The cyclic voltammetry (CV) curves of oxides generated from solid-solution MXenes mainly showed behavior similar to that of the BVO/nANO composite, with some notable exceptions.

Z. Shen et al. [69] were able to synthesize $(Ti_{0.5}V_{0.5})_3C_2$ by exfoliating its MAX phase (i.e., $(Ti_{0.5}V_{0.5})_3AlC_2$). They comprehensively assessed the catalytic effect of as-synthesized MXenes on the Mg hydrogen storage process. The synthesized $(Ti_{0.5}V_{0.5})_3C_2$ had a typical layer shape and demonstrated a higher catalytic activity than Ti_3C_2. For the MgH$_2$–10wt%, the apparent activation energy was determined to be 77.3 kJ mol^{-1}, which was around half that of the pristine. The TiVAlC MAX phase was etched by S. Yazdanparast and colleagues [70] using a combination of HF and

TABLE 4.4

All the Compositions of Solid-Solution MXenes

Solid-Solution MXenes					
M_2X n = 1		M_3X_2 n = 2		M_4X_3 n = 3	
$(M', M'')_2C$ (on M Site)	**$(M'_y M''_{1-y})_2C$ (on M Site)**	**$(M', M'')_3C_2$ (on M Site)**	**$(M'_y M''_{1-y})_3C_2$ (on M Site)**	**$(M', M'')_4C_3$: (on M Site)**	**$(M'_y M''_{1-y})_4C_3$ (on M Site)**
$(Ti, V)_2C$, $(Ti, Zr)_2C$, $(Ti, Nb)_2C$, $(Ti, Mo)_2C$, $(Ti, Hf)_2C$, $(Ti, Ta)_2C$, $(V, Mn)_2C$, $(Cr, Ti)_2C$, $(Cr, V)_2C$, $(Cr, Mn)_2C$, $(Nb, V)_2C$, $(Nb, Zr)_2C$, $(Mo, V)_2C$ (Total 66 combinations can be made using M' and M'')	$(Ti_{1/2}V_{1/2})_2C$, $(Ti_{1/2}Nb_{1/2})_2C$, $(Ti_{1-y}V_y)_2C$, TiNbC, $(Ti_{1-y}Nb_y)_2C$, $(V_{1-y}Nb_y)_2C$ (y=0.2, 0.4, 0.6, and 0.8) $(V_{1-x}Ti_x)_2C$, $Nb_y V_{2-y}C$ (y=0.4, 0.8, 1.2, 1.6), $Ti_y Nb_{2-y}C$ (0.4, 0.8, 1.2, 1.6) **$M_2(C_y N_{1-y})$: (on X site)** $Ti_2(C_{0.5}N_{0.5})$ (Total 12 combinations can be made for solid solution on X site using M)	$(Ti, V)_3C_2$, $(Ti, Nb)_3C_2$, $(Ti, Hf)_3C_2$, $(Ti, Ta)_3C_2$, $(Ti, W)_3C_2$, $(V, Ta)_3C_2$, $(Cr, V)_3C_2$, $(Zr, Ti)_3C_2$, $(Zr, Nb)_3C_2$, $(Zr, Hf)_3C_2$, $(Zr, Ta)_3C_2$, $(Nb, Ta)_3C_2$, $(Hf, Nb)_3C_2$, $(Hf, Ta)_3C_2$, $(Ta, Sc)_3C_2$ (Total 66 combinations can be made using M' and M'').	$(Ti_{1/2}V_{1/2})_3C_2$, $(Cr_{1/2}V_{1/2})_3C_2$ **$M_3(C_y N_{1-y})_2$: (on X site)** $Ti_3(C, N)_2$ or $Ti_3(C_{1/2}N_{1/2})_2$, Ti_3CN (Total 12 combinations can be made for solid solution on X site using M: Sc, Y, Ti, Zr, Hf, V, Nb, Ta, Cr, Mo, W, Mn) **$(M'_a M''_b M'''_c) C_2$: (on M site)** (Ti V Cr) C_2	$(Sc, Mo)_4C_3$, $(Sc, Ta)_4C_3$, $(Sc, W)_4C_3$, $(Ti, V)_4C_3$, $(Ti, Nb)_4C_3$, $(Ti, Ta)_4C_3$, $(Cr, V)_4C_3$, $(Cr, Ti)_4C_3$, $(Zr, Hf)_4C_3$, $(Zr, Nb)_4C_3$, $(Zr, Ta)_4C_3$, $(Nb, V)_4C_3$, $(Nb, Zr)_4C_3$, $(Nb, Hf)_4C_3$, $(Nb, Ta)_4C_3$, $(Nb, W)_4C_3$, $(Mo, V)_4C_3$, $(Ta, V)_4C_3$ (Total 66 combinations can be made using M' and M'')	$(Mo_{1/2}T_{1/2})_4C_3$, $(Nb_{1/2}V_{1/2})_4C_3$, $(Mo_z V_{1-z)4}C_3$ (z= 0.25, 0.375, 0.5, and 0.075), $(Ti_{0.5}Ta_{0.5})_4C_3$, $(Nb_{4/5}Ti_{1/5})_4C_3$, $(Nb_{4/5}Zr_{1/5})_4C_3$, $Nb_{3.5}Ta_{0.5}C_3$, $Nb_{3.9}W_{0.1}C_3$ **$(M'_a M''_b M'''_c M^{IV}_{1-a-b-c})_4C_3$: (on M site)** (Ti V Nb Mo) C_3 (Ti V Cr Mo) C_3

M: Sc, Ti, V, Cr, Mn, Y, Zr, Nb, Mo, Hf, Ta, W; **X**: C, N; T_x: O, S, Se, Te, F, Cl, Br, I, H

Source: Experimentally synthesized MXenes are represented in red, and theoretically predicted MXenes are represented in black.

LiF/HCl. For large-scale delamination, the material was further intercalated by using TBAOH. The TiVC was observed to be O- and F-terminated, which is favorable for energy-storage applications. The compositions of all of the theoretically predicted and experimentally synthesized solid-solution MXenes are given in Table 4.4.

4.2.4 OTHER CLASSIFICATIONS AND LATEST ADVANCEMENTS

According to the latest research and advancements, a number of studies are being carried out by researchers to expand the structural and compositional range of MXenes. S. K. Nemani et al. [71] presented two high-entropy materials, $TiVCrMoC_3T_x$ and $TiVNbMoC_3T_x$, along with their MAX phases, that is, $TiVCrMoAlC_3$ and $TiVNbMoAlC_3$. The authors investigated the possibility of synthesizing these high-entropy MAX phases and calculated their formation energies using first-principle calculations. They also demonstrated that when three transition metals are used instead of four, two distinct MAX phases can emerge under synthesis conditions. Similarly, a new MAX phase, that is, $Ti_2V_{1-y}Cr_yAlC_2$, was successfully synthesized along with its associated MXenes for the first time [72]. The structural stability was confirmed by using synchrotron X-ray pair distribution functions (PDFs), DFT, and aberration-corrected (AC)-scanning transmission electron microscopy (STEM). This structure shows a solid solution of Ti, V, and Cr in the three atomic layers. Additionally, the $Ti_2V_{0.9}Cr_{0.1}C_2T_x$ monolayer demonstrated a notable gravimetric capacitance when used as the electrode of a supercapacitor.

B. C. Wyatt et al. [73] conducted a study using DFT calculations, Rietveld refinement, and electron imaging techniques to validate the preferred ordering of Mo and Nb in the M-layers. They also demonstrated the fabrication and electrocatalytic behavior of $(Mo_{2+\alpha}Nb_{2-\alpha})AlC_3$ phases, with $0 \leq \alpha \leq 0.3$. Another study revealed that the HER activity of Mo-based DTM $Mo_2MC_2O_2$ MXenes is better than that of Cr-based $Cr_2M''C_2O_2$, indicating that $Mo_2VC_2O_2$ and $Mo_2NbC_2O_2$ exhibit higher HER activity [74]. Z. Zeng et al. [75] performed DFT computations to examine the HER activity of 64 O-terminated carbonitrides with a formula $M'_2M''CNO_2$. The study found that 11 M'_2MCNO_2-MXene candidates had better HER performance than Pt. Additionally, Ti_2VCNO_2, Ti_2NbCNO_2, and Mo_2TiCNO_2 were identified as more stable choices based on the stability screening. For the first time, $Ti_xTa_{(4-x)}C_3$ MXene was successfully produced by R. Syamsai et al. [76] by Al etching of the MAX phase. This material was observed to exhibit 97% capacity retention with high reversible discharge capacity (459 mAh g^{-1}).

Half-metallic pure and functionalized $ScCr_2C_2$ and $YCr_2C_2H_2$ possess high Curie temperatures and wide bandgaps [77]. The magnetic tunnel junction based on $ScCr_2C_2$ displayed an extremely intriguing high tunnel magnetoresistance ratio. Additionally, $ZrCr_2C_2$, $TiCr_2C_2$, and $ZrCr_2C_2(OH)_2$ are predicted to be antiferromagnetic semiconductors with high Néel temperatures and moderate bandgaps. The spin-dependent transport features of $ScCr_2C_2F_2$-based van der Waals (vdW) magnetic tunnel junctions (MTJs) were extensively studied by Z. Cui et al. [78]. In these MTJs, 1T-MoS_2 served as the electrode, $ScCr_2C_2F_2$ as the spin-filter tunnel barrier, and 2H-MoS_2 as the tunnel barrier. They discovered that the spin-up electrons mostly determined the transmission behavior in the parallel configuration state. The outcomes suggested potential uses of this material in spintronics. Another research reported the substitution of transition metals on $Mo_2TiC_2O_2$ and $Cr_2TiC_2O_2$ monolayers and observed better thermodynamic stability and HER performance of as-modified MXenes [79]. M. Zhou et al. [80] used MoWC and $MoWCO_2$ as anodes in sodium-ion batteries and theoretically studied the influence of Mo and W on their electrochemical behavior. A new Na adsorption layer resulted from the activation and transfer of this excess charge to the first adsorption layer. These materials outperformed the Mo_2C and W_2C in terms of battery performance.

To better understand the electrochemical behavior of MoWC MXene, which is widely used as an effective electrode material in lithium-ion batteries, V. Mehta et al. [81] used a first-principle approach. Using phonon dispersion calculations, shape optimization, and AIMD, the thermodynamic stability was ascertained. The monolayer exhibited strong electronic conduction, according to the electronic structural calculations. Additionally, it was noted that the MoWC monolayer had a low diffusion barrier of 0.029 eV on the Mo-layer side and 0.040 eV on the W-layer side. Furthermore, compared to Mo_2C and W_2C, their superior electrochemical performance was demonstrated by high storage capacity (670 mAh g^{-1}). The electronic, optical, thermodynamic, and structural stability characteristics of novel YScX MXenes were investigated by Z. Amoudeh et al. [82]. 2D YScC and YScN were observed to have structural stability to be synthesized experimentally. Using local-density approximation (LDA), generalized gradient approximation (GGA), and GGA + HSE06 approximations, total density of states (DOS) and band structure computations revealed that these materials are metallic in nature.

W. L. Chang et al. [83] investigated the stability and electrical characteristics of $ScYCT_2$ (T = F or OH) using first-principle calculations. Based on this, their semiconductor characteristics were identified, and the Slack model and Boltzmann transport theory were used to examine their thermoelectric properties. The findings demonstrated the strong thermoelectric qualities of the MXene structures under investigation. Specifically, the n-type $ScYC(OH)_2$ has a maximum ZT value greater than 3 and the largest power factor at 900 K, 0.072 W mK^{-2}. The topochemical transformation of $(TiNb)_5C_4T_x$ and $(TiTa)_5C_4T_x$ was reported by M. Downes et al. [84]. The resultant MXenes underwent structural analysis, thermal and optical property characterization, and delamination into single-layer flakes. In doing so, a family of M_5AX_4 phases and associated MXenes were established. With the help of theoretical predictions, these materials were created experimentally, paving the way for intriguing new uses for MXenes.

Energy storage and conversion systems can utilize two different metals arranged in either an in-plane or out-of-plane orientation, forming a sandwich-like structure. However, due to the large number of compositions that require further computational or experimental research, DTM MXene research is still in its early stages. Despite the fact that over 50 of them are theoretically stable, only a few ordered DTM compositions have been realized analytically. Even in theoretical studies, the structures and properties of only a few ordered DTM MXenes have been analyzed. Incorporating solid-solution DTM MXenes widens the range of potential compositions and allows for the creation of materials with adjustable qualities. This gives us the ability to design 2D materials that have optimized properties for specific purposes. Theoretical studies on how transition metals affect MXene characteristics can guide future research on the synthesis of new MXenes using theoretical principles [32, 33].

Recent advancements in MAX phase precursors have opened up new opportunities for DTM MXenes. Currently, there are ordered and solid-solution DTM MAX phases that have not yet been transformed into their corresponding MXenes. Additionally, no research has been conducted on DTM MXenes derived from nitride or carbonitride MAX phases, even though several mono-M nitrides and carbonitrides have been successfully synthesized. So far, the focus has been solely on DTM MXenes derived from C-based MAX phases. It is expected that DTM nitride/carbonitride MXenes would exhibit promising properties due to nitrogen or carbon/nitrogen as the X element. Nitrides and carbonitrides have exhibited unique properties in energy storage applications compared to their carbide counterparts. Further research should investigate the effect of nitrogen and carbon/nitrogen on the structural stabilities and properties of DTM MXenes [32, 33].

REFERENCES

[1] Naguib, M., M.W. Barsoum, and Y. Gogotsi, *Ten years of progress in the synthesis and development of MXenes.* Advanced Materials, 2021. **33**(39): p. 2103393.

[2] Champagne, A. and J.-C. Charlier, *Physical properties of 2D MXenes: from a theoretical perspective.* Journal of Physics: Materials, 2020. **3**(3): p. 032006.

[3] Ronchi, R.M., J.T. Arantes, and S.F. Santos, *Synthesis, structure, properties and applications of MXenes: current status and perspectives.* Ceramics International, 2019. **45**(15): p. 18167–18188.

[4] Anasori, B. and Û.G. Gogotsi, *2D Metal Carbides and Nitrides (MXenes).* Vol. 416. 2019, Springer.

[5] Wang, Y., et al., *MXenes: focus on optical and electronic properties and corresponding applications.* Nanophotonics, 2020. **9**(7): p. 1601–1620.

[6] Verger, L., et al., *MXenes: an introduction of their synthesis, select properties, and applications.* Trends in Chemistry, 2019. **1**(7): p. 656–669.

[7] Firouzjaei, M.D., et al., *MXenes: the two-dimensional influencers.* Elsevier, 2022: p. 100202.

[8] Pogorielov, M., et al., *MXenes—A new class of two-dimensional materials: structure, properties and potential applications.* Nanomaterials, 2021. **11**(12): p. 3412.

[9] Chen, N., W. Yang, and C. Zhang, *Perspectives on preparation of two-dimensional MXenes.* Science and Technology of Advanced Materials, 2021. **22**(1): p. 917–930.

[10] Tang, X., et al., *2D metal carbides and nitrides (MXenes) as high-performance electrode materials for Lithium-based batteries.* Advanced Energy Materials, 2018. **8**(33): p. 1801897.

[11] Naguib, M., et al., *Two-dimensional nanocrystals produced by exfoliation of Ti_3AlC_2.* Advanced Materials, 2011. **23**(37): p. 4248–4253.

[12] Tang, M., et al., *Surface terminations of MXene: synthesis, characterization, and properties.* Symmetry, 2022. **14**(11): p. 2232.

[13] Ashton, M., et al., *Predicted surface composition and thermodynamic stability of MXenes in solution.* The Journal of Physical Chemistry C, 2016. **120**(6): p. 3550–3556.

[14] Meng, Q., et al., *Theoretical investigation of zirconium carbide MXenes as prospective high capacity anode materials for Na-ion batteries.* Journal of Materials Chemistry A, 2018. **6**(28): p. 13652–13660.

[15] Bai, Y., et al., *Dependence of elastic and optical properties on surface terminated groups in two-dimensional MXene monolayers: a first-principles study.* RSC Advances, 2016. **6**(42): p. 35731–35739.

[16] Zhang, H., et al., *Computational studies on the structural, electronic and optical properties of graphene-like MXenes (M_2CT_2, M = Ti, Zr, Hf; T = O, F, OH) and their potential applications as visible-light driven photocatalysts.* Journal of Materials Chemistry A, 2016. **4**(33): p. 12913–12920.

[17] Zha, X.-H., et al., *Promising electron mobility and high thermal conductivity in Sc_2CT_2 (T = F, OH) MXenes.* Nanoscale, 2016. **8**(11): p. 6110–6117.

[18] Frey, N.C., et al., *Prediction of synthesis of 2D metal carbides and nitrides (MXenes) and their precursors with positive and unlabeled machine learning.* ACS Nano, 2019. **13**(3): p. 3031–3041.

[19] Jadav, R.P., et al., *Structural stability and electronic properties of 2D MXene $Hf_3C_2F_2$ monolayer by density functional theory approach.* Biointerface Research in Applied Chemistry, 2022. **13**(2): p. 152.

[20] Zhao, S., et al., *Effect of pinholes in Nb_4C_3 MXene sheets on its electrochemical behavior in aqueous electrolytes*. Electrochemistry Communications, 2022. **142**: p. 107380.

[21] Bin, X., et al., *Self-assembling delaminated $V_4C_3T_x$ MXene into highly stable pseudo-capacitive flexible film electrode for supercapacitors*. Advanced Materials Interfaces, 2022. **9**(15): p. 2200231.

[22] Mishra, A., et al., *Isolation of pristine MXene from Nb_4AlC_3 MAX phase: a first-principles study*. Physical Chemistry Chemical Physics, 2016. **18**(16): p. 11073–11080.

[23] Deysher, G., et al., *Synthesis of Mo_4VAlC_4 MAX phase and two-dimensional Mo_4VC_4 MXene with five atomic layers of transition metals*. ACS Nano, 2019. **14**(1): p. 204–217.

[24] Zhao, X., et al., *Screening MXenes for novel anode material of lithium-ion batteries with high capacity and stability: a DFT calculation*. Applied Surface Science, 2021. **569**: p. 151050.

[25] Zhan, C., et al., *Computational screening of MXene electrodes for pseudocapacitive energy storage*. The Journal of Physical Chemistry C, 2018. **123**(1): p. 315–321.

[26] Teh, S. and H.-T. Jeng, *Magnetoelastic and magnetoelectric coupling in two-dimensional nitride MXenes: a density functional theory study*. Nanomaterials, 2023. **13**(19): p. 2644.

[27] Mehta, V., et al., *Assessment of Mo_2N monolayer as Li-ion battery anodes with high cycling stability*. Materials Today Communications, 2021. **26**: p. 102100.

[28] Mehta, V., et al. *First principles study of Mo_2N monolayer as potential anode material for Na-ion batteries*. in *AIP Conference Proceedings*. 2020, AIP Publishing.

[29] Ranjan, B., G.K. Sharma, and D. Kaur, *Rationally synthesized Mo_2N nanopyramids for high-performance flexible supercapacitive electrodes with deep insight into the Na-ion storage mechanism*. Applied Surface Science, 2022. **588**: p. 152925.

[30] Liu, H., et al., *Two-dimensional V_2N MXene monolayer as a high-capacity anode material for lithium-ion batteries and beyond: first-principles calculations*. ACS Omega, 2022. **7**(21): p. 17756–17764.

[31] Venkateshalu, S., et al., *New method for the synthesis of 2D vanadium nitride (MXene) and its application as a supercapacitor electrode*. ACS Omega, 2020. **5**(29): p. 17983–17992.

[32] Hong, W., et al., *Double transition-metal MXenes: atomistic design of two-dimensional carbides and nitrides*. Mrs Bulletin, 2020. **45**(10): p. 850–861.

[33] Venkateshalu, S., et al., *Ordered double transition metal MXenes*. ChemNanoMat, 2022. **8**(11): p. e202200320.

[34] Ahmed, B., A.E. Ghazaly, and J. Rosen, *i-MXenes for energy storage and catalysis*. Advanced Functional Materials, 2020. **30**(47): p. 2000894.

[35] Thore, A. and J. Rosén, *An investigation of the in-plane chemically ordered atomic laminates $(Mo_{2/3}Sc_{1/3})_2AlC$ and $(Mo_{2/3}Y_{1/3})_2AlC$ from first principles*. Physical Chemistry Chemical Physics, 2017. **19**(32): p. 21595–21603.

[36] Persson, I., et al., *Tailoring structure, composition, and energy storage properties of MXenes from selective etching of in-plane, chemically ordered MAX phases*. Small, 2018. **14**(17): p. 1703676.

[37] Tao, Q., et al., *Two-dimensional $Mo_{1.33}C$ MXene with divacancy ordering prepared from parent 3D laminate with in-plane chemical ordering*. Nature Communications, 2017. **8**(1): p. 14949.

[38] Qin, L., et al., *High-performance ultrathin flexible solid-state supercapacitors based on solution processable $Mo_{1.33}C$ MXene and PEDOT:PSS*. Advanced Functional Materials, 2018. **28**(2): p. 1703808.

[39] Liu, Y., et al., *$Mo_{1.33}C$ MXene-assisted PEDOT:PSS hole transport layer for high-performance bulk-heterojunction polymer solar cells*. ACS Applied Electronic Materials, 2019. **2**(1): p. 163–169.

[40] Halim, J., et al., *Synthesis of two-dimensional $Nb_{1.33}C$ (MXene) with randomly distributed vacancies by etching of the quaternary solid solution $(Nb_{2/3}Sc_{1/3})_2AlC$ MAX phase*. ACS Applied Nano Materials, 2018. **1**(6): p. 2455–2460.

[41] Meshkian, R., et al., *W-based atomic laminates and their 2D derivative $W_{1.33}C$ MXene with vacancy ordering*. Advanced Materials, 2018. **30**(21): p. 1706409.

[42] Chen, L., et al., *Theoretical prediction and synthesis of $(Cr_{2/3}Zr_{1/3})_2AlC$ i-MAX phase*. Inorganic Chemistry, 2018. **57**(11): p. 6237–6244.

[43] Tao, Q., et al., *Atomically layered and ordered rare-earth i-MAX phases: a new class of magnetic quaternary compounds*. Chemistry of Materials, 2019. **31**(7): p. 2476–2485.

[44] Jin, D., et al., *Computational screening of 2D ordered double transition-metal carbides (MXenes) as electrocatalysts for hydrogen evolution reaction*. The Journal of Physical Chemistry C, 2020. **124**(19): p. 10584–10592.

[45] Huang, Z.-Q., et al., *Large-gap topological insulators in functionalized ordered double transition metal carbide MXenes*. Physical Review B, 2020. **102**(7): p. 075306.

[46] Feng, X., T. Bai, and B. Xiao. *Prediction of surface termination preference of out-of-plane ordered double-transition metal MXenes (o-MXenes) from first-principles calculations*. Journal of Physics: Conference Series, 2022. (IOP Publishing).

[47] Anasori, B., et al., *Two-dimensional, ordered, double transition metals carbides (MXenes)*. ACS Nano, 2015. **9**(10): p. 9507–9516.

[48] Si, C., et al., *Quantum spin Hall phase in $Mo_2M_2C_3O_2$ (M = Ti, Zr, Hf) MXenes*. Journal of Materials Chemistry C, 2016. **4**(48): p. 11524–11529.

[49] Anasori, B., et al., *Experimental and theoretical characterization of ordered MAX phases Mo_2TiAlC_2 and $Mo_2Ti_2AlC_3$*. Journal of Applied Physics, 2015. **118**(9).

[50] Khaledialidusti, R., A.K. Mishra, and A. Barnoush, *Atomic defects in monolayer ordered double transition metal carbide $(Mo_2TiC_2T_x)$ MXene and CO_2 adsorption*. Journal of Materials Chemistry C, 2020. **8**(14): p. 4771–4779.

[51] Meshkian, R., et al., *Theoretical stability and materials synthesis of a chemically ordered MAX phase, Mo_2ScAlC_2, and its two-dimensional derivate Mo_2ScC_2 MXene*. Acta Materialia, 2017. **125**: p. 476–480.

[52] Liu, H., et al., *Bare Mo-based ordered double-transition metal MXenes as high-performance anode materials for aluminum-ion batteries*. The Journal of Physical Chemistry C, 2020. **124**(47): p. 25769–25774.

[53] Zahra, S.A., et al., *Two-dimensional double transition metal carbides as superior bifunctional electrocatalysts for overall water splitting*. Electrochimica Acta, 2022. **434**: p. 141257.

[54] Kim, H., et al., *Thermoelectric properties of two-dimensional molybdenum-based MXenes*. Chemistry of Materials, 2017. **29**(15): p. 6472–6479.

[55] Vovusha, H., et al., *Sensing of sulfur containing toxic gases with double transition metal carbide MXenes*. Materials Today Chemistry, 2023. **30**: p. 101543.

[56] Anasori, B., et al., *Control of electronic properties of 2D carbides (MXenes) by manipulating their transition metal layers*. Nanoscale Horizons, 2016. **1**(3): p. 227–234.

[57] Caffrey, N.M., *Prediction of optimal synthesis conditions for the formation of ordered double-transition-metal MXenes (o-MXenes)*. The Journal of Physical Chemistry C, 2020. **124**(34): p. 18797–18804.

[58] Yang, J., et al., *Tuning magnetic properties of $Cr_2M_2C_3T_2$ (M = Ti and V) using extensile strain*. Computational Materials Science, 2017. **139**: p. 313–319.

[59] Yang, J., et al., *Tunable electronic and magnetic properties of $Cr_2M'C_2T_2$ (M' =Ti or V; T = O, OH or F)*. Applied Physics Letters, 2016. **109**(20).

[60] Jing, Z., et al., *Superior thermoelectric performance of ordered double transition metal MXenes: $Cr_2TiC_2T_2$ (T = −OH or −F)*. The Journal of Physical Chemistry Letters, 2019. **10**(19): p. 5721–5728.

[61] Sun, W., Y. Xie, and P.R. Kent, *Double transition metal MXenes with wide band gaps and novel magnetic properties.* Nanoscale, 2018. **10**(25): p. 11962–11968.

[62] Dong, L., et al., *Rational design of two-dimensional metallic and semiconducting spintronic materials based on ordered double-transition-metal MXenes.* The Journal of Physical Chemistry Letters, 2017. **8**(2): p. 422–428.

[63] Siriwardane, E.M., et al., *Engineering magnetic anisotropy and exchange couplings in double transition metal MXenes via surface defects.* Journal of Physics: Condensed Matter, 2020. **33**(3): p. 035801.

[64] Yang, J., et al., *Two-dimensional Nb-based M_4C_3 solid solutions (MXenes).* Journal of the American Ceramic Society, 2016. **99**(2): p. 660–666.

[65] Wang, L., et al., *Adjustable electrochemical properties of solid-solution MXenes.* Nano Energy, 2021. **88**: p. 106308.

[66] Ingason, A.S., M. Dahlqvist, and J. Rosén, *Magnetic MAX phases from theory and experiments; a review.* Journal of Physics: Condensed Matter, 2016. **28**(43): p. 433003.

[67] Pinto, D., et al., *Synthesis and electrochemical properties of 2D molybdenum vanadium carbides—solid solution MXenes.* Journal of Materials Chemistry A, 2020. **8**(18): p. 8957–8968.

[68] Andrade, M.A., et al., *Synthesis of 2D solid-solution (Nb y V_2–y) CT x MXenes and their transformation into oxides for energy storage.* ACS Applied Nano Materials, 2023. **6**(18): p. 16168–16178.

[69] Shen, Z., et al., *A novel solid-solution MXene $(Ti_{0.5}V_{0.5})_3C_2$ with high catalytic activity for hydrogen storage in MgH2.* Materialia, 2018. **1**: p. 114–120.

[70] Yazdanparast, S., et al., *Synthesis and surface chemistry of 2D TiVC solid-solution MXenes.* ACS Applied Materials & Interfaces, 2020. **12**(17): p. 20129–20137.

[71] Nemani, S.K., et al., *High-entropy 2D carbide MXenes: $TiVNbMoC_3$ and $TiVCrMoC_3$.* ACS Nano, 2021. **15**(8): p. 12815–12825.

[72] Ma, W., et al., *A new $Ti_2V_{0.9}Cr_{0.1}C_2Tx$ MXene with ultrahigh gravimetric capacitance.* Nano Energy, 2022. **96**: p. 107129.

[73] Wyatt, B.C., et al., *Design of atomic ordering in $Mo_2Nb_2C_3T$ x MXenes for hydrogen evolution electrocatalysis.* Nano Letters, 2023. **23**(3): p. 931–938.

[74] Li, N., et al., *Double transition metal carbides MXenes (D-MXenes) as promising electrocatalysts for hydrogen reduction reaction: Ab initio calculations.* ACS Omega, 2021. **6**(37): p. 23676–23682.

[75] Zeng, Z., et al., *Computational screening study of double transition metal carbonitrides $M'_2M''CNO_2$-MXene as catalysts for hydrogen evolution reaction.* npj Computational Materials, 2021. **7**(1): p. 80.

[76] Syamsai, R., et al., *Double transition metal MXene $(Ti_xTa_{4-x}C_3)$ 2D materials as anodes for Li-ion batteries.* Scientific Reports, 2021. **11**(1): p. 688.

[77] Zhang, Y., et al., *Computational design of double transition metal MXenes with intrinsic magnetic properties.* Nanoscale Horizons, 2022. **7**(3): p. 276–287.

[78] Cui, Z., et al., *Giant tunneling magnetoresistance in two-dimensional magnetic tunnel junctions based on double transition metal MXene $ScCr_2C_2F_2$.* Nanoscale Advances, 2022. **4**(23): p. 5144–5153.

[79] Jin, D., et al., *Single transition metal atom stabilized on double metal carbide MXenes for hydrogen evolution reaction: a density functional theory study.* Journal of Physics D: Applied Physics, 2022. **55**(44): p. 444002.

[80] Zhou, M., et al., *Excellent double metal MXenes MoWC anode: the synergistic effect of molybdenum and tungsten transition metal.* Vacuum, 2023. **213**: p. 112152.

[81] Mehta, V., et al., *Ultralow diffusion barrier of double transition metal MoWC monolayer as Li-ion battery anode.* Journal of Materials Science, 2022. **57**(23): p. 10702–10713.

[82] Amoudeh, Z., P. Amiri, and A. Aliakbari, *The new solid solution of double transition metal MXenes: atomistic modeling of two-dimensional YScX (X = C and N)*. Solid State Sciences, 2023. **144**: p. 107306.

[83] Chang, W.-L., et al., *Thermoelectric properties of two-dimensional double transition metal MXenes: ScYCT$_2$ (T = F, OH)*. Journal of Physics and Chemistry of Solids, 2023. **176**: p. 111210.

[84] Downes, M., et al., *M$_5$X$_4$-a family of MXenes*. ACS Nano, 2023. **17**: p. 17158–17168.

5 Layered Heterostructure MXenes for Energy Applications

5.1 THEORETICAL CALCULATIONS

Elemental doping or substitution has emerged as one of the most effective ways for precisely controlling the characteristics of MXenes by modifying their atomic structures. Several synthetic approaches have been developed so far for synthesizing MXenes, but the process may introduce defects that allow for elemental doping or substitution [1]. When heteroatoms are introduced into MXenes, it inevitably causes structural and electrical distortion, which in turn alters the bandgap, electroconductivity, Fermi level, and charge transport of the MXenes. With the help of different dopants of specific sizes and bonding arrangements, the properties of MXenes can be made advantageous for energy storage [2]. Both experimental and theoretical methods are being explored for the formation and functionalization of doped MXenes. MXenes can be doped with heteroatoms like O, S, N, or P in the desired positions either during the MAX stage or after exfoliation, and the results can be categorized as M-, X-, or T-doped. Halides can replace T elements, and transition metals can be added to the doped MXenes to occupy any or both M and X positions. The resultant materials offer great potential to be utilized for a vast range of applications due to their unique elemental composition and the availability of stoichiometry within the structural configuration [3].

Theoretical simulations are vital to fully recognize the properties and applications of hetero-MXenes. Due to recent advancements in density functional theory (DFT) and low-cost computation, it is now possible to generate reliable data from basic physical equations. However, there are certain method limitations, for example, the inability to accurately predict van der Waals interactions and bandgap values. To address these limitations, the hybrid functional method employing the Heyd–Scuseria–Ernzerhof (HSE06) functional is commonly used, as well as empirical corrections such as DFT-D2, DFT-D3 (Grimme), and DFT+U [1, 3]. The DFT calculations of M doping introduce an alternate d orbital that affects the basic properties of MXenes, such as variations in layer thickness or lattice parameters, bandgap, and mechanical properties. The associated changes in the interatomic distance of MXene layers and its effect on the related lattice values and layer thickness are presented by the DFT study on X doping (i.e., Si, N, B, or S). First-principle simulations have also been used to predict the structural and electronic properties of functionalized surfaces (T = F, OH, O, S, and Cl) for a range of MXenes [1, 3]. The next sections will outline the variety of approaches that can strategically and successfully introduce dopants (metals, halides, nitrogen, sulfur, etc.) into the desired M, X, and T positions of MXenes. It will also include examples of the preparation of doped MXenes, their accomplishments, and their importance in developing new materials, all based on recent literature.

5.1.1 M-Doped MXenes

It is well understood that the d orbitals of M atoms give intriguing characteristics to MXenes. For example, research has shown that bandgaps increase when M changes in the following order: $Hf_2CO_2 > Zr_2CO_2 > Ti_2CO_2$. It is therefore expected that the modification of transition metals will result in a significant change to its inherent properties. The MXene family, which consists of sandwiched arrangements of two distinct transition metals, has grown recently due to the development of double-transition metal (DTM) MXenes. These hetero-M-MXenes, which include ordered DTM MXenes, ordered divacancy, and solid solution, can be thought of as typical hetero-M-MXenes since they contain more than one type of transition metal [1]. Several studies can be found on the theoretical screening of M-doped MXenes. Using first-principle computations, J. Yang et al. [4] examined the magnetic and electronic properties of $Cr_2M'C_2T_2$ structures (where M': Ti, or V). The findings showed that depending on the bond coupling interactions and choice of M' and T, $Cr_2M'C_2T_2$ can be ferromagnetic ($Cr_2VC_2O_2$, $Cr_2VC_2F_2$, and $Cr_2VC_2(OH)_2$), nonmagnetic ($Cr_2TiC_2O_2$), antiferromagnetic ($Cr_2TiC_2F_2$ and $Cr_2TiC_2(OH)_2$), and either a semiconductor or a metal.

T. L. Tan et al. [5] explored the stability of eight MXene alloys, i.e., $(V_{1-x}Mo_x)_3C_2$, $(Nb_{1-x}V_x)_3C_2$, $(Nb_{1-x}Mo_x)_3C_2$, $(Ta_{1-x}Mo_x)_3C_2$, $(Ti_{1-x}Ta_x)_3C_2$, $(Ti_{1-x}Nb_x)_3C_2$, $(Ti_{1-x}Mo_x)_3C_2$, and $(Ti_{1-x}V_x)_3C_2$, with $0 \leq x \leq 1$. They developed fast-to-compute interactions using the cluster expansion method, which allowed them to search through the stability and formation energies of millions of structural compositions. Based on Monte Carlo simulations, the results showed that for the Mo-rich MXenes, Mo atoms preferably occupy the outer layers of $(M1_{1-x}Mo_x)_3C_2$ and ordering retains its stability at high temperatures. When Ti is alloyed with Nb or Ta, it prefers the surface layers with the Ti–C–Nb–C–Ti sequence. However, across all the compositions, solid solutions have resulted when Ti is alloyed with V. $(Nb_{1-x}V_x)_3C_2$ phase, which exhibits solid-solution states at synthesis temperatures and tends to separate at lower temperatures. To prevent the atomic aggregation associated with conventional transition-metal (TM) modification, Z. Chen et al. [6] substituted TM atoms into MXenes. The DFT simulations were used to examine the hydrogen evolution reaction (HER) activity of M_2CO_2 (where M = Ti, Hf, V, etc.). The outcomes demonstrated that the substitution of TM atoms may alter the differential HER Gibbs free energies and conductivities of MXenes.

Using Vienna Ab initio simulation package (VASP) code, V. Mehta et al. [7] reported the electrical and magnetic characteristics of pure and Co-doped Mo_2C by spin-polarized DFT-based computations. According to the computed results, the ferromagnetism of a compound is produced by the dopant at the Mo site, while the pure Mo_2C is determined to be nonmagnetic. It has been determined that the system's total magnetic moment is 1.2 μB, rising to 2.03 μB when the dopant concentration increases from 3% to 8%, respectively. Using first-principle calculations, C. Cheng et al. [8] examined the catalytic oxidation of CO on Cu_3 cluster-decorated pristine MXene (Cu_3/p-Mo_2CO_2) and defective monolayers (Cu_3/d-Mo_2CO_2). Through the use of molecular dynamics simulations, geometry distortion, and energy analysis, the stability of the developed catalysts was thoroughly proven. Cu_3/d-Mo_2CO_2 was discovered to have good stability and reactivity, suggesting that it could be an excellent catalyst for CO oxidation. The Cu_3 cluster's capacity to donate and take electrons was controlled by its active sites, which functioned as an electron reservoir.

By using DFT, C. Cheng et al. [9] methodically examined the structural stability and catalytic properties of both pristine and defective Pd-doped Mo_2CO_2 monolayers with an oxygen vacancy. They noticed that the Pd/OV-Mo_2CO_2 combination functions well as a monodispersed atomic catalyst due to the oxygen vacancy's (OV's) ability to stabilize the Pd dopant. C. Si et al. [10] projected that $Mo_2MC_2O_2$ (M = Ti, Hf, or Zr) are robust quantum spin Hall (QSH) insulators based on first-principle calculations. The topological gap (at Γ point) was observed to be contributed by the atomic spin-orbit coupling (SOC) of transition metal. This is a useful feature that sets it apart from typical scenarios in which the topological gap is significantly smaller than the atomic SOC. As a result, $Mo_2MC_2O_2$ exhibited significant gaps with varying M atoms, ranging from 0.1 to 0.2 eV, which were adequate to achieve room-temperature QSH effects. DFT calculations were performed by L. Li et al. [11] to screen many transition metal atoms (i.e., Zr, Mo, Hf, Ta, W, Re, and Os) confined in a vacancy of $Mo_2TiC_2O_2$ MXene nanosheets. The Zr-doped single-atom catalyst (i.e., $Mo_2TiC_2O_2$-ZrSA) was observed to exhibit the highest HER activity and the lowest barrier of the potential-determining step (i.e., 0.15 eV). Moreover, the formation energy of $Mo_2TiC_2O_2$-ZrSA was significantly lower than that of the synthesized $Mo_2TiC_2O_2$-PtSA catalyst, indicating its possibility of the experimental preparation.

J. Yang and colleagues [12] examined the 2D Sc_2CT_2 structures, and it was observed that the precursor, that is, Sc_2AlC, can be doped with TMs when ScAl is abundant. TM-doped $ScCO_2$ were observed to be p-type semiconductors, while the ensuing TM-doped $Sc_2C(OH)_2$ were n-type semiconductors. They also observed the magnetic properties of all Cr- and Mn-doped Sc_2CT_2 structures. Using first-principle methods of DFT, D. Yang et al. [13] investigated MXenes X_2CO_2 (X = Ti, Sc, Hf, and Zr) as NO and CO gas sensor material. According to the calculations, Sc_2CO_2 is sensitive to NO molecules because of their chemical interaction and significant 0.303 e charge transfer, which would cause a shift in current that would be observed. More significantly, the application of external strain improved the contact and demonstrated the potential of Sc_2CO_2 as a material for gas capture. Mn doping considerably enhances the adsorption of CO on Sc_2CO_2 due to high adsorption energy (i.e., −0.85 eV). The theoretical calculations of M-doped MXenes are summarized in Table 5.1.

5.1.2 X-Doped MXenes

Despite its significant influence on the characteristics of materials, the doping of X element has been less simulated than M-doped MXenes [1]. Some of the theoretical studies of X-doped MXenes are reviewed here. Using the DFT calculations, B. Ding et al. [26] reported the doping effect of heteroatom X (X = N, S, P, etc.) on the

TABLE 5.1

Theoretical Calculations of M-doped MXenes

M-Doped MXenes	Heteroelements	Theoretical Calculation Methods	Properties/ Applications	Ref.
$(M'_{1-x}M''_x)_3C_2$	M': Ti, V, Nb, Ta M'': Mo, Nb, Ta, V	Perdew, Burke, and Ernzerhof (PBE) functional	Structure–stability relationship	[5]
M_2CO_2 (M: Ti, V, Hf, Zr, and Ta)	3d, 4d, and 5d TM	Generalized gradient approximation (GGA)-PBE	HER	[6]
Mo_2C	Co	GGA-PBE	Electronic and magnetic	[7]
Mo_2CO_2	Cu_3	GGA-PBE	CO oxidation reaction	[8]
Mo_2CO_2	Pd	GGA-PBE	CO oxidation reaction	[9]
$Mo_2MC_2O_2$	M: Ti, Zr, or Hf	GGA-PBE/HSE06	Electronic	[10]
$Mo_2TiC_2O_2$	3d, 4d, and 5d TM	GGA-PBE/DFT-D3	NRR	[11]
Sc_2CT_2 (T: O, OH, or F)	TM: Ti, V, Cr, and Mn	GGA-PBE	Magnetic and electronic	[12]
M_2CO_2 (M: Sc, Ti, Zr, and Hf)	Mn	GGA-PBE	CO adsorption	[13]
V_2CO_2	Fe, Co, and Ni	GGA-PBE	HER	[14]
Nb_2CT_2	Pt/Pd	GGA+U/DFT-D3	ORR/OER	[15]
$Cr_2M'C_2T_2$ (T: O, OH, or F)	M': Ti or V	GGA-PBE	Electronic and magnetic	[4]
Cr_2CO_2	TM: Ag, Au, Co, Fe, Ir, Mn, Mo, Ni, Pd, and Ru	GGA-PBE/DFT-D3	HER/OER	[16]
Cr_2CO_2	TM: Ni, Co	GGA-PBE/DFT-D3	HER	[17]
$MWCO_2$	M: Cr, Mo	GGA-PBE/DFT-D2	HER	[18]
$MVCO_2$	M: Ti, Zr, Nb, Hf, and Ta	GGA-PBE/DFT-D2	HER	[18]
Ti_2C	V	GGA+U	Mechanical	[19]
Ti_2CO_2	Sc, V	GGA	Mechanical and electronic	[20]
Ti_2CO_2	V, Sc	GGA-PBE	Electronic	[21]
Ti_3C_2	TM: Fe, Co, Ni, Cu, Zn, Ru, Rh, etc.	GGA-PBE	Stability	[22]
$Ti_3C_2O_2$	TM: Fe	GGA-PBE	NRR	[23]
$Ti_{n+1}C_nT_x$ (T_x: O, F)	Pt	GGA-PBE	ORR	[24]
Ti_4N_3	Sc, V, Zr	GGA-PBE	Electronic and magnetic	[25]

HER activity of M_2C MXene (M = Mo and Ti). Compared to the X-doped bare M_2C, the X-doped functionalized material demonstrated superior HER catalytic activity. According to the computed Gibbs free energies (ΔG_H), N-doped Ti_2CO_2 exhibited better electrocatalytic activity than Pt. Furthermore, N-doped Ti_2CO_2 has better electrical conductivity than bare Ti_2CO_2. E. Balci et al. [27] theoretically replaced

C atoms in the Sc_2CF_2 MXene monolayer by different elements, that is, Si, Ge, Sn, N, S, F, and B atoms. They observed that doped Sc_2CF_2 monolayers are semiconductors with bandgap values ranging from 0:55 eV to 0:24 eV.

Using DFT, E. Balci et al. [28] also examined the electrical structures of Ge- and Si-doped monolayers of $Sc_2C(OH)_2$ MXene. Using molecular dynamics simulations, the parameters of doped structures and dynamical stability were examined. Band inversion was detected by the projection of electronic orbitals. At TRends and Indices for Monitoring data (TRIM) locations, symmetry analysis was carried out, and the parities were found. Z2 theory has shown that the doped $Sc_2C(OH)_2$ MXene is topologically invariant. Bandgap openings were observed when SOC was incorporated. In a theoretical study, R. Z. Zhang et al. [29] examined the doped Ti_2CF_2 MXenes' structural, electrical, and optical characteristics by incorporating F, N, B, S, Sn, Si, and Ge atoms. According to the binding energy and cohesive energy analysis, $N\text{-}Ti_2CF_2$ exhibit the most stable structure, but the stability of Si-, Ge-, and Sn-doped Ti_2CF_2 MXenes was not observed to be favorable. All of the investigated doped monolayers showed metallic character, suggesting that the doped atom does not affect the electronic characteristics. Additional bands surround the Fermi energy if a Si, Sn, or Ge atom fills the carbon vacancy of pure material.

To reveal the nitrogen doping mechanism in Ti_3C_2 MXene, C. Lu et al. [30] conducted a thorough investigation using theoretical simulation and experimental characterization. In $Ti_3C_2T_x$, three potential locations for the nitrogen dopants were identified: surface absorption (for −O), function substitution (for −OH), and lattice substitution (for carbon). Additionally, the findings of the electrochemical tests verified that the three different types of nitrogen dopants are beneficial for raising the Ti_3C_2 electrode's specific capacitance and that the underlying causes can be identified. In another study, DFT and DFT+U calculations were used to determine the lithium adsorption on bare and functionalized Ti_3CN MXenes [31]. The Li diffusion barriers for Ti_3CNT_2 range in value from 0.2 eV to 0.3 eV. Based on their findings, it can be concluded that Li adsorbs on the surface of carbon for Ti_3CNT_2 (in the presence of functional groups) and prefers to adsorb on the nitrogen for bare Ti_3CN. A simple and controlled method for creating N-doped $Ti_3C_{1.6}N_{0.4}$ and $Ti_3C_{1.8}N_{0.2}$ flakes by utilizing an insitu nitrogen solution and etching method was described by Y. Tang et al. [32]. The addition of nitrogen to $Ti_3C_{1.6}N_{0.4}$ flakes results in enhanced wettability for more accessible sites, faster charge transfer during an electrochemical process, and more exposed active sites. Consequently, the $Ti_3C_{1.6}N_{0.4}$ catalyst showed improved electrocatalytic characteristics for oxygen evolution reaction (OER). The theoretical calculations of X-doped MXenes are summarized in Table 5.2.

5.1.3 T-DOPED MXENES

The characteristics of MXenes are known to be significantly influenced by surface terminations, and several computations have been done using the most prevalent terminations (−F, −O, and −OH) [1]. As the MXene surface groups (T_x) are much stronger, regulated and tunable covalent surface modifications are difficult to tune to desired uses after the synthesis process. But more recently, a more secure approach free of fluoride has been created, which presents many possibilities for improving

TABLE 5.2

Theoretical Calculations of X-doped MXenes

X-doped MXenes	Heteroelements	Theoretical Calculation/ Synthesis Methods	Properties/ Applications	Ref.
M_2CT_2 (M: Ti, Mo)	X: N, B, P, S	GGA-PBE	HER	[26]
Sc_2CF_2	Si, Ge, Sn, F, S, N, B, B+N	GGA-PBE	Electronic	[27]
$Sc_2C(OH)_2$	Si, Ge	GGA-PBE	Electronic	[28]
Ti_2CF_2	Si, Ge, Sn, F, B, N, S	GGA-PBE	Electronic and optical	[29]
$Ti_3C_2T_x$ (T_x: F, OH, O)	N	GGA-PBE	Supercapacitors	[30]
Ti_3CNT_2 (T_x: F, OH, O)	N	GGA-PBE/DFT-D2	Lithium-ion batteries	[31]
$Ti_3C_{1.8}N_{0.2}$ and $Ti_3C_{1.6}N_{0.4}$	N	In situ nitrogen solid solution followed by etching (HCl + LiF)	OER	[32]

MXene preparation and enabling the addition of more adjustable halide terminations. When molten salt etching is used, MXenes can be transformed into the surface terminations that are easily adjustable to the preferred composition with remarkable consistency. This has made it simple to add halides other than fluorine, that is, iodine, bromine, or chlorine, with up to 100% coverage, thereby removing undesired hydrolysis and oxidation. Although fluoride termination bonds exhibit good stability, those of chlorides and bromides exhibit much greater liability, offering the possibility of easy substitution with –NH groups, oxygen, sulfur, selenium, and tellurium to the desired extent [3].

These surface groups provide researchers with an unmatched scalable method for the commercial manufacture of MXenes, as well as a potentially greener chemistry procedure with much greater flexibility in customizing the surface chemistry of MXenes. This is in addition to enable the researchers to avoid the harmful hydrofluoric acid (HF) etching process [3]. There are several examples of T-doped MXenes; for example, to examine the uses and applications of MXenes as supercapacitor electrodes and other electrical devices, Y. Xin et al. [33] used ab initio DFT calculations to study the work functions and quantum capacitances of bare and functionalized $Nb_{n+1}C_nZ_2$ (where Z = F, O, OH, and OCH_3). They showed that except for the Nb_2C sheet, the functional groups have very little effect at negative electrodes even while they significantly reduce the quantum capacitances of bare MXenes at positive electrodes. The work function of the MXene sheets was observed to decrease with the addition of OCH_3 and OH groups and increase with the addition of F and O atoms.

The energy storage features and structural stabilities of Li-decorated Zr_2C and Zr_2CX_2 were studied theoretically by J. Zhu et al. [34]. They replaced the OH, O, and

F groups with S during the synthesis and showed that an exchange reaction is feasible to get around the crucial disadvantages of those groups. As a result, Zr_2CS_2 exhibited a significantly lower diffusion barrier but a similar specific capacity to Zr_2CO_2. S. Zheng et al. [35] theoretically examined the catalytic activity of $B-Mo_2CO_2$ and $B-W_2CO_2$ MXenes. The limiting potentials of these MXenes, which were 0.20 and 0.24 V, respectively, showed exceptional catalytic behavior. Significantly, a strong B-to-N bonding would substantially impede the conversion of $*NH_2$ to $*NH_3$. This shows that optimal performance with a moderate electron donation can be attained because the TM in the MXene substrate may suitably tune such an electron-donation effect.

Using DFT modulations, V. Shukla et al. [36] investigated the potential of V_2NS_2 and Ti_2NS_2 as anode materials for lithium- or sodium-ion batteries. On both surfaces, high (> −2 eV) Li/Na ion adsorption energies were discovered to be connected to substantial charge transfer. It is interesting to note that this ion intercalation can extend up to many layers, hence giving the substrate a higher specific capacity. In particular, it was discovered that Li-ion batteries will benefit more from these two 2D materials (V_2NS_2 and Ti_2NS_2), whose estimated theoretical capacities are 308.28 mAh g^{-1} and 299.52 mAh g^{-1}, respectively. Using the DFT technique, J. Yang et al. [37] studied 20 different S-functionalized MXenes with the structural formula M_2XS_2 (where X = C, N). The findings of molecular dynamic simulations (at 1000 K) and phonon frequency analysis indicated that a number of compounds are dynamically stable and can be synthesized, that is, V_2CS_2, Nb_2CS_2, Ta_2CS_2, Hf_2CS_2, and Cr_2NS_2. It suggests that S atoms may also terminate MXenes, a phenomenon that is dependent on the synthesis procedure.

An asymmetric functionalized double MXene (Cr_2TiC_2FCl) with opposite spin directions in the conduction and valence band around the Fermi level was presented by J. He et al. [38] by using DFT calculations. It was observed that the gate voltage can control the spin orientation of Cr_2TiC_2FCl, resulting in a change from bipolar antiferromagnetic semiconductor (BAFS) to half-metal antiferromagnets (HMAFs). Furthermore, as the chemical environment of Cr atom differed, the Cr_2TiC_2FCl with varying F/Cl concentrations was observed to display the BAFS property. Likewise, J. He et al. [39] predicted the characteristics of Janus Cr_2C MXene (i.e., asymmetrically functionalized), represented as Cr_2CXX' (where X, X' = H, F, Cl, Br, and OH). These materials can function as zero magnetization bipolar magnetic semiconductors because their conduction and valence bands are composed of opposite spin channels. For materials including Cr_2CHCl, Cr_2CFCl, Cr_2CHF, Cr_2CFOH, and Cr_2CClBr, a Néel temperature of up to 400 K has been discovered.

In another study, TM carbonitrides (M_3CN) were studied by B. Huang et al. [40]. They also examined the charge transfer, TM-modified, surface-terminated, and bare M_3CNO_2 structural topologies and ΔG_H (i.e., adsorption free energy). According to the computed data, all bare materials showed substantial binding between the H atom and catalysts, and only Nb_3CNO_2 and Ti_3CNO_2 were observed to have the capability for HER catalysis. M. Li et al. [41] synthesized Ti_2ZnC, V_2ZnC, Ti_3ZnC2, and Ti_2ZnN by replacing the Al element in their corresponding MAX phase precursors and the Zn from molten salt, that is, $ZnCl_2$. When sufficient $ZnCl_2$ was used,

Cl-terminated MXenes like Ti_2CCl_2 and $Ti_3C_2Cl_2$ were obtained (due to their strong Lewis acidity). Similarly, the effective synthesis of MXenes with different surface groups (i.e., O, NH, S, Cl, Se, Br, and Te) and pristine MXenes was presented by V. Kamysbayev et al. [42]. The MXenes exhibited unique electrical and structural characteristics. Nb_2C MXenes, for instance, demonstrated superconductivity (dependent on surface groups). The theoretical calculations of T-doped MXenes are summarized in Table 5.3.

TABLE 5.3

Theoretical Calculations of T-doped MXenes

T-Doped MXenes	Heteroelements	Theoretical Calculation/ Synthesis Methods	Properties/ Applications	Ref.
$Nb_{n+1}C_nT_2$	H, O, F, OCH3	GGA-PBE	Work function	[33]
Zr_2CT_2	O, F, S	GGA-PBE	Li-ion batteries	[34]
M_2CO_2 (M: Mo, W, etc.)	B	PBE/DFT-D3	NRR	[35]
M_2NT_2 (M: V, Ti)	S	GGA-PBE	Li-/Na-ion batteries	[36]
M_2XT_2 (M: Sc, Ti, V, Cr, Zr, etc. and X: C, N)	S	GGA-PBE	Stability and electronic	[37]
$Cr_2TiC_2T_2$	F, Cl	HSE06	Magnetic	[38]
Cr_2CTT'	H, OH, F, Cl, Br	GGA-PBE/DFT+U	Magnetic and electronic	[39]
M_3CNO_2 (M: Cr, Hf, Mo, Nb, Ta, Ti, V, Zr)	O, OH	GGA-PBE	HER	[40]
Ti_2CT_2 and $Ti_3C_2T_2$	Cl	GGA-PBE	Stability	[41]
$M_{n+1}C_nT_x$ (M: Ti, Mo, Nb, V)	O, NH, S, Cl, Se, Br, Te	Using molten inorganic salts	Structural and electronic	[42]
Ti_2CT_2	S	GGA-PBE/DFT-D2	Magnesium-ion batteries	[43]
Ti_2CT_2	O, F, S	GGA-PBE	Li–S batteries	[44]
$Ti_3C_2T_2$	N, O, F, S, Cl	GGA+U/DFT-D3	Li–S batteries	[45]
$Ti_3C_2T_2$	O, S, Se, Te	GGA-PBE/DFT-D2	Li-ion batteries	[46]
$Ti_3C_2T_2$	O, H, OH	HF etching	CO_2 and H_2 dehydrogenation	[47]
$Ti_3C_2T_x$	F, OH, O	HF etching	HER	[48]
$Ti_3C_2T_x$	F, O	LiF + HCl etching	CO_2 and H_2O reaction	[49]
$Ti_3C_2T_2$	O, S	GGA-PBE	Na-ion batteries	[50]
$Ti_3C_2T_x$	O, Cl	Lewis acid etching	Li-ion storage	[51]
$Ti_3C_2T_x$	O, F, OH	HF etching	Stability	[52]

5.2 HETERO-MXENES

The elemental property allows us to distinguish between nonmetal and metal dopants in MXenes. The in situ and ex situ techniques can be used to produce the elemental doping or substitution of MXenes from an experimental perspective [1]. For the in situ synthesis, the dopant is added in the MXene precursor (e.g., by sintering), and the doped MXenes then undergo selective etching and exfoliation. The primary outcomes of this method are substitutions at M and X, which vary according to the heteroatom dopant and TM that are employed. Ex situ strategies have generally entailed post-synthesis adjustments (e.g., via heat treatment, hydro/solvothermal modification, or plasma modification), thereby allowing the substitution at X and/or T. A more modern ex situ technique uses molten salts to etch the 3D MAX phase while also providing easily adjustable termination groups (T), which make it simple to dope the MXenes at T by substitution [3]. Figure 5.1 [3] represents the types of heteroatoms doping of MXenes.

This section provides a detailed discussion of the experimental doping/substituting outcomes of several MXenes, including metal and nonmetal (i.e., N, S, and P).

5.2.1 HETERO-N-MXENES

Nitrogen is one of the primary components of bare nitride MXenes, that is, $M_{n+1}N_n$. However, due to the higher formation energy and less stability of the M-N bond,

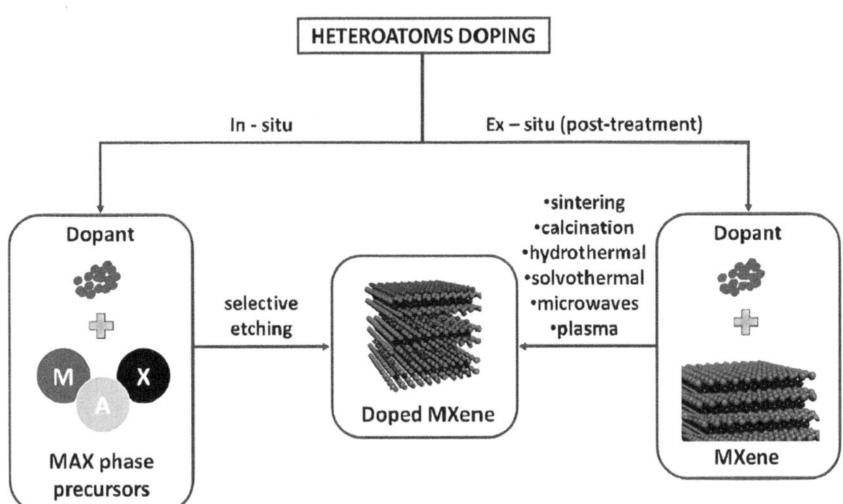

FIGURE 5.1 Types of heteroatoms doping of MXenes.

Source: Reprinted from [3] *Progress in Materials Science*, 139, A. Dey et al., Doped MXenes—A new paradigm in 2D systems: Synthesis, properties and applications, 101166, Copyright 2023, with permission from Elsevier. The source of material Guoxing Qu, Yang Zhou, Tianli Wu, et al. 2018. Phosphorized MXene-phase Molybdenum Carbide as Earth-Abundant Hydrogen Evolution Electrocatalyst. *ACS Appl. Energy Mater.*, 1, 12, 7206–7212. Copyright 2018, The American Chemical Society is also acknowledged.

nitride MXenes are difficult to produce using a conventional wet-etching technique. Fortunately, in situ and ex situ methods have been used to produce nitrogen doping and substitution, which further modify the characteristics of MXenes and expands their applicability [1]. Since the atomic radius of nitrogen is similar to that of carbon, it is a preferred pentavalent doping material for a variety of carbon-based nanostructures. It is commonly recognized that N doping can cause the introduction of oxygen vacancies and/or interstitials, which can modify the bandgap in addition to changing the electrical and optical characteristics. For the first time, Naguib et al. (2012) demonstrated an in situ synthesis of N-doped MXenes by including dopant N atoms into the preparative mixture used to produce the MAX phase precursor [3]. Since then, there have been several papers published in the field of N-doped MXenes, which have grown significantly.

Y. Y. Chen et al. [53] used phosphomolybdic acid (PMo_{12}) as a source for P and Mo to synthesize N, P-Mo_2C@C nanocrystals and to start the polymerization of polypyrrole (PPy). The molecular presence of PMo_{12} facilitates the creation of nanospheres that resemble pomegranates, with a porous carbon shell as the peel, and uniformly distributed Mo_2C nanocrystals functioning as seeds. The matrix of this nanostructure and carbon shell effectively prevents Mo_2C nanocrystals from aggregating, facilitates the transportation of electrons, and provides catalytic active sites. These are just a few of the favorable features this nanostructure provided for hydrogen evolution application. J. Q. Chi et al. [54] synthesized porous core-shell N-Mo_2C@C with a conductive substrate, porous nanostructures, and N-doping using inorganic–organic precursor (i.e., MoO_4^{2-}/aniline-pyrrole or MoO_4^{2-}-polymer). Because MoO_4^{2-}-polymer contains aniline-pyrrole, it inhibits the rapid growth and aggregation of Mo_2C, forming a porous structure that exhibits more exposed active sites. The ultrathin carbon shell that results from the carbonization of MoO_4^{2-}-polymer, on the other hand, is produced on the surface of nanospheres and can optimize electrical structures and significantly increase the charge transfer rate.

Ultrathin N-Mo_2C nanosheets (NSs) were also highlighted by J. Jia et al. [55] as very effective platinum-free HER electrocatalysts. In-depth research was done on the crystal phase and structural transformation between MoO_2 NSs (thickness of about 1.1 nm) and N-Mo_2C NSs (thickness of around 1.0 nm). Theoretical studies revealed that the N doping, structure, and particular crystalline phase of Mo_2C provide more exposed active sites. An entirely new kind of N-doped two-dimensional (2D) Nb_2CT_x MXene was synthesized by R. Liu et al. [56]. After thermally reacting with urea, N-doped MXene had a nitrogen concentration of 4.5%. When nitrogen is added to MXene NSs, the c-lattice parameter of Nb_2CT_x MXene increases to 34.78 Å from 22.32 Å. Furthermore, MXene nanosheets with nitrogen included exhibit improved electrochemical performance. Researchers R. G. Mendes et al. [57] studied the stability of $Mo_2Ti_2C_3$ in the presence of electron radiation. $Mo_2Ti_2C_3$ was discovered to be fairly stable throughout the first two minutes of radiation. But then there are structural alterations that lead to more drastic and quick rearrangements.

Y. Yoon et al. [58] presented a method for nitriding Ti_2CT_x NSs using sodium amide ($NaNH_2$) to convert it into an active electrocatalyst. Ti-Nx formed a chemical bond on the surface of Ti_2CT_x, upon the addition of $NaNH_2$, and this reaction occurred at 500 °C. In another study, a sacrificial template technique was used to synthesize

N-Ti$_3$C$_2$T$_x$ with consistent nitrogen doping and well-defined porosity structure [59]. This N-doped MXene was used as an anode in a 3D-printed sodium-ion hybrid capacitor. The N-doped carbon-decorated composites (Ti$_3$C$_2$T$_x$@NC) were also created by T. Zhao et al. [60]. They used an in situ self-polymerization of dopamine, followed by a carbonization procedure. The characterizations showed that a distinct 3D composit nanostructure was created and that NC was equally decorated on the surface and interlayer of MXene sheets. This kind of nanostructure can give the NC layer a large surface area and effectively prevent the Ti$_3$C$_2$T$_x$ sheets from restacking. More importantly, it can give the composites greater pseudocapacitance and strong conductivity. A simple method for creating Ti$_3$C$_2$T$_x$ MXenes using ultrasonication and controlled N-doping was described by M. Han et al. [61]. To further increase the HER activity, the surface of MXenes can be modified by forming TiAN chemical bonds at an ideal ultrasonic temperature.

W. Bao et al. [62] produced N-Ti$_3$C$_2$T$_x$ NSs with strong chemical and physical co-adsorption of polysulfides, by thermally annealing negatively charged MXene flakes and positively charged melamine, which were subsequently used as a host for Li–S batteries. The nitrogen-doping approach resulted in a large pore volume and a well-defined porous structure, while also allowing heteroatoms to be introduced into MXene NSs. Likewise, N- and O-doped C@Ti$_3$C$_2$ composites were synthesized [63] using an in situ polymerization carried out in annealing atmospheres containing Ar or a mixture of Ar and NH$_3$. The doped concentrations change depending on the reaction environment, which affects how well the electrode material performs electrochemically. A simple technique for creating N-doped delaminated Ti$_3$C$_2$/rGO hybrid films that are freestanding and binder-free was devised by Yang et al. [64] for use as high-performance supercapacitor electrodes. Z. Zhang et al. [65] developed a NC N–Ti$_3$C$_2$/Fe$_2$O$_3$, by solvent-free thermal degradation of a precursor. Using Ti$_3$C$_2$ MXene as the starting material, Q. Xu et al. [66] produced unique full-color MXene quantum dots (MQDs) by adjusting the hydrogen bonds. When ammonia water (S- or N-doped) and sodium thiosulfate were doped, the emission of the resulting MQDs showed the whole light spectrum, spanning from blue to orange light. In another study, a unique N-MX-CoS$_2$ nanohybrid was created by C. Yang et al. [67] using a one-step in situ sulfidation technique. Fast sulfur electrochemistry redox reaction is facilitated by the wide layered structure and strong electrical conductivity of the MXene substrate. The summary of hetero-N-MXenes and their corresponding synthesis methods is presented in Table 5.4.

5.2.2 Hetero-S-MXenes

As sulfur has a large covalent radius, which is expected to enhance the interlayer distance of S-MXenes and effect the inherent properties of materials, sulfur is excellent in this regard [3, 89]. Hetero-S-MXenes can also be produced using in situ and ex situ doping techniques [1]. Mo$_2$C NSs, measuring 1.0–1.2 nm in thickness, were produced by carburization and solvent exfoliation, as shown by H. Ang et al. [90]. These NSs had a significant specific surface area (122–153 m^2 g^{-1}) and were single crystalline. Compared to bulk-layered Mo$_2$C, this particular NS shape offers a shorter electron transport channel during HER, which causes the overpotential to

TABLE 5.4

Hetero-N-MXenes by Various Dopants and Synthesis Methods

Hetero-N-MXenes	Dopants	Synthesis Methods	Properties/ Applications	Ref.
N, P-Mo$_2$C@C	Phosphomolybdic acid/ pyrrole	Carbothermal/ polymerization	HER	[53]
N-Mo$_2$C@NC	Aniline + pyrrole	Carbonization	HER	[54]
N-Mo$_2$C NSs	Dicyandiamide	Calcine	HER	[55]
N-Nb$_2$CT$_x$	Urea	Hydrothermal	Li-ion batteries	[56]
Mo$_2$Ti$_2$C$_3$	N-doped graphene	HF etching followed by vacuum-assisted filtration	Stability	[57]
N-Ti$_2$CT$_x$	Sodium amide	Heat treatment	HER	[58]
N-Ti$_2$CT$_x$	Cyanamide	Annealing	Supercapacitors	[68]
N-Ti$_3$C$_2$	Urea	Hydrothermal	Supercapacitors	[69]
N-Ti$_3$C$_2$	Urea	Carbonization	Supercapacitors	[70]
N-Ti$_3$C$_2$T$_x$	Ammonia	Annealing	Supercapacitors	[71]
N-Ti$_3$C$_2$T$_x$	Ammonia	Heat treatment	HER	[72]
N-Ti$_3$C$_2$	Urea	Hydrothermal	Supercapacitors	[73]
N-Ti$_3$C$_2$T$_x$	Melamine formaldehyde	Annealing	Na-ion hybrid capacitors	[59]
Ti$_3$C$_2$T$_x$@NC	Dopamine	Carbonization	Supercapacitors	[60]
N-Ti$_3$C$_2$T$_x$	Ammonia	Ultrasonic treatment	HER	[61]
N-Ti$_3$C$_2$T$_x$	Melamine formaldehyde	Annealing	Li–S batteries	[62]
N, O-C@Ti$_3$C$_2$	Ammonium citrate	Sintering	Supercapacitors	[63]
N-Ti$_3$C$_2$/rGO	Urea	Hydrothermal	Supercapacitors	[64]
N-Ti$_3$C$_2$/Fe$_2$O$_3$	Cyanamide	Thermal treatment	Li-ion batteries	[65]
N, S-Ti$_3$C$_2$ QDs	Na$_2$S$_2$O$_3$/NH$_3$·H$_2$O	Hydrothermal	Light-emitting diodes (LEDs)	[66]
N-MX-CoS$_2$	Thiourea	Heat treatment	Li–S batteries	[67]
N-Ti$_3$C$_2$/C	2-Methylimidazole + PVP	Annealing	Li–Sulfur batteries	[74]
Ti$_3$CN	AlN	Sintering	Mode-locked lasers	[75]
N-Ti$_3$C$_2$	Urea/ monoethanolamine	Solvothermal	Supercapacitors	[76]
MCoNPCNSs	2-Methylimidazole + PVP	Carbonization	Li–S batteries	[77]
Ti$_{3-x}$CNT$_y$/ CNT	AlN	Sintering	Potassium–metal batteries	[78]
Fe$_3$O$_4$@Ti$_3$C$_2$/ CNFs	DMF	Annealing	Li-ion batteries	[79]
N-TiO$_2$/TiN/ Ti$_3$C$_2$T$_x$	Hexamethylenetetramine	Hydrothermal	Supercapacitors	[80]
N-Ti$_3$C$_2$T$_x$	Ammonium chloride	Annealing	Capacitive deionization (CDI)	[81]
P, N-Ti$_3$C$_2$T$_x$	Melamine formaldehyde	Annealing	Li–S batteries	[82]

(Continued)

TABLE 5.4 (Continued)

Hetero-N-MXenes	Dopants	Synthesis Methods	Properties/ Applications	Ref.
N-Ti$_3$C$_2$@CNT	Melamine	Pyrolysis	Li–S batteries	[83]
N-Ti$_3$C$_2$ QDs	Diethylenetriamine	Solvothermal	Detector	[84]
N-Ti$_3$C$_2$T$_x$	Ammonium fluoride	Calcination	Supercapacitors	[85]
N-Ti$_3$C$_2$T$_x$	Ammonium bicarbonate	Plasma treatment	OER	[86]
N-Ti$_3$C$_2$T$_x$/P	EDTA/RP	Annealing	Li-ion batteries	[87]
N-Ti$_3$C$_2$T$_x$ NSs	Melamine formaldehyde	Annealing	Supercapacitors	[88]

drop by 15 mV. 2D MoS$_2$-on-MXene (Mo$_2$TiC$_2$T$_x$) heterostructures were produced by C. Chen et al. [91] using the in situ sulfidation of MXene. The outcomes proved that these heterostructures exhibit metallic characteristics. In addition, MXene promotes improved Li and Li$_2$S adsorption throughout the conversion and intercalation processes. These features resulted in the steady lithium-ion storage performance of the heterostructures.

Two simple methods were developed by D. Wang et al. [92] to create distinct Mo$_2$C/C hybrid electrocatalysts that are dual-doped with N, P, and N, S, and that are highly active HER catalysts. To obtain a greater synergistic effect, co-doping of N, P, and N, S includes substituting C atoms in the carbon matrix and in the Mo$_2$C crystals as well. This can further increase the HER performance of Mo$_2$C catalyst in comparison to specific N doping. According to expectations, the resulting dual-doped Mo$_2$C catalyst exhibited better activity of each active site, a larger active surface area, and lower electrochemical resistance. A specially designed sulfur host that can immobilize Li$_2$S$_x$ via a dual chemisorption process was reported by X. Wang et al. [93]. The MXene matrix and polydopamine (PDA) topcoat made up the new sulfur host. PDA retains Li$_2$S$_x$ through the polar–polar contact, while MXene uses the Lewis acid–base process to generate a strong Ti-S bond. The novel cathode, which benefited from double chemisorption, showed an initial capacity of 1,001 mAh g^{-1} at 0.2 C, with a capacity retention of 65% over 1,000 cycles.

In another study, a strong Ti$_3$C$_2$T$_x$/S conductive paper integrating the mechanical strength, exceptional conductivity, and distinct chemisorption of LiPSs from MXene NSs was described by H. Tang et al. [94]. Significantly, frequent cycling starts the in situ development of a thick layer of sulfate complex on the surface of MXenes. This layer functions as a protective barrier, thereby preventing LiPS shuttling and enhancing sulfur utilization. A freeze-drying method for functionalizing MXene NSs using in situ sulfur doping was described by W. Bao et al. [95]. The resultant S-Ti$_3$C$_2$T$_x$ NSs were used as the electrode in Na–S batteries that were operated at room temperature. With sodium polysulfides, the S–Ti$_3$C$_2$T$_x$ matrix exhibits strong polarity, which limits the diffusion of sodium polysulfides. C. Du et al. [96] developed the first S@TiO$_2$/Ti$_2$C nanoarchitecture as cathodes for Li–S batteries by embedding TiO$_2$ hollow nanospheres enclosing sulfur inside Ti2C interlayers. The S@TiO$_2$/Ti$_2$C composites showed outstanding electrochemical properties in their prepared state. The summary of hetero-S-MXenes and their corresponding synthesis methods are presented in Table 5.5.

TABLE 5.5

Hetero-S-MXenes by Various Dopants and Synthesis Methods

Hetero-S-MXenes	Dopants	Synthesis Methods	Properties/ Applications	Ref.
N, S-Mo$_2$C	Thioacetamide	Carburization	HER	[90]
MoS$_2$-Mo$_2$TiC$_2$T$_x$	Sulfur	Heat treatment	Li-ion batteries	[91]
N, S-Mo$_2$C@C	Dicyandiamide/ thioacetamide	Hydrothermal	HER	[92]
S-Ti$_3$C$_2$@PDA	Sulfur	Solution mixing	Li–S batteries	[93]
S-Ti$_3$C$_2$T$_x$	Sulfur	Vapor deposition	Li–S batteries	[94]
CT-S-Ti$_3$C$_2$	Sulfur	Heat treatment	Na-ion batteries	[97]
S-Ti$_3$C$_2$T$_x$	Sulfur	Sintering	Na–S batteries	[95]
S-TiO$_2$/Ti$_2$C	Sulfur	Mechanically mixing	Li–S batteries	[96]
S-Ti$_3$C$_2$T$_x$	Sodium thiosulfate	Sulfur-template method	Li-ion batteries	[98]
Ru$_{SA}$-N-S-Ti$_3$C$_2$T$_x$	Thiourea	Annealing	HER	[99]
S-Ti$_3$C$_2$T$_x$	Na$_2$S·9H$_2$O	Solution soaking	Na-ion batteries	[100]
S-Ti$_3$C$_2$T$_x$	Thiourea	Heat treatment	Na-ion batteries	[101]
S-TiO$_2$@C	Sulfur/Ti$_3$C$_2$	Oxidation	HER	[102]
N, S-Ti$_3$C$_2$	Thiourea	Carbonization	Supercapacitors	[103]
Ti$_3$C$_2$/N, S-G	Thiourea	Ultra-sonication	Supercapacitors	[104]
S-Ti$_3$C$_2$T$_x$	Thiourea	Heat-treatment	Room temperature gas sensing	[105]
S-TCD/TCS	Sulfur	Heat treatment	Li–S batteries	[106]

5.2.3 Hetero-P-MXenes

It is verified that the electronegativity of phosphorus is less than that of carbon, nitrogen, and sulfur, suggesting that P-doping can efficiently change the electronic structure of P-doped MXenes and give them inherently better qualities [1]. Phosphorus doping in MXenes has been more commonly employed as a co-dopant with nitrogen and has not been thoroughly investigated as a solo dopant. It is anticipated that it will affect the fundamental properties of the material and increase the interlayer distance of MXenes, similar to large covalent radius of sulfur and hyper-valency. Ex situ routes have been used in the majority of reported P-doped MXene synthesis [3]. For example, the large surface area and 3D conductive network of 3D V$_2$C multilayer structures allowed X. Guo et al. [107] to set up a 3D channel for electrolyte storage and quick charge transfer, bringing the electrolyte and electrode into complete contact. MXene's special structure, which can efficiently withstand volume expansion, allows it to prevent NiCo-LDH NS accumulation and break up during the intercalation or deintercalation process.

J. Q. Chi et al. [108] used MoO$_4^{2-}$@polymer spheres as precursors for developing a unique dual N, P-doped core–shell nanostructure represented as Mo$_2$C/MoP@NPC. First, using an annealing process, porous core–shell Mo$_2$C@NC nanospheres were created, with ultrafine Mo$_2$C serving as the core and ultrathin NC serving as the shell. Second, Mo$_2$C/MoP@NPC was produced via a high-temperature phosphidation reaction while retaining an intact spherical-like morphology. To successfully alter the electrical arrangement of nanostructured Mo_2C and achieve a noble metal-free HER electrocatalyst, Z. Shi et al. [109] developed a controlled phosphorus doping. Using MoOx–phytic acid–polyaniline hybrids with controllable precursors, simple pyrolysis under inert flow produced a sequence of hierarchical nanowires referred to as $P-Mo_2C@C$. In a study by G. Qu et al. [110], a straightforward phosphorization process added both phosphorus and oxygen to Mo_2CT_x MXenes. When compared to pure Mo_2CT_x MXenes, the phosphorized MXenes demonstrated an enhanced HER electrocatalytic performance, accompanied by a notable reduction in overpotential (> 100 mV at 10 mA cm^{-2}). According to theoretical calculations, P, O-Mo_2CT_x exhibit excellent hydrogen adsorption and a metallic band structure, which enhance conductivity and electrocatalytic kinetics, respectively.

According to Y. Tang et al. [111], the annealing environment significantly modifies the performance of P-doped Mo_2C nanodots in the HER and changes their structure when they are hybridized with Ti_3C_2 MXene flakes. It was found that when the nanohybrids are annealed in N_2, they partially oxidize, but when they are annealed in Ar, NH_3, and H_2/Ar, they maintain their comparable nanostructures. A novel nanohybrid, known as $P-Mo_2C/Ti_3C_2@NC$, was described by Y. Tang et al. [112], and it exhibited excellent dispersion and limited growth due to the conductive intrinsic anchoring sites of Ti_3C_2 matrix. By using a nonmetallic electron donor to achieve tunable interfacial chemical doping, that is, heat treatment using triphenylphosphine as a phosphorous source in 2D vanadium carbide, Yoon et al. [113] were able to overcome several of the obstacles that researchers had been facing. This technique resulted in the substitution of phosphorus at the basal plane with controlled chemical compositions. R. Meng et al. [114] developed black phosphorus quantum dot (BPQD)/titanium carbide nanosheet (TNS) composite anodes using BPQDs as battery and TNSs as pseudocapacitive components. In particular, BPQDs anchored on the TNSs have reduced stress during cycling and enhanced conductivity, making them suitable for use as battery-type components that allow for stable and high-capacity energy storage.

$FePS_3$ is a common ternary metal phosphor-sulfide. Y. Ding et al. [115] generated 2D ultrathin $FePS_3$ NSs by ultrasonic exfoliation. After that, in situ mixing of ultrathin MXene and $FePS_3$ NSs led to the effective synthesis of the unique 2D/2D $FePS_3@$ MXene composite. $FePS_3$ nanosheets@MXene hybrids that are produced as a result have superior surface and interfacial charge transfer capabilities due to large specific surface area and electronic conductivity. Moreover, $FePS_3$ nanosheets@MXene composite's special heterojunction allows it to facilitate Na^+ diffusion and reduce the significant volume shift that occurs throughout the cyclic process, improving its capacity to store salt. A new hybridization of 1D $NiCo_2S_4$ hollow nanotubes and 2D Ti_3C_2-MXene NSs was created by W. Wu et al. [116] owing to a favorable electrostatic interaction between the nanotubes (positively charged) and the MXene (negatively charged). The electrode displayed exceptional electrochemical performance for supercapacitors by combining the high pseudocapacitance of $NiCo_2S_4$ with the strong metallic conductivity of Ti_3C_2-MXenes. The summary of hetero-P-MXenes and their corresponding synthesis methods are presented in Table 5.6.

TABLE 5.6

Hetero-P-MXenes by Various Dopants and Synthesis Methods

Hetero-P-MXenes	Dopants	Synthesis Methods	Properties/ Applications	Ref.
P-V$_2$C/NiCo-LDH	Sodium hypophosphite	Calcination	Li-ion batteries	[107]
Mo$_2$C/MoP@ NPC	Sodium hypophosphite	Calcination	HER	[108]
P-Mo$_2$C@C	Phytic acid	Pyrolysis	HER	[109]
P-Mo$_2$CT$_x$	Red phosphorous	Annealing	HER	[110]
P-Mo$_2$C	Phosphomolybdic acid	Solution mixing	HER	[111]
P-Mo$_2$C/Ti$_3$C$_2$@NC	Phosphomolybdic acid	Annealing	HER	[112]
P-V$_2$CT$_x$	Triphenyl phosphine	Heat treatment	HER	[113]
BPQDs/Ti$_3$C$_2$	BP	Solution mixing	Li–Na storage	[114]
FePS$_3$@Ti$_3$C$_2$T$_x$	Fe, P, S	Solution mixing	Na-ion batteries	[115]
P–Ti$_3$C$_2$@NiCo$_2$S$_4$	Dopamine hydrochloride	Solution mixing	Supercapacitors	[116]

5.2.4 Hetero-M-MXenes

Despite the widespread simulation of MXenes doped with different transition metals, the calculations have primarily addressed the substitution of M sites, and the absence of synthetic techniques continues to pose difficulties for experimental fabrications. As previously stated, an in situ method is used to synthesize the MAX phases of stoichiometric MXenes with ordered double-M. Consequently, several stoichiometric MXenes, including Mo$_2$TiC$_2$, Cr$_2$TiC$_2$, Mo$_2$ScC$_2$, Cr$_2$VC$_2$, and Mo$_2$M$_2$C$_3$ (where M might be Ti, V, or Nb), have been developed [1]. A thorough analysis of experiments and first-principle computations demonstrated that alloying Ti can greatly improve the exfoliation of $(V_{1-x}Ti_x)_2AlC$ precursor, as reported by J. Zhou et al. [117]. Additionally, scanning electron microscopy (SEM) and XRD X-ray diffraction (XRD) analysis displayed the multilayered morphology of exfoliated MXene (i.e., $(V_{1-x}Ti_x)_2C$). Another method for preparing solid-solution $(V_x, Ti_{1-x})_2C$ MXenes (x = 0, 0.3, 0.5, 0.7, and 1) by etching their corresponding MAX precursor was described by Y. Wang et al. [118]. When utilized as Li-ion battery (LIB) anodes, the exfoliated MXenes showed good electrochemical characteristics due to the presence of heteroatoms on their M site.

A non-oxide-based substitution between Pt and Nb$_2$CT$_x$ MXene was presented by Z. Li et al. [119]. It is possible to reduce the surface terminations of the 2D carbide, and a Pt–Nb surface alloy forms at 350 °C. The CO adsorption of this alloy is not as strong as that of monometallic platinum. The kinetics of the water–gas shift showed that, in contrast to bulk niobium carbide, the reactive metal-support interactions (RMSI) stabilizes the material and forms the interfaces of the alloy–MXene with a higher H$_2$O activation. According to D. A. Kuznetsov et al. [120], a two-step synthesis can be used to obtain a Mo$_2$CT$_x$:Co phase from a Co-doped bulk material (i.e., β-Mo$_2$C:Co). The first step involved intercalating gallium to yield Mo$_2$Ga$_2$C:Co, and the second step involved treating the Ga phase with HF to remove it. The HER activity of the doped material was observed to be increased significantly when compared to the undoped Mo_2CT_x catalyst, indicating the positive impact of cobalt substitution on the redox characteristics. A technique for considerably increasing the electrochemical properties of $Ti_3C_2T_x$ MXenes through surface modification and cation intercalation was given by J. Li et al. [121]. Following K^+ intercalation and the removal OH^-/F^-, MXene sheets showed a notable improvement in gravimetric capacitance (about 211% of the original), and the intercalation pseudocapacitance was observed to be three times higher than that of the pristine MXene. The lowest terminating surface group concentration and extensive interlayer voids of Ti_3C_2 are responsible for this enhanced electrochemical performance.

By employing acid etching and alkaline intercalation, X. Zhu et al. [122] synthesized 2D accordion-like alk-Ti_3C_2. The distinct morphology and surface chemistry properties of alk-Ti_3C_2 electrode allowed it to show a noticeably better electrochemical response when compared to the Ti_3C_2-modified electrode. The measurement of trace metal ions involved the optimization of key operating parameters, such as deposition potential, deposition time, and pH. Under ideal circumstances, it demonstrated a high sensitivity that outperformed the majority of published values, with detection limits for Pb(II), Cd(II), Hg(II), and Cu(II). Additionally, the mutual interference of target metal ions was investigated, and it was found that Pb(II) deposited preferentially in the presence of the other three metal ions and that Cd(II) enhanced Hg(II) sensitivity. This technique provides a fresh approach to heavy metal detection with MXene materials. A bifunctional catalyst for ORR and OER processes based on 3D MXenes ($Ti_3C_2T_x$) combined with N-$CoSe_2$ was demonstrated by Z. Zeng et al. [123].

In another study, the optical, magnetic, and structural characteristics of pure and La-$Ti_3C_2T_x$ MXenes were reported by M. Iqbal et al. [124]. The parent MXene's c-lattice parameters (c = 19.2Å), which were determined using X-ray diffraction data, differ somewhat from the c-lattice parameters computed for La-MXene (i.e., c = 18.3Å). The layers of MXene contract perpendicular to the planes when La^{+3} ions are doped, although the in-plane lattice parameters marginally expand. The bandgap of the MXene was calculated to be 1.06 eV, but after doping with La^{+3} ions, it increases to 1.44 eV, demonstrating its strong semiconducting nature. Y. Dall'Agnese et al. treated Ti_3C_2 by delamination or intercalation to show how surface chemistry affects capacitive performance [125]. O-groups were responsible for the increase in capacitance that electrochemical testing revealed. Electrodes with a specific surface area of only 98 $m^2\ g^{-1}$ produced an exceptionally high capacitance of 415 F cm^{-3} at 5 A g^{-1}. For delaminated MXene sheets, values as high as 520 F cm^{-3} were measured at 2 mV s^{-1}. Vanadium-doped $Ti_3C_2T_x$ 2D NSs were hydrothermally synthesized by Z. W. Gao et al. [126] to adjust the interaction between MXene and the alkali metals in the neutral electrolyte. The summary of hetero-M-MXenes and their corresponding synthesis methods are summarized in Table 5.7.

5.3 ENERGY APPLICATIONS OF HETERO-MXENES

Energy storage and conversion are necessary for transforming the available resources into sustainable energy, given the energy limitations caused by the overuse of fossil fuels. The stability and activity of the electrocatalyst greatly influence the electrochemical system, with a reduced overpotential of the catalyst allowing for a faster

TABLE 5.7

Hetero-M-MXenes by Various Dopants and Synthesis Methods

Hetero-M-MXenes	Dopants	Synthesis Methods	Properties/ Applications	Ref.
$V_2C@Sn$	$SnCl_4$	Alkalization	Li-ion batteries	[127]
Sn^{4+}-Ti_3C_2	$SnCl_4$	Solution mixing	Li-ion batteries	[128]
$Mo_2Ti_2C_3T_x$	Ti	Sintering	Thermoelectric	[129]
$(V_{1-x}Ti_x)_2C$	Ti	Sintering	Li-ion batteries	[117]
$(V_xTi_{1-x})_2C$	Ti	Sintering	Li-ion batteries	[118]
$PtNP/Nb_2CT_x$	$Pt(NO_3)_2(NH_3)_4$	Impregnation/annealing	Catalyst	[119]
Mo_2CT_x:Co	$Co(NO_3)_2.6H_2O$	Carburization	HER	[120]
KOH-$Ti_3C_2T_x$	KOH	Intercalation/calcination	Supercapacitors	[121]
KOH-$Ti_3C_2T_x$	KOH	Solution mixing	Supercapacitors	[122]
N-$CoSe_2$/3D $Ti_3C_2T_x$	$Co_2(CO)_8$	Solution mixing	ORR/OER/zinc– air batteries	[123]
La^{3+}-Ti_3C_2	La-nitrate	Coprecipitation	Magnetic	[124]
K^+-$Ti_3C_2T_x$	KOH/KOAc	Delamination	Supercapacitors	[125]
V-$Ti_3C_2T_x$	NH_4VO_3	Hydrothermal	Supercapacitors	[126]
$Ti_3C_2T_x$/PtNP	H_2PtCl_6	Reduction	Sensor	[130]
Gd^{3+}-Ti_3C_2	$Gd(NO_3)_3.6H_2O$	Coprecipitation	Spintronics	[131]
CTAB-Sn^{4+}-Ti_3C_2	$SnCl_4.5H_2O$	Solution mixing	Li-ion capacitors	[132]
Pt/e-TAC	H_2PtCl_6	Reduction	ORR	[133]
KOH-$Ti_3C_2T_x$	KOH	Annealing/calcination	Supercapacitors	[134]
K^+-$Ti_3C_2T_x$	KOH	Hydrothermal	Thermoelectric	[135]
$Ni_{1-x}Co_x@$ $Ti_3C_2T_x$	Ni/Co-LDHs	Reduction	HER	[136]
$Ti_3C_2T_x$/PtNP	H_2PtCl_6	Reduction	HER/ORR	[137]

rate of reaction. MXenes have recently revealed intriguing properties in the domain of catalytic energy conversion due to their exceptional conductivities and surface terminations. By elemental doping, MXenes can be given more active sites, making them ideal for a variety of energy conversion systems such as HERs, NRRs, and OERs, among others [1, 3]. As mentioned earlier, several theoretical studies have demonstrated the effectiveness of MXene-based catalysts supported by single atoms in enhancing the activity of HER. These findings have influenced the exploration and utilization of MXenes doped with single atoms for electrocatalytic oxygen evolution/ reduction [3].

A simple and universal method utilizing ammonia heat treatment, by adding a nitrogen heteroatom to improve HER performance of $Ti_3C_2T_x$ MXenes, was described by T. A. Le et al. [72]. According to the experimental results, N-$Ti_3C_2T_x$ exhibited significantly better HER performance than pure MXene. Similarly, one of the Mo-based catalysts, that is, the pomegranate-like $Mo_2C@C$ nanospheres, showed exceptional catalytic activity in 1 M KOH with low overpotential (47 mV at 10 mA cm^{-2}) [53]. J. Q. Chi et al. [54] stated that the optimized $Mo_2C@NC$ has the ability to demonstrate improved HER performance and long-term stability in both alkaline and acidic solutions by adjusting the MoO_4^{2-} content and carbonization temperature.

In another study, N-Mo$_2$C NSs exhibited HER activity at 10 mA cm^{-2}, long-term stability (with an onset potential of −48.3 mV), and an overpotential of 99 mV [55]. According to Y. Yoon et al. [58], N-Ti$_2$CT$_x$ showed strong HER activity at 10 mA cm^{-2}, when utilized as electrocatalytic materials. Compared to pristine-Ti$_2$CT$_x$, the values are more than three times less (i.e., −645 mV vs. NHE). In 0.5 M H$_2$SO$_4$, the as-synthesized sample demonstrated outstanding endurance, signifying strong catalytic activity toward the HER.

Utilizing the NH$_3$/Ar plasma treatment approach, X. Chen et al. [87] produced nitrogen-doped, few-layered Ti$_3$C$_2$T$_x$ MXene suitable for electrocatalysis. They observed that the produced Ti$_3$C$_2$T$_x$-N$_6$ had the highest nitrogen concentration (1.57 wt%) and the lowest HER overpotential (119.17 mV) when the ammonia-to-argon volume ratio was 6:1. In another study, Ti$_3$C$_2$T$_x$ MXene as effective solid support was utilized to host N ruthenium single-atom catalyst and S-doped RuSA, and the resulting catalyst showed a low overpotential of 76 mV (at 10 mA cm^{-2}) [99]. The dual N, P-doped carbon matrix has the ability to stop the corrosion of Mo$_2$C/MoP nanoparticle during the electrocatalytic process along with enhancing the electroconductivity of the catalysts. The resulting Mo$_2$C/MoP@NPC, which only requires an overpotential of 160 mV, demonstrated durability and good catalytic activity when employed as an HER cathode in acids [108]. Y. Tang et al. [112] reported on a novel nanohybrid P-Mo$_2$C/Ti$_3$C$_2$@NC and observed that the catalyst outperformed many nanohybrids in terms of HER activity, by exhibiting an overpotential of 177 mV (at 10 mA cm^{-2}), long-term stability over 60 hours in the acidic electrolyte, and fast reaction kinetics (57.3 mV dec^{-1}). MXene-based 3D structures can provide an excellent support for dispersing active chemicals in their pores. To enhance their stability, it may be useful to cover the surface of the MXene with an oxidation-resistant layer [138].

Rechargeable batteries are another of the most popular power sources because of their low weight and extended cycling life, particularly for portable electronics. Since the second year of their discovery, 2D MXenes have been used as battery electrodes, and the literature on the subject is constantly expanding in terms of both theoretical calculations and experimental research. Many efforts have been undertaken to improve the capacity of MXenes using different techniques such as surface modification, elemental doping or substitution, and hybrid composites. Notably, it has been demonstrated that elemental doping and substitution techniques effectively improve the performance of MXenes for a variety of metal-ion and metal–air batteries. Regarding the applications, rechargeable batteries have been the primary focus of hetero-P- and hetero-S-MXenes, whereas half of the applications of hetero-M- and hetero-N-MXenes have been concentrated on supercapacitor electrodes [1]. For example, the N-doped MXene NSs acquired a high sulfur loading of 5.1 mg cm^{-2} and demonstrated efficient electrochemical characteristics for Li–S batteries. These characteristics include a high reversible capacity, that is, 1,144 mAh g^{-1} (at 0.2°C) and cycling stability, that is, 610 mAh g^{-1} (after 1,000 cycles) [62].

In another study, the crumpled N–Ti$_3$C$_2$/Fe$_2$O$_3$ nanocomposite improved the electron transport and brought about significant volume change of the active material. The nanocomposites showed strong cycling performance, that is, 549 mAh g^{-1} after 400 cycles (at 2 A g^{-1}), quick charge/discharge ability, and high reversible capacity (1,065 mAh g^{-1} at 100 mA g^{-1}) [65]. A unique N-doped MXene-CoS$_2$ nanohybrid was synthesized by C. Yang et al. [67], and it successfully immobilized and converted LiPSs, addressing the drawbacks of Li–S batteries. An exceptional rate performance, high initial specific capacity, and cycling stability were all displayed by the cell. Using $Ti_3C_2T_x$ MXene NSs as the starting point, J. Wang et al. [77] created MXene-based Co, N co-doped porous carbon nanosheets (CNSs) as sulfur hosts by allowing the bimetallic zeolite imidazole framework to self-assemble in situ before calcining and etching the material. Co, N co-doping greatly improves the adsorption to polysulfides while also considerably accelerating sulfur cathode kinetics. As a result, in the first cycle of charge/discharge process, the electrode reaches a high capacity of 1340.2 mAh g^{-1}, and after 1,000 cycles at 1 C, it retains an extraordinary value of 914.7 mAh g^{-1}.

Similarly, P-NTC (porous N-doped Ti_3C_2) was observed to be an active electrocatalyst for Li–S chemistry [82]. After over 1,200 cycles, the as-fabricated cathode retained a low-capacity decline (about 0.033% per cycle). A conductive $Ti_3C_2T_x$/S paper with the mechanical strength, exceptional conductivity, and distinct chemisorption of LiPSs from MXene NSs was reported [94]. It was observed that after 1,500 cycles, the $Ti_3C_2T_x$/S paper showed a retention of 0.014%, which is the lowest figure for Li–S batteries that have been reported to date. S@TiO_2/Ti_2C composites [96] demonstrated the first discharge-specific capacity (1,408.6 mAh g^{-1} at 0.2 C) at a high sulfur content of 78.4 wt%. After 200 cycles, the battery retained a capacity of 464.0 at 2 C and 227.3 mAh g^{-1} at 5 C, respectively. The synergistic impact between the conductivity and adsorption of TiO_2 for active sulfur was primarily responsible for the increased electrochemical properties of composites. MXene-based $Ti_3C_2T_x$ nanodot-interspersed $Ti_3C_2T_x$ NS (TCD-TCS) was introduced by Z. Xiao et al. [106] to achieve the conversion and spatial immobilization of high-loaded sulfur species. A small sulfur loading (1.8 mg cm^{-2}) presented an almost theoretical discharge capability for the TCD-TCS/S electrode. Notably, at a high sulfur loading of 13.8 mg cm^{-2}, high capacity (13.7 mAh cm^{-2}) and volumetric capacity (1,957 mAh cm^{-3}) were attained simultaneously.

R. Meng et al. [114] reported the dual-model energy storage (DMES) mechanism for Li-ion and Na-ion batteries, by developing BPQD/TNS composite anodes using BPQDs as battery and Ti_3C_2 NSs as pseudocapacitive components. $FePS_3$ @ MXene hybrids can also enhance the specific surface area and electronic conductivity, thereby ensuring superior charge transfer capabilities [115]. Thus, after 90 cycles, the ultrathin MXene-coated few-layered $FePS_3$ NSs offer an outstanding reversible capacity of 676.1 mAh g^{-1} (at 100 mA g^{-1}). In addition to presenting the promising anode material with superior properties for Na-ion batteries, this work offered an innovative procedure for creating 2D/2D materials. When employed as LIB anodes, solid-solution $(V_x, Ti_{1-x})_2C$ MXenes demonstrated good electrochemical properties [118]. The electrochemical properties were optimized for x = 0.5, that is, $(V_{0.5}, Ti_{0.5})_2C$, which displayed the highest reversible capacity of 204.9 mAh g^{-1}. In addition, Zn–air batteries fitted with the newly designed N-CoSe$_2$/3D MXene air cathode demonstrated higher power/energy densities and longer cycling lives (more than 500 cycles) when compared to combined Pt/C and RuO_2 batteries [123].

MXenes have the potential to be excellent electrode materials for highly effective pseudocapacitors. However, due to repulsion interactions, they are only suitable as negative electrodes and are not able to interact well with anionic species. Researchers are extensively studying MXene and MXene-based composite materials as supercapacitor electrode materials. It has been discovered that electrodes produced from MXenes exhibit outstanding performance. When heteroatoms like nitrogen are added, their specific capacitance is significantly increased. In conclusion, doped MXenes, specifically those doped with nitrogen, are ideal anode materials for pseudocapacitors. Further, quick ion exchange interactions with MXenes allow for high-rate capabilities. There are a lot of important MXene materials available; therefore, more optimization will result in excellent electrode materials for storage applications. Fabricating 3D composite structures based on MXenes or overlaying MXene structures with bimetallic hybrid compounds is a feasible strategy to increase the electrode capacity [3]. For example, Y. Yoon et al. [68] stated that $N\text{-}Ti_2CT_x$ MXenes are excellent electrochemical capacitor electrodes due to the synergistic effect of heteroatom compositions and delaminated structures. The resultant characteristics include high capacitance and stable long cyclic performance at high current density.

As a potential electrode material for supercapacitors, Y. Wen et al. [71] developed and optimized $N\text{-}Ti_3C_2T_x$. The resulting doped MXene materials showed significantly higher electrochemical capacitances than the pure materials (i.e., 34 F g^{-1} and 52 F g^{-1} in 1 M H_2SO_4 and 1 M $MgSO_4$, respectively). Another $Ti_3C_2T_x$@NC-2 composite demonstrated a high specific capacitance of 442.2 F g^{-1} at 1 A g^{-1} [60]. This is 281% greater than that of $Ti_3C_2T_x$. To elaborate, $Ti_3C_2T_x$@NC demonstrated outstanding cycling stability, maintaining 91.9% of capacitance after 5,000 cycles and 92.5% of capacitance retention (at 10 A g^{-1}). N, O co-doped C@Ti_3C_2 composites were observed to exhibit 250.6 F g^{-1} at 1 A g^{-1}, with 94% capacitance retention (after 5,000 cycles) [63]. Moreover, the composite-based symmetric supercapacitor exhibited an energy density of 10.8 Wh kg^{-1} at a power density of 600 W kg^{-1}. The $N\text{-}Ti_3C_2$/rGO hybrid film demonstrated exceptional long-term electrochemical stability, as seen by its high specific capacitance of 247 F g^{-1} in 6M KOH and the capacitive retention after 10,000 cycles. Additionally, a flexible symmetric device was developed using such films, and it produces a power density of 3,738.7 W kg^{-1} (at 11.1 Wh kg^{-1}) and an energy density of 15.7 Wh kg^{-1} (at 309.3 W kg^{-1}) in PVA/H_2SO_4 gel [64].

An efficient electrode with significantly enhanced gravimetric capacitance and superior cycling stability was shown by B. Yang et al. [80]. The optimized TiO_2/TiN/$Ti_3C_2T_x$ electrodes for nitrogen-doped intercalation were evaluated in LiOH, KOH, H_2SO_4, Li_2SO_4, and Na_2SO_4 electrolytes. With a cycling stability of 85.8% after 10,000 charge–discharge cycles and a high capacitance of 361 F g^{-1}, the electrode in the H_2SO_4 electrolyte demonstrated the best performance. Electrodes based on the N surface-modified $Ti_3C_2T_x$ film also exhibit significantly improved electrochemical performances [85]. In 1 M Li_2SO_4 electrolyte, the NS-Ti_3C_2 NSs showed notable cycling stability (90.1%) and a high specific capacitance (175 F g^{-1}). The increased surface areas and more mesopores are responsible for the NS-Ti_3C_2's notable improvement in electrochemical performance; in fact, N and S doping can improve ion electrolyte intercalation, produce more electrical charge, and increase pseudocapacitance [103]. The impact of adding N-, S-graphene at different ratios on electrochemical properties was observed by using galvanostatic charge–discharge (GCD) and cyclic voltammetry (CV) analyses [104].

Despite the adaptable chemistry of MXenes that makes their qualities adjustable for a range of applications, there are some difficulties, and more studies in doped/heterostructure MXenes are required. Unfortunately, the study of MXenes is restricted to only a few MXenes due to a shortage of precursor materials. As a result, the synthesis of new MAX precursors will allow for the use of previously unexplored MXenes. Machine learning has been successful in predicting characteristics such as bandgaps and activities like cytotoxicity. Therefore, the use of this technique is suggested in future research in this era. Although practical implementation of hetero-MXene-based devices is still in the early stages, they hold great potential for multifunctional solutions in more reliable electronic devices, energy storage, and environmentally friendly energy [3].

5.4 LATEST RESEARCH ON HETEROSTRUCTURE MXENES

In 2022, Y. Fan et al. [139] reported on the latest developments in MXene crystal formation using the chemical vapor deposition (CVD) technique. They made several MXene heterostructures with both lateral and vertical spatial orientations, including graphene/tungsten carbide (WC) for electrocatalysis and graphene/α-Mo$_2$C for superconductivity. Similar to this, in 2023, R. A. B. John et al. [140] offered a thorough investigation of hetero-MXenes used in gas-sensing applications. They emphasized the potential of these heterostructures to improve gas-sensing performance while addressing the lack of knowledge regarding the sensor mechanisms of different heterostructures, such as MXene/metal oxide, MXene/carbonaceous, MXene/noble-metal, MXene/polymer, and MXene/metal chalcogenide heterostructures. The composite of MXenes with other photocatalysts, like g-C$_3$N$_4$, perovskite materials (Cs$_3$Bi$_2$Br$_9$, FAPbBr$_3$, CsPbBr$_3$, and Cs$_2$AgBiBr$_6$), LDHs (NiAl, Co–Co, and Co$_2$Al$_{0.95}$La$_{0.05}$), bi-based photocatalysts (BiOX (X = Cl, Br, I), Bi$_2$XO$_6$ (X = W, Mo), and hybrid Bi$_2$O$_2$SiO$_3$), and metal oxides and metal sulfides (TiO$_2$, CeO$_2$, InVO$_4$, CdS, Cd$_{0.2}$Zn$_{0.8}$S, and ZnIn$_2$S$_4$) are some of the reports presented by T. Amrillah et al. [141]. Additionally, they noted that more than two material combinations can be used in MXene-based nanocomposite photocatalysts. Examples of such combinations are TiO$_2$/C$_3$N$_4$/Ti$_3$C$_2$, meso-TiO$_2$@ZnIn$_2$S$_4$/Ti$_3$C$_2$, CdS/Ti$_3$C$_2$/g-C$_3$N$_4$, g-C$_3$N$_4$/Bt/Ti$_3$C$_2$, and BiOIO$_3$/g-C$_3$N$_4$/Ti$_3$C$_2$. Each component of these nanocomposites has a specific function, such as the photocatalyst TiO$_2$, the co-catalyst ZnIn$_2$S$_4$, and the electron-trapping agent C$_3$N$_4$.

By using an in situ polymerization and carburization technique, X. Wang et al. [142] produced N, S-Mo$_2$C-Mo/C nanorods containing a lot of active heterointerfaces and defect sites. The combined actions of Mo$_2$C and N, S-doped carbon greatly improved the HER activity. According to J. L. Hart et al. [143], annealing MXenes in a low-power Ar$^+$ O$_2$ plasma increases $-$O functionalization while producing little secondary phase formation. Using two MXenes, Ti$_2$CT$_x$ and Mo$_2$TiC$_2$T$_x$, they utilized this approach and demonstrated that the higher $-$O concentration in both cases increases the electrical resistance and decreases the electron count of the surface transition metal. We demonstrate that the O content of Mo$_2$TiC$_2$O$_x$ may be reversibly changed by sequential vacuum and plasma annealing. A novel approach for the alteration of V$_2$CT$_x$ by thermal treatment with molten salt was proposed by W. Jiang et al. [144]. Through a substitution reaction, S heteroatoms were added to V_2CT_x via the unique approach, which involved dissolving Li_2S in molten salts of LiCl and KCl. The S-doped V_2CT_x ($MS-S-V_2CT_x$) cathode was then electrochemically charged and discharged in situ to produce surface V_2O_5. After undergoing extensive cycling stability testing for 3,000 cycles, the constructed $Zn/MS-S-V_2CT_x$ battery demonstrated a high reversible discharge capacity of 411.3 mAh g^{-1} (at 0.5 A g^{-1}) and an 80% capacitance retention.

As potential catalysts for ORR, Y. Chen et al. [145] constructed and reported heterostructures of eight distinct transition metals (represented as $M-N4-Gr/V_2C$). The results of the computations demonstrated the thermodynamic stability of all catalysts, except $Zn-N4-Gr/V_2C$. Higher activity toward the ORR was demonstrated by $Co-N4-Gr/V_2C$ and $Ni-N4-Gr/V_2C$ with low overpotentials. A 2D Nb_2C/MoS_2 heterostructure that surpasses pristine Nb_2C in both linear and nonlinear optical performance was rationally designed and synthesized in situ by Y. Wang et al. [146]. The Nb_2C/MoS_2 inherited the majority of Nb_2C and MoS_2 in absorption at different wavelengths, leading to broadband increased optical absorption, as indicated by the agreement of results between experimental and theoretical studies. They also achieved nonlinear optical modulation in the near-infrared; the nonlinear absorption coefficient of Nb_2C/MoS_2 is more than double that of pristine Nb_2C. A novel heterostructure electrocatalyst, represented as $S-ML-Nb_4C_3T_x$, was developed using a hydrothermal approach, according to a paper by F. Wu et al. [147]. This electrocatalyst showed exceptional HER catalytic activity in a 1.0 M KOH, with a reduced overpotential of 118 mV (at 10 mA cm^{-2}). This was in contrast to the pristine $ML-Nb_4C_3T_x$ catalyst. Furthermore, the catalyst has a consistent electrochemical durability in 1.0 M KOH for up to 24 hours.

N-MXene, also known as wrinkled, flexible, and layered nitrogen-doped $Ti_3C_2T_x$, was synthesized by S. Y. Liao et al. [148] using cyanamide as the N-dopant. At the proper weight ratio (i.e., 2.5 wt.%), the obtained electrodes enable improved electrochemical performance when the wrinkled N-MXene is applied as a conductive addition to $LiNi_{0.8}Co_{0.1}Mn_{0.1}O_2$. The intercalation of Li ions from the multilayer, flexible, and wrinkled N-MXene is attributed to the additional capacity of 226.9 mAh g^{-1} (at 20 mA g^{-1}), provided by the 2.5-N-MXene. Aqueous production of N-doped MXene-TiO_2 hybrid anode materials were reported by U. Alli et al. [149]. After 100 cycles, its initial specific energy capacity (305 mAh g^{-1}) was improved to 369 mAh g^{-1} (at 0.1 C), with a capacitance retention of 99.7%. A unique iodine-doped MXene was proposed by W. Yu et al. [150] to enhance the performance of electrodes in Li-S batteries. It was observed that I-MXenes can quicken the reaction kinetics and efficiently immobilize LiPS through Ti–S bonds. Heteroatom-doped MXene (B, $S-Ti_3C_2T_x$) NSs were developed [151] as effective electrocatalysts for HER. With its high surface area, the synthesized B, $S-Ti_3C_2T_x$ demonstrated better catalytic activity in acidic conditions.

In another study, ammonium citrate was used as a source of nitrogen in a modified two-step multielement method [152] to increase the degree of heteroatom doping. The resultant nitrogen/sulfur co-doped MXenes showed remarkable rate capability, superior cycling stability, and increased gravimetric capacitance (495 F g^{-1} at 1 A g^{-1}). By annealing MXene NSs in sodium hypophosphite, Y. Wen et al. [153] created unique

P-$Ti_3C_2T_x$ NSs in an easy and controlled manner. For the first time, the as-obtained P-$Ti_3C_2T_x$ displayed efficient electrochemical properties as the electrode material. L. Cai et al. [154] used the etching and the electrostatic self-assembly technology to develop hollow Co-ZIF particles (HCF) and layered Ti_3CNT_x flakes. Electromagnetic wave (EMW) attenuation could be enhanced by the synergistic effect of improved dielectric loss and interface polarization when combined with impedance matching. Thus, this straightforward assembling approach expands the viable candidates for trivial and highly effective microwave absorbers and provides a new avenue for the layered materials. B. Sarfraz et al. [155] created Cl-terminated MXenes ($Ti_3C_2Cl_2$) by using a molten salt method in which the corresponding MAX phase was reacted with $CuCl_2$ salt and then thermally treated for 5 hours at 550°C. Y. Tian et al. [156] used a simpler method than the intricate post-doped procedure to produce in situ O-doped $Ti_3C_2T_x$ NSs. Additionally, electrochemical performances showed that the pure $Ti_3C_2T_x$ electrode cannot match the higher capacity of the O-$Ti_3C_2T_x$−0.05 film electrode, which can deliver 306.0 C g^{-1}.

An iodine-terminated MXene was effectively synthesized by S. Gong et al. [157] using a simple Lewis-acidic etching technique, and its performance was thoroughly examined for supercapacitors. The I-Ti_3C_2 MXene with pseudocapacitor property showed a much better specific capacitance than HF-$Ti_3C_2T_x$ MXene, due to the presence of iodine. Even at 50 A g^{-1}, the I-Ti_3C_2 MXene demonstrated impressive long-term cycling performance, allowing for the preservation of 91% of specific capacitance over 100,000 cycles. The formation of P-Ti_3C_2-MW was observed to be directed by the MW absorption ability of MXene, phytic acid, and water, as well as the homogeneous mixing of the MXene and doping source through electrostatic attraction [158]. With an enlarged interlayer spacing (i.e., 1.18 nm), the electrode enables improved pseudocapacitance, which facilitates optimized electronic conductivity, an abundance of redox active sites, intercalation of ions, etc. A novel 0D/2D heterostructure of ZnO/N-Ti_3C_2 with a high nitrogen-doping level was designed [159], using glycine as the nitrogen precursor through a simple heat treatment. A Cl-terminated MXene ($Ti_3C_2Cl_2$) with a low diffusion barrier and a large Cl capacity was also synthesized and reported [160] to be used as a cathode material for chloride-ion hybrid capacitors. As a result, it displayed better rate capability, that is, 153.1 mAh g^{-1} at 1 A g^{-1} and 121.2 mAh g^{-1} at 2 A g^{-1}. In another study, a protective heterogeneous layer made of ZnS on Zn anode and conductive S-doped 3D MXene was produced by Y. An et al. [161]. Sulfur-doped 3D MXene can reduce volume change, reduce local current density, and homogenize the distribution of electric field. ZnS can quicken Zn^{2+} migration, encourage homogeneous Zn^{2+} distribution, and limit side reactions.

REFERENCES

[1] Gao, L., et al., *Hetero-MXenes: theory, synthesis, and emerging applications.* Advanced Materials, 2021. **33**(10): p. 2004129.

[2] Lei, Y.-J., et al., *Tailoring MXene-based materials for sodium-ion storage: synthesis, mechanisms, and applications.* Electrochemical Energy Reviews, 2020. **3**: p. 766–792.

[3] Dey, A., et al., *Doped MXenes—a new paradigm in 2D systems: synthesis, properties and applications.* Progress in Materials Science, 2023: p. 101166.

[4] Yang, J., et al., *Tunable electronic and magnetic properties of Cr_2M' C_2T_2 (M' = Ti or V; T = O, OH or F).* Applied Physics Letters, 2016. **109**(20).

[5] Tan, T.L., et al., *High-throughput survey of ordering configurations in MXene alloys across compositions and temperatures.* ACS Nano, 2017. **11**(5): p. 4407–4418.

[6] Chen, Z., et al., *Transition metal atoms implanted into MXenes (M_2CO_2) for enhanced electrocatalytic hydrogen evolution reaction.* Applied Surface Science, 2020. **509**: p. 145319.

[7] Mehta, V., K. Tankeshwar, and H.S. Saini. *Ab-initio study of electronic and magnetic properties of Co-doped Mo_2C monolayer.* in *AIP Conference Proceedings.* 2018, AIP Publishing.

[8] Cheng, C., et al., *Cu_3-cluster-doped monolayer Mo_2CO_2 (MXene) as an electron reservoir for catalyzing a CO oxidation reaction.* ACS Applied Materials & Interfaces, 2018. **10**(38): p. 32903–32912.

[9] Cheng, C., et al., *Single Pd atomic catalyst on Mo_2CO_2 monolayer (MXene): unusual activity for CO oxidation by trimolecular Eley–Rideal mechanism.* Physical Chemistry Chemical Physics, 2018. **20**(5): p. 3504–3513.

[10] Si, C., et al., *Large-gap quantum spin Hall state in MXenes: d-band topological order in a triangular lattice.* Nano Letters, 2016. **16**(10): p. 6584–6591.

[11] Li, L., et al., *Theoretical screening of single transition metal atoms embedded in MXene defects as superior electrocatalyst of nitrogen reduction reaction.* Small Methods, 2019. **3**(11): p. 1900337.

[12] Yang, J., et al., *Investigation of magnetic and electronic properties of transition metal doped Sc_2CT_2 (T = O, OH or F) using a first principles study.* Physical Chemistry Chemical Physics, 2016. **18**(18): p. 12914–12919.

[13] Yang, D., et al., *Sc_2CO_2 and Mn-doped Sc_2CO_2 as gas sensor materials to NO and CO: a first-principles study.* Physica E: Low-dimensional Systems and Nanostructures, 2019. **111**: p. 84–90.

[14] Ling, C., et al., *Transition metal-promoted V_2CO_2 (MXenes): a new and highly active catalyst for hydrogen evolution reaction.* Advanced Science, 2016. **3**(11): p. 1600180.

[15] Kan, D., et al., *Rational design of bifunctional ORR/OER catalysts based on Pt/Pd-doped Nb_2CT_2 MXene by first-principles calculations.* Journal of Materials Chemistry A, 2020. **8**(6): p. 3097–3108.

[16] Cheng, Y., et al., *Nanostructure of Cr_2CO_2 MXene supported single metal atom as an efficient bifunctional electrocatalyst for overall water splitting.* ACS Applied Energy Materials, 2019. **2**(9): p. 6851–6859.

[17] Cheng, Y.-W., et al., *Transition metal modification and carbon vacancy promoted Cr_2CO_2 (MXenes): a new opportunity for a highly active catalyst for the hydrogen evolution reaction.* Journal of Materials Chemistry A, 2018. **6**(42): p. 20956–20965.

[18] Ling, C., et al., *Searching for highly active catalysts for hydrogen evolution reaction based on O-terminated MXenes through a simple descriptor.* Chemistry of Materials, 2016. **28**(24): p. 9026–9032.

[19] Chakraborty, P., et al., *Manipulating the mechanical properties of Ti_2C MXene: effect of substitutional doping.* Physical Review B, 2017. **95**(18): p. 184106.

[20] Feng, L., et al., *Structures and mechanical and electronic properties of the Ti_2CO_2 MXene incorporated with neighboring elements (Sc, V, B and N).* Journal of Electronic Materials, 2017. **46**: p. 2460–2466.

[21] Zhou, Y., et al., *Current rectification induced by V-doped and Sc-doped in Ti_2CO_2 devices.* Computational Materials Science, 2017. **138**: p. 175–182.

[22] Gao, Y., et al., *Functionalization Ti_3C_2 MXene by the adsorption or substitution of single metal atom.* Applied Surface Science, 2019. **465**: p. 911–918.

[23] Luo, H., et al., *A theoretical study of Fe adsorbed on pure and nonmetal (N, F, P, S, Cl)-doped $Ti_3C_2O_2$ for electrocatalytic nitrogen reduction.* Nanomaterials, 2022. **12**(7): p. 1081.

[24] Liu, C.-Y. and E.Y. Li, *Termination effects of Pt/v-Ti$_{n+1}$C$_n$T$_2$ MXene surfaces for oxygen reduction reaction catalysis.* ACS Applied Materials & Interfaces, 2018. **11**(1): p. 1638–1644.

[25] Zhou, T., et al., *Atomic vacancy defect, Frenkel defect and transition metals (Sc, V, Zr) doping in Ti$_4$N$_3$ MXene nanosheet: a first-principles investigation.* Applied Sciences, 2020. **10**(7): p. 2450.

[26] Ding, B., et al., *Uncovering the electrochemical mechanisms for hydrogen evolution reaction of heteroatom doped M$_2$C MXene (M = Ti, Mo).* Applied Surface Science, 2020. **500**: p. 143987.

[27] Balcı, E., Ü.Ö. Akkuş, and S. Berber, *Band gap modification in doped MXene: Sc 2 CF 2.* Journal of Materials Chemistry C, 2017. **5**(24): p. 5956–5961.

[28] Balcı, E., Ü.Ö. Akkuş, and S. Berber, *Doped Sc$_2$C(OH)$_2$ MXene: new type s-pd band inversion topological insulator.* Journal of Physics: Condensed Matter, 2018. **30**(15): p. 155501.

[29] Zhang, R.-Z., H.-L. Cui, and X.-H. Li, *First-principles study of structural, electronic and optical properties of doped Ti$_2$CF$_2$ MXenes.* Physica B: Condensed Matter, 2019. **561**: p. 90–96.

[30] Lu, C., et al., *Nitrogen-doped Ti$_3$C$_2$ MXene: mechanism investigation and electrochemical analysis.* Advanced Functional Materials, 2020. **30**(47): p. 2000852.

[31] Chen, X., et al., *Proposing the prospects of Ti$_3$CN transition metal carbides (MXenes) as anodes of Li-ion batteries: a DFT study.* Physical Chemistry Chemical Physics, 2016. **18**(48): p. 32937–32943.

[32] Tang, Y., et al., *The effect of in situ nitrogen doping on the oxygen evolution reaction of MXenes.* Nanoscale Advances, 2020. **2**(3): p. 1187–1194.

[33] Xin, Y. and Y.-X. Yu, *Possibility of bare and functionalized niobium carbide MXenes for electrode materials of supercapacitors and field emitters.* Materials & Design, 2017. **130**: p. 512–520.

[34] Zhu, J., et al., *S-functionalized MXenes as electrode materials for Li-ion batteries.* Applied Materials Today, 2016. **5**: p. 19–24.

[35] Zheng, S., et al., *Electrochemical nitrogen reduction reaction performance of single-boron catalysts tuned by MXene substrates.* The Journal of Physical Chemistry Letters, 2019. **10**(22): p. 6984–6989.

[36] Shukla, V., et al., *Modelling high-performing batteries with MXenes: the case of S-functionalized two-dimensional nitride MXene electrode.* Nano Energy, 2019. **58**: p. 877–885.

[37] Yang, J., et al., *Stability and electronic properties of sulfur terminated two-dimensional early transition metal carbides and nitrides (MXene).* Computational Materials Science, 2018. **153**: p. 303–308.

[38] He, J., et al., *Cr$_2$TiC$_2$-based double MXenes: novel 2D bipolar antiferromagnetic semiconductor with gate-controllable spin orientation toward antiferromagnetic spintronics.* Nanoscale, 2019. **11**(1): p. 356–364.

[39] He, J., et al., *High temperature spin-polarized semiconductivity with zero magnetization in two-dimensional Janus MXenes.* Journal of Materials Chemistry C, 2016. **4**(27): p. 6500–6509.

[40] Huang, B., et al., *Insights into the electrocatalytic hydrogen evolution reaction mechanism on two-dimensional transition-metal carbonitrides (MXene).* Chemistry—A European Journal, 2018. **24**(69): p. 18479–18486.

[41] Li, M., et al., *Element replacement approach by reaction with Lewis acidic molten salts to synthesize nanolaminated MAX phases and MXenes.* Journal of the American Chemical Society, 2019. **141**(11): p. 4730–4737.

[42] Kamysbayev, V., et al., *Covalent surface modifications and superconductivity of two-dimensional metal carbide MXenes.* Science, 2020. **369**(6506): p. 979–983.

[43] Wang, Y., et al., *Achieving superior high-capacity batteries with the lightest Ti_2C MXene anode by first-principles calculations: overarching role of S-functionate (Ti_2CS_2) and multivalent cations carrier.* Journal of Power Sources, 2020. **451**: p. 227791.

[44] Liu, X., et al., *Anchoring effects of S-terminated Ti_2C MXene for lithium-sulfur batteries: a first-principles study.* Applied Surface Science, 2018. **455**: p. 522–526.

[45] Wang, D., et al., *A general atomic surface modification strategy for improving anchoring and electrocatalysis behavior of $Ti_3C_2T_2$ MXene in lithium–sulfur batteries.* ACS Nano, 2019. **13**(10): p. 11078–11086.

[46] Li, D., et al., *Chalcogenated-$Ti_3C_2X_2$ MXene (X = O, S, Se and Te) as a high-performance anode material for Li-ion batteries.* Applied Surface Science, 2020. **501**: p. 144221.

[47] Hou, T., et al., *Modulating oxygen coverage of $Ti_3C_2T_x$ MXenes to boost catalytic activity for HCOOH dehydrogenation.* Nature Communications, 2020. **11**(1): p. 4251.

[48] Jiang, Y., et al., *Oxygen-functionalized ultrathin $Ti_3C_2T_x$ MXene for enhanced electrocatalytic hydrogen evolution.* ChemSusChem, 2019. **12**(7): p. 1368–1373.

[49] Persson, I., et al., *How much oxygen can a MXene surface take before it breaks?* Advanced Functional Materials, 2020. **30**(47): p. 1909005.

[50] Meng, Q., et al., *The S-functionalized Ti_3C_2 Mxene as a high capacity electrode material for Na-ion batteries: a DFT study.* Nanoscale, 2018. **10**(7): p. 3385–3392.

[51] Li, Y., et al., *A general Lewis acidic etching route for preparing MXenes with enhanced electrochemical performance in non-aqueous electrolyte.* Nature Materials, 2020. **19**(8): p. 894–899.

[52] Tao, Y., et al., *Surface group-modified MXene nano-flake doping of monolayer tungsten disulfides.* Nanoscale Advances, 2019. **1**(12): p. 4783–4789.

[53] Chen, Y.-Y., et al., *Pomegranate-like N, P-doped $Mo_2C@C$ nanospheres as highly active electrocatalysts for alkaline hydrogen evolution.* ACS Nano, 2016. **10**(9): p. 8851–8860.

[54] Chi, J.-Q., et al., *Porous core-shell N-doped $Mo_2C@C$ nanospheres derived from inorganic-organic hybrid precursors for highly efficient hydrogen evolution.* Journal of Catalysis, 2018. **360**: p. 9–19.

[55] Jia, J., et al., *Ultrathin N-doped Mo_2C nanosheets with exposed active sites as efficient electrocatalyst for hydrogen evolution reactions.* ACS Nano, 2017. **11**(12): p. 12509–12518.

[56] Liu, R., et al., *Nitrogen-doped Nb_2CT_x MXene as anode materials for lithium-ion batteries.* Journal of Alloys and Compounds, 2019. **793**: p. 505–511.

[57] Mendes, R.G., et al., *In situ N-doped graphene and Mo nanoribbon formation from $Mo_2Ti_2C_3$ MXene monolayers.* Small, 2020. **16**(5): p. 1907115.

[58] Yoon, Y., et al., *Enhanced electrocatalytic activity by chemical nitridation of two-dimensional titanium carbide MXene for hydrogen evolution.* Journal of Materials Chemistry A, 2018. **6**(42): p. 20869–20877.

[59] Fan, Z., et al., *3D printing of porous nitrogen-doped Ti_3C_2 MXene scaffolds for high-performance sodium-ion hybrid capacitors.* ACS Nano, 2020. **14**(1): p. 867–876.

[60] Zhao, T., et al., *Dopamine-derived N-doped carbon decorated titanium carbide composite for enhanced supercapacitive performance.* Electrochimica Acta, 2017. **254**: p. 308–319.

[61] Han, M., et al., *Efficient tuning the electronic structure of N-doped Ti-based MXene to enhance hydrogen evolution reaction.* Journal of Colloid and Interface Science, 2021. **582**: p. 1099–1106.

[62] Bao, W., et al., *Facile synthesis of crumpled nitrogen-doped MXene nanosheets as a new sulfur host for lithium–sulfur batteries.* Advanced Energy Materials, 2018. **8**(13): p. 1702485.

[63] Pan, Z. and X. Ji, *Facile synthesis of nitrogen and oxygen co-doped $C@Ti_3C_2$ MXene for high performance symmetric supercapacitors.* Journal of Power Sources, 2019. **439**: p. 227068.

[64] Yang, L., et al., *Freestanding nitrogen-doped d-Ti$_3$C$_2$/reduced graphene oxide hybrid films for high performance supercapacitors*. Electrochimica Acta, 2019. **300**: p. 349–356.

[65] Zhang, Z., et al., *Highly-dispersed iron oxide nanoparticles anchored on crumpled nitrogen-doped MXene nanosheets as anode for Li-ion batteries with enhanced cyclic and rate performance*. Journal of Power Sources, 2019. **439**: p. 227107.

[66] Xu, Q., et al., *Hydrochromic full-color MXene quantum dots through hydrogen bonding toward ultrahigh-efficiency white light-emitting diodes*. Applied Materials Today, 2019. **16**: p. 90–101.

[67] Yang, C., et al., *In situ N-doped CoS$_2$ anchored on MXene toward an efficient bifunctional catalyst for enhanced lithium-sulfur batteries*. Chemical Engineering Journal, 2022. **427**: p. 131792.

[68] Yoon, Y., et al., *A strategy for synthesis of carbon nitride induced chemically doped 2D MXene for high-performance supercapacitor electrodes*. Advanced Energy Materials, 2018. **8**(15): p. 1703173.

[69] Yang, C., et al., *Flexible nitrogen-doped 2D titanium carbides (MXene) films constructed by an ex situ solvothermal method with extraordinary volumetric capacitance*. Advanced Energy Materials, 2018. **8**(31): p. 1802087.

[70] Yang, C., et al., *Improved capacitance of nitrogen-doped delaminated two-dimensional titanium carbide by urea-assisted synthesis*. Electrochimica Acta, 2017. **225**: p. 416–424.

[71] Wen, Y., et al., *Nitrogen-doped Ti$_2$C$_2$T$_x$ MXene electrodes for high-performance supercapacitors*. Nano Energy, 2017. **38**: p. 368–376.

[72] Le, T.A., et al., *Synergistic effects of nitrogen doping on MXene for enhancement of hydrogen evolution reaction*. ACS Sustainable Chemistry & Engineering, 2019. **7**(19): p. 16879–16888.

[73] Tang, Y., et al., *Synthesis of nitrogen-doped two-dimensional Ti$_3$C$_2$ with enhanced electrochemical performance*. Journal of the Electrochemical Society, 2017. **164**(4): p. A923.

[74] Jiang, G., et al., *In-situ decoration of MOF-derived carbon on nitrogen-doped ultrathin MXene nanosheets to multifunctionalize separators for stable Li-S batteries*. Chemical Engineering Journal, 2019. **373**: p. 1309–1318.

[75] Jhon, Y.I., et al., *Metallic MXene saturable absorber for femtosecond mode-locked lasers*. Advanced Materials, 2017. **29**(40): p. 1702496.

[76] Yang, C., et al., *Methanol and diethanolamine assisted synthesis of flexible nitrogen-doped Ti$_3$C$_2$ (MXene) film for ultrahigh volumetric performance supercapacitor electrodes*. ACS Applied Energy Materials, 2019. **3**(1): p. 586–596.

[77] Wang, J., et al., *MXene-based Co, N-Co-doped porous carbon nanosheets regulating polysulfides for high-performance lithium–sulfur batteries*. ACS Applied Materials & Interfaces, 2019. **11**(42): p. 38654–38662.

[78] Tang, X., et al., *MXene-based dendrite-free potassium metal batteries*. Advanced Materials, 2020. **32**(4): p. 1906739.

[79] Guo, Y., et al., *MXene-encapsulated hollow Fe$_3$O$_4$ nanochains embedded in N-doped carbon nanofibers with dual electronic pathways as flexible anodes for high-performance Li-ion batteries*. Nanoscale, 2021. **13**(8): p. 4624–4633.

[80] Yang, B., et al., *Nitrogen doped intercalation TiO$_2$/TiN/Ti$_3$C$_2$T$_x$ nanocomposite electrodes with enhanced pseudocapacitance*. Nanomaterials, 2020. **10**(2): p. 345.

[81] Amiri, A., et al., *Porous nitrogen-doped MXene-based electrodes for capacitive deionization*. Energy Storage Materials, 2020. **25**: p. 731–739.

[82] Song, Y., et al., *Rational design of porous nitrogen-doped Ti$_3$C$_2$ MXene as a multifunctional electrocatalyst for Li–S chemistry*. Nano Energy, 2020. **70**: p. 104555.

[83] Wang, J., et al., *Rational design of porous N-Ti$_3$C$_2$ MXene@CNT microspheres for high cycling stability in Li–S battery*. Nano-Micro Letters, 2020. **12**: p. 1–14.

[84] Feng, Y., et al., *Solvothermal synthesis of in situ nitrogen-doped Ti_3C_2 MXene fluorescent quantum dots for selective Cu^{2+} detection.* Ceramics International, 2020. **46**(6): p. 8320–8327.

[85] Tian, Y., et al., *Surface nitrogen-modified 2D titanium carbide (MXene) with high energy density for aqueous supercapacitor applications.* Journal of Materials Chemistry A, 2019. **7**(10): p. 5416–5425.

[86] Chen, X., et al., *Tunable nitrogen-doped delaminated 2D MXene obtained by NH_3/Ar plasma treatment as highly efficient hydrogen and oxygen evolution reaction electrocatalyst.* Chemical Engineering Journal, 2021. **420**: p. 129832.

[87] Zhang, S., et al., *Vapor deposition red phosphorus to prepare nitrogen-doped $Ti_3C_2T_x$ MXenes composites for lithium-ion batteries.* The Journal of Physical Chemistry Letters, 2019. **10**(21): p. 6446–6454.

[88] Yu, L., et al., *Versatile N-doped MXene ink for printed electrochemical energy storage application.* Advanced Energy Materials, 2019. **9**(34): p. 1901839.

[89] Prakash, N.J. and B. Kandasubramanian, *Nanocomposites of MXene for industrial applications.* Journal of Alloys and Compounds, 2021. **862**: p. 158547.

[90] Ang, H., et al., *Hydrophilic nitrogen and sulfur Co-doped molybdenum carbide nanosheets for electrochemical hydrogen evolution.* Small, 2015. **11**(47): p. 6278–6284.

[91] Chen, C., et al., *MoS_2-on-MXene heterostructures as highly reversible anode materials for lithium-ion batteries.* Angewandte Chemie International Edition, 2018. **57**(7): p. 1846–1850.

[92] Wang, D., et al., *N, P (S) Co-doped Mo_2C/C hybrid electrocatalysts for improved hydrogen generation.* Carbon, 2018. **139**: p. 845–852.

[93] Wang, X., et al., *A robust sulfur host with dual lithium polysulfide immobilization mechanism for long cycle life and high capacity Li-S batteries.* Energy Storage Materials, 2019. **16**: p. 344–353.

[94] Tang, H., et al., *A robust, freestanding MXene-sulfur conductive paper for long-lifetime Li-S batteries.* Advanced Functional Materials, 2019. **29**(30): p. 1901907.

[95] Bao, W., et al., *Boosting performance of Na—S batteries using sulfur-doped $Ti_3C_2T_x$ MXene nanosheets with a strong affinity to sodium polysulfides.* ACS Nano, 2019. **13**(10): p. 11500–11509.

[96] Du, C., et al., *Embedding S@ TiO_2 nanospheres into MXene layers as high rate cyclability cathodes for lithium-sulfur batteries.* Electrochimica Acta, 2019. **295**: p. 1067–1074.

[97] Luo, J., et al., *Atomic sulfur covalently engineered interlayers of Ti_3C_2 MXene for ultrafast sodium-ion storage by enhanced pseudocapacitance.* Advanced Functional Materials, 2019. **29**(10): p. 1808107.

[98] Zhao, Q., et al., *Flexible 3D porous MXene foam for high-performance lithium-ion batteries.* Small, 2019. **15**(51): p. 1904293.

[99] Ramalingam, V., et al., *Heteroatom-mediated interactions between ruthenium single atoms and an MXene support for efficient hydrogen evolution.* Advanced Materials, 2019. **31**(48): p. 1903841.

[100] Sun, S., et al., *Hybrid energy storage mechanisms for sulfur-decorated Ti_3C_2 MXene anode material for high-rate and long-life sodium-ion batteries.* Chemical Engineering Journal, 2019. **366**: p. 460–467.

[101] Li, J., et al., *Improved sodium-ion storage performance of $Ti_3C_2T_x$ MXenes by sulfur doping.* Journal of Materials Chemistry A, 2018. **6**(3): p. 1234–1243.

[102] Yuan, W., et al., *Laminated hybrid junction of sulfur-doped TiO_2 and a carbon substrate derived from Ti_3C_2 MXenes: toward highly visible light-driven photocatalytic hydrogen evolution.* Advanced Science, 2018. **5**(6): p. 1700870.

[103] Yang, C., et al., *Nitrogen and sulfur Co-doped 2D titanium carbides for enhanced electrochemical performance.* Journal of the Electrochemical Society, 2017. **164**(9): p. A1939.

[104] Kaewpijit, P., J. Qin, and P. Pattananuwat. *Preparation of MXene/N, S-doped graphene electrode for supercapacitor application.* IOP Conference Series: Materials Science and Engineering, 2019. (IOP Publishing).

[105] Shuvo, S.N., et al., *Sulfur-doped titanium carbide MXenes for room-temperature gas sensing.* ACS Sensors, 2020. **5**(9): p. 2915–2924.

[106] Xiao, Z., et al., *Ultrafine Ti_3C_2 MXene nanodots-interspersed nanosheet for high-energy-density lithium–sulfur batteries.* ACS Nano, 2019. **13**(3): p. 3608–3617.

[107] Guo, X., et al., *Constructing P-doped self-assembled V 2 C MXene/NiCo-layered double hydroxide hybrids toward advanced lithium storage.* Materials Advances, 2023. **4**(6): p. 1523–1533.

[108] Chi, J.-Q., et al., *Nitrogen, phosphorus dual-doped molybdenum-carbide/molybdenum-phosphide-@-carbon nanospheres for efficient hydrogen evolution over the whole pH range.* Journal of Colloid and Interface Science, 2018. **513**: p. 151–160.

[109] Shi, Z., et al., *Phosphorus-Mo$_2$C@carbon nanowires toward efficient electrochemical hydrogen evolution: composition, structural and electronic regulation.* Energy & Environmental Science, 2017. **10**(5): p. 1262–1271.

[110] Qu, G., et al., *Phosphorized MXene-phase molybdenum carbide as an earth-abundant hydrogen evolution electrocatalyst.* ACS Applied Energy Materials, 2018. **1**(12): p. 7206–7212.

[111] Tang, Y., et al., *Phosphorus-doped molybdenum carbide/MXene hybrid architectures for upgraded hydrogen evolution reaction performance over a wide pH range.* Chemical Engineering Journal, 2021. **423**: p. 130183.

[112] Tang, Y., et al., *Synergistically coupling phosphorus-doped molybdenum carbide with MXene as a highly efficient and stable electrocatalyst for hydrogen evolution reaction.* ACS Sustainable Chemistry & Engineering, 2020. **8**(34): p. 12990–12998.

[113] Yoon, Y., et al., *Precious-metal-free electrocatalysts for activation of hydrogen evolution with nonmetallic electron donor: chemical composition controllable phosphorous doped vanadium carbide MXene.* Advanced Functional Materials, 2019. **29**(30): p. 1903443.

[114] Meng, R., et al., *Black phosphorus quantum dot/Ti_3C_2 MXene nanosheet composites for efficient electrochemical lithium/sodium-ion storage.* Advanced Energy Materials, 2018. **8**(26): p. 1801514.

[115] Ding, Y., et al., *Facile synthesis of FePS$_3$ nanosheets@MXene composite as a high-performance anode material for sodium storage.* Nano-Micro Letters, 2020. **12**: p. 1–12.

[116] Wu, W., et al., *Hierarchical materials constructed by 1D hollow nickel–cobalt sulfide nanotubes supported on 2D ultrathin MXenes nanosheets for high-performance supercapacitor.* Ceramics International, 2020. **46**(8): p. 12200–12208.

[117] Zhou, J., et al., *Ti-enhanced exfoliation of V$_2$AlC into V$_2$C MXene for lithium-ion battery anodes.* Ceramics International, 2017. **43**(14): p. 11450–11454.

[118] Wang, Y., et al., *Preparation of $(V_x, Ti_{1-x})2C$ MXenes and their performance as anode materials for LIBs.* Journal of Materials Science, 2019. **54**: p. 11991–11999.

[119] Li, Z., et al., *Reactive metal–support interactions at moderate temperature in two-dimensional niobium-carbide-supported platinum catalysts.* Nature Catalysis, 2018. **1**(5): p. 349–355.

[120] Kuznetsov, D.A., et al., *Single site cobalt substitution in 2D molybdenum carbide (MXene) enhances catalytic activity in the hydrogen evolution reaction.* Journal of the American Chemical Society, 2019. **141**(44): p. 17809–17816.

[121] Li, J., et al., *Achieving high pseudocapacitance of 2D titanium carbide (MXene) by cation intercalation and surface modification.* Advanced Energy Materials, 2017. **7**(15): p. 1602725.

[122] Zhu, X., et al., *Alkaline intercalation of Ti_3C_2 MXene for simultaneous electrochemical detection of Cd (II), Pb (II), Cu (II) and Hg (II).* Electrochimica Acta, 2017. **248**: p. 46–57.

[123] Zeng, Z., et al., *Bifunctional N-CoSe$_2$/3D-MXene as highly efficient and durable cathode for rechargeable Zn—air battery.* ACS Materials Letters, 2019. **1**(4): p. 432–439.

[124] Iqbal, M., et al., *Co-existence of novel ferromagnetic/anti-ferromagnetic phases in two-dimensional Ti$_3$C$_2$ MXene.* Arxiv Preprint Arxiv:1907.12588, 2019.

[125] Dall'Agnese, Y., et al., *High capacitance of surface-modified 2D titanium carbide in acidic electrolyte.* Electrochemistry Communications, 2014. **48**: p. 118–122.

[126] Gao, Z.W., W. Zheng, and L.Y.S. Lee, *Highly enhanced pseudocapacitive performance of vanadium-doped MXenes in neutral electrolytes.* Small, 2019. **15**(40): p. 1902649.

[127] Wang, C., et al., *Atomic Sn^{4+} decorated into vanadium carbide MXene interlayers for superior lithium storage.* Advanced Energy Materials, 2019. **9**(4): p. 1802977.

[128] Luo, J., et al., *Sn^{4+} ion decorated highly conductive Ti$_3$C$_2$ MXene: promising lithium-ion anodes with enhanced volumetric capacity and cyclic performance.* ACS Nano, 2016. **10**(2): p. 2491–2499.

[129] Kim, H., et al., *Thermoelectric properties of two-dimensional molybdenum-based MXenes.* Chemistry of Materials, 2017. **29**(15): p. 6472–6479.

[130] Lorencova, L., et al., *Highly stable Ti$_3$C$_2$T$_x$ (MXene)/Pt nanoparticles-modified glassy carbon electrode for H$_2$O$_2$ and small molecules sensing applications.* Sensors and Actuators B: Chemical, 2018. **263**: p. 360–368.

[131] Rafiq, S., et al., *Novel room-temperature ferromagnetism in Gd-doped 2-dimensional Ti$_3$C$_2$T$_x$ MXene semiconductor for spintronics.* Journal of Magnetism and Magnetic Materials, 2020. **497**: p. 165954.

[132] Luo, J., et al., *Pillared structure design of MXene with ultralarge interlayer spacing for high-performance lithium-ion capacitors.* ACS Nano, 2017. **11**(3): p. 2459–2469.

[133] Xie, X., et al., *Surface Al leached Ti$_3$AlC$_2$ as a substitute for carbon for use as a catalyst support in a harsh corrosive electrochemical system.* Nanoscale, 2014. **6**(19): p. 11035–11040.

[134] Zhang, X., et al., *Surface modified MXene film as flexible electrode with ultrahigh volumetric capacitance.* Electrochimica Acta, 2019. **294**: p. 233–239.

[135] Liu, P., et al., *Surface termination modification on high-conductivity MXene film for energy conversion.* Journal of Alloys and Compounds, 2020. **829**: p. 154634.

[136] Du, C.F., et al., *Synergy of Nb doping and surface alloy enhanced on water–alkali electrocatalytic hydrogen generation performance in Ti-based MXene.* Advanced Science, 2019. **6**(11): p. 1900116.

[137] Filip, J., et al., *Tailoring electrocatalytic properties of Pt nanoparticles grown on Ti$_3$C$_2$T$_x$ MXene surface.* Journal of the Electrochemical Society, 2019. **166**(2): p. H54.

[138] Kang, Z., et al., *Recent progress of MXenes and MXene-based nanomaterials for the electrocatalytic hydrogen evolution reaction.* Journal of Materials Chemistry A, 2021. **9**(10): p. 6089–6108.

[139] Fan, Y., et al., *Recent advances in growth of transition metal carbides and nitrides (MXenes) crystals.* Advanced Functional Materials, 2022. **32**(16): p. 2111357.

[140] John, R.A.B., et al., *Gas-sensing mechanisms and performances of MXenes and MXene-based heterostructures.* Sensors, 2023. **23**(21): p. 8674.

[141] Amrillah, T., et al., *MXene-based photocatalysts and electrocatalysts for CO$_2$ conversion to chemicals.* Transactions of Tianjin University, 2022. **28**(4): p. 307–322.

[142] Wang, X., et al., *Interfacial engineering of N, S-doped Mo$_2$C-Mo/C heterogeneous nanorods for enhanced alkaline hydrogen evolution.* Applied Surface Science, 2023. **614**: p. 156276.

[143] Hart, J.L., et al., *Termination-property coupling via reversible oxygen functionalization of MXenes.* ACS Nanoscience Au, 2022. **2**(5): p. 433–439.

[144] Jiang, W., et al., *Molten salt thermal treatment synthesis of S-doped V$_2$CT$_x$ and its performance as a cathode in aqueous Zn-ion Batteries.* ACS Applied Materials & Interfaces, 2022. **14**(12): p. 14482–14491.

[145] Chen, Y., et al., *Rational design of M–N$_4$–Gr/V$_2$ C heterostructures as highly active ORR catalysts: a density functional theory study.* RSC Advances, 2022. **12**(23): p. 14368–14376.

[146] Wang, Y., et al., *2D Nb$_2$CT$_x$ MXene/MoS$_2$ heterostructure construction for nonlinear optical absorption modulation.* Opto-Electronic Advances, 2023: p. 220162-1-220162-11.

[147] Wu, F., et al., *S-doped multilayer niobium carbide (Nb$_4$C$_3$Tx) electrocatalyst for efficient hydrogen evolution in alkaline solutions.* International Journal of Hydrogen Energy, 2022. **47**(39): p. 17233–17240.

[148] Liao, S.-Y., et al., *Wrinkled and flexible N-doped MXene additive for improving the mechanical and electrochemical properties of the nickel-rich LiNi$_{0.8}$Co$_{0.1}$Mn$_{0.1}$O$_2$ cathode.* Electrochimica Acta, 2022. **410**: p. 139989.

[149] Alli, U., et al., *In-situ continuous hydrothermal synthesis of TiO$_2$ nanoparticles on conductive N-doped MXene nanosheets for binder-free Li-ion battery anodes.* Chemical Engineering Journal, 2022. **430**: p. 132976.

[150] Yu, W., et al., *Immobilization and kinetic acceleration of lithium polysulfides by iodine-doped MXene nanosheets in lithium—sulfur batteries.* The Journal of Physical Chemistry C, 2022. **126**(27): p. 10986–10994.

[151] Tekalgne, M.A., et al., *Efficient electrocatalysts for hydrogen evolution reaction using heteroatom-doped MXene nanosheet.* International Journal of Energy Research, 2023. **2023**.

[152] Yang, F., et al., *Synthesis of nitrogen-sulfur co-doped Ti$_3$C$_2$T$_x$ MXene with enhanced electrochemical properties.* Materials Reports: Energy, 2022. **2**(1): p. 100079.

[153] Wen, Y., et al., *A temperature-dependent phosphorus doping on Ti$_3$C$_2$T$_x$ MXene for enhanced supercapacitance.* Journal of Colloid and Interface Science, 2021. **604**: p. 239–247.

[154] Cai, L., et al., *Etching engineering and electrostatic self-assembly of N-doped MXene/ hollow Co-ZIF hybrids for high-performance microwave absorbers.* Chemical Engineering Journal, 2022. **434**: p. 133865.

[155] Sarfraz, B., et al., *HF free greener Cl-terminated MXene as novel electrocatalyst for overall water splitting in alkaline media.* International Journal of Energy Research, 2022. **46**(8): p. 10942–10954.

[156] Tian, Y., et al., *In situ oxygen doped Ti$_3$C$_2$T$_x$ MXene flexible film as supercapacitor electrode.* Chemical Engineering Journal, 2022. **446**: p. 137451.

[157] Gong, S., et al., *Iodine-functionalized titanium carbide MXene with ultra-stable pseudocapacitor performance.* Journal of Colloid and Interface Science, 2022. **615**: p. 643–649.

[158] Gupta, N., et al., *Microwave-assisted rapid synthesis of titanium phosphate free phosphorus doped Ti$_3$C$_2$ MXene with boosted pseudocapacitance.* Journal of Materials Chemistry A, 2022. **10**(29): p. 15794–15810.

[159] Jiang, D., et al., *One-pot synthesis of ZnO quantum dots/N-doped Ti$_3$C$_2$ MXene: tunable nitrogen-doping properties and efficient electrochemiluminescence sensing.* Chemical Engineering Journal, 2022. **430**: p. 132771.

[160] Zhu, L., et al., *Pseudocapacitance of Cl-terminated MXene nanosheets for efficient chloride-ion hybrid capacitors.* Energy & Fuels, 2023. **37**(7): p. 5607–5612.

[161] An, Y., et al., *Rational design of sulfur-doped three-dimensional Ti$_3$C$_2$T$_x$ MXene/ZnS heterostructure as multifunctional protective layer for dendrite-free zinc-ion batteries.* ACS Nano, 2021. **15**(9): p. 15259–15273.

6 Synthesis Mechanisms

The properties and structures of MXenes, which are used for various applications, depend on the synthetic techniques used for their fabrication [1]. After successfully synthesizing the first multilayer $Ti_3C_2T_x$ using hydrofluoric (HF) acid, various alternative methods have been discovered and applied to obtain new compositions. Generally, there are two primary approaches that is, top-down and bottom-up, that are being used by researchers to synthesize 2D MXenes. In top-down techniques, bulk materials are exfoliated by selective MAX phase etching to produce monolayer or multilayer films. On the other hand, bottom-up approaches start with atoms and build up to a layered structure. It has been observed that the properties of MXenes are influenced by the precursor materials and surface modification techniques. Therefore, it is crucial to closely examine and compare various techniques in order to identify their benefits and limitations and to outline the essential steps involved in the synthesis process [2, 3]. There are two types of top-down exfoliation: mechanical and chemical. Initially, graphene was separated using adhesive tapes, which is a standard form of mechanical exfoliation. However, the M-A bonds in MXene structures are metallic or covalent in nature, which makes the mechanical exfoliation of these materials very difficult. Therefore, MXenes are chemically exfoliated from their corresponding MAX phase precursors, by selective etching of strong interlayer bonds [3].

On the other hand, bottom-up approaches use different methods to produce high-quality MXene films, which can be utilized in various applications. Due to hydrophilicity and stability of MXenes in aqueous solutions, different solution-processing techniques have also been developed [1]. Initially, wet chemical methods were used for developing multilayered nanoflakes of various MXenes, such as V_2CT_x, Nb_2CT_x, Ti_3CNT_x, and Ti_2CT_x. In 2013, researchers utilized an organic solvent [dimethyl sulfoxide (DMSO)] to delaminate multilayered Ti_3C_2, TiNbC, and Ti_3CN into single flakes, which displayed photothermal properties with high electronic conductivity. Tetrabutylammonium hydroxide (TBAOH), tetrapropylammonium hydroxide (TPAOH), and tetramethylammonium hydroxide (TMAOH) have also been employed successfully to delaminate multilayered MXenes. As synthetic techniques have evolved, ammonium bifluoride (NH_4HF_2) salt has also been employed for etching. Additionally, an in situ LiF/HCl method was developed to avoid the direct use of hazardous and corrosive HF. Furthermore, minimally intensive layer delamination (MILD) was also developed for the delamination of MXenes after acid treatment using only shaking (without sonication) to produce large-sized yet high-quality MXene flakes [1]. The year-by-year development in the synthesis of MXenes is presented in Fig. 6.1 [1].

DOI: 10.1201/9781003465768-6

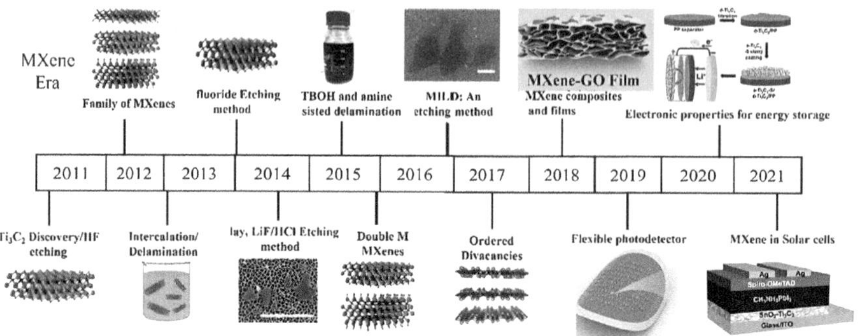

FIGURE 6.1 Development in the synthesis of MXenes.

Source: Reprinted from [1] Israt Ali, Muhammad Faraz Ud Din and Zhi-Gang Gu.: MXenes Thin Films: From Fabrication to Their Applications. *Molecules* 2022, 27, 4925. Open access.

6.1 TOP-DOWN STRATEGIES

MXenes are commonly produced through top-down synthesis, which involves the selective etching of A-layer from bulk MAX precursors, due to weaker M-A bonds, using various etching reagents without affecting the M-X bonds. In addition to the MAX phases, non-MAX precursors have also been exfoliated for the synthesis of different MXene compositions with specific properties. The composition of MXenes from their corresponding MAX phases after etching of A-layer is shown in Fig. 6.2 [4]. There are several phases involved in the top-down synthesis methods of MXenes. First, the precursor material is dipped in an etching reagent for a certain period of time, which selectively removes A element layers. After that, the treated material is converted into discrete nanolayers by sonication, which produces 2D MXenes with ultra-thin structure. For sonication, probe/tip or bath ultrasonic devices can be used. Probe sonicators are suitable for producing smaller-sized flakes, while bath sonication is more effective in producing larger flakes of MXenes [2].

Power, frequency, and amplitudes are the most important sonication parameters. The amplitude of the irradiation has a substantial impact on the intensity of the ultra-sonication. However, high amplitude vibrations negatively influence the stability of MXenes due to an increase in collisions between precursor particles and delaminated MXenes. As a result, an appropriate amplitude value must be chosen, and particles must be diluted to decrease their concentration. Additionally, the power must be sufficient to cause bubble formation. As compared to the effect of amplitudes, higher frequencies positively influence the stability of MXenes by increasing the quantity of uniform-sized bubbles. This facilitates the prevention of collisions between particles and weakens the van der Waals forces between layers. Generally, these sonication frequencies range from 6 kHz to 40 kHz; however, high-quality flakes have been observed above 20 kHz, with 40 kHz being the most typically used [2].

The etching procedure also greatly influences the properties and potential uses of the resultant materials. For example, even trace amounts of HF in the synthesized

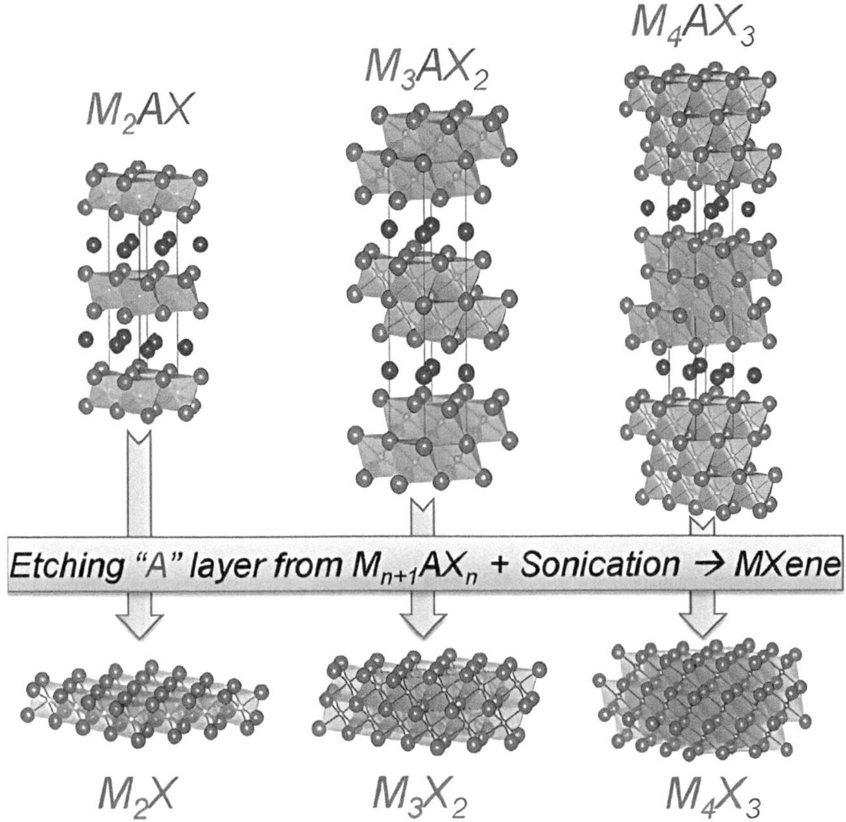

FIGURE 6.2 Composition of MXenes from corresponding MAX phases after etching of A-layer.

Source: [4] Naguib M., Mochalin V. N., et al.: 25th anniversary article: MXenes: a new family of two-dimensional materials. *Adv. Mater*. 2014, 26, 992–1005. Copyright WILEY-VCH Verlag GmbH & Co. KGaA, Weinheim. Reproduced with permission.

material would be hazardous to biological beings and can cause toxicity in live cells. As a result, while HF is not commonly utilized to create MXenes for medicinal purposes, it can be employed in other disciplines such as catalysis, light detectors, and energy storage devices. However, due to certain limitations, a variety of new etching agents, that is, alkaline solutions, $ZnCl_2$, and molten and fluoride salts, have been developed to replace HF [2]. Among all the fluoride-based etchants, NH_4HF_2 salt has been observed to produce MXenes with most stable characteristics. It is important to note that F-based etchants result into functionalized MXenes with −O, −OH, or −F groups, whereas fluoride-free etching techniques result in MXenes decorated with non-F groups [5, 6]. MXenes synthesized by all of these etching methods are summarized in the next sections.

6.1.1 Acid Etching

The earliest known etching agent for synthesizing MXenes is HF, an unsafe chemical requiring specific safety precautions. Exfoliation of A-layers from the MAX precursor replaces these sites with surface groups with hydrophilicity [5, 7]. A schematic representation of the formation of MXenes by HF etching is shown in Fig. 6.3 [4]. A few basic processes are required to prepare MXenes using concentrated HF solution. First, the precursor is gradually mixed with an HF solution. The mixing should be done gradually to avoid the formation of bubbles due to the exothermic reaction between Al and HF. This results in the formation of a multilayered structure with hydrogen bonds and van der Waals forces between the layers. The unreacted precursor, residual HF, and reaction byproducts can be removed by washing several times with deionized water [2].

The concentration of HF varies (5–50 wt.%) in an aqueous solution, as does the MAX phase immersion period, and alternative precursor materials can be used [2]. More MAX phases are being employed to create MXenes as HF etching procedures improve. MAX phases include Ti_2AlC, Nb_2AlC, V_2AlC, TiNbAlC, $(Ti_{0.5}V_{0.5})_2AlC$, and others. Ti_3AlC_2, Ti_3AlCN, Ti_3SiC_2, $(V_{0.5}Cr_{0.5})_3AlC_2$, $(Ti_{0.5}V_{0.5})_3AlC_2$, Mo_2TiAlC_2, and other 312 series of MAX phases are possible. The 413 series of MAX phases include V_4AlC_3, Nb_4AlC_3, Ta_4AlC_3, $Mo_2Ti_2AlC_3$, $(Nb_{0.8}Ti_{0.2})_4AlC_3$, and $(V_{0.5}Nb_{0.5})_4AlC_3$. Among these structures, researchers used HF to etch the Si layer when they etched Ti_3SiC_2. To generate Ti_3C_2, HF can be used with various oxidants, that is, $KMnO_4$, HNO_3, $FeCl_3$, and $(NH_4)_2S_2O_8$, to etch the Si layer. In addition to MAX phases,

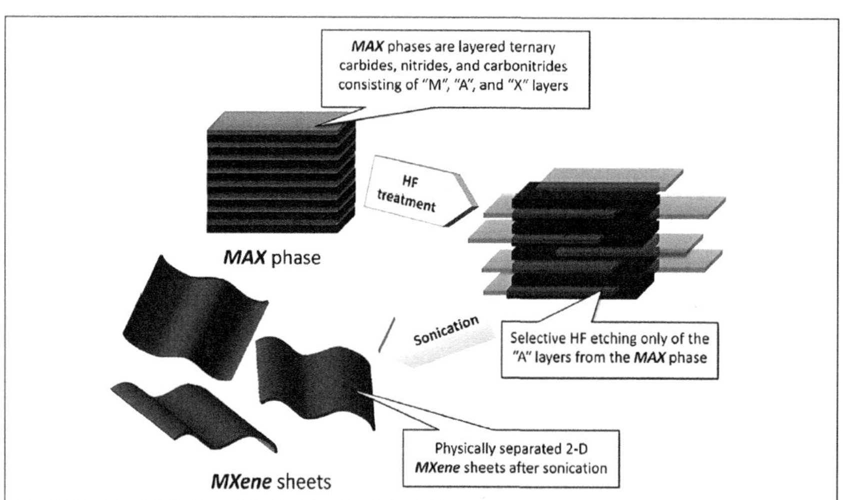

FIGURE 6.3 Schematic representation of formation of MXenes.

Source: [4] Naguib M., Mochalin V. N., et al.: 25th anniversary article: MXenes: a new family of two-dimensional materials. *Adv. Mater.* 2014, 26, 992–1005. Copyright WILEY-VCH Verlag GmbH & Co. KGaA, Weinheim. Reproduced with permission. The source of material from Michael Naguib, Olha Mashtalir, Joshua Carle, et al. 2012. Two-Dimensional Transition Metal Carbides. *ACS Nano*, 6, 2, 1322–1331. Copyright 2012, American Chemical Society is also acknowledged.

non-MAX phases such as Mo_2Ga_2C, $Zr_3Al_3C_5$, and $Hf_3[Al\,(Si)]_4C_6$ can also be developed [8]. The etching conditions vary corresponding to different MAX phases. It should be noted that when concentrations of HF etchant vary, the morphology of synthesized materials also varies, between accordion-like or bulk-like morphologies. Therefore, rather than just SEM images, other characterization techniques, that is, EDS and XRD, should also be used to determine the degree of removal of Al [5, 7].

To obtain high-quality MXenes, the etching process must be carefully controlled. The size of the material being etched, etching solution concentration, etching duration, and the temperature during etching all play a crucial role in the quality of the resulting MXenes. MXenes developed by HF etching have transparent and evenly spaced layers. However, poor quality MXenes are often obtained when the etching parameters are not properly established. For instance, if the HF concentration is too low, the etching duration is too short, or the temperature is too low, the A-layer may not be completely removed, resulting in incomplete MXene formation. When the concentration of HF is high, or the etching duration is too long, or the temperature is too high, the material structure being etched can be damaged, leading to structural defects. Moreover, HF is a highly corrosive acid solution, and its extensive use in etching can not only harm human beings but also pollute the environment [8].

6.1.2 HF Etching

In 2013, O. Mashtalir et al. [9] published an extensive study to explain the factors (i.e., etching duration, temperature, and particle size) that affect the kinetics of Al etching in a 50% HF solution by using Ti_3AlC_2 as a MAX precursor. Different characterization techniques were used to analyze the structure and morphology of the exfoliated material. Furthermore, the surface group (−OH) or adsorbed water was removed by heating it in a vacuum. F. Chang et al. [10] used pressure-less argon shielding synthesis procedures to synthesize Ti_3AlC_2 powders. The original powders of Ti, Al, and graphite were ball-milled for 5 hours at 300 rpm at a molar ratio of 3:1.1:1.9. The mixture was then sintered for 2 hours at 1350 °C in a flowing Ar environment. Magnetic stirring (at room temperature) of 2.5 gas-prepared Ti_3AlC_2 particles and 40% HF solution developed graphene-like products. The resultant suspension was then filtered and rinsed with distilled water and 100% ethanol multiple times. In another study, the effect of synthesis conditions on the electrochemical properties of same material was studied [11]. It was observed that $Ti_3C_2T_x$ etched in 6 M HF exhibits the value of capacitance that is two times greater than the that of the one etched in 15 M HF solution.

M. Naguib et al. [12] immersed the powders of Ti_2AlC, Ti_3AlCN, Ta_4AlC_3, $(V_{0.5},Cr_{0.5})_3AlC_2$, and $(Ti_{0.5},Nb_{0.5})_2AlC$ in different concentrations of HF solution (for 10–72 hours). The resistance of the MXene sheets was found to be equivalent to that of multilayer graphene. In another investigation, it was observed that in addition to the etching of Al layer, a small quantity of MXene can also be dissolved in HF solutions. For this purpose, Ti_3AlCN powders were etched in 30% HF for 18 hours followed by washing with deionized water, vacuum filtration, and drying at room temperature. As a result, weight loss during etching was observed to be 20% [13]. Similarly, M. Iqbal et al. [14] effectively developed Ti_3C_2 by selective etching of

Ti_3AlC_2 using 39% HF for 66 hours. The resultant suspension was rinsed with deionized water (to reach a pH of 6). The suspension result was filtered and washed with 100% ethanol using a sterile membrane. In addition to Al layers, Ga layers have also been observed to be removed by HF etching. In brief, Mo_2CT_x was created by removing the Ga atoms from its parent ternary carbide, Mo_2Ga_2C, using HF etching [15]. Another work showed the stability and catalytic activity of Mo_2CT_x in the water gas shift reaction with high selectivity (>99%) toward H_2 and CO_2, synthesized by etching away Ga from generated Mo_2Ga_2C using a modified literature approach [16].

Furthermore, Mo-based double transition metal (DTM) MXenes have also been observed to be synthesized by this etching route. For example, $Mo_2TiC_2T_x$, synthesized by HF etching, demonstrated efficient hydrogen evolution reaction (HER) catalytic behavior in both neutral or acidic solutions [17]. Similarly, Mo_2TiAlC_2 or $Mo_2Ti_2AlC_3$ powders were also etched at different times (48 and 90 hours, respectively) by using 50% HF at 55°C to obtain their corresponding multilayered MXenes. The solutions were stirred with a Teflon-coated magnetic bar [18]. The desalination performance of $Mo_{1.33}C$ MXene with divacancy, synthesized via HF etching of $(Mo_{2/3}Sc_{1/3})_2AlC$, in seawater concentrations was presented [19]. A. S. Etman et al. [20] synthesized $Mo_{1.33}CT_z$ and described a one-step method for making composite films based on $Mo_{1.33}CT_z$ and $Ti_3C_2T_z$ MXenes. In 1 M H_2SO_4, these films demonstrated high electronic conductivity, flexibility, and high capacitance value. After 17,000 cycles, the capacitance retention was 96%. J. Halim et al. [21] demonstrated that purity of the sample can be increased (by 70%) by using $(Mo_{2/3}Y_{1/3})_2AlC$ i-MAX. Furthermore, using a modified etching method, they were able to create a $Mo_{1.33}CT_z$ MXene with outstanding structural quality, while increasing the yield by a factor of 6 over prior efforts. The crystal structure of $(Mo_{2/3}Y_{1/3})_2AlC$ i-MAX is shown in Fig. 6.4 [21].

M. Naguib et al. [22] reported the HF etching and Li-absorption capability of Nb_2CT_x and V_2CT_x. To develop Nb_2CT_x material, its corresponding MAX phase was treated in 50% HF for 90 hours. The photocatalytic HER activity of a hybrid $Nb_2O_5/C/Nb_2C$ composite was also presented [23]. The $Nb_2O_5/C/Nb_2C$ composites were created by oxidizing Nb_2CT_x with CO_2 in a single process. After mild oxidation, Nb_2O_5 developed homogeneously on Nb_2C, along with some amorphous

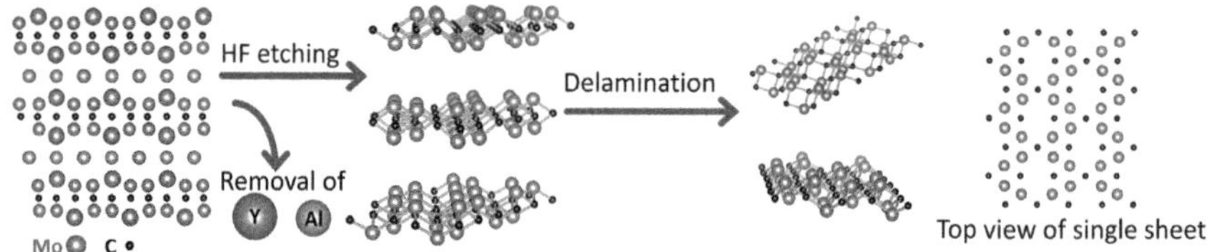

FIGURE 6.4 Crystal structure of $(Mo_{3/2}Y_{1/3})_2AlC$. Etching and removal of Y and Al atoms result in multilayered $Mo_{1.33}CT_z$, which is subsequently delaminated into single flakes.

Source: [21] Halim J., Etman A. S., Elsukova A., et al. 2021. Tailored synthesis approach of $(Mo_{2/3}Y_{1/3})_2AlC$ i-MAX and its two-dimensional derivative $Mo_{1.33}CT_z$ MXene: Enhancing the yield, quality, and performance in supercapacitor applications. *Nanoscale*, 13, 311–319. Reproduced with permission from the Royal Society of Chemistry.

carbon. Similarly, H. Xiang et al. [24] demonstrated a cancer-therapeutic technique using M_2C MXene, which induces cancer cell death using free radicals generated by thermo-labile initiators, which are not dependent on oxygen. In other studies, Nb_2C MXene, synthesized via HF etching, was used as mode lockers and photoelectrochemical-type photodetectors [25], as well as in intestinal systems to protect and stimulate cell durability [26].

In further studies, S. Zhao et al. [27] described the incorporation of Li into a 2D $Nb_4C_3T_x$ MXene produced by HF-etching of Nb_4AlC_3. Furthermore, they discovered that the anode's charge/discharge capacity increases with cycling. Y. Liu et al. [28] also created a 2D $Nb_4C_3T_x$ by chemical exfoliation and ball-milling it into MgH_2. The hydrogen storage kinetics of the activated MgH_2–5 wt% $Nb_4C_3T_x$ were superior. X. Li et al. [29] used HF as the etching agent in a wet chemical etching technique to create V_2CT_x powder. By using V_2CT_x MXene as the cathode, they also showed a zinc hybrid-ion battery with an extraordinary capacity improvement even after 18,000 cycles. J. Zhou et al. [30] reported the synthesis of 2D vanadium carbide V_4C_3 in HF at 55 °C (Fig. 6.5 [30]).

It was also reported that HF-etched V_4C_3 MXene exhibits excellent electrochemical properties in (1M H_2SO_4), that is, high capacitance (209 F g^{-1} at 2 mV s^{-1}), high rate of performance, and 97.23% capacitance retention after 10,000 cycles [31]. J. Bai

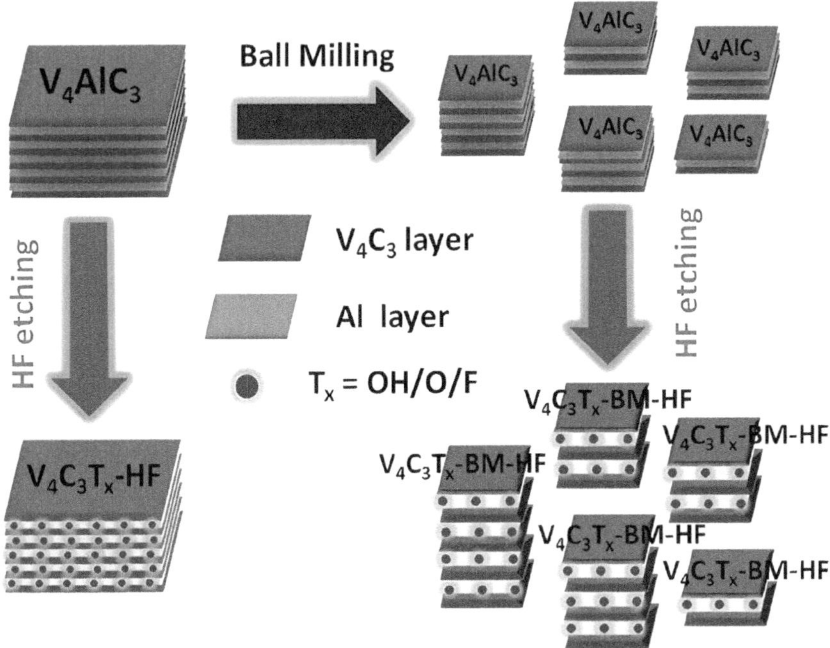

FIGURE 6.5 Schematic showing the two different processes of $V_4C_3T_x$ MXene.

Source: Reprinted from [30] *Chemical Engineering Journal*, 373, Jiafeng Zhou, Shuai Lin, Yanan Huang, Peng Tong, et al., Synthesis and lithium ion storage performance of two-dimensional V_4C_3 MXene, 203–212, Copyright 2019, with permission from Elsevier.

et al. [32] developed a simple approach for producing $V_4C_3/MoS_2/C$ nanohybrids. In this method, V_4C_3-MXene was intercalated by MoS_2 nanosheets (NSs) with thin C coating. As a result, this nanohybrid provided a wide range of properties, including large surface area, fast charge carrier transfer, quick reaction kinetics, high electronic conductivity, and structural stability. Furthermore, HF etching can also be used to synthesize MXenes from their corresponding non-MAX phase precursors. For example, J. Zhou et al. [33] reported for the first time the synthesis of Zr-containing carbide by selective exfoliation of Al-C units from layered ternary $Zr_3Al_3C_5$. Similarly in another study, the synthesis of conductive and flexible $Hf_3C_2T_z$ was presented by etching of a layered parent $Hf_3[Al(Si)]_4C_6$ molecule (Fig. 6.6 [34]) [34]. A Si substitutional solution on Al sites substantially decreased the interfacial adhesion between HfC and Al(Si)C sublayers within the parent compound's unit cell, allowing for further selective etching. First-principle density functional simulations were used to extensively investigate the underlying mechanism.

6.1.3 NON-HF ETCHING/FLUORIDE-FREE ETCHING

MXenes can be produced using a safe process that involves HF-containing etchants like NH_4HF_2 or HF-forming etchants like $LiF + HCl$. Additionally, cations such as Li^+ and NH^{4+} along with water molecules can be introduced into the multilayer structure of MXenes during synthesis. This results in MXenes with a higher c-lattice constant than HF-etched MXenes. The newly produced MXenes can be easily delaminated

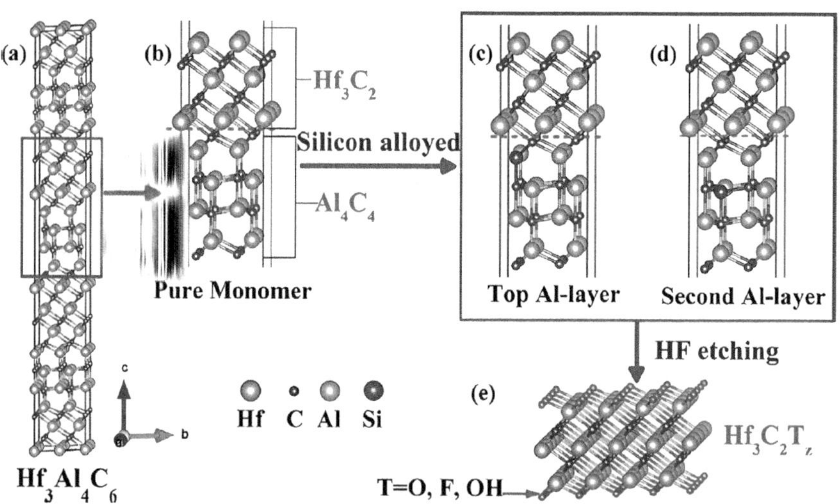

FIGURE 6.6 (a) Pure $Hf_3Al_4C_6$; (b) $Hf_3Al_4C_6$ monomer in its supercell. (c, d) Silicon alloying in the $Hf_3Al_4C_6$ monomer. Two different replacement positions: (c) in the top Al layer and (d) in the second Al layer. (e) Side view of $Hf_3C_2T_z$ MXene.

Source: Reprinted with permission from [34] Jie Zhou, Xianhu Zha, Xiaobing Zhou, et al. 2017. Synthesis and Electrochemical Properties of Two-Dimensional Hafnium Carbide. *ACS Nano*, 11, 3841–3850. Copyright 2017, American Chemical Society.

into few-layer MXene NSs through sonication, which simplifies the intercalation process. This technology is a significant advancement in the preparation of MXenes as it now requires only a single step [35]. These methods were observed to produce MXenes with higher conductivities and fewer structural defects. Furthermore, there is a wide range of fluoride salts that can be utilized, that is, NaF, LiF, KF, CsF, CaF_2, and FeF_3, with concentrations of HCl from 6 to 12 M. In addition, $NaHF_2$, KHF_2, and NH_4HF_2 are the recently discovered bifluoride salts that can be utilized as etchants. After removing the A-layer during etching, $-NH_3$ and $-NH_4$ can also be intercalated between the layers, which results in the formation of delaminated MXenes. The primary byproducts of the process were identified as $(NH_4)_3AlF_6$ and NH_4AlF_4. Furthermore, during high temperatures and pressures, NH_4F and NaOH solutions can also be produced [36, 37].

Etching MAX phases using LiF/HCl resulted in the production of MXenes such as Ti_2C, Ti_3CN, Cr_2TiC_2, and $(Nb_{0.8}Zr_{0.2})_4C_3$. The produced Ti_3C_2 showed no nanoscale defects, and this mixture was found to be milder as an etchant than pure HF. Additionally, non-MAX phases such as Mo_2C were also etched to obtain MXenes. As a result, MXenes with ideal characteristics, that is, superior quality, large lateral size, and high yield, may be obtained by employing the approach. Researchers have employed this process to synthesize different nitride structures such as V_2C, Ti_2N, and Ti_4N_3 [8]. Furthermore, the non-HF etching or fluoride-free etching techniques for MXenes include alkali etching, electrochemical etching, and molten salt etching. These etching methods are being used by the researchers on a large scale to avoid the hazards produced by fluoride-based etching techniques and to increase the characteristics of MXenes for corresponding applications.

6.1.4 ETCHING WITH FLUORIDE SALTS

Researchers have been exploring alternative methods for producing HF, as direct use of it has been proven to be hazardous to biological beings. One such method involves using a low-cost in situ combination of HCl and HF, which is now commonly used for producing MXenes. Fluoride salts, such as NaF, LiF, KF, CaF_2, and NH_4F, have been observed to be much milder than HF. This is evidenced by the creation of MXene flakes that have much larger transverse dimensions and lack the nanoscale flaws that are typically found when using HF [2, 38]. $Ti_3C_2T_x$ was synthesized by improving on the clay technique [39]. Researchers used the MILD approach to minimize the extensive processing as it was required for the delamination of $Ti_3C_2T_x$. In brief, the etchant for this method comprised a HCl/HF solution, which was further used to exfoliate Ti_3AlC_2 at 35 °C for 24 hours. The etching process was followed by washing the product with deionized water and centrifuging at 3,500 rpm, to obtain pH 6. The electrodes composed of 2D Ti_3C_2 clay, formed by this approach, have been observed to display high volumetric capacitance values [40, 41]. Furthermore, the electronic properties of $Ti_3C_2T_x$ flakes, synthesized via the MILD approach, were also studied for field-effect transistors (FETs) [42].

A. Miranda et al. [43] produced $Ti_3C_2T_x$ particles by mixing Ti_3AlC_2 powders into an ice-cold HCl/LiF solution. The $Ti_3C_2T_x$ flakes obtained after rinsing with LiCl, HCl, and deionized water and hoover filtering are comparable to those of other

MXenes. Likewise, a number of studies followed the same synthesis method and used as-synthesized MXenes for production of filtered seawater [44], for highly efficient gas separation [45]. F. Liu et al. [46] selectively etched Ti_3AlC_2 and Ti_2AlC in HCl with various fluoride salts, including KF, LiF, NaF, and NH_4F. Urea, DMSO, or ammonium hydroxide was used to delaminate the as-prepared Ti_2C. S. Kajiyama et al. [47] revealed the importance of chemical composition and surface termination groups in the $Ti_{n+1}C_nT_x$ MXene. Because of its excellent electronic conductivity and chemical durability against synthetic chemicals (HF or LiF-HCl), the molecule with n = 2, $Ti_3C_2T_x$, is the most investigated MXene. However, because of its lower molecular weight, Ti_2CT_x (n = 1) has a substantially higher theoretical gravimetric capacitance. In another study, researchers synthesized Ti_2CT_x MXene thin films and studied its properties, that is, sheet resistance, thickness, etc., which could make it a potential material for transparent conductive electrode application [48]. The schematic illustration of Ti_2CT_x MXene thin films and plasma treatment is given in Fig. 6.7 [48].

Ti_3CNT_x can be synthesized using Ti_3AlCN by etching it in HF. The resulting material can then be delaminated into 2D MXene layers by intercalating with hydrazine or TBAOH. Alternatively, Ti_3CNT_x can be synthesized using LiF/HCl solution, which in situ forms HF without any need for an intercalant. While Ti_3CNT_x has similar properties to $Ti_3C_2T_x$, it has substantially lower conductivity. Moreover, MXenes synthesized through the LiF/HCl method have greater electronic conductivity than those synthesized through HF etching [49]. In another study, multilayered Ti_3CNT_x, synthesized by using the LiF/HCl method, exhibited a morphology with low percentage of restacked NSs, high rate of performance, better cycling performance, and

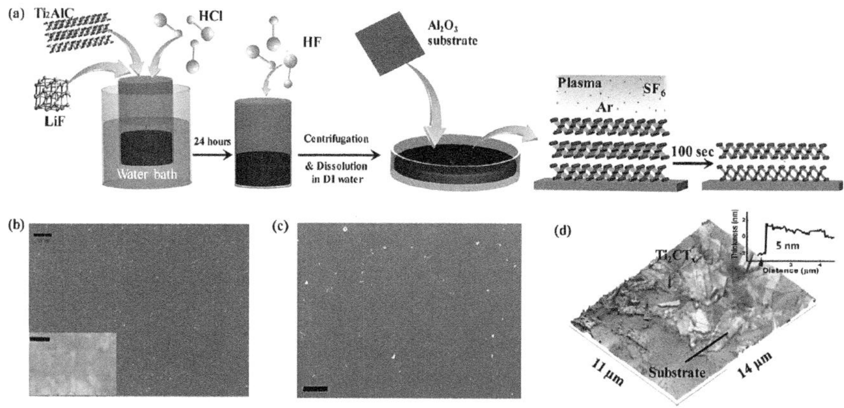

FIGURE 6.7 (a) Schematic illustration of the synthesis of Ti_2CT_x thin film and plasma treatment. (b) Optical microscopy (OM) image of Ti_2CT_x thin film on Al_2O_3. (c) SEM image of Ti_2CT_x thin film. (d) AFM image of Ti_2CT_x thin film.

Source: Reprinted with permission from [48] Yajie Yang, Sima Umrao, Shen Lai, et al. 2017. Large-Area Highly Conductive Transparent Two-Dimensional Ti_2CT_x Film. *J. Phys. Chem. Lett.*, 8, 859–865. Copyright 2017, American Chemical Society.

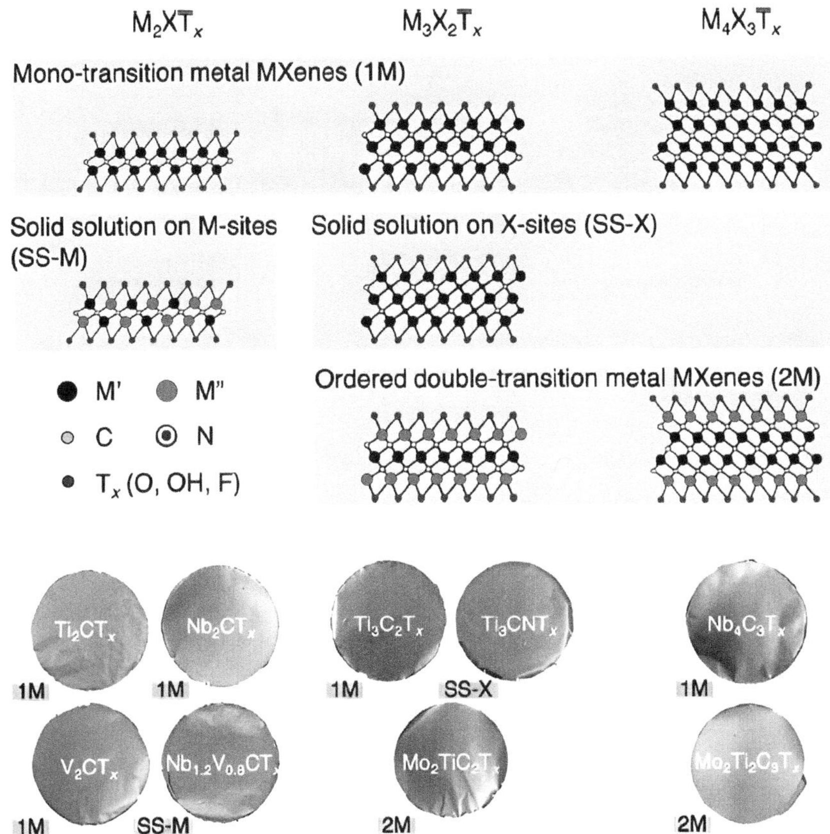

FIGURE 6.8 Structures and compositions of as-synthesized MXenes.

Source: Reprinted with permission from [51] Meikang Han, Christopher Eugene Shuck, Roman Rakhmanov, et al. 2020. Beyond $Ti_3C_2T_x$: MXenes for Electromagnetic Interference Shielding. *ACS Nano*, 14, 5008–5016. Copyright 2020, American Chemical Society.

excellent charge storage capacities [50]. Furthermore, a thorough analysis of electromagnetic interference (EMI) shielding properties of 16 distinct MXenes (Fig. 6.8 [51]) was presented by M. Han et al. [51]. They used different synthesis methods to produce films with thicknesses ranging from nanometers to micrometers. In micrometer-thick sheets, all MXenes displayed an effective EMI shielding (>20 dB).

N. H. Attanayake et al. [52] used LiF/HCl solutions to synthesize the colloidal suspension of d-Mo_2C and d-Ti_3C_2 from their corresponding precursors, namely, Mo_2Ga_2C and Ti_3AlC_2, respectively. They also investigated the CO_2 reduction electrocatalytic behavior of as-synthesized MXenes. In another study, selective etching of Mo_2Ga_2C (using HF and/or LiF/HCl) was carried out to produce delaminated Mo_2CT_x flakes [53], which were further investigated for photonic tumor hyperthermia [54]. Furthermore, lithium- and sodium-ion storage of pillared Mo_2TiC_2, synthesized using the amine-assisted method, was studied and compared to non-pillared

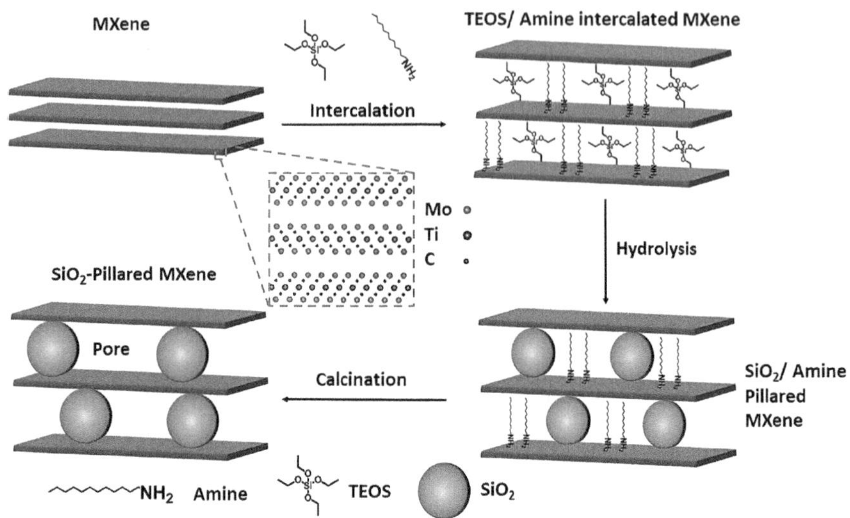

FIGURE 6.9 Schematic representation of the amine-assisted silica pillaring method.

Source: [55] Philip A. Maughan, Luc Bouscarrat, et al. 2021. Pillared Mo_2TiC_2 MXene for high-power and long-life lithium and sodium-ion batteries. *Nanoscale Adv.*, 3, 3145–3158. Reproduced with permission from the Royal Society of Chemistry.

materials. As a result, these MXenes exhibited large surface area (as shown in Fig. 6.9 [55]), high-rate capabilities, and high capacitance values [55].

Y. Guan et al. [56] reported milder etching of V_2AlC powders by using LiF/HCl solution to synthesize the corresponding V_2C MXene. Researchers demonstrated a broad technique, which involves an acid/alkali treatment, for the synthesis of MXenes to improve their potassium-ion storage. The resulting $K-V_2C$ were observed to exhibit a higher capacity of 195 mAh g^{-1}, compared to that of pure V_2C, along with a better rate performance [57]. Furthermore, the synthesis and characterization of $(V_{2/3}Sc_{1/3})_2AlC$, a novel laminated i-MAX phase, were also reported using fluoride-free methods [58]. To synthesize V_2C, selective etching of V_2AlC powder was carried out by stirring it in a mixture of HCl/LiF in an argon atmosphere for 48 hours [59]. The leftover black slurry was rinsed with deionized water several times to obtain a pH above 6. The hazy solution was treated with 5% TMAOH by mass, and the solution was agitated at room temperature for 18 hours to make it homogeneous (as shown in Fig. 6.10 [59]).

In addition to V-based MXenes, J. Xiao et al. [60] synthesized Nb_2CT_x by etching Nb_2AlC powders in a mixture of HCl/LiF solution at 60 °C for 90 hours. The resulting Nb_2CT_x-based electrode demonstrated excellent electrochemical properties, which can be further enhanced by adding carbon nanotubes (CNT) as a conductive agent. The application of as-synthesized Nb_2CT_x as a radioprotectant in scavenging free radicals against infrared (IR) was also reported [61]. Similarly, by using fluoride salts, X. Zou et al. [62] presented a straightforward approach for synthesizing ultrathin Cr_2CT_x from its Cr_2AlC phase (Fig. 6.11 [62]). The electrochemical tests

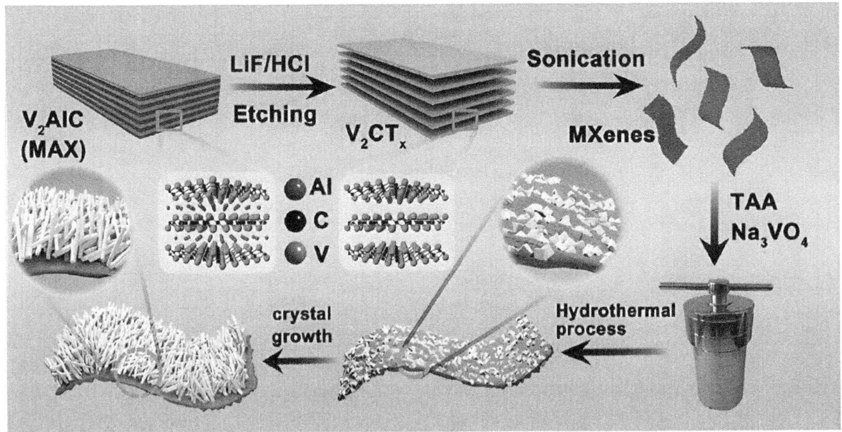

FIGURE 6.10 Schematic representation of the preparation of composite.

Source: Reprinted with permission from [59] Zhenguo Wang, Ke Yu, Yu Feng, et al. 2019. VO_2(p)-V_2C (MXene) Grid Structure as a Lithium Polysulfide Catalytic Host for High-Performance Li–S Battery. *ACS Appl. Mater. Interfaces*, 11, 44282–44292. Copyright 2019, American Chemical Society.

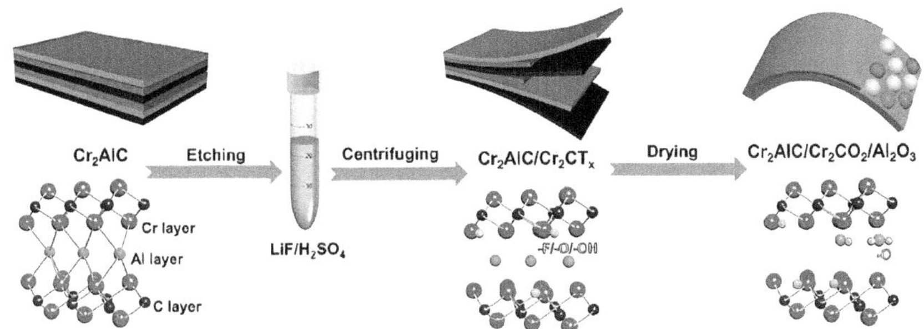

FIGURE 6.11 Schematic illustration of the etching process from Cr_2AlC MAX to Cr_2CT_x MXene.

Source: Reprinted from [62] *Materials Today Energy*, 20, Xinshu Zou, Hao Liu, Hao Xu, Xueyun Wu, et al., A simple approach to synthesis Cr_2CT_x MXene for efficient hydrogen evolution reaction, 100668. Copyright 2021, with permission from Elsevier.

revealed that the manufactured Cr_2CT_x composite with minimum Al_2O_3 exhibits better HER performance, which outperformed the catalytic activity of Pt sheets at high current density.

6.1.5 ETCHING WITH AMMONIUM HYDROFLUORIDE

The process for preparing MXenes using NH_4HF_2 etchant is the same as the process for preparing HF-etched MXenes. However, NH_4HF_2-etched MXenes have a larger interplanar spacing of 24.9 Å compared to 19.5 Å for HF-etched MXenes. Additionally, NH_4HF_2-etched MXenes exhibit higher heat resistance, withstanding

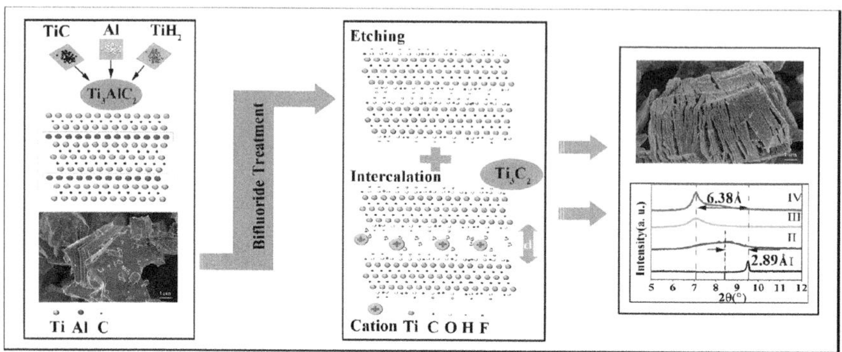

FIGURE 6.12 Schematic illustration of the exfoliation of Ti_3AlC_2 by using bifluoride salts.

Source: Reprinted from [64] *Materials & Design*, 114, Aihu Feng, Yun Yu, Yong Wang, et al., Two-dimensional MXene Ti_3C_2 produced by exfoliation of Ti_3AlC_2, 161–166. Copyright 2017, with permission from Elsevier.

temperatures up to 500°C in air and 900°C in argon, compared to 350°C in air and 550°C in argon for HF-etched MXenes. This is due to their substantially greater lattice distance. Thus, a considerable amount of NH_4^+ is adsorbed onto the Ti_3C_2 surfaces, preventing Ti_3C_2 from being directly exposed to O_2, thereby reducing the oxidation process. Furthermore, the absorbed NH_4^+ would volatilized as NH_3 by expelling extra heat, increasing Ti_3C_2's thermal stability [5]. The MXenes synthesized by using these salts also exhibit a wide range of characteristics and applications. For example, the MAX precursor of $Ti_3C_2T_z$ MXene was exfoliated in an Ar-filled glove box, by using NH_4HF_2 at 35 °C for 196 hours. The resultant MXene displayed double capacity when employed in sodium-ion batteries [63]. In another study, Ti_3AlC_2 was exfoliated by a mixture of bifluoride salts, that is, NH_4HF_2, KHF_2, and $NaHF_2$, to develop corresponding MXenes exhibiting wider interplanar spacing [64]. The detailed exfoliation process of the MAX phase and morphology of the resulting MXene are illustrated in Fig. 6.12 [64].

T. Wang et al. [65] developed a hybrid material based on TiO_2 (derived from the bifluoride salt) and Ti_3C_2 after selectively exfoliating the MXene with $(NH_4)_2TiF_6$. The thermal stability of hybrids was evaluated utilizing thermogravimetric and differential thermal studies. Ti_3AlC_2 interacted with $(NH_4)_2TiF_6$ for 24 hours at 60°C to generate hybrids combined with NH_4TiOF_3 crystals, alongside the synthesis of Ti_3C_2 MXene. After reaction with H_3BO_3, NH_4TiOF_3 crystals were converted into TiO_2 and ultimately into Ti_3C_2/TiO_2 hybrids.

Furthermore, Ti_3C_2 films synthesized by this etching method were reported to exhibit high transmission of vis-to-IR light [66]. It was observed that around 19-nm-thick MXene films have metallic conductivity (down to 100 K) along with 90% transmission of Vis-to-IR light. As temperature decreased below 100 K, the resistivity of these films increased, thereby facilitating negative magnetoresistance. A. Feng et al. [67] studied the thermal stability and wider interplanar spacing of Ti_3C_2 (up to 24.9 Å), synthesized by etching its MAX precursor with NH_4HF_2, by using a differential thermal analyzer (DTA) and thermogravimetry (TG) analysis.

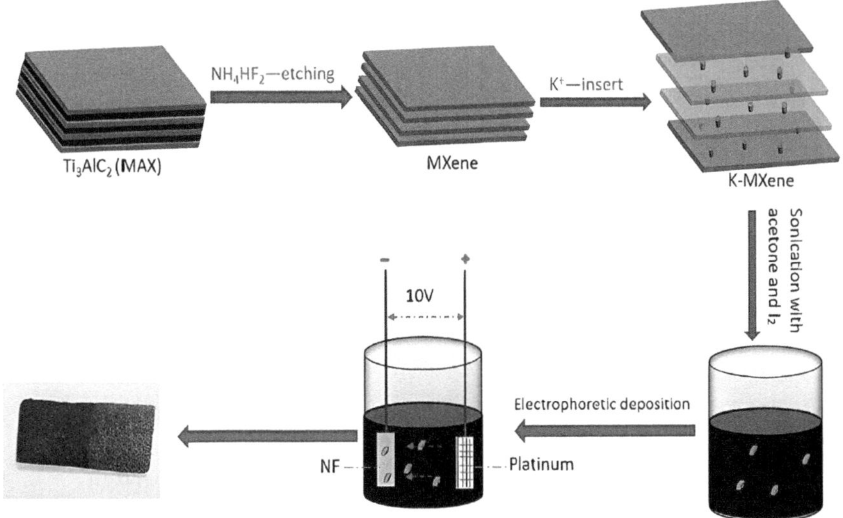

FIGURE 6.13 Schematic diagram illustrating the synthesis of binder-free K-MXene electrode on NF.

Source: [68] Li J., Liu Y., Xu Fang., et al.: K⁺ intercalation of NH₄HF₂-exfoliated Ti₃C₂ MXene as binder-free electrodes with high electrochemical capacitance. *Phys. Status Solidi A* 2020, 217, 1900806. Copyright WILEY-VCH Verlag GmbH & Co. KGaA, Weinheim. Reproduced with permission.

J. Li et al. [68] employed a two-step technique to create a layer structure for Ti_3C_2, with high interlayer spacing (Fig. 6.13 [68]). First, the standard Ti_3C_2 MXene was prepared using the moderate NH_4HF_2 etching agent. Second, KOH was added to the resulting solution to intercalate the MXene with K^+. The K^+ intercalation was observed to significantly enhance the Ti_3C_2 (002) crystal planar spacing from 10.8 Å to 12.4 Å.

6.1.6 Alkali Etching

Due to the strong binding capacity of alkali with Al, it was found to be a potential etching agent for creating MXenes from their MAX phases. Consequently, several alkali-based techniques were employed for the synthesis [69]. For example, O- and OH-functionalized $Ti_3C_2T_x$ was synthesized hydrothermally by an alkali-assisted approach [70] (as shown in Fig. 6.14 [70]). This path was inspired by the Bayer method, which is commonly employed in the refinement of bauxite. Multilayered MXenes were synthesized by treating the MAX precursor with NaOH at 270°C, under Ar atmosphere. The process was carried out by varying the temperature and concentration, which resulted in the formation of MXene with 92 wt.% purity. They also employed the as-synthesized MXene as an electrode material in 1 M H_2SO_4, which displayed better gravimetric capacitance value (314 F g⁻¹) than the one produced by HF etching.

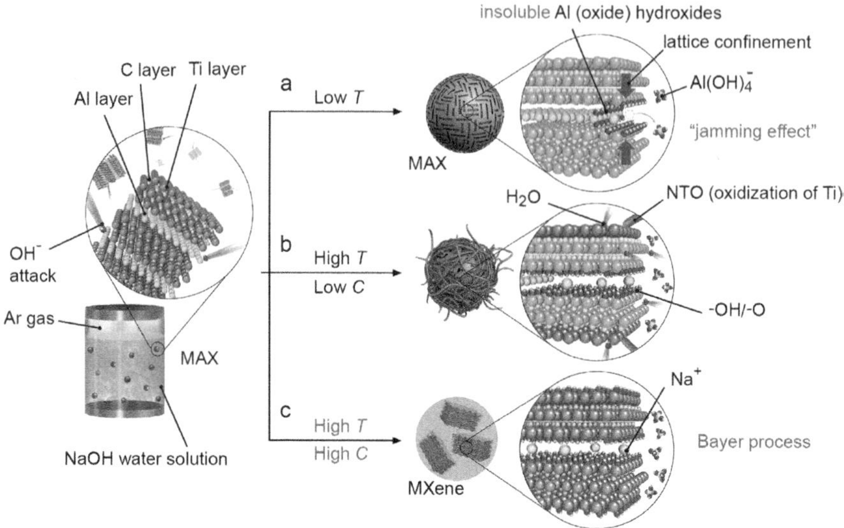

FIGURE 6.14 Schematic of the reaction between Ti_3AlC_2 and NaOH water solution under different conditions. (a) At low temperature, Al (oxide) hydroxides block the exfoliation of Al. (b) At high temperatures, some Al (oxide) hydroxides dissolve in lower concentration of NaOH. (c) High temperatures and NaOH concentrations will dissolve the Al (oxide) hydroxides in NaOH.

Source: [70] Li T., Yao L., Liu Q., Gu J., et al.: Fluorine-Free Synthesis of High-Purity $Ti_3C_2T_x$ (T=OH, O) via Alkali Treatment. *Angewandte Chemie* 2018, 57, 6115–6119. Copyright WILEY-VCH Verlag GmbH & Co. KGaA, Weinheim. Reproduced with permission.

In another study, the adsorption behavior of $Ti_3C_2T_x$ was modified by tuning its surface groups and interlayer spacing [71]. These properties were modified by treating the MXene with a hot alkaline solution, as illustrated in Fig. 6.15 [71]. Three alkali solutions were used for this purpose, that is, LiOH, NaOH, and KOH. As a result, F-functionalized surfaces were converted to OH-functionalized ones, along with a 29% increase in the interlayer distance. Furthermore, LiOH- and NaOH-based MXenes were observed to exhibit better adsorption properties in methylene blue (MB) dye.

Similarly, the synthesis of OH-terminated $Ti_3C_2(OH)_2$ NSs by utilizing KOH solution for etching of the precursor in the presence of water was reported [72]. The process of synthesis is illustrated in Fig. 6.16 [72]. As a result, a simple washing process resulted in the formation of desired NSs by quickly replacing the Al layer of the Ti_3AlC_2 phase with hydroxide. Atomic force microscopy indicated that the lateral size of the synthesized NSs was several micrometers. Interestingly, the exfoliated $Ti_3C_2(OH)_2$ NSs can be restacked after drying. When the restacked components were redistributed in water, they easily re-delaminated after several hours of shaking.

It was also observed that after etching in fluoride salts, when MXenes are treated with DHT (di(hydrogenated tallow)benzyl methyl ammonium chloride), the lithium cations between the layers of these materials become organophilic, which facilitates

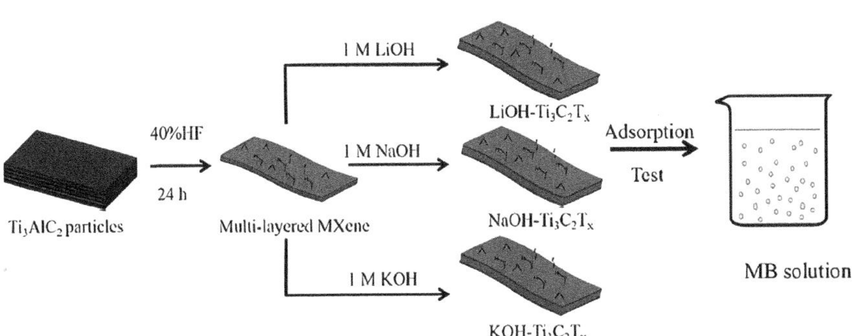

FIGURE 6.15 Schematic illustration of the synthesis and adsorption test of MXenes, using hot alkaline solutions.

Source: Reprinted from [71] *Materials Chemistry and Physics*, 206, Zheng Wei, Zhang Peigen, Tian Wubian, et al., Alkali treated $Ti_3C_2T_x$ MXenes and their dye adsorption performance, 270–276. Copyright 2018, with permission from Elsevier.

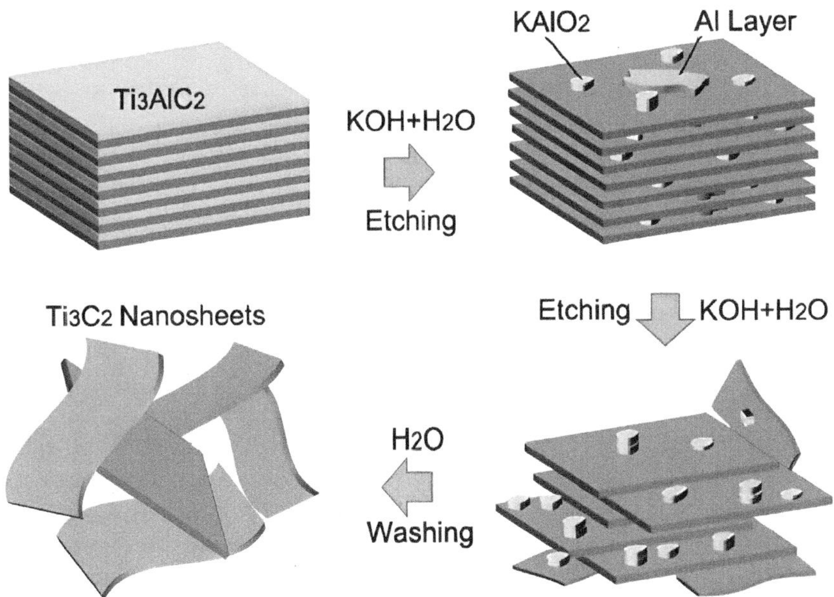

FIGURE 6.16 Schematic illustration of the etching and exfoliation of Ti_3AlC_2.

Source: Reprinted with permission from [72] Gengnan Li, Li Tan, Yumeng zhang, et al. 2017. Highly Efficiently Delaminated Single-Layered MXene Nanosheets with Large Lateral Size. *Langmuir*, 33, 9000–9006. Copyright 2017, American Chemical Society.

the formation of colloidal suspensions [73]. Under ambient circumstances, fast cation exchange occurs, which results in oxide-free and well-dispersed 2D flakes.

6.1.7 Molten Salt Etching

It has been observed that a replacement reaction between molten salt, that is, $ZnCl_2$, and MAX phases generates a sequence of Zn-based precursors, by replacing the A-layer with Zn, and Cl-functionalized MXenes [74]. By using this method, novel Zn-based precursors were synthesized, including V_2ZnC, Ti_2ZnC, Ti_3ZnC_2, and Ti_2ZnN. Further exfoliation of Ti_2ZnC_2 and Ti_3ZnC_2 provided Cl-functionalized MXenes, that is, Ti_2CCl_2 and $Ti_3C_2Cl_2$, due to the strong Lewis acidity exhibited by $ZnCl_2$. These findings suggested that replacing A-site elements with molten salts facilitates the analysis of thermodynamically instable MAX phases, which are difficult to be synthesized using a traditional approach. Similarly, Ti_3C_2 was produced by immersing Ti_3SiC_2 in $CuCl_2$ molten salt at 750°C. The synthesized MXene was further washed with ammonium persulfate (APS) and represented as $MS\text{-}Ti_3C_2T_x$. Furthermore, these MXenes were observed to be efficient electrode materials for hybrid devices, as they displayed high-rate performance in aqueous electrolytes along with high lithium-ion storage capacity values [75].

In another study, water-dispersible $Ti_3C_2T_z$ NSs were developed by using SnF_2 as a molten salt etchant [76]. This salt diffuses between the MAX layers, separates the layers, and produces Sn and AlF_3 as byproducts. In addition, Li^+-, K^+-, and Mg^{2+}-intercalated $Ti_3C_2T_z$ can also be synthesized by using aqueous salt solutions, that is, LiCl, KCl, and $MgCl_2$, respectively [77]. M. Ghidiu et al. [78] observed the dehydration/hydration and ion exchange behavior of Li-intercalated Ti_3C_2 MXene, synthesized by exfoliating its MAX precursor using LiCl (as illustrated in Fig. 6.17 [78]). The MXenes produced via molten salt etching display efficient electrochemical properties in metal-ion batteries. For example, H. Dong et al. [79] described the Lewis acidic etching of Nb_2CT_x MXene. The developed MXene exhibited a high lithium-ion storage capacity value of up to 330 mAh g^{-1} at 0.05 A g^{-1}, surpassing the highest capacity (205 mAh g^{-1}) of $Ti_3C_2T_x$ MXene produced using molten salt. A capacity of 80 mAh g^{-1} at 10 A g^{-1} was also obtained with high-rate performance. The aqueous dispersibility of Nb_2CT_z NSs was also observed by molten salt etching and adding hydroxyl surface groups with a KOH wash [80].

The electrocatalytic activity of V-Mo bimetallic nitridene solid solution ($V_{0.2}Mo_{0.8}N_{1.2}$) was reported, synthesized using a catalytic molten-salt approach [81]. The characterizations revealed that a monomer assembly enables the dissolution of V by lowering the growth barrier of the nitridene. Furthermore, theoretical calculations confirmed that the electronic structure can be optimized by doping of V, which facilitates the formation of hydrogen by fast proton coupling. J. Chen et al. [82] reported the open-air rapid synthesis of MXenes utilizing a molten salt-shielded synthesis (MS_3) approach. In this method, exfoliation was carried out by using Lewis-acid salts and a eutectic salt mixture (with a low melting point) served as a reaction medium, which prevents the oxidation of MXenes at increased temperatures. The MS_3 technique was used to successfully synthesize Ti_2CT_x, $Ti_3C_2T_x$, Ti_3CNT_x, and $Ti_4N_3T_x$. The synthesized MS_3-MXenes showed outstanding electrochemical characteristics

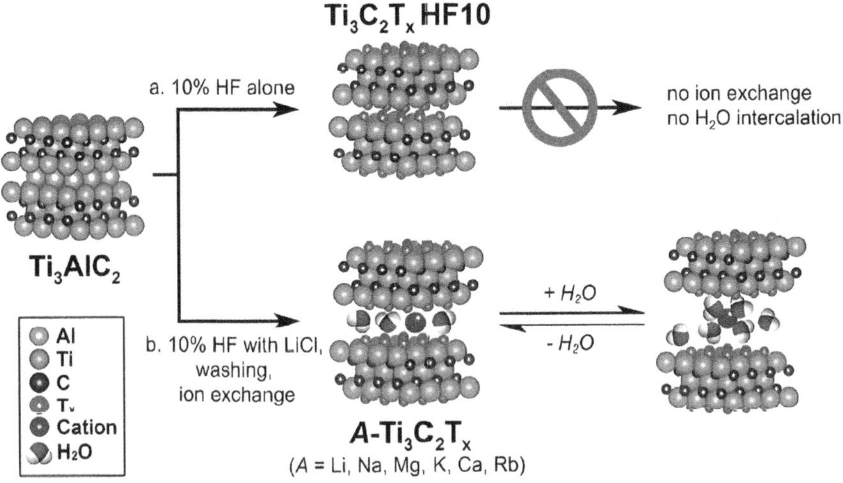

FIGURE 6.17 Ti$_3$AlC$_2$ phase etched with (a) 10% HF alone to yield Ti$_3$C$_2$T$_x$ HF10. This material does not show intercalation of cations or water. When etched with (b) 10% HF in the presence of LiCl, a variety of intercalated A-Ti$_3$C$_2$T$_x$ is produced (where A is an intercalated cation).

Source: Reprinted with permission from [78] Michael Ghidiu, Joseph Halim, Sankalp Kota, et al. 2016. Ion-Exchange and Cation Solvation Reactions in Ti$_3$C$_2$ MXene. Chem. Mater., 28, 10, 3507–3514. Copyright 2016, American Chemical Society.

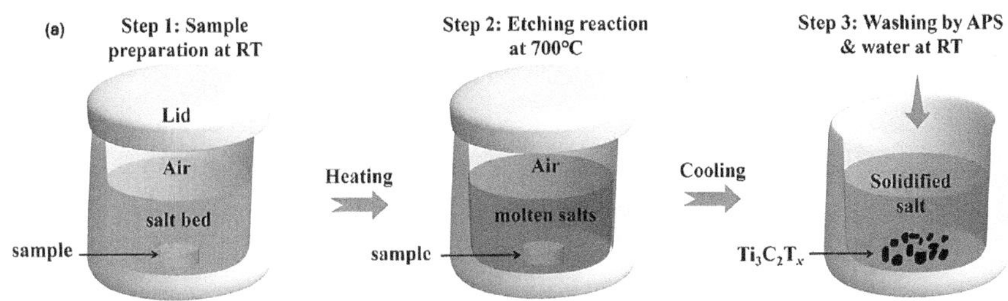

FIGURE 6.18 Scheme of the MS$_3$ of Ti$_3$C$_2$T$_x$ MXene in air. The sample contains NaCl/KCl and Ti$_3$AlC$_2$ phase mixture, and the salt bed is a NaCl/KCl mixture with a 1:1 mole ratio mixed with CuCl$_2$ etchant.

Source: [82] Chen J., Jin Q., Li Y., et al.: Molten Salt-Shielded Synthesis (MS$_3$) of MXenes in Air. *Energy Environ. Mater*. 2023, 6, e12328. Copyright Wiley. Reproduced with permission.

for high-rate Li-ion storage when utilized as negative electrodes. The MS$_3$ mechanism in air is illustrated in Fig. 6.18 [82].

6.1.8 Electrochemical Etching

Electrochemical etching is another approach that can be used to exfoliate the MAX precursors of MXenes. In this process, an electrolyte is significantly used, as it influences not only the ultimate yield but also the microstructure and surface chemical characteristics of the resulting MXenes. The halogens strongly interact with aluminum; therefore, Cl-based electrolytes are most commonly used. To develop O- and OH-functionalized $Ti_3C_2T_x$, researchers used a binary electrolyte of 1M NH_4Cl + 0.2M TMAOH to etch the Ti_3AlC_2 phase. According to the researches, OH concentration greatly influences the etching parameters. The reaction efficiency was great when the value was 0.2 M, and only 10 hours were required. The mentioned technique was later enhanced, and a thermal-assisted electrochemical etching was established. This technique is comparatively safe and milder, as it uses dilute HCl electrolyte instead of tetramethylammonium ion, which is hazardous [36].

In addition to the Ti-based materials, this approach has also synthesized V_2CT_x and Cr_2CT_x MXenes, in contrast to other etching techniques. Since this technique is cost-effective and carried out under ambient conditions, its environmental stability can be increased by optimizing the selection of electrolyte [36]. W. Sun et al. [83] reported that Ti_2CT_x can be etched into a three-layered structure, from Ti_2AlC, using an aqueous electrolyte of HCl (Fig. 6.19 [83]). This etching technique can synthesize O-, Cl-, and OH-functionalized MXenes, as it is a fluoride-free etching method. The three-layered structure comprises the CDC, unetched MAX, and MXene. Bath sonication can further extract MXenes from this three-layer structure.

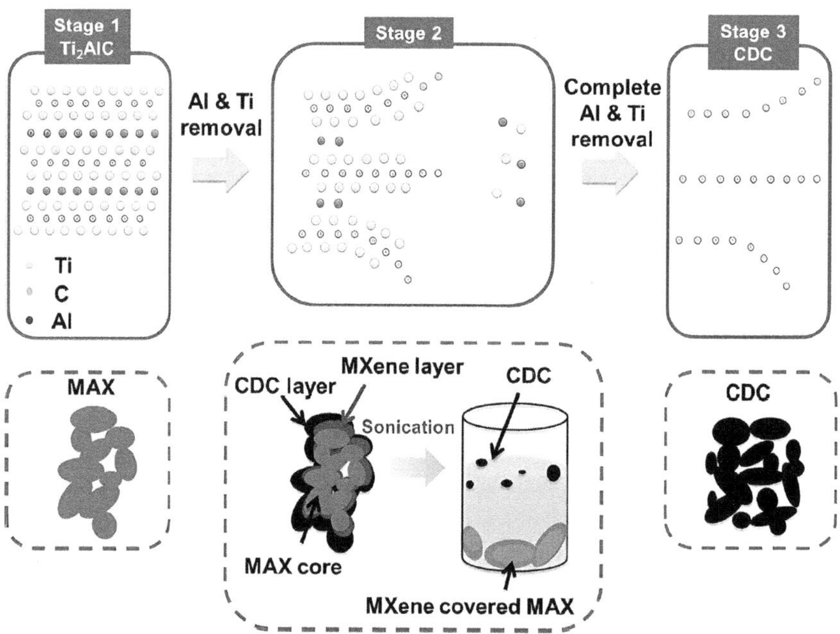

FIGURE 6.19 Proposed mechanism for electrochemical etching of Ti_2AlC in a HCl aqueous electrolyte.

Source: [83] Sun W., Shah S. A., Chen Y., et al. 2017. Electrochemical etching of Ti_2AlC to Ti_2CT_x (MXene) in low-concentration hydrochloric acid solution. *J. Mater. Chem. A*, 5, 21663–21668. Reproduced with permission from the Royal Society of Chemistry.

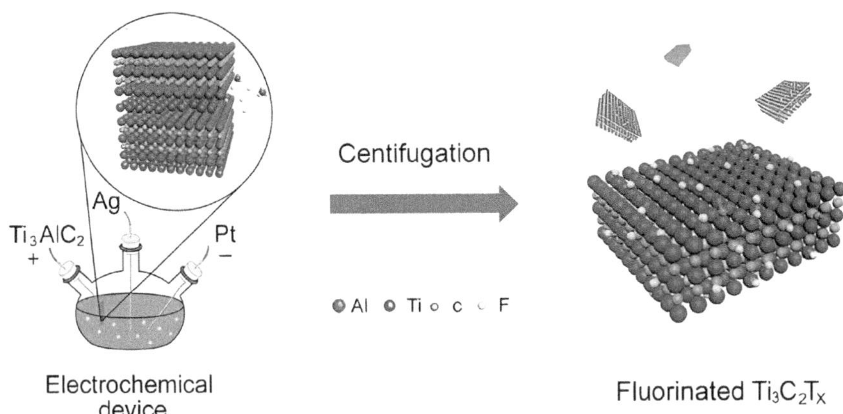

FIGURE 6.20 Schematic illustration of the synthesis and exfoliation of MXene.

Source: Reprinted from [86] *Ceramics International*, 47, Teng Yin, Yan Li, Renheng Wang, et al., Synthesis of Ti₃C₂Fₓ MXene with controllable fluorination by electrochemical etching for lithium-ion batteries applications, 28642–28649. Copyright 2021, with permission from Elsevier.

S. Yang et al. [84] demonstrated an effective, fluoride-free electrochemical method for the delamination of Ti_3C_2 in a binary aqueous electrolyte. In this method, Cl-ions enable the rapid etching of Al by interrupting the Ti-Al bonds. Intercalation of NH_4OH was observed to be important in expanding the edges of the etched anode and in aiding the underlying surface etching. Under ambient circumstances, the anode (Ti_3AlC_2) was entirely etched in a short period (e.g., 5 hours). More than 90% of the exfoliated materials were single- or double-layered with an average lateral size of more than 2 m, which was bigger than the sizes obtained by the traditional HF etching procedure. In another study, Ti_2C was directly synthesized from its elements, that is, Ti, C, and Al, by using a one-pot molten salt electrochemical etching process [85]. T. Yin et al. [86] presented a green and mild exfoliation process for synthesizing $Ti_3C_2F_x$ and controlling the degree of fluorides on its surface (see Fig. 6.20 [86]). A non-aqueous ionic liquid solution was used as an electrolyte, based on [BMIM][PF6]. The resultant MXene was F-functionalized by using the TiF_3 and CF groups.

S. Pang et al. [87] synthesized V_2CT_x, Cr_2CT_x, and Ti_2CT_x using thermal-assisted electrochemical etching. Furthermore, cobalt ion-doped MXenes showed substantially improved HER and oxygen evolution reaction (OER) activity, revealing their multifunctionality equivalent to commercialized catalysts. As a proof-of-concept, X. Li et al. [88] used V_2AlC as the cathode and 21 M LiTFSI + 1 M $Zn(OTf)_2$ solution as the electrolyte to develop a MXene battery. As a result, the V_2AlC inside the battery was electrochemically etched to be MXene. Another work [89] effectively generated V_2C MXene with a quasi-2D structure by etching the V_2AlC powders with NaF and HCl at 90°C. The effects of time on the exfoliation process were investigated and defined. During the exfoliation process, a significant degree of conversion (>90wt% V_2C) was attained. The primary impurities were $Na_5Al_3F_{14}$ and V_2AlC. In LIBs, as-prepared V_2C MXene was investigated as an electrode material.

6.1.9 Topochemical Synthesis

M. Downes et al. [90] reported the successful topochemical transformation of two new MAX-phases, $(Ti, Ta)_5AlC_4$ and $(Ti, Nb)_5AlC_4$, into $(Ti, Ta)_5C_4T_x$ and $(Ti, Nb)_5C_4T_x$ MXenes. The resultant MXenes were delaminated and their structural, optical, and thermal properties were evaluated. This study resulted in the formation of a new family of MAX phases and MXenes. Another topochemical synthesis was carried out to develop $Ti_3C_2T_x$ from its corresponding MAX phase. The influence of concentration, temperature, and duration of etching was also evaluated [91]. Similarly, in the presence of alkali and alkaline-earth metal chlorides, oxidation of V_2CT_x in H_2O_2 was reported to produce chemically pre-intercalated bilayered vanadium oxides (BVOs) with distinctive 2D shape and better electrochemical stability. HCl and HF were mixed with deionized water in a volume ratio of 12:12:6 (60 mL total) for topochemical synthesis [92]. The V_2AlC powder was agitated (using a Teflon stir bar) at 300 rpm for 120 hours. After that, the slurry was centrifuged at 3,500 rpm for 5 minutes. The washing procedure was continued until the pH reached 6. The electrochemical characteristics of δ-$Li_xV_2O_5nH_2O$ and δ-$Mg_xV_2O_5nH_2O$ electrodes generated from V_2CT_x were investigated in Li-ion cells.

Recently, a new class of these materials has been discovered, named as MBene (i.e., 2D metal borides), which can be synthesized from their corresponding bulk Mo_2AlB_2 phase by using a low-temperature topochemical synthesis [93]. A stable MoAlB compound was chemically deintercalated at room temperature to remove Al. As a result, destabilized $MoAl1$-xB phase was developed, which was further precipitated (to isolate the grains) and annealed at 600°C. This resulted in additional deintercalation of Al to crystallize Mo_2AlB_2. Further heating led to topotactic disintegration into bulk-scale $Mo_2AlB_2AlO_x$ nanolaminates containing 13-nm-thick Mo_2AlB_2 NSs interleaved by 13-nm-thick amorphous aluminum oxide (as illustrated in Fig. 6.21 [93]).

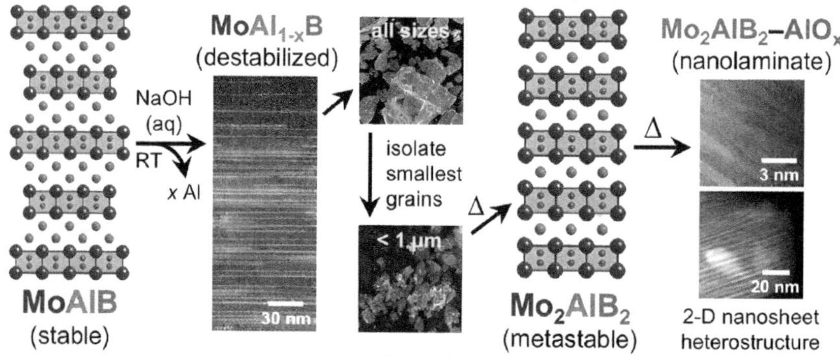

FIGURE 6.21 Topochemical synthesis route of Mo_2AlB_2–AlO_x.

Source: Reprinted with permission from [93] Lucas T. Alameda, Robert W. Lord, Jordan A. Barr, et al. 2019. Multi-Step Topochemical Pathway to Metastable Mo_2AlB_2 and Related Two-Dimensional Nanosheet Heterostructures. *J. Am. Chem. Soc.*, 141, 10852–10861. Copyright 2019, American Chemical Society.

This method was also used to develop a hybrid TiN NM@C, which displayed efficient electrochemical properties for lithium–sulfur batteries, due to the rapid charge transfer and availability of active sites [94]. The resultant hybrid electrode material showed better rate capabilities than that obtained by TiN NPs. L. T. Alameda et al. [95] demonstrated that MBene along with different metastable phases (Mo-Al-B) can be developed by stepwise topochemical deintercalation of Al from single MoAlB crystals. A high-resolution microscopic examination indicated that as Al is deintercalated, stacking defects emerge in MoAlB, and with the increase in Al removal, the density of these defects increases. This study discovered four new intergrowth phases in nanoscale regions, that is, Mo_2AlB_2, $Mo_3Al_2B_3$, $Mo_4Al_3B_4$, and $Mo_6Al_5B_6$, with significant densities of stacking defects. The crystal structures of representative MAX phases are shown in Fig. 6.22 [95].

Similarly, lithium ions were deintercalated from layered LiNiB polymorphs by using a topochemical synthesis, carried out at room temperature in the presence of water, dilute HCl, or air. This resulted in the formation of different crystal structures of metastable $Li_{0.5}NiB$ borides, which contain lithium stored between NiB layers [96]. They demonstrated that Li deintercalation occurs via a "zip-lock" process, resulting in the condensing of single [NiB] layers into multilayers coupled by covalent bonds, yielding structural fragments with different compositions, that is, $Li[NiB]_2$ and $Li[NiB]_3$. A C/Fe_3C/Fe hybrid with a quasi-MXene structure was also reported to be synthesized topochemically, which displayed better electrochemical

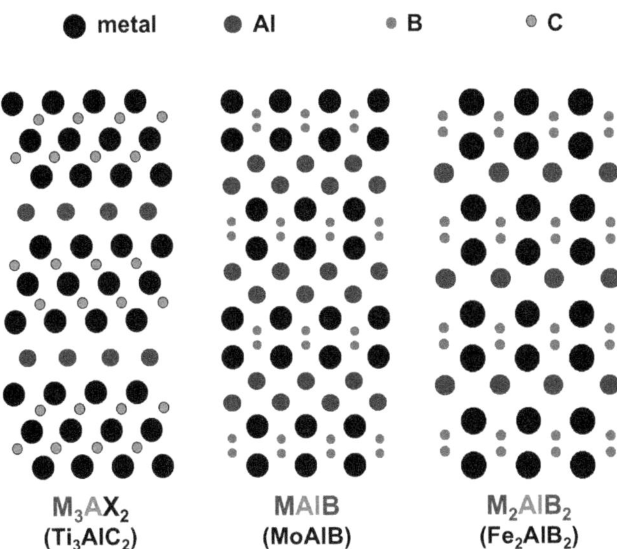

FIGURE 6.22 Crystal structures of representative MAX phases, MAlB-type MAB phases, and M_2AlB_2-type MAB phases.

Source: Reprinted with permission from [95] Lucas T. Alameda, Parivash Moradifar, et al. 2018. Topochemical Deintercalation of Al from MoAlB: Stepwise Etching Pathway, Layered Intergrowth Structures, and Two-Dimensional MBene. *J. Am. Chem. Soc.*, 140, 28, 8833–8840. Copyright 2018, American Chemical Society.

behavior [97]. In another study, the precursor for MBenes, that is, Mo_2AlB_2, was synthesized topochemically by removing one Al layer from MoAlB [98]. A theoretical study showed that this process follows a sequence of stages to change from MoAlB to Mo_2AlB_2 and $Mo_4Al_3B_4$ (stages I and II, respectively.). To generalize, topochemical synthesis can be utilized to synthesize MXenes, MBenes, and other 2D materials (such as MoS_2 by sulfurization of MoO_3 MPs [99]), with enhanced electrochemical and electrocatalytic properties.

6.2 BOTTOM-UP STRATEGIES

Another approach to synthesize 2D MXenes is a bottom-up strategy, which starts with atoms and builds up a layered structure by depositing multilayered epitaxial films. The top-down methods have been observed to reduce the sizes of MXenes along with the formation of structural defects. In contrast, bottom-up techniques have the ability to develop structural compositions with fewer defects and large transverse dimensions [2, 100]. These approaches use small organic or inorganic molecules/elements that are combined with a 2D layered structure through crystal formation. MXenes with modified morphologies, sizes, and surface groups can be developed by using these methods. However, the presence of multicomponent element layers in MXenes somehow limits the use of these approaches, which need to be studied in detail. Furthermore, while top-down synthesis cannot modify the surface groups accurately, MXenes can be easily functionalized to provide specific groups for specialized applications due to their extensive chemistry. As a result, reliable surface termination determination is required to investigate the inherent properties of MXenes. Therefore, an essential orientation for the advancement of bottom-up preparation procedures is considerably advantageous for creating MXenes with predefined properties [101].

A wide range of bottom-up approaches to synthesize MXenes have been reported so far, including solution processing techniques, chemical vapor deposition (CVD), hydrothermal/solvothermal synthesis, spray coating, dip coating, drop casting, hot press techniques, vacuum-assisted filtration (VAF), and plasma-enhanced pulsed-laser deposition (PEPLD). MXenes synthesized by all of these bottom-up techniques are summarized in the next sections.

6.2.1 CHEMICAL VAPOR DEPOSITION

Several deposition processes, that is, CVD, photo-deposition, electrodeposition, atomic layer deposition (ALD), etc., have been used to fabricate different MXene-based composites. During the CVD process, volatile precursors react on the surface of the substrate, which results in thin film formation. On the other hand, to avoid the contact of precursor material, ALD splits the reaction process in two parts. This method can control the deposition on an atomic scale due to constrained and self-reliant growth of ALD films. MXene hybrids composed of transition metal (oxy) hydroxides (TMOs), carbon-based materials, and transition metal phosphides could be synthesized using electrodeposition procedures. Similarly, metallic NPs (such as Pt and Cu) can be deposited on MXene surfaces by using photo-deposition. Although

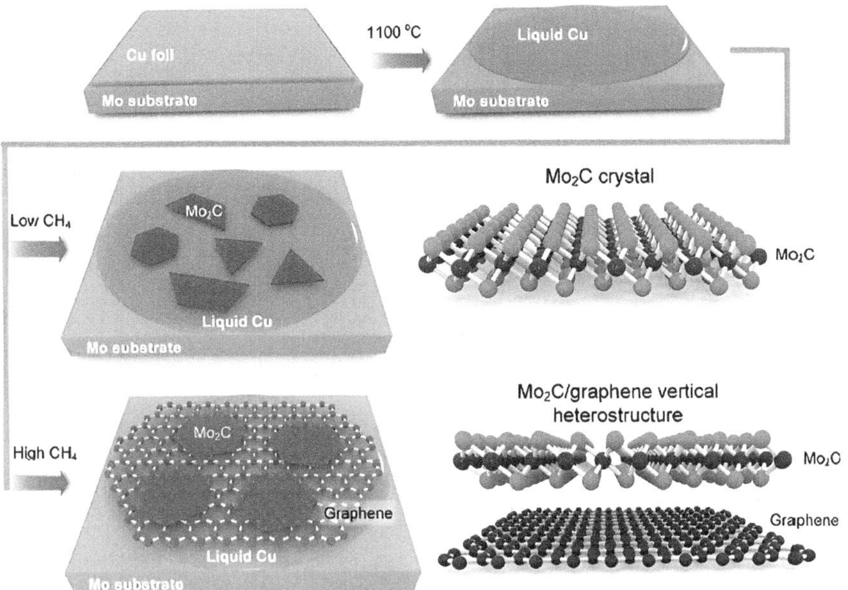

FIGURE 6.23 CVD growth process of Mo_2C/graphene heterostructure.

Source: [103] Geng D., Zhao X., Chen Z., et al.: Direct Synthesis of Large-Area 2D Mo_2C on In Situ Grown Graphene. *Adv. Mater.* 2017, 29, 1700072. Copyright WILEY-VCH Verlag GmbH & Co. KGaA, Weinheim. Reproduced with permission.

deposition techniques exhibit many advantages over other methods, the increased cost of production makes this method less desirable [102].

In the CVD method, high-quality Mo_2C thin films with a homogeneous structure can be developed on graphene by using a Mo-Cu alloy catalyst. Within the heterostructure, graphene serves as a diffusion barrier to the phase-segregated Mo, which allows the growth of Mo_2C thin films of nanometer thickness. D. Geng et al. [103] demonstrated molten copper-catalyzed CVD to synthesize Mo_2C/graphene thin films. At 1,100 °C, liquid Cu was used as a catalyst on Mo substrate, H_2 as a reducing agent, and CH_4 as a carbon-based precursor. Fig. 6.23 [103] depicts the growth process schematically. The as-grown thin films exhibited a considerably lower onset voltage for HER than Mo_2C-only electrodes because of their more effective transfer kinetics.

In another study, similar parameters were used to grow α-Mo_2C crystals on Mo substrate, except the temperature used was above 1,085°C [104]. The resultant high-quality crystals were a few nanometers thick and exhibited superconducting transition characteristics along with a significant anisotropy with magnetic field orientation. It was observed that the superconducting behavior of the as-synthesized crystal was thickness-dependent. Z. Kang et al. [105] used Mo_2C graphene, synthesized via the CVD method, to produce a hybrid Mo_2C-Gr/$Sb_2S_{0.42}Se_{2.58}$/TiO_2, which was further utilized to develop a photodetector. In this device, Mo_2C-Gr served as a hole-transporting film with transparency (Fig. 6.24 [105]).

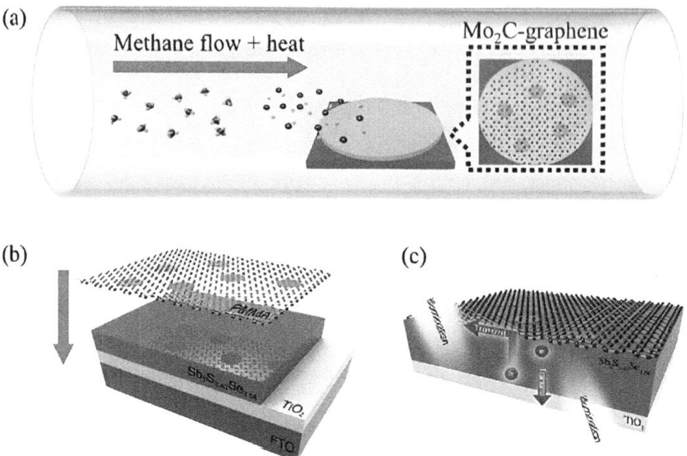

FIGURE 6.24 (a) CVD method to grown Mo_2C graphene; (b) transfer of the Mo_2C-Gr layer; and (c) self-driven two-sided photodetector.

Source: Reprinted from [105] Kang Z., Zheng Z., Wei H., Zhang Z., et al. Controlled Growth of an Mo_2C–Graphene Hybrid Film as an Electrode in Self-Powered Two-Sided Mo_2C–Graphene/$Sb_2S_{0.42}Se_{2.58}$/ TiO_2 Photodetectors. *Sensors*. 2019; 19(5):1099. Open access.

After successful CVD synthesis of Mo-based thin films or crystals, a number of studies developed these materials by changing the reaction parameters and reported their applications in various fields. For example, W. Sun et al. [106] used liquid Au substrate to develop Mo_2C crystals and Mo_2C/graphene heterostructures. According to the antiferromagnetic (AFM) analysis, a reducing growth of MXenes was observed on this substrate, which indicated the growth of some crystals below the level of Au substrate. Furthermore, HER catalytic activity of these heterostructures was also observed. In another study, MXene-CNT nanocomposites were developed using a catalytic-CVD approach, and their corresponding dye-degradation activity was reported for rhodamine B (RhB) [107]. The influence of reaction parameters on the formation of crystals and heterostructures was also observed [108]. According to the results, the growth rate of Mo_2C or Mo_2C/graphene was observed to be dependent on methane flow, the presence of hydrogen, and growth temperature.

Studies have shown that α-Mo_2C flakes can be analyzed by using scanning tunneling microscopy/spectroscopy (STM/S) [109]. The scanning tunneling spectroscopy (STS) technique revealed increased superconductivity with lower correlation length, despite the presence of defects. S. Chaitologou et al. [110] used liquid Sn-Cu alloys to construct thin Mo_2C domains and Mo_2C/graphene heterostructures, which exhibit micrometer-size and semi transparency at lower temperatures. No Sn-related defects were found, making the alloy an acceptable growth substrate. The fluctuation in the graphene layer strain revealed the vertical contact between graphene and Mo_2C using Raman spectroscopy. The resultant structures were observed to be efficient

HER electrocatalysts. The morphology and structural anisotropy of CVD-grown α-Mo_2C crystals were determined using angle-resolved polarized Raman spectroscopy (ARPRS) [111]. In addition, a highly sensitive photodetector was developed by using hybrid features MoS_2 and Mo_2C [112]. The resultant device displayed broad spectral response and high sensitivity, thereby outperforming the performance of conventional 2D photodetectors.

6.2.2 Physical Vapor Deposition

S. Rouhi et al. [113] investigated thin copper films formed via physical vapor deposition (PVD) onto molybdenum precursor substrates, which resulted in better surface coverage of MXenes. The precursor substrates were thoroughly cleaned before the deposition of a copper thin layer using a rotational thermal evaporation technique to ensure immaculate conditions. A Bruker Dektak XT profilometer was used to measure the thickness of the resultant layers. Following that, the copper-coated precursor was placed in a horizontal CVD furnace and purged with hydrogen (H_2) before heating. The methane (CH_4) flow rate was kept constant during the 30-minute procedure, while the CH_4/H_2 flow rate was 0.25. Following that, the sample was swiftly removed by interrupting the methane flow. The methane flow rate remained steady during the 30-minute procedure, while the CH_4/H_2 flow rate stayed constant at 0.25. When the operation was finished, the sample was removed quickly by stopping the methane flow.

Characterization techniques such as μ-Raman spectroscopy and energy dispersive X-ray (EDX) validated the method's greater MXene coverage. The findings revealed that even thin copper layers of 50 nm offer 100% MXene coverage, whereas increasing the thickness of the copper catalyst film causes shifts in the μ-Raman spectral peaks. Furthermore, increasing the growing temperature encourages the formation of MXene–graphene hybrids. Topographies using AFM demonstrated that the thickness of the catalyst is critical in determining the production and characteristics of the resultant MXene thin films. This research is essential to the evolution of MXene synthesis processes with regulated structures and characteristics. The combined PVD–CVD technique employs thin copper films generated using PVD, solving the difficulty of troublesome dissolution during transfer, which is a key impediment to the practical application of 2D materials.

The fabrication of thin films with modified characteristics is gaining popularity. This necessitates appropriate deposition processes and an appropriate control diagnosis. As a PVD process, ion beam sputtering (IBS) can meet technological challenges. It is mostly due to the spatial separation of ion production in an ion source, sputtered particle formation at targets, and thin film growth on substrates in IBS (in contrast to other PVD processes). Thus, the deposition conditions can be tuned by adjusting ion beam properties (i.e., intensity, energy, and ion species) and/or target system parameters (ion incidence or emission angles). A new low-energy ion facility (LEIF) system has just been installed in the nuclear physics institute's (NPI) CANAM research infrastructure. The targets were positioned on a cooled Cu holder during the IBS process. The phase synthesis was promoted by fixing the substrates at elevated temperatures. P. Horak et al. [114] produced several thin films of MAX (and MXene) phases using the LEIF technology, and the radiation endurance of the films was investigated using ion bombardment.

In another study, thin films of single crystalline Ti_2AlC MAX were developed by using the PVD method at low temperatures, and their mechanical and structural properties were investigated [115]. This MAX phase was developed by employing a methodology of layer-by-layer deposition along with a comparatively lower deposition rate. Epitaxial development was studied by using different characterization techniques, that is, high-resolution transmission electron microscopy (HR-TEM), Raman spectroscopy, and XRD. Furthermore, nanoindentation and nano-scratch experiments were used to investigate the tribological and mechanical properties. Similarly, L. Groner et al. [116] oriented and annealed double layers of Ti-AlN on Al_2O_3 substrate to synthesize thin films of polycrystalline Ti_2AlN. SEM, XRD, electron backscatter diffraction (EBSD), and Raman spectroscopy demonstrated that the multilayer system was successfully transformed into a polycrystalline and dense Ti_2AlN coating with a thickness of 2.7 m.

6.2.3 Hydrothermal Method

The hydrothermal or solvothermal method is a cost-effective way of synthesizing a wide range of materials. The structures and morphologies of synthesized materials depend on the reaction parameters, that is, liquid solvent, mineralizer, and high pressure and temperature. The only difference between these two methods is the choice of solvent; organic solvents are used in the solvothermal method, whereas hydrothermal methods use water as a solvent. The chemical reaction is carried out in a sealed autoclave, which is heated at high temperature and pressure (up to the critical point of the solvent). This results in the formation of the supercritical fluid phase, which includes both liquid and gas properties with higher viscosities. This method can dissolve chemical compounds that are insoluble at lower temperatures. However, corrosion by hydroxyl or water molecules is one of the major disadvantages of this approach, which can cause some unexpected alterations [102]. Due to efficient synthesis conditions, different structural compositions of MXenes have been synthesized by using these methods.

For example, F. Han et al. [117] presented a hydrothermal-assisted intercalation (HAI) approach to increase the yield of MXene sheets by 74%. This approach successfully aids reagent diffusion and intercalation, improving upcoming delamination; in the meantime, an antioxidant is used to shield $Ti_3C_2T_x$ from oxidation throughout the process. This process can easily synthesize high-quality $Ti_3C_2T_x$ sheets. Figure 6.25 [117] depicts the entire synthesis process. These $Ti_3C_2T_x$ 2D sheets exhibit interesting applications in supercapacitors due to high conductivity and considerable terminated capabilities, yielding a high capacitance of 482 F g^{-1}. Furthermore, because of the comparatively lengthy bleaching relaxation time, the ultrafast carrier dynamics studies of $Ti_3C_2T_x$ sheets signal a possible application in photocatalysis.

V. T. Quyen et al. [118] utilized a hydrothermal method to develop TiO_2@Ti_3C_2 nanoflowers. This hybrid material displayed photocatalytic efficiency toward RhB, thereby outperforming the activity of commercial TiO_2 and pristine Ti_3C_2. Furthermore, even after the fifth degrading cycle, the composite exhibited good

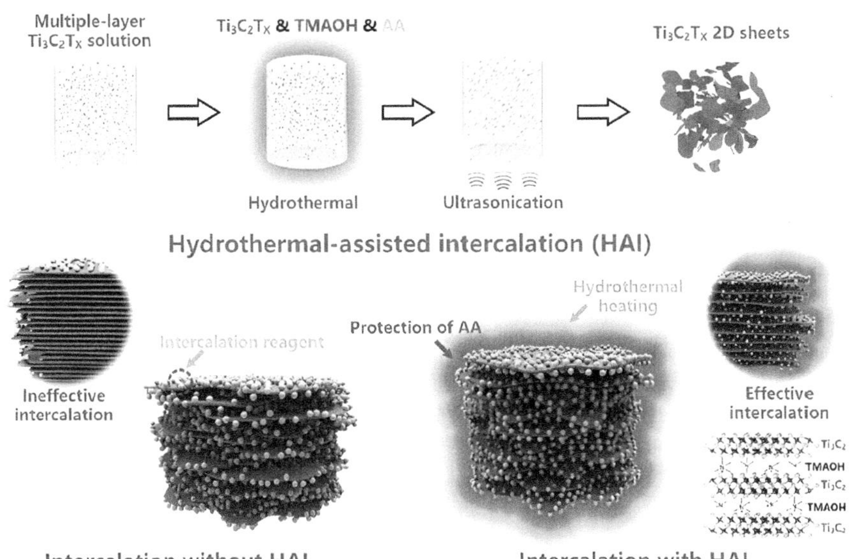

FIGURE 6.25 Schematic illustration of HAI.

Source: Reprinted with permission from [117] Fei Han, Shaojuan Luo, Luoyuan Xie, et al. 2019. Boosting the Yield of MXene 2D Sheets via a Facile Hydrothermal-Assisted Intercalation. *ACS Appl. Mater. Interfaces*, 11, 8443–8452. Copyright 2019, American Chemical Society.

recyclability. This shows their effectiveness for the treatment of organic contaminants in wastewater. Q. Xue et al. [119] described a simple hydrothermal approach for producing water-soluble, monolayered Ti_3C_2 MXene quantum dots (MQDs). The reaction temperature can be used to control their average size. Due to substantial quantum confinement, these MQDs exhibited excitation-dependent photoluminescence (PL) spectra (with around 10% quantum yields). In addition, these materials were used to mark RAW264.7 cells as a biocompatible multicolor cellular imaging probe, demonstrating their considerable potential for biomedical and optical applications. MQDs synthesized at high temperatures (150°C) were also tested for their ability to detect zinc ions.

The hydrothermally synthesized $Ti_3C_2T_x$ was also observed to exhibit surface terminations and efficient electrochemical properties for supercapacitors [120]. It was discovered that the reaction parameters, that is, duration, NH_4F concentration, and temperature, greatly influence the structure and morphology of the material. MXene-based composite materials can also be synthesized by using this approach. For example, M. Xu et al. [121] developed a petal-like MoS_2/MXene composite, which was further utilized as a cathode for Mg-ion batteries. The combination of layered sulfide structure and high conductivity of MXene displayed high-capacity values (165 mAh g^{-1} at 50 mA g^{-1}) as well as high-rate performance at 200 mA g^{-1}. In addition to the electrochemical properties, the composite materials also exhibit excellent dye adsorption in sewage treatment (Fig. 6.26 [122]). Phytic acid

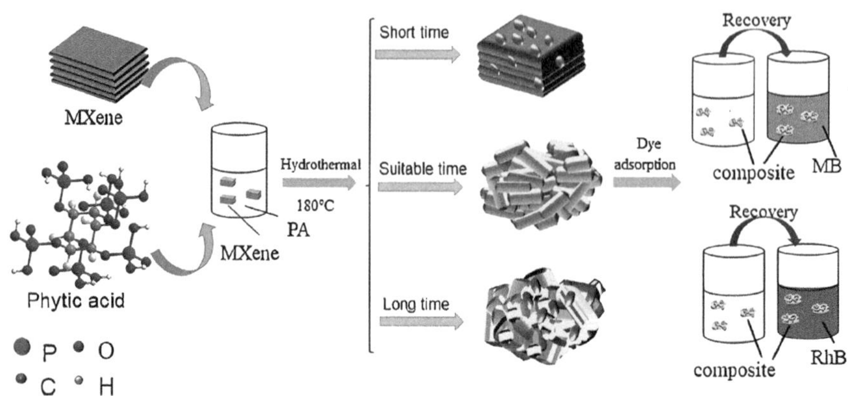

FIGURE 6.26 Schematic illustration of the PA-MXene synthesized hydrothermally using the dye adsorption process.

Source: Reprinted from [122] *Colloids and Surfaces A*, 589, Chong Cai, Ran Wang, Shufeng Liu, et al., Synthesis of self-assembled phytic acid-MXene nanocomposites via a facile hydrothermal approach with elevated dye adsorption capacities, 124468. Copyright 2020, with permission from Elsevier.

(PA)-MXene composites were characterized, and it was discovered that additional rod-like $Ti(OH)PO_4$ components developed during the process. Furthermore, this composite displayed better adsorption properties for RhB and MB, compared to the pristine MXene.

A one-step hydrothermal technique was used to successfully manufacture MXene/$NiFe_2O_4$ nanocomposites with varying concentrations of Ti_3C_2 [123]. The morphologies, structural compositions, and electromagnetic properties were characterized using SEM, XRD, vector network analyzer, and vibrating sample magnetometer. MXenes exhibit a large surface area; therefore, as the concentration of MXene increases, $NiFe_2O_4$ crystallizes from the surface to the interlayer, resulting in a decrease in grain aggregation. W. Kong et al. [124] used TiCN powder as the precursor in a simple and ecologically friendly hydrothermal technique to create MXene-like TiCN QDs. As demonstrated in Fig. 6.27 [124], the as-prepared QDs have fluorescence characteristics along with high water dispersibility.

It was also observed that the hydrothermal process can also utilize $NaBF_4$ and HCl to synthesize h-Ti_3C_2, which are low-toxicity etching chemicals [125]. When compared to the MXenes synthesized by HF etching (t-Ti_3C_2), h-Ti_3C_2 displayed larger interlayer spacing, larger specific surface area, and higher lattice parameter. Apart from the Ti-based MXenes, this method can also be utilized to produce Nb- and V-based materials, that is, Nb_2C and V_2C MXenes. For example, V_2C-based QDs were hydrothermally synthesized and used to develop a white laser, which resulted in a broadband nonlinear scattering system with enhanced gain [126]. X. Wang et al. [127] described a hydrothermal method for developing a NiCoAl-LDH/$V_4C_3T_x$ heterostructure. The hydrothermally synthesized heterostructure displayed high-rate performance and high specific capacity value in 1 M KOH, that is, 627 C g^{-1} at 1 A g^{-1}.

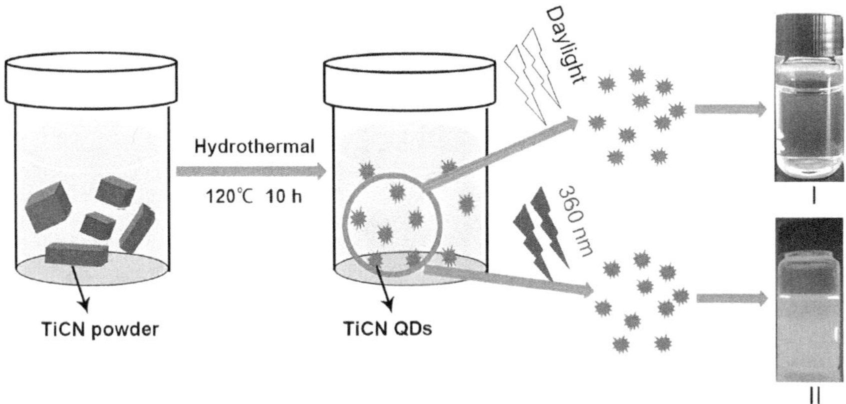

FIGURE 6.27 Schematic illustration of hydrothermal synthesis of TiCN QDs with blue fluorescence emission.

Source: Reprinted from [124] *Inorganic Chemistry Communications*, 105, Wenhan Kong, Yusheng Niu, Mengli Liu, et al., One-step hydrothermal synthesis of fluorescent MXene-like titanium carbonitride quantum dots, 151–157. Copyright 2019, with permission from Elsevier.

6.2.4 SOLVOTHERMAL METHOD

Yan, F. et al. [128] used a solvothermal method to synthesize nitrogen-doped MQDs from Ti_3C_2 precursors, and o-phenylenediamine was used as the source of nitrogen. These N-MQDs were observed to be scattered quasi-spherical NPs comprising O, C, and N elements, which displayed a 5.42% of quantum yield and a yellow fluorescence (580 nm). Another study [129] developed $Ti_3C_2T_x$ MQDs by using different solvents and analyzed their solvent-dependent optical properties. Under UV irradiation (365 nm), DMSO produced white PL, while dimethylformamide (DMF) and ethanol produced blue PL (as illustrated in Fig. 6.28 [129]). Interestingly, among the metal ions tested, Fe^{3+} had a stronger quenching effect in DMF than the other two. Furthermore, for the first time, electrochemiluminescence of MQDs was detected employing persulfate as a co-reactant.

Cui, G. et al. [130] described the microwave absorption performance and design of sandwich-like $CuS/Ti_3C_2T_x$ MXene composites, synthesized by a solvothermal approach. The results showed that the composite exhibits a hetero-layered architecture with a higher microwave absorption ability. Electrostatic self-assembly and solvothermal methods were used to create 2D CdS/2D MXene heterostructures [131]. Figure 6.29 [131] depicts a schematic example of the synthesis of 2D CdS/2D $Ti_3C_2T_x$ nanocomposites. The composite photocatalysts outperformed the pristine CdS NSs in terms of hydrogen-evolving efficiency and robustness. Furthermore, the charge-transport mechanism was revealed using density functional theory (DFT) computations.

In another study, Co_3O_4 particles were evenly distributed on the surface and interior layers of the Ti_3C_2 sheets, thereby preventing restacking and forming an organized composite structure [132]. The structural morphologies of these nanocomposites

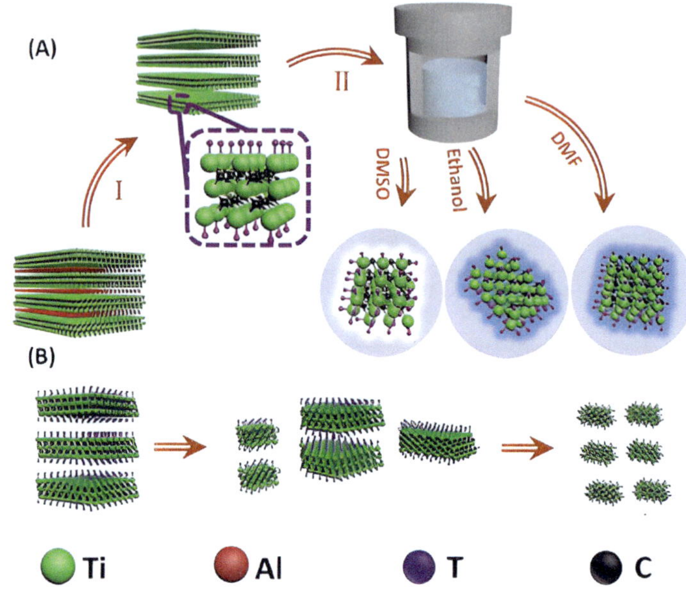

FIGURE 6.28 Scheme of the preparation of MQDs with white/blue-emitting solvent PL via the solvothermal treatment in different solvents of DMF, DMSO, or ethanol.

Source: [129] Xu G., Niu Y., Yang X., et al.: Preparation of $Ti_3C_2T_x$ MXene-Derived Quantum Dots with White/Blue-Emitting Photoluminescence and Electrochemiluminescence. *Adv. Optical Mater*. 2018, 6, 1800951. Copyright WILEY-VCH Verlag GmbH & Co. KGaA, Weinheim. Reproduced with permission.

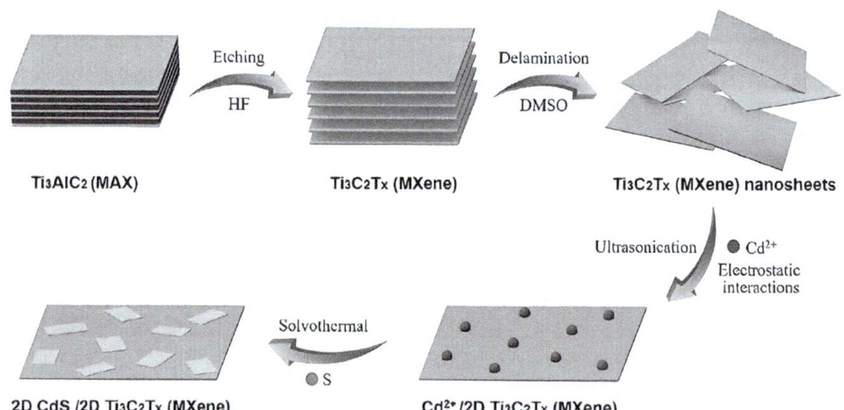

FIGURE 6.29 Schematic illustration of the synthesis of 2D CdS/2D Ti3C2Tx nanocomposites.

Source: [131] Ding M., Xiao R., Zhao C., et al.: Evidencing Interfacial Charge Transfer in 2D CdS/2D MXene Schottky Heterojunctions toward High-Efficiency Photocatalytic Hydrogen Production. *Sol. RRL* 2020, 2000414. Copyright WILEY-VCH Verlag GmbH. Reproduced with permission.

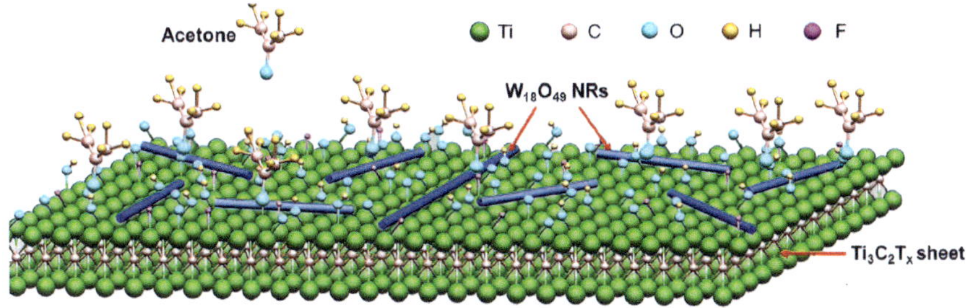

FIGURE 6.30 Schematic of the reaction between acetone and W18O49/Ti3C2Tx composite.

Source: Reprinted from [135] *Sensors and Actuators: B. Chemical*, 304, Shibin Sun, Mingwei Wang, Xueting Chang, et al., $W_{18}O_{49}/Ti_3C_2T_x$ MXene nanocomposites for highly sensitive acetone gas sensor with low detection limit, 127274. Copyright 2020, with permission from Elsevier.

were characterized by different techniques, that is, SEM, TEM, XRD, EDX, etc. Furthermore, their catalytic behavior was also tested for dye degradation of MB and RhB. By using cetyltrimethylammonium bromide-assisted solvothermal synthesis, a simple one-step technique for growing TiO_2 QDs on ultrathin $Ti_3C_2T_x$ NSs was established [133]. The resultant $TiO_2@Ti_3C_2T_x/S$ nanocomposite showed better rate capability and long-term cyclability as a cathode, revealing a new avenue for quick and stable lithium–sulfur batteries. J. Chen et al. [134] successfully synthesized $TiO_2/MXene$ composites using an in situ solvothermal technique in which the shape of TiO_2 was changed by varying concentrations. $Ti_3C_2T_x$, a co-catalyst with electronic storage properties, was also used to boost photocatalytic degradation activity. Similarly, another study [135] demonstrated the fabrication of $W_{18}O_{49}/Ti_3C_2T_x$ composites (Fig. 6.30 [135]) using a simple solvothermal technique. According to the results, composites displayed a very low limit of detection of 170 ppb acetone, fast response and recovery rates, and long-term stability.

$ZnO/Ti_3C_2T_x$ nanocomposites synthesized using a simple solvothermal technique demonstrated an improved NO_2 gas sensor [136]. Apart from Ti-based materials and composites, Mo- and Nb-based MXenes were also synthesized using the solvothermal method. For example, a new NiS/Mo_2CT_x hybrid [137] was developed using chemical etching (NH_4F/HCl solution) followed by solvothermal reactions. The inherent structure and electrochemical properties of an HER electrocatalyst were also examined experimentally. As a result, this composite exhibited large electrochemical surface areas with abundance of active sites along with an improved electrolysis process. Similarly, CdS/Nb_2CT_x composite was created, which facilitated the electron–hole separation and charge transfer in CdS, thereby enhancing its photocatalytic efficiency [138].

6.2.5 PEPLD Method

In this process, a high-temperature plasma is created by focusing the target's surface with a high-power pulsed laser, which then expands and emits to deposit the thin layers on the substrate. Standard pulsed laser deposition (PLD) systems use PEPLD

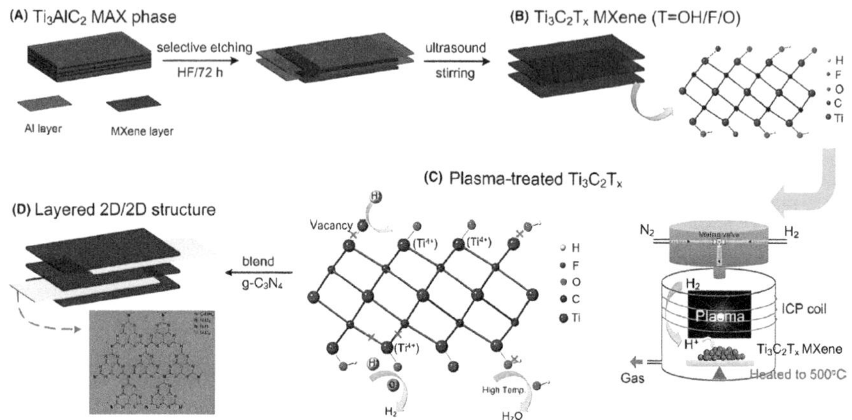

FIGURE 6.31 (A) Structure of Ti_3AlC_2; (B) etching and delamination of layered $Ti_3C_2T_x$; (C) plasma treatment of $Ti_3C_2T_x$; and (D) intercalation of 2D g-C_3N_4 and formation of layered composite.

Source: Reprinted with permission from [139] Xu F., Zhang D., Liao Y., et al. 2020. Synthesis and photocatalytic H_2-production activity of plasma-treated $Ti_3C_2T_x$ MXene modified graphitic carbon nitride. *Journal of the American Ceramic Society*, 103, 849–858. Copyright 2020, the American Ceramic Society.

with plasma enhancement to facilitate the synthesis of high-quality material at lower temperatures [8]. The plasma processing technique is gaining popularity among researchers as a possible method for increasing the photocatalytic activity of catalysts. For example, in an N_2/H_2 atmosphere, F. Xu et al. [139] selected 2D $Ti_3C_2T_x$ as a cocatalyst of g-C_3N_4 for plasma treatment (500°C) (Fig. 6.31 [139]). The surface of layered $Ti_3C_2T_x$ MXene has a greater fraction of Ti-O surface groups as a result of plasma treatment, particularly for Ti^{4+}. An increased photocatalytic activity of the g-C_3N_4/p$Ti_3C_2T_x$ composite with a sandwich-like structure was observed. The increased photocatalytic activity is due to the presence of Ti^{4+} as a result of plasma treatment, which can collect photoinduced electrons from g-C_3N_4 and promote electron and hole separation following visible light irradiation.

$Ti_3C_2T_x$ MXenes exhibit high conductivity, high transparency, solution processability, and variable work function, which has recently piqued the curiosity of its uses in perovskite solar cells (PSCs). However, it has been observed that PSCs based on MXenes display lower performances than TiO_2- or SnO_2-based counterparts. Some crucial concerns concerning the MXene/perovskite interface remain unresolved. J Wang et al. [140] employed a room-temperature solution technique followed by oxygen plasma treatment to create an electron transport layer in PSCs using $Ti_3C_2T_x$ MXene. The surface groups of MXenes and their corresponding influence on the characteristics were determined by using several characterization techniques. It was observed that oxygen plasma treatment generates randomly distributed Ti-O bonds on the surface by disrupting portions of TiC bonds. These modified surface terminations can vary the work functions and increase the electron transport. The resultant material was also analyzed to determine the related perovskite morphology.

The plasma in the PEPLD was created by altering a typical PLD system to include a high-voltage electrode at the methane appearance of chamber that may ionize the gas entering into the chamber. A power generator supplied a DC voltage of 500 V. F. Zhang et al. [141] successfully fabricated ultrathin Mo_2C films at 700°C using a PEPLD method. After growth, the sample was allowed to cool down naturally in a vacuum. Without any etching or transfer processes, transport characteristics can be measured directly on the grown sample. The superconductive Mo_2C films were identified as single crystals with a seldom reported fcc structure using extensive structural studies. Furthermore, in this system, high upper critical magnetic fields were detected. In another study, controlled development of superconductive Mo_2C thin films in PLD on the sapphire (0001) substrate was reported [142]. Recently, this method has also been used to create epitaxial Cr_2AlC MAX phase thin films on MgO(111) and Al_2O_3(0001) substrates at 600°C [143].

6.2.6 Magnetron Sputtering

It has been observed that yarn-based sensors exhibit lower conductivity and sensing stability. Therefore, the hydrophilicity of polyurethane (PU) was improved by its surface engineering, which facilitates the deposition of MXene NSs [144]. Figure 6.32 [144] illustrates the magnetron sputtering process along with the formation of yarn-based sensors. The magnetron sputtering process introduced Ag NPs, which significantly improved the electric conductivity. However, the sensing stability was improved (over 15,000 cycles) due to the encapsulation of the polydimethylsiloxane (PDMS) layer.

Studies have shown that epitaxial thin films of Ti_2AlC and Nb_2AlC phases on a sapphire substrate can also be synthesized via magnetron sputtering [145]. These films were further chemically etched (in LiF/HCl solutions), which provided corresponding Li-intercalated MXene films. As a result, temperature-dependent fast ion transport, optical properties, and high metallic conductivity were displayed by these

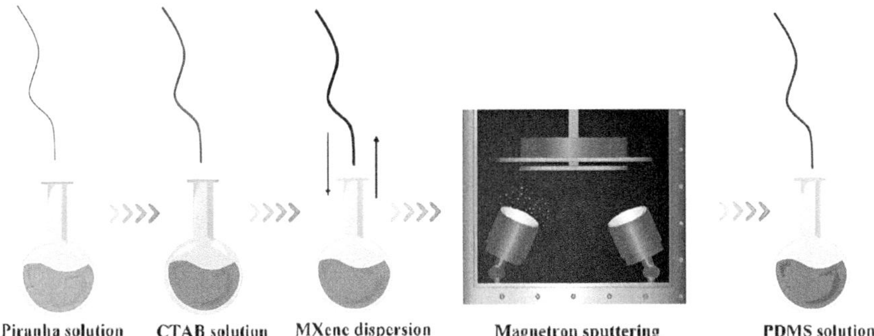

FIGURE 6.32 Fabrication of conductive strain-sensing yarns.

Source: Reprinted from [144] Zhang S, Xu J. PDMS/Ag/MXene/Polyurethane Conductive Yarn as a Highly Reliable and Stretchable Strain Sensor for Human Motion Monitoring. *Polymers*. 2022; 14(24):5401. Open access.

films. Similarly, G. Li et al. [146] used magnetron sputtering to produce few-layer 2D Nb_2C NSs. The Z-scan measurements (with a pump source operating at 1.0 m) revealed that Nb_2C exhibits third-order nonlinear optical response, by displaying remarkable light modulation capabilities. The saturable absorption produced a two times greater nonlinear absorption coefficient than the one produced by reverse saturable absorption processes. In another study, thin films of Mo_2Ga_2C were deposited by employing DC magnetron sputtering from three elemental targets (i.e., Mo, Ga, and C) [147]. Depositions were carried out at 560°C on MgO substrates that had been ultrasonically cleaned for 10 minutes using ethanol, acetone, and isopropanol.

Q. Chen et al. [148] used magnetron sputtering to create semiconductive MXene Sc_2CO_x. The composition of sample was determined using XRD and X-ray photoelectron spectroscopy. The bandgap of the Sc_2CO_x sample was evaluated using UV-Vis and PL spectroscopy. Recently, analysis of the structural morphology and different characteristics of thin-film Cr_2GeC and Cr_2-xMn_xGeC phases, synthesized via magnetron sputtering, was carried out [149]. The elemental ratio of Cr:Ge:C was observed to influence the formation of these phases. Cr_2GeC films with thicknesses of more than 40 nm were composed of well-developed crystallites with metallic optical and transport characteristics. The columnar structure of crystallites could explain the hopping conduction seen in the Cr_2-xMn_xGeC film. Calculations using a two-band model revealed that the best-synthesized Cr_2GeC film had high carrier concentrations of N and P, and mobility m, indicating transport capabilities similar to single-crystal material.

6.2.7 TEMPLATE METHOD

Q. Zhang et al. [150] employed a sacrificial template method to synthesize porous MXenes by using polystyrene (PS) microspheres. This method prevents the restacking of MXene layers for quick charge transport. Figure 6.33 [150] illustrates the complete synthesis process. For the synthesis process, a homogeneous solution was obtained by dissolving PS in deionized water followed by ultrasonication and addition of MXene suspension (the PS/MXene weight ratio was 1:2). The resulting PS–MXene mixture was then rapidly stirred under bath ultrasonication for 2 hours with H_2PtCl_6 solution. The amount of H_2PtCl_6 solution added to the $Ti_3C_2T_x$ colloidal suspension determined the Pt to $Ti_3C_2T_x$ ratio. A small amount of ethanol was used as reducing agent, which facilitates the reduction of residual $PtCl_6$ to Pt. After completing the reaction, PS particles were removed by using the toluene solution.

In another study, to fabricate a MXene@rGO hybrid, first 3D MXene nanospheres were synthesized using a template method. For this purpose, positively charged PS was used as template, which resulted in MXenes with an average diameter of 300 nm [151]. As a result, the 3D composite prevented the restacking of MXenes, provided a large surface area, and reduced the influence of volume expansion. Similarly, Co-LDH/MXene hybrid was developed by this method, which displayed a high reversible capacity for lithium-ion batteries [152]. A 2D $Ti_3C_2T_x$/carbon composite was synthesized by Z. Hu et al. [153], and this delivered an ultralow overpotential of 1.38 V at 0.2 A g^{-1} via a self-sacrificial templating approach (Fig. 6.34 [153]). The heterostructure, which inherited $Ti_3C_2T_x$ MXene's high catalytic performance and

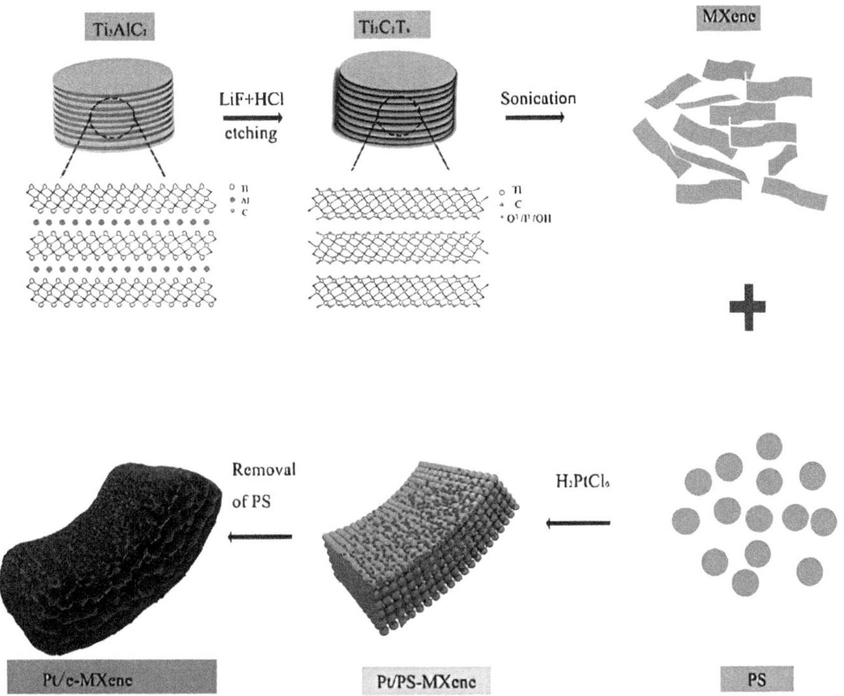

FIGURE 6.33 Synthesis process of the Pt/e-MXene composite.

Source: Reprinted from [150] *Journal of Electroanalytical Chemistry*, 894, Qiang Zhang, Yuxuan Li, Ting Chen, et al., Fabrication of 3D interconnected porous MXene-based PtNPs as highly efficient electrocatalysts for methanol oxidation, 115338. Copyright 2021, with permission from Elsevier.

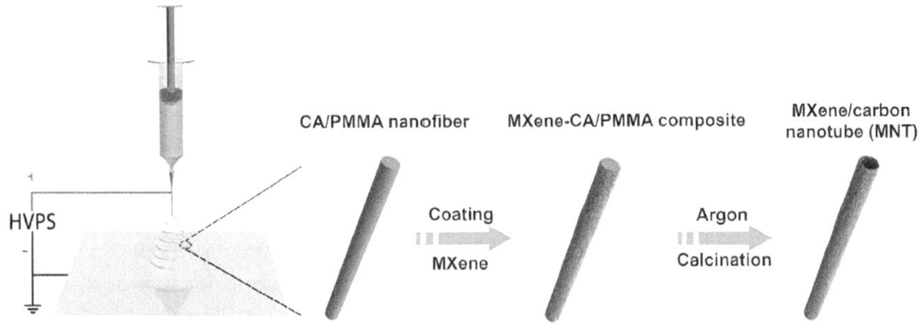

FIGURE 6.34 Schematic illustration of synthetic route for MXene/Carbon Nanotubes (MNT).

Source: Reprinted with permission from [153] Zhe Hu, Yaoyi Xie, Deshuang Yu, et al. 2021. Hierarchical $Ti_3C_2T_x$ MXene/Carbon Nanotubes for Low Overpotential and Long-Life Li-CO$_2$ Batteries. *ACS Nano*, 15, 8407–8417. Copyright 2021, American Chemical Society.

carbon material's outstanding stability, promoted CO_2 adsorption and accelerated the decomposition of lithium carbonate, as demonstrated by DFT calculations.

Studies have shown that template synthesis can also be used along with another technique to improve the synthesis conditions. For example, to develop $Co_3O_4/Al_2O_3@$ $Ti_3C_2T_x$ composite, MXenes were synthesized using a self-sacrificial template and a hydrothermal process [154]. Co_3O_4 as the active center component and Al_2O_3 as the addition were equally dispersed on $Ti_3C_2T_x$ MXene NSs in a three-dimensional $Co_3O_4/Al_2O_3@Ti_3C_2T_x$ MXene composite. The 3D structure provided more active sites and gas diffusion channels, both of which improved gas-sensing performance. In addition to Ti-based materials, Mo_2C was also observed to be synthesized by using a salt-assisted template method [155]. During the synthesis process, KCl serves as a template, which prevents the coarsening of material during melting. It was observed that thickness and HER activity of the as-synthesized material are inversely related.

6.2.8 SOLUTION-PROCESSING METHOD

Due to the hydrophilicity of functionalized MXenes, MXene-based composites can easily be synthesized by solution-processing methods [102]. These approaches readily convert stable dispersions into uniform MXene films, with appropriate adherence to the target substrate. This type of dispersion is known as ink, and it typically comprises additives (e.g., surfactants and binders) to improve the processability and formation of films. Additives damage the electronic characteristics of films/structures in most circumstances and should be eliminated at the end of the synthesis process. However, because MXenes have an abundance of surface functional groups, it is easier to formulate additive-free inks compared to other materials. The type and concentration of the surface groups primarily determine the characteristics, stability, and formation of MXene inks [156]. A guideline map for solution processing of colloidal MXenes is presented in Fig. 6.35 [156].

Viscosity in mPa s versus Throughput graph depicts the printing and coating techniques resolution, conductivity, film thickness, and film transparency. Dip coating with viscosity from 9 to 20, 3 for conductivity, 1 for film thickness and 5 for film transparency, and low throughput. Spin coating with viscosity from 80 to 110, 3 for conductivity, 1 for film thickness and 4 for film transparency, with low throughput. Inkjet printing with low to medium throughput, viscosity from 1 to 10, and resolution 4, conductivity 3, film thickness 1 and film transparency 4. Spray coating aerosol jet printing has medium throughput, viscosity from 1 to 1000, resolution 5, conductivity 4, film thickness 3 and film transparency 3. Blade coating has low to medium throughput, viscosity from 500 to 1000, conductivity 5, film thickness 4, and film transparency 1. Flexographic printing Gravure printing has high to very high throughput, viscosity from 90 to 1000, resolution 3, conductivity 3, film thickness 2 and film transparency 4. Screen printing has high throughput, viscosity from 1300 to 10000, resolution 2, conductivity 4, film thickness 4 and film transparency 1. Extrusion printing has low to medium throughput, viscosity from 500 to 10000, resolution 1, conductivity 4, film thickness 5 and film transparency 1. A scale depicts 1 to 5 with 1 lowest and 5 highest.

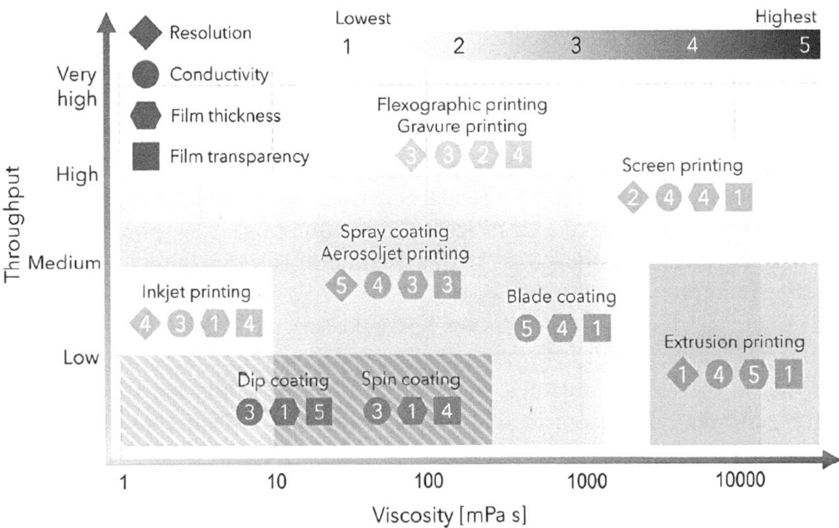

FIGURE 6.35 Guideline map for solution processing of colloidal MXenes.

Source: Reprinted from [156] *Materials Today*, 48, Sina Abdolhosseinzadeh, Xiantao Jiang, Han Zhang, et al., Perspectives on solution processing of two-dimensional MXenes, 214–240. Copyright 2021, with permission from Elsevier.

In a study by S. Sun et al. [157], S-Ti_3C_2 was synthesized by combining the solution-processing technique with electrostatic attraction. The resultant material displayed long cycling life, quick sodium-ion diffusion, and high-rate capability for sodium-ion batteries. Similarly, MXene-based composites can also be synthesized by this method. For example, $FePS_3$ NSs@MXene composite was developed, which displayed higher surface area and electronic conductivity. This resulted in efficient charge storage capabilities and the capacity to increase sodium-ion diffusion, with high reversible capacity (676.1 mAh g^{-1} at 100 mA g^{-1}) after 90 cycles [158]. Furthermore, CTABSn(IV)@Ti_3C_2 was developed using a simple liquid-phase pillaring process, by combining the polymer and intercalated ions with pillared MXene. Due to the pillaring effect, the lithium-ion capacitor developed using this composite exhibited high energy and power densities [159]. These examples show that the MXene-based composites and hybrids synthesized via solution-processing methods display enhanced electrochemical properties. Different solution-based coating and printing techniques are summarized below with the help of recent literature on the synthesis of MXenes.

6.2.9 Spray Coating

The higher values of conductivity require defect-free coatings. A number of films have been formed on various substrates using the CVD technique. However, the higher fabrication costs and reduction in size generate commercialization issues.

Furthermore, owing to inter-flake distance and defects, spin coating and electrodeposition processes also have limitations. To address these challenges, the spray-coating method (SCM) evolved as a new solution-based process for producing defect-free films. In this method, controlled and uniform thin films can be synthesized by spraying the MXene solution repeatedly on the substrate. The viscosity of synthesized films depends upon the substrate type, spray speed, flow, etc. To fabricate the devices, researchers always need a reliable approach to obtain outstanding chemical and electrical properties from large-scale MXene thin films. The spray-coating process has also been investigated for the production of large-scale thin films without affecting their properties [1].

Y. Cheng et al. [160] used a thermal spray-coating method to uniformly deposit highly conductive $Ti_3C_2T_x$ NSs on polydimethylsiloxane (PDMS). It was discovered that at increasing pressures, the contact area of the conductive channel can also expand due to bioinspired microspines, resulting in increased pressure sensor-sensing capability (as illustrated in Fig. 6.36 [160]). Similarly, A. Sarycheva et al. [161] demonstrated wireless communication RF devices based on metallic MXene produced by a single-step spray-coating method. Because of its transparency and flexibility, a smooth PET sheet was used as a spraying substrate. Before deposition, PET substrates were cleaned by sonication in detergent solution, deionized water, and ethanol (limited to 3 minutes in all situations). Substrates were also cleaned with O_2 plasma after drying with compressed air to remove residual pollution and enhance the hydrophilicity of the surface. Plasma cleaning was performed with a 4-sccm O_2 gas flow at 50W for 5 minutes. Following that, the chamber was cleaned with Ar gas.

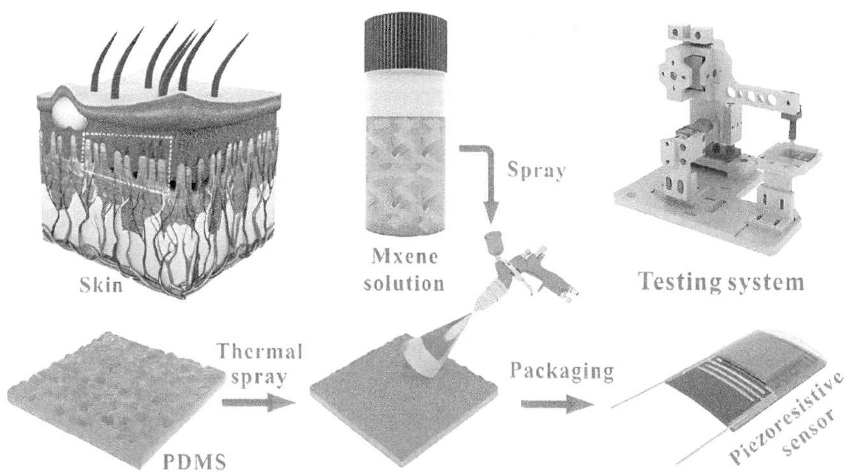

FIGURE 6.36 Schematic illustration of spray coating and piezoresistive activity of $Ti_3C_2T_x$ NSs.

Source: Reprinted with permission from [160] Yongfa Cheng, Yanan Ma, Luying Li, et al. 2020. Bioinspired Microspines for a High-Performance Spray $Ti_3C_2T_x$ MXene-Based Piezoresistive Sensor. *ACS Nano*, 14, 2, 2145–2155. Copyright 2020, American Chemical Society.

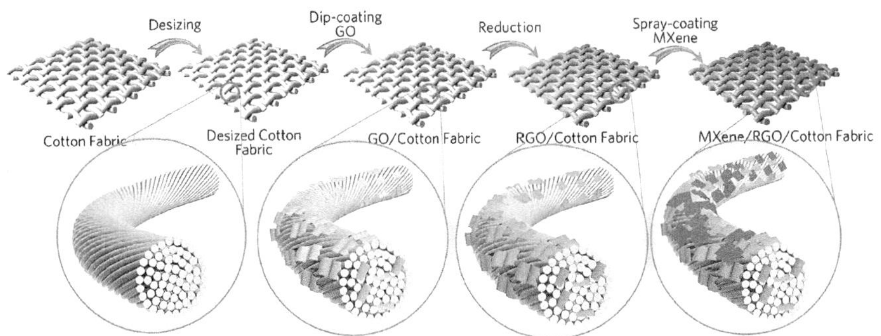

FIGURE 6.37 Spray-coating and dip-coating strategies to fabricate reduced RGO/Ti$_3$C$_2$T$_x$-decorated cotton fabrics.

Source: Reprinted from [163] Materials and Design, 200, Xianhong Zheng, Wenqi Nie, Qiaole Hu, et al., Multifunctional RGO/Ti$_3$C$_2$T$_x$ MXene fabrics for electrochemical energy storage, electromagnetic interference shielding, electrothermal and human motion detection, 109442. Copyright 2021, with permission from Elsevier.

W. Chen et al. [162] revealed that Ti$_3$C$_2$T$_x$ layer, synthesized using a scalable spray-coating process, can be combined with Ag NW to improve its environmental and mechanical stabilities along with EMI shielding performance. The resultant composite film was observed to exhibit better EMI performance (34 dB) in comparison to the Ag NW film of the same density. In another study, dip-coating and spray-coating methods were used together to develop a MXene/RGO/cotton fabric composite [163]. The complete synthesis process is illustrated in Fig. 6.37 [163]. It was observed that this hybrid of electroactive materials exhibits flexibility, a hydrophilic surface, high electrical conductivity, and excellent electrochemical performance for all-solid-state supercapacitors (ASSSs) with one of the highest values of specific capacitances, that is, 383.3 F g^{-1}.

Spray-coating is a cost-effective approach, and it has found wide applicability in thin-film electronics. This method is also observed to be compatible with roll-to-roll processing and inkjet printing, which are large-area processing techniques. Apart from the electrochemical properties, MXenes synthesized by this method can also be used to fabricate metal oxide thin-film transistors. For example, Ti$_3$C$_2$-based large area source were developed using a solution-processing method [164]. It was observed that a good ohmic contact can be formed due to the correspondence of band edges of metal oxides (i.e., SnO and ZnO) with the work function of MXene. M. Zhao et al. [165] developed 2D MXene/graphene heterostructured papers by spray-assisted layer-by-layer assembly of Ti$_3$C$_2$T$_x$ MXene and reduced graphene oxide (rGO) nanosheets (Fig. 6.38 [165]). When utilized as anode materials for Na-ion storage, these films outperformed rGO films and pure MXenes in terms of rate performance, cycling stability, and capacity. The Ti$_3$C$_2$T$_x$ MXene–20 wt% rGO heterostructured films attain capacities of 600 mAh g^{-1} and have an outstanding cycling stability.

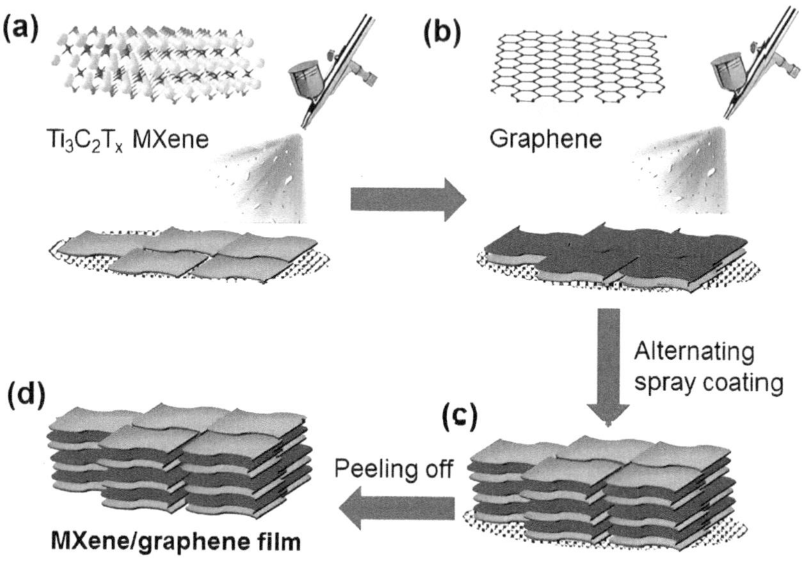

FIGURE 6.38 Schematic showing the manufacturing of 2D MXene/graphene heterostructured films using a spray-assisted layer-by-layer assembly.

Source: [165] Anasori B., Gogotsi Y., Ren C. E., et al.: Scalable Manufacturing of Large and Flexible Sheets of MXene/Graphene Heterostructures. *Adv. Mater. Technol.* 2019, 1800639. Copyright WILEY-VCH Verlag GmbH & Co. KGaA, Weinheim. Reproduced with permission.

6.2.10 Dip Coating

Another approach for producing thin-film MXenes with reduced roughness and tunable thickness is the dip-coating method (DCM). In this process, thin films are developed by immersing the substrate in the MXene solution. The nature and surface modification of the substrate, concentration of the MXene solution, and dipping time are the parameters that influence the film production process in the DCM. For adequate layer deposition, the substrate must be immersed in the MXene solution for nearly 5 minutes. To create uniform and defect-free films, cleaning the substrate and multiple dipping cycles are required. Due to less contact with air, a coating shield is observed to be provided on MXenes. Thin films of MXenes synthesized by this approach can be utilized in various applications, including flame retardancy, micro-supercapacitors, and EMI shielding. A pristine or functionalized substrate is dipped into the MXene-containing solution in the DCM. An extremely homogeneous thin coating developed on various substrates after numerous dipping cycles in and out [1]. For example, Ti_3C_2 thin films, synthesized by dip coating, exhibited good electrical and optoelectronic properties [166]. Previously, unreported automated scalpel engraving was used to create transparent and semitransparent MXene micro-supercapacitors in a single step. MXene-based devices demonstrated remarkable high-rate storing capacities without the use of external current collectors by combining transparent conductive electrode (TCE) and pseudocapacitive properties.

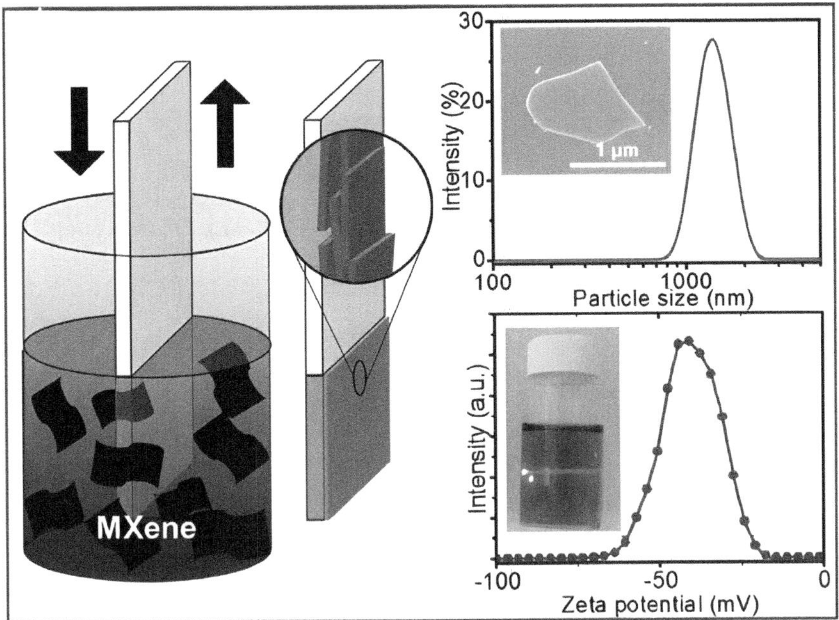

FIGURE 6.39 Dip coating of optically transparent Ti_3C_2 film with typical distribution profiles for particle size and zeta potential.

Source: [166] Gogotsi Y., Kurra N., et al.: Automated Scalpel Patterning of Solution Processed Thin Films for Fabrication of Transparent MXene Microsupercapacitors. *Small* 2018, 1802864. Copyright WILEY-VCH Verlag GmbH & Co. KGaA, Weinheim. Reproduced with permission.

Schematic image of the synthesis and structural morphologies of the optically transparent Ti_3C_2 film are shown in Fig. 6.39 [166].

A. Kumar et al. [167] reported that posttreating the 3D-printed electrodes with electroactive compounds can improve their electrochemical behavior. For this purpose, $Ti_3C_2T_x$ and TMDs (e.g., $MoSe_2$, WSe_2, WS_2, and MoS_2) were dip-coated onto nanocarbon electrodes to investigate their HER catalytic activity. The resultant electrode materials displayed better HER performance. Similarly, this method can also be used to synthesize thin and transparent MXene films [168]. They used UV-vis-NIR spectroscopy to analyze the electrochromic behavior of $Ti_3C_2T_x$. The results revealed that in an acidic electrolyte, the absorption peak shifts reversibly by 100 nm, thereby displaying 12% change in transmittance. This solution-processing method was observed to synthesize MXene-based polymer composites. L. Li et al. [169] developed MXene/PVA/PU sponge composites by employing solution blending along with the dip-coating method. The step-by-step synthesis process is schematically illustrated in Fig. 6.40 [169]. According to the results, thermal stability and anti-dripping performance of polymer matrices were observed to be increased significantly due to dip-coated MXenes.

Similarly, several studies reported the synthesis of MXene-based materials by using a dip-coating technique, and the materials were further utilized for various

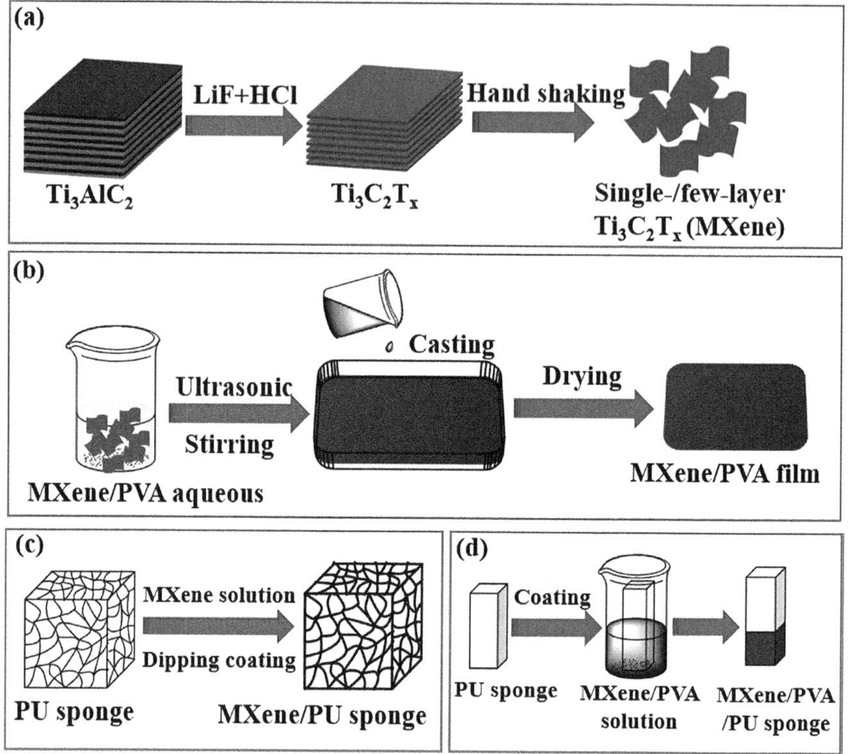

FIGURE 6.40 (a) Etching and delamination of MXene NSs, (b) preparation of PVA/MXene film, (c) PU/MXene sponge, and d) PU-PVA/MXene sponge.

Source: Reprinted from [169] *Composites Part A*, 127, Lei Li, Xiaoya Liu, Jianfeng Wang, et al., New application of MXene in polymer composites toward remarkable anti-dripping performance for flame retardancy, 105649. Copyright 2019, with permission from Elsevier.

applications. To mention a few examples, MXene-sponge was created by inserting insulating polyvinyl alcohol (PVA) NWs for piezoresistive sensors [170]. A MXene/cellulose nanocomposite paper was developed, which was observed to be thermally and electrically conductive [171], and the coating of $Ti_3C_2T_x$ MXene on a PT (PAN@ TiO_2) film substrate exhibited a highly conductive framework [172]. In another study, hybrids of Ni flower/MXene were created using electrostatic self-assembly and dip-coating adsorption on the surface of melamine foam (MF) [173]. The resulting Ni/MXene-MF displayed better heat insulation, low density, flame retardancy, and low density. The preparation process is presented in Fig. 6.41 [173]. The strong electromagnetic wave absorption resulted in interface polarization, magnetic and dielectric loss, multiple attenuation, and outstanding impedance matching. S. Yao et al. [174] proposed a unique fine silk yarn dip-coated in an aqueous solution of $Ti_3C_2T_x$ MXene to achieve low electrical resistance ($\gg 25.6$ Ohms cm^{-1}). They discovered that both solution concentration and soaking time had a substantial influence on yarn performance by varying two coating parameters as the design of experiment (DoE) components.

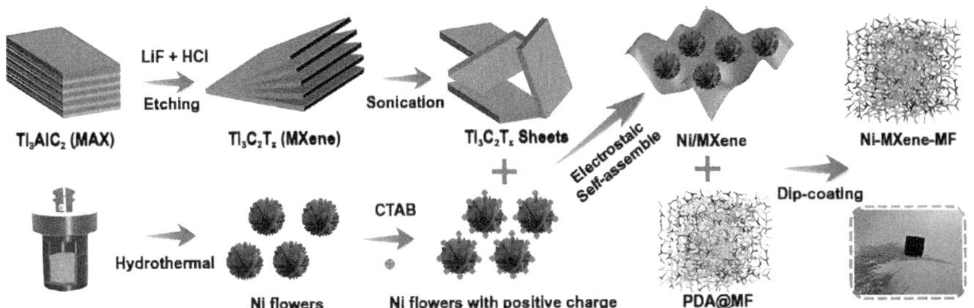

FIGURE 6.41 Schematic illustration of the preparation process for Ni/MXene-MF.

Source: Reprinted with permission from [173] Springer Nature. *Nano-Micro Letters*, Haoran Cheng et al. Ni Flower/MXene-Melamine Foam Derived 3D Magnetic/Conductive Networks for Ultra-Efficient Microwave Absorption and Infrared Stealth. 14:63, copyright 2022.

6.2.11 Spin Casting

Spin coating is a widely adopted technique to produce thin films of different materials. The process involves high-speed spinning of a substrate while spreading a solution on its surface. The surface tension of the liquid and centripetal force combines to distribute the liquid uniformly on the surface of substrate. After removal of the solvent, thin films can be obtained, with the thickness ranging from a few nm to μm. The spin coating method has also been used to create MXene thin films, in which the nature of substrate depends on how the MXene components are employed [1]. For example, L. Yang et al. [175] reported the synthesis of Ti$_3$C$_2$T$_x$ NSs for low-temperature PSCs, in which the colloidal solutions were spin-coated onto indium tin oxide (ITO) substrates. The surface of Ti$_3$C$_2$T$_x$ films was treated with UV-ozone to enhance the interface properties of the Ti$_3$C$_2$T$_x$/perovskite junction, which resulted in the formation of oxide-like (Ti-O) bonds. Similarly, thick and freestanding V$_2$CT$_x$ films were developed by spin-casting the concentrated colloidal solution of the MXene on a glass substrate, followed by vacuum annealing at 150°C [176].

The spin-casted MXene films have been observed to exhibit optical transmittance and high electrical conductivities. In a recent study, photodetectors were developed by using Ti$_3$C$_2$-based electrodes, which outperformed the performance of typical Au electrodes. These electrodes were developed by spin-casting the aqueous suspensions of MXenes on GaAs substrate, followed by treating with acetone [177]. The devices based on these electrodes (MXene and Au) were observed to exhibit comparable internal electric fields and Schottky barrier heights. However, the MXene-based devices demonstrated much greater quantum efficiencies and responsivities, compared to that of Au-based devices, resulting in an improved detectivity and dynamic range, as well as equivalent sub-nanosecond reaction rates. Figure 6.42 [177] illustrates the synthesis process for MXene–GaAs–MXene photodetector devices.

Layered structure (a) with T i, A l, C and T$_2$ under etching and exfoliation leads to separated layers (b). (c) U V light passes over a layer with mask and photoresist over a thick layer. Developing (d) leads to two layers with two rectangular structures. Spin casting (e) with MXenes leads to MXenes deposited thick sheet layered structure,

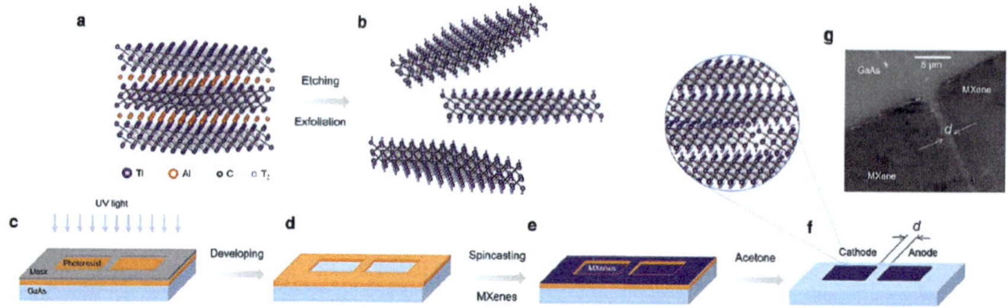

FIGURE 6.42 Fabrication process and energy band diagrams for MXene–GaAs–MXene (MX–S–MX) photodetector devices.

Source: [177] Montazeri K., Currie M., Verger L., et al.: MXene Photodetectors: Beyond Gold: Spin-Coated Ti$_3$C$_2$-Based MXene Photodetectors. *Adv. Mater.* 2019, 31, 1903271. Copyright WILEY-VCH Verlag GmbH & Co. KGaA, Weinheim. Reproduced with permission.

which reacts with acetone and leads to step (f). A circular structure of molecules in the cathode, and the anode is adjacent to the cathode. The space labelled d is between the cathode and anode. (g) A microscopic view of the MXene with gap between them labelled d.

The studies have shown that mostly Ti-based MXene films have been synthesized by using the spin-casting approach. However, these films have shown superior applications in PSCs, photovoltaic devices, micro supercapacitors (MSCs), and piezoelectric devices, due to high conductivity and structural properties. For example, conductive spin-casted Ti$_3$C$_2$T$_x$ films [178] were combined with SnO$_2$ electron transport layer (ETL) to produce a SnO$_2$-Ti$_3$C$_2$ colloidal suspension, which displayed efficient characteristics in low-temperature PSCs [179]. Similarly, low-cost and scalable on-chip MSCs were also developed by creating MXene-based electrodes via cold laser cutting, followed by spin coating on Si/SiO$_2$ substrates [180]. In another study, flexible and freestanding MXene films were synthesized by inserting the MXene particles into hydrophobic poly(vinylidene fluoride) trifluoro ethylene (P(VDFTrFE)) [181]. These films were used to develop highly sensitive piezoresistive pressure sensors with long-term stability, quick reaction time, and no responsivity when exposed to air (even after 20 weeks). Furthermore, the developed sensors could be utilized to detect human physiological signals such as knee bending and cheek bulging, as well as for speech recognition.

It has been observed that on a macroscopic scale, the spin-casted films exhibit the conductivity that is comparable to that of 2D sheets. For example, high optical quality and thin Ti$_3$C$_2$ films were developed via spin-casting, which exhibited more than 97% transmission of visible light along with high electric conductivity (at 6,500 S cm^{-1}) [182], thereby outperforming the conductivity values of other solution-processed 2D materials. The schematic illustration of the crystal structure and deposition of Ti$_3$C$_2$T$_x$ film on different substrates is shown in Fig. 6.43 [182]. M. Zhang et al. [183] developed uniform MXene films by fine-tuning the production parameters and Ti$_3$C$_2$ concentration in several solutions. Three alternative processing methods, including mechanical exfoliation, wafer dipping, and spin-coating technology, were thoroughly explored. Furthermore, optimized dispersion and concentration

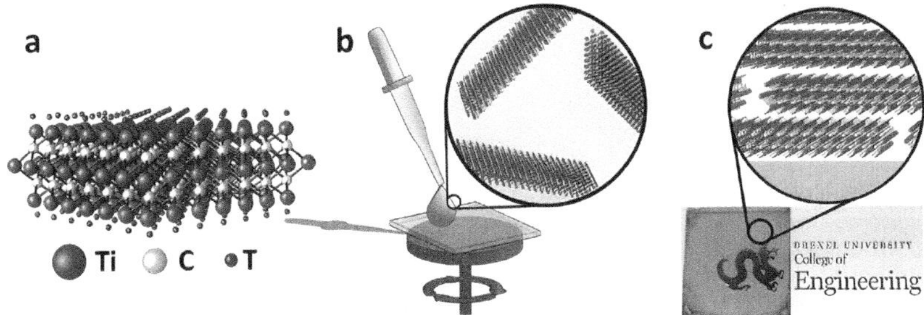

FIGURE 6.43 Schematic of (a) the crystal structure of $Ti_3C_2T_x$ NSs. (b) Deposition of $Ti_3C_2T_x$ onto a variety of substrates by spin casting. (c) Spin-cast $Ti_3C_2T_x$ film on glass.

Source: [182] Dillon A. D., Ghidiu M. J., Krick A. L., et al.: Highly Conductive Optical Quality Solution-Processed Films of 2D Titanium Carbide. *Adv. Funct. Mater.* 2016, 26, 4162–4168. Copyright WILEY-VCH Verlag GmbH & Co. KGaA, Weinheim. Reproduced with permission.

of Ti_3C_2, along with spinning speed, were also investigated. Following that, OM, XRD, SEM, and TEM measurements were used to explore and analyze the physical properties of MXene films.

6.2.12 DROP-CASTING AND ADSORPTION

Drop-casting and adsorption are the nonreactive ways of self-assembly that rely on electrostatic/van der Waals forces. This technique is suitable for synthesizing materials that do not require high-temperature treatments. Generally, MXene/C-based hybrids are synthesized by this method, which include MXene/CNTs, MXene/GO, and MXene/rGO hybrids [102]. For example, $Ti_3C_2T_x$/C composite was synthesized by using a drop-casting technique, which displayed efficient electrochemical behavior in asymmetric supercapacitors [184]. To synthesize the composite material, a carbon fabric was washed with ethanol, acetone, and deionized water, followed by blowing with N_2 gas. The MXene dispersion was then drop-cast onto the carbon cloth before hardening in an oven at 50°C. This procedure was performed numerous times to achieve the desired mass loading.

Similarly, W. Chen et al. [185] reported a drop-casting method, with high speed and uniformity, to deposit Ti_3C_2 on a clean wiper (CW, a typical textile) while also alleviating the resulting self-restacking phenomenon of Ti_3C_2 (shown in Fig. 6.44 [185]). For the synthesis process, a CW was placed on a hotplate (at 60°C), and a colloidal solution of MXene was added by using a pipette gun with a row of tips. For uniform distribution, the solution was rolled with a hydrophobic polytetrafluoroethylene (PTFE) rod for 5 minutes until the water was fully evaporated. This procedure was repeated numerous times to guarantee that the desired amount of Ti_3C_2 was loaded. A thin PTFE layer was used between the CW and the hotplate to prevent contamination. The resulting Ti_3C_2/CW composites were indicated as CWM-X, where "X" represents the number of drop-casting rounds.

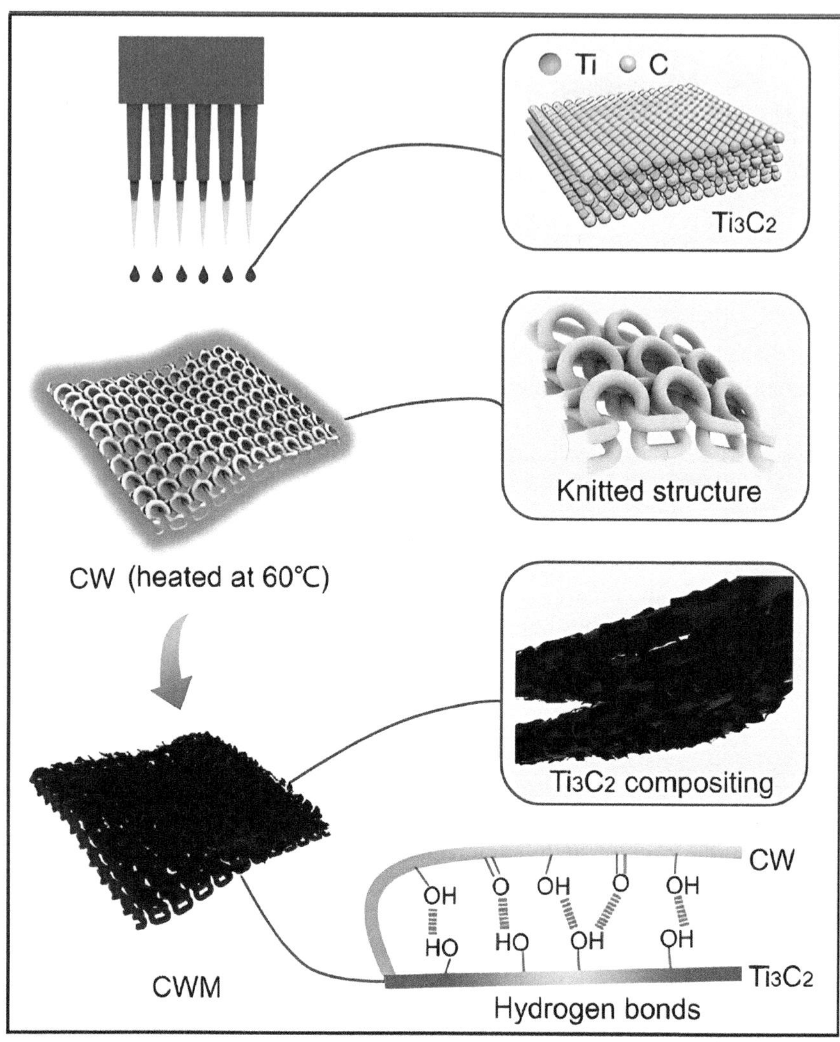

FIGURE 6.44 Schematic illustration on preparing the clean wiper loaded MXene (CWM).

Source: Reprinted from [185] *Chemical Engineering Journal*, 423, Weimin Chen, Min Luo, Kai Yang, et al., MXene loaded onto clean wiper by a dot-matrix drop-casting method as a free-standing electrode for stretchable and flexible supercapacitors, 130242. Copyright 2021, with permission from Elsevier.

In another study, drop-casting of delaminated MXene (d-V_2CT_x) was carried out for a room temperature gas-sensor device [186]. M. K. Singh et al. [187] reported the drop-casting of $Ti_3C_2T_x$ layer on a fabric-based material, developed by growing CoS on carbon cloth (CC) fibers. Due to the combined effect of hydrophilicity, high conductivity, and redox active sites, the resultant electrode $Ti_3C_2Tx/CoS@CC$ showed an excellent augmentation in supercapacitor characteristics when compared to its components. Similarly, to make composite solvent resistant nanofiltration

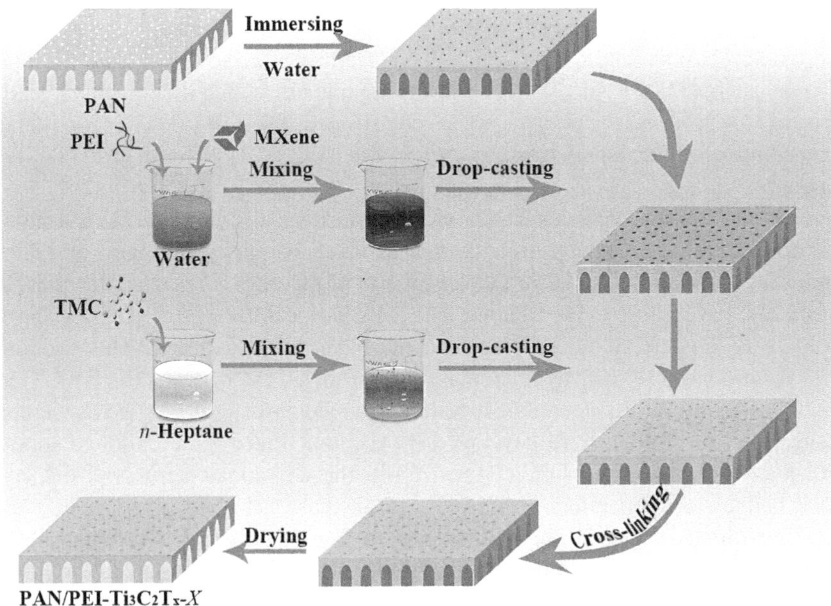

FIGURE 6.45 Synthesis process of the PAN/PEI-Ti$_3$C$_2$T$_x$-X.

Source: Reprinted from [188] *Journal of Membrane Science*, 515, Xiaoli Wu, Lan Hao, Jiakui Zhang, et al., Polymer-Ti$_3$C$_2$T$_x$ composite membranes to overcome the trade-off in solvent resistant nanofiltration for alcohol-based system, 175–188. Copyright 2016, with permission from Elsevier.

(SRNF) membranes, Ti$_3$C$_2$T$_x$ was well-dispersed into a PEI matrix to create a hybrid active layer, which was then drop-cast on top of polyacrylonitrile (PAN) ultrafiltration membrane that served as a support layer. The detailed synthesis process is illustrated in Fig. 6.45 [188]. PVA and MXenes were mixed in another study using stirring and ultrasonication [189]. Drop-casting was used to generate PVA-MXene nanocomposite thin films. This approach allows for a wider scale of attainable perovskite-molecule composite (PMC) thin films.

6.2.13 MOLECULAR INTERCALATION

After obtaining multilayered MXenes by etching procedures, different intercalants can be introduced between MXene layers by simply shaking or ultrasonication, to obtain mono- or few-layered materials. In 2013, an ultrasonic treatment was carried out to delaminate the first multilayered Ti$_3$C$_2$ by using DMSO as intercalants. However, due to its high boiling point, DMSO can replace the active groups on the top layer of the surface of MXenes. Up till now, researchers have introduced a wide range of organic and inorganic intercalants, including hydrazine hydrate (HM), urea, and DMF. Furthermore, different ions, that is, Li$^+$, Na$^+$, K$^+$, etc., can also be intercalated between MXene layers [8]. Later, to produce additional MXene delamination without being limited to Ti_3C_2, other MXenes, such as Nb_2CT_z, were intercalated by using isopropylamine dissolved in water. Different organic bases, such as choline hydroxide, TBAOH, and n-butylamine, have been reported to delaminate additional MXenes such as V_2CT_z and Ti_3CNT_z. These organic bases increase the interlayer distance of MXenes, and the mixture is then centrifuged. TBAOH is later used to delaminate additional MXenes such as $Mo_2TiC_2T_z$, $Mo_2Ti_2C_3T_z$, $Ti_4N_3T_z$, and Mo_2CT_z. Aryl diazonium salts were also used as intercalants for Ti_3C_2. First, the sodium ion is injected into multilayered material, and then the surface is modified with sulfanilic acid diazonium ions. The approach has the advantage of quickly removing aryl diazonium salts and improving the surface chemistry of MXenes for new applications [8].

Studies have shown that delaminated MXenes exhibit better electrochemical behavior as compared to non-delaminated materials. For example, O. Mashtalir et al. [190] delaminated Nb_2CT_x by intercalating isopropylamine between the Nb_2CT_x layers, followed by mild sonication in water. After delamination, CNT/MXene composite paper electrodes were developed by vacuum filtering the colloidal solution with a CNT-containing solution. As a result, the delaminated material displayed much better electrochemical performance than those achieved using multilayered (non-delaminated) material. In another study, delaminated V_2C/CNT composite was utilized as a promising cathode for an aqueous zinc-ion hybrid supercapacitor [191]. Similarly, polyvinylpyrrolidone (PVP) was used to synthesize Sn(IV)-intercalated Ti_3C_2 MXene [192]. XRD and TEM studies showed that Sn^{4+} was successfully inserted into an alkalization-intercalated Ti_3C_2 (alk-Ti_3C_2) matrix. As a result, the nanocomposites demonstrated better reversible volumetric capacity. In comparison to sodium ion, intercalation of high-valency aluminum ions was observed to increase the cyclic stability and Li-ion storage capacity of $Ti_3C_2T_x$ anode [193]. This process enlarges the ion transfer channel, which becomes broadened after charging, thereby improving the ionic conductivity. In another study, fatty diamines and aromatic diamines of different molecular weights were inserted as pillars between MXene interlayers using a one-step amination procedure to restrict self-stacking and generate variably enlarged interlayer spacings with increased antioxidant stability [194]. The XRD analysis revealed that this intercalation increased the interlayer spacing from 1.23 to 1.40 nm, which resulted in an electrolyte-accessible surface area, improved charge-transport characteristics, and facilitated Zn^{2+} ion storage.

6.2.14 Electrospray

S. Cho et al. [195] developed $Ti_3C_2T_x$-based binder-free electrodes for supercapacitors using an electrospray deposition (ESD) process (as illustrated in Fig. 6.46 [195]). In a controlled atmosphere, an aerosolized MXene solution was deposited, which resulted in the evaporation of the solvent in the aerosolized droplets. This quick evaporation of the solvent caused capillary forces and initiated the crumpling process. Using this method, they homogeneously deposited crumpled MXene flakes onto a Ti-foil current collector, resulting in a crumpled 3D structure with no additives, further synthetic procedures, or binders. As a result, a porous structure evolved within the electrode between MXene flakes, resulting in the creation of channels for fast ion diffusion. These ion diffusion channels, when combined with binder-free networks

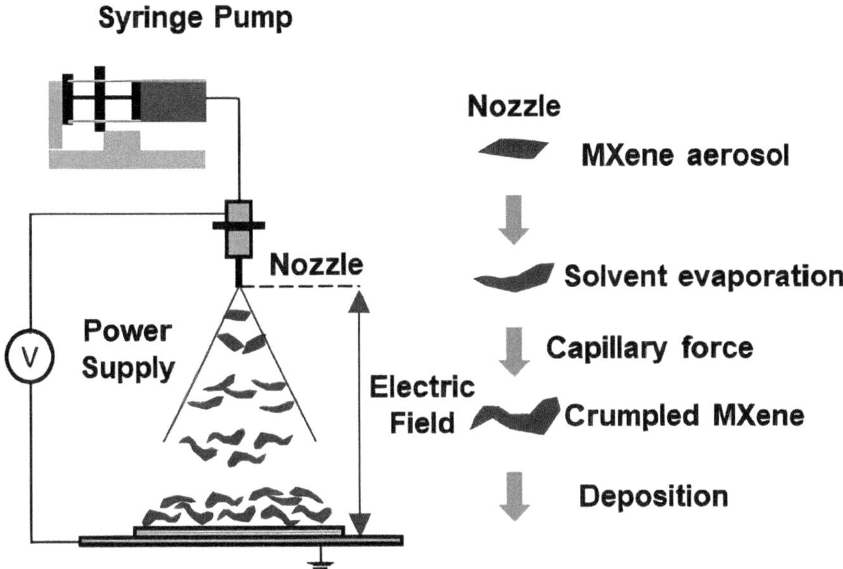

FIGURE 6.46 ESD technique to fabricate binder-free porous $Ti_3C_2T_x$ electrodes.

Source: [195] Cho S., Kim D. Y., and Seo Y.: Binder-Free High-Performance MXene Supercapacitors Fabricated by a Simple Electrospray Deposition Technique. *Adv. Mater. Interfaces* 2020, 2000750. Copyright WILEY-VCH Verlag GmbH. Reproduced with permission.

of MXene sheets, significantly increased the rate capability of electrodes. The cyclic voltammetry (CV) curves revealed the capacitance retention of these electrodes in KOH (even at 10,000 mV s^{-1}). In H_2SO_4, the electrode displayed a specific capacitance of 400 F g^{-1} along with 85% capacitance retention at high scan rates.

In another study, researchers studied the visible-light-driven photocatalytic activity of Ti_3C_2 cocatalyst combined with Ag/Ag_3PO_4, synthesized via electrostatic self-assembly [196]. The resultant hybrid exhibited photocatalytic activity toward MO and Cr(VI), which was observed to be dependent on the concentration of the MXene under irradiation of visible light. Similarly, a $Ti_3C_2T_x$ MXene-engineered membrane was developed, which displayed strong water repellency and photothermal effect. To develop PM-poly(vinylidene fluoride) (PVDF) membranes, MXene NSs were inserted onto the C-PVDF layer by using an electrospray technique [197]. The electrospray engineering method consists mostly of three steps: charged droplet creation, coulombic explosion, and phase separation.

6.2.15 Vacuum-Assisted Filtration

VAF is the simplest method for producing homogenous MXene thin films. A stable MXene dispersion was generated in this approach, and it was filtered using a suction filtering apparatus. After filtration, the MXene adheres to the surface of the porous membrane and creates a thick coating. The concentration of the material along with the solvent and filter membrane decides the thickness of the film. On the surface of the filter membrane, the formation of the film is greatly influenced by van der Waals forces. The MXene films synthesized by this method can be employed in various applications. Although the VAF method is simple to use, it requires a significant amount of time and effort to make a film. To speed up the process, a small amount of alkali is added to Ti_3C_2 MXene before filtration, which reduces the filtering time. Another key component of the VAF-fabricated MXene thin films is the peeling of the entire film for patterning to gain improved applications [1].

A. Iqbal et al. [198] used the VAF method to synthesize $Ti_3C_2T_x$ and Ti_3CNT_x freestanding films. After filtration, the films were annealed at various increasing temperatures for 6 hours in an argon atmosphere. The shielding capability of carbonitride was produced by thermal annealing, which resulted in high absorption of electromagnetic waves. Similarly, d-$Ti_3C_2T_x$/cellulose nanofiber (CNF) composite was developed using the VAF approach. The resulting material exhibited flexible and ultrathin structure with nacre-like lamellar morphology [199], as illustrated in Fig. 6.47 [199]. In another study, thin $Ti_3C_2T_x$ NSs were developed on PVDF filter membranes by using the VAF method [200]. The resultant MXene-based membranes, based on ion charge and hydration radius, exhibited differential salt sieving and fast water flux. These membranes are suitable for separation applications, as they display hydrophilic surfaces, flexibility, high electrical conductivity, and mechanical strength.

W. Xin et al. [201] showed that vacuum-assisted filtered MXene–CNFs–silver (MCS) composite membranes exhibit electromagnetic shielding properties of 50.7 dB and high electrical conductivity of 588.2 S m^{-1}. Similarly, due to aqueous dispersion and negative surface charges, a colloidal combination of MnO_2 and Ti_3C_2 sheets displayed superior conductivity [202]. A hybrid film was created by molecularly stacking the constituent sheets using a simple VAF approach on a cellulose acetate membrane, as shown schematically in Fig. 6.48 [202].

6.2.16 Green Synthesis Method

J. Iqbal et al. [203] used NaOH as an etching agent to develop a more ecologically friendly technique to synthesize MXenes. Furthermore, magnetic characteristics were induced during the etching process, which resulted in the development of a magnetic composite. Different structures (denoted as M-Ti_2CT_x) could be generated by carefully manipulating etching parameters (such as concentration and time). Magnetic nanostructures having distinct physicochemical properties (i.e., a large number of binding sites) were used to adsorb radioactive Sr^{2+} and Cs^+ from various matrices, including deionized, salt, and tap water. When compared to corrosive acid-etched MXenes, these materials were shown to be particularly stable in the aqueous state, acquiring a unique structure with $-O$ groups. The removal efficiencies of Sr^{2+} and Cs^+ from M-Ti_2CT_x were determined using traditional batch adsorption methods. M-Ti_2CT_x-AIII was observed to outperform other M-Ti_2CT_x phases in terms of adsorption performance. A. Wojciechowska et al. [204] presented a simple, low-cost, and entirely green method for limiting the potential cytotoxicity of delaminated MXenes by exploiting interactions between the surface of the MXene and a collagen.

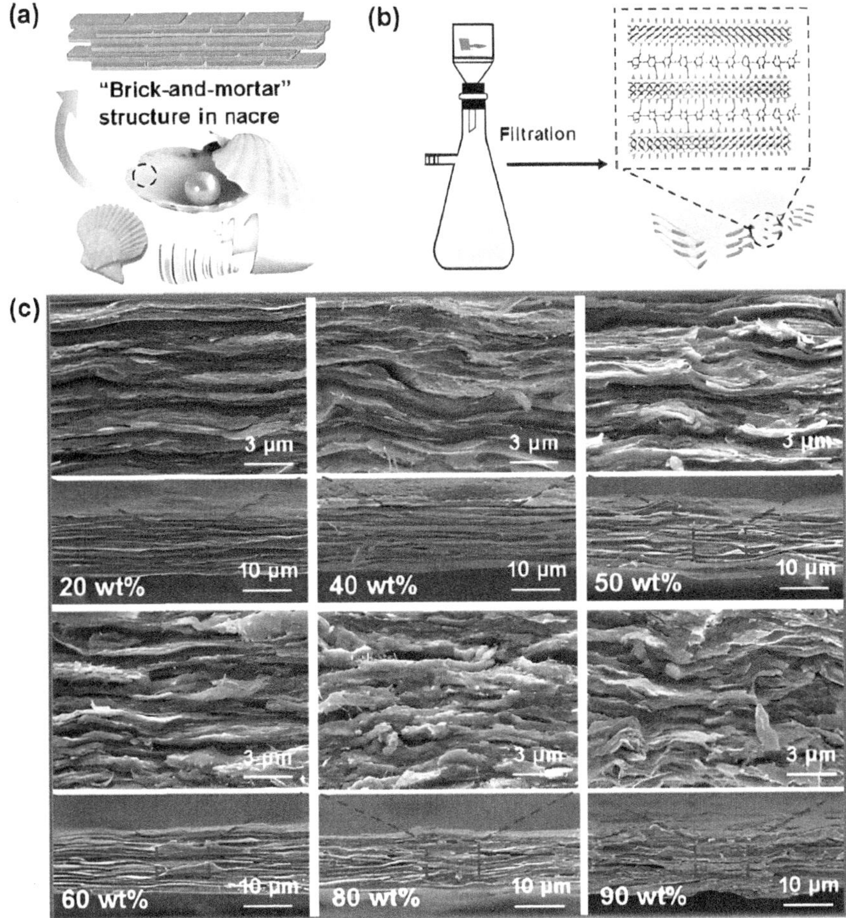

FIGURE 6.47 (a) "Brick-and-mortar" structure in nacre. (b) Preparation of the d-Ti$_3$C$_2$T$_x$/ CNF composite paper via a vacuum filtration method. (c) SEM images of the d-Ti$_3$C$_2$T$_x$/CNF composite paper.

Source: Reprinted with permission from [199] Wen-Tao Cao, Fei-Fei Chen, Ying-Jie Zhu, et al. 2018. Binary Strengthening and Toughening of MXene/Cellulose Nanofiber Composite Paper with Nacre-Inspired Structure and Superior Electromagnetic Interference Shielding Properties. *ACS Nano*, 12, 4583–4593. Copyright 2018, American Chemical Society.

It was observed that adsorption/desorption of collagen can easily be controlled by utilizing in situ zeta potential along with dynamic light scattering (DLS) measurements. The results showed that the surface of MXenes displayed electrostatically induced sensitivity toward collagen.

According to L. Xiu et al. [205], the synthesis of MXenes by using Lewis's acid in molten salt provides a safer and more efficient method for MXene manufacture. Extending the spectrum of parent MAX phases allows for more flexibility in modifying the capabilities and features of MXenes. More importantly, it enabled the

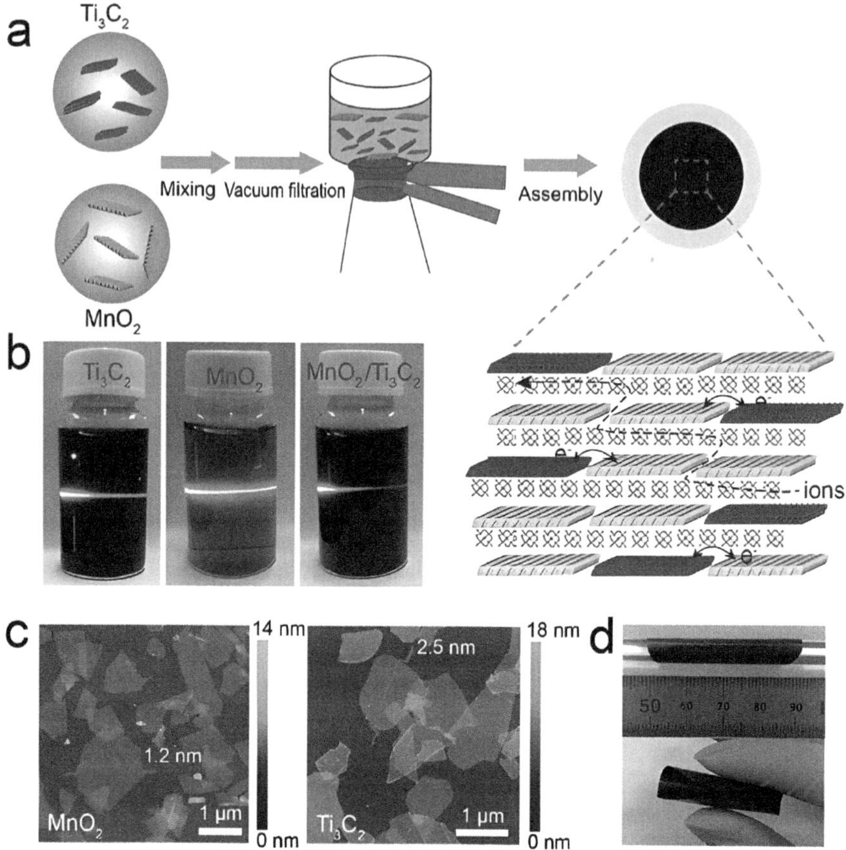

FIGURE 6.48 (a) Fabrication of the MnO$_2$/Ti$_3$C$_2$ hybrid film by VAF. (b) Digital photographs of the MnO$_2$ and Ti$_3$C$_2$ colloids, and the MnO$_2$/Ti$_3$C$_2$ mixture.

Source: [202] Liu W., Wang Z., Su Y., et al.: Molecularly Stacking Manganese Dioxide/Titanium Carbide Sheets to Produce Highly Flexible and Conductive Film Electrodes with Improved Pseudocapacitive Performances. Adv. Energy Mater. 2017, 1602834. Copyright WILEY-VCH Verlag GmbH & Co. KGaA, Weinheim. Reproduced with permission.

creation of novel MXenes that are difficult to produce using other approaches. Without a doubt, if energy usage could be decreased further, this general strategy could offer considerable promise in the scalable manufacturing of MXenes with tunable chemical and structural compositions. R. Ramrez et al. [206] presented a study in which MQDs were prepared by making a suspension of MAX phase in water. To obtain permanent dispersion, the suspension was sonicated, transferred to a quartz cuvette closed with a septum, and bubbled through a needle (for 15 minutes) with argon. The second harmonic of a Q switched Nd:YAG laser was used to irradiate the dispersion. A power meter was used to monitor laser power, while the probe head was placed in the sample holder. Similarly, reduced Ti$_3$C$_2$T$_x$ was developed by using L-ascorbic acid treatment, which is a green approach carried out at room temperature. The resultant material displayed oxidation stability and high electrical conductivity [207].

6.2.17 Sol-gel Technique

Another bottom-up approach is the sol-gel method, which is widely being used to synthesize MXene-based composites. Y. Huang et al. [208] used a sol-gel technique to develop Li_3VO_4 (LVO)/$Ti_3C_2T_x$ MXene composites. The resultant composite displayed better electrochemical performance than commercial graphite anodes and bare LVO. Figure 6.49 [208] depicts a schematic representation of the LVO/$Ti_3C_2T_x$ MXene composite preparation. Similarly, J. Wang et al. [209] used a blade-coating approach with a sol-gel conversion step to create large-scale, layered MXene/ANF (amarid nanofiber) sheets inspired by natural nacre, which exhibited excellent mechanical properties.

MXene-based aerogel fibers are widely used in flexible electronic devices due to their multi-mesoporous characteristics. A pure $Ti_3C_2T_x$ aerogel fiber was developed by employing a simple dynamic sol-gel spinning, followed by supercritical CO_2 drying [210]. The resultant aerogel fiber displayed an ultrahigh electronic conductivity mesoporous structure. In addition to MXenes, MAX phases can also be synthesized using this method. J. P. Siebert et al. [211] described a solution-based synthesis method for producing MAX phase Cr_2GaC from a solution of citric acid and metal nitrates. This solution-processable precursor combination can be easily be fabricated onto substrates, as evidenced by the creation of hollow carbon microspheres adorned with Cr_2GaC particles (as illustrated in Fig. 6.50 [211]).

J. Sinclair et al. [212] described an alternate sol-gel-based method for preparing nearly single-phase V_2PC. The adaptability of sol-gel chemistry was further proved by changing the gel-building ingredient from citric acid to another carbon source.

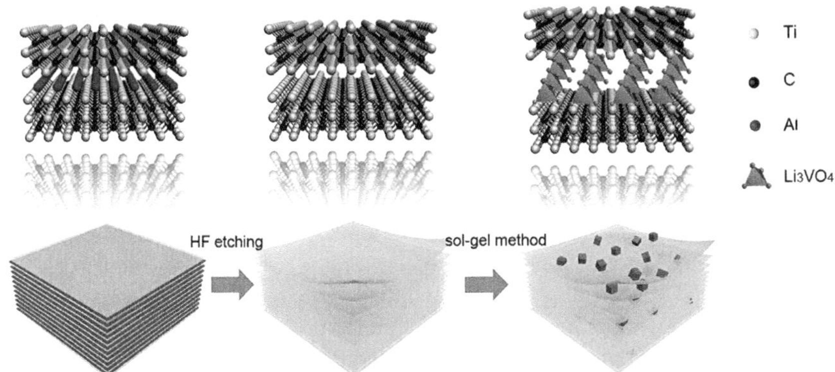

FIGURE 6.49 Schematic illustration of LVO/$Ti_3C_2T_x$ MXene composite preparation.

Source: [208] Chen P., Huang Y., Liu N., Ma Y., et al. 2019. A safe and fast-charging lithium-ion battery anode using MXene supported Li_3VO_4. *J. Mater. Chem. A*,7, 11250–11256. Reproduced with permission from the Royal Society of Chemistry.

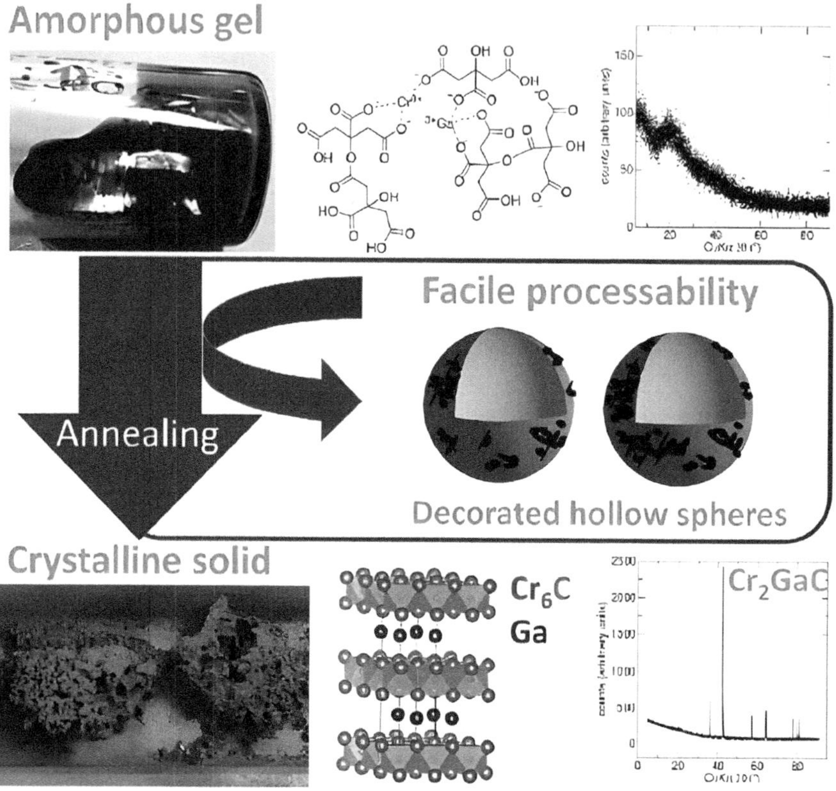

FIGURE 6.50 Schematic showing the transformation of the amorphous gel into a crystalline MAX phase solid.

Source: [211] Siebert J. P., Bischoff L., Lepple M., et al., 2019. Sol-gel based synthesis and enhanced processability of MAX phase Cr$_2$GaC. *J. Mater. Chem. C*,7, 6034–6040. Reproduced with permission of the Royal Society of Chemistry.

DFT computations confirmed the experimentally determined structural characteristics and demonstrated that V$_2$PC is metallic in nature. H. Zhang et al. [213] successfully suppressed voltage hysteresis by using a nonstoichiometric synthesized approach for acquiring N$_{3.5}$MTP. Furthermore, as-fabricated N$_{3.5}$MTP grains were grown in situ on MXene-rGO aerogels using a sol-gel synthesis technique, yielding N$_{3.5}$MTP@ MXene-rGO. This aerogel, as an outstanding conductive matrix, can efficiently reduce particle agglomeration and ameliorate lattice volume change during sodium-ion intercalation/deintercalation. An essentially different approach was used in another study: a series of controlled compositions of TiO$_2$-MXene composite were prepared by mixing sol-gel-derived TiO$_2$ nanopowder with the Ti$_3$C$_2$T$_x$ component, which was obtained by HF etching of the corresponding MAX phase [214].

Z. Wu et al. [215] reported sol-gel, freeze drying, and heat treatment methods to develop silane-modified MXene and polybenzazole (F-MP) nanocomposite

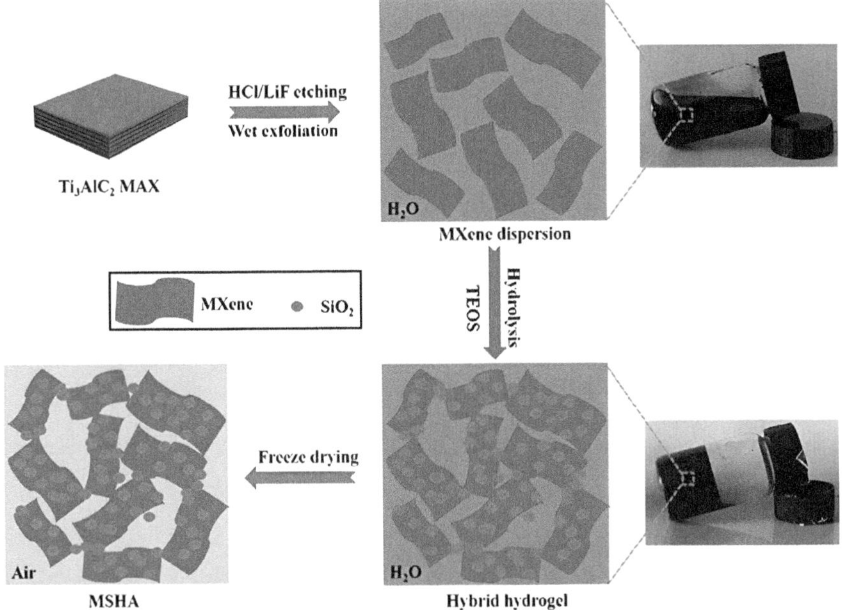

FIGURE 6.51 Synthesis process of MXene/SiO$_2$ hybrid aerogel.

Source: Reprinted from [216] Zhou J, Pei Z, Sui Z, Liang Y, Xu X, Li Y, Li Y, Qiu J, Chen Q. Hierarchical Porous and Three-Dimensional MXene/SiO$_2$ Hybrid Aerogel through a Sol-Gel Approach for Lithium–Sulfur Batteries. *Molecules*. 2022; 27(20):7073, Open access.

aerogels. After a strain of 50% compression, the optimized aerogel displayed low density, good surface hydrophobicity (water contact angle of nearly 141°), and stale mechanical toughness. Furthermore, even after a 60-second flame attack, this aerogel demonstrated consistent flame resistance and structural integrity. Figure 6.51 [216] shows how a new porous material with a high surface area, MXene/SiO$_2$ hybrid aerogel, was created using sol-gel and freeze-drying processes. The composite containing a combination of 2D and 3D material in the hierarchical porous hybrid aerogel was used not only as a sulfur-based cathode but also as an efficient functional separator.

R. Tian et al. [217] used a sol-gel approach along with a thermal treatment to control the production of phases in mesoporous iron oxide. They discovered that mixed α/γ-Fe$_2$O$_3$ phases have improved acetone detection performance. Furthermore, when α/γ-Fe$_2$O$_3$ is coupled with Ti$_3$C$_2$T$_x$, agglomeration of Fe$_2$O$_3$ nanoparticles can be prevented, thereby increasing the absorption sites. In another study, sol-gel auto combustion, etching, and mechanical blending were used to create Ti$_3$C$_2$T$_x$-MXene, Ba$_{1.8}$Sr$_{0.2}$Co$_2$Fe$_{11.9}$Pr$_{0.1}$O$_{22}$, and their composites [218]. A sol-gel approach was used to create polyacrylamide (PAM) composite hydrogels with Ti$_3$C$_2$T$_x$-MXene (PAM/MXene) [219]. MXene integration at a low level of 0.3 wt% in hydrogel resulted in a significantly compact 3D porous structure (Fig. 6.52 [219]).

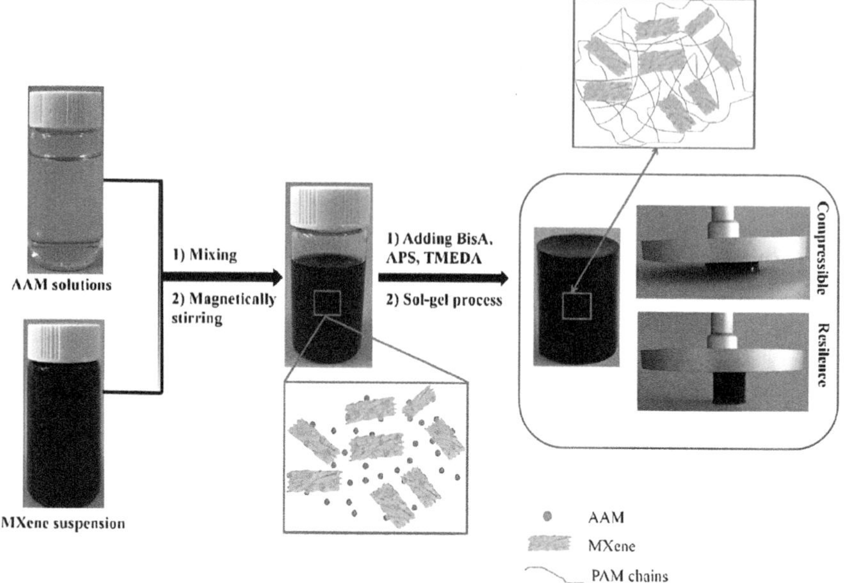

FIGURE 6.52 Schematic illustration of the composite hydrogels of PAM with $Ti_3C_2T_x$-MXene.

Source: Reproduced from [219] Chen F., Wang R., Chen H., Lu H. 2021. Preparation of polyacrylamide/ MXene hydrogels as highly-efficient electro-adsorbents for methylene blue removal. *Polymer-Plastics Technology and Materials*, 1568–1584. With permission from Taylor & Francis Group.

6.2.18 Hot Press Technique

MXenes have higher thermal stability and a lower degradation temperature than polymeric compounds. In this approach, MXene and a melted polymer matrix are blended to form a dispersion using a stirrer of desired intensity under temperature control; finally, this melt is hot-pressed to generate the final composite. This approach has various advantages, including being versatile, solvent-free, cost-effective, and environmentally friendly. As a result, this can be used for scaled product manufacturing. This method is useful for hydrophobic polymers, which cannot be synthesized using solution-based techniques [102]. R. Syamsai et al. [220] used a hot-pressing process to create V_4C_3-delaminated MXene from its MAX precursor, V_4AlC_3. In summary, this process consisted of a two-step synthesis (Fig. 6.53 [220]). The parental MAX phase was originally synthesized by utilizing vanadium, aluminum, and carbon (4:1.5:3) elemental metals. The elemental elements' stoichiometry weights were ball-milled for 20 hours at a ball to powder ratio of 1:10. The powder was put onto a graphite punching die and heat-pressed at 1,500°C at 1 ton pressure with a ramping rate of 5°C/min under a high vacuum of 5×10^{-6} bar.

In another study, dense MXene/PI films were developed by the interaction of MXenes with PI macromolecules [221]. The MXene/PAA foam, synthesized via a freeze-drying method, was hot-pressed to obtain the required film. A complete synthesis process is illustrated in Fig. 6.54 [221]. Layered structure MXenes facilitate

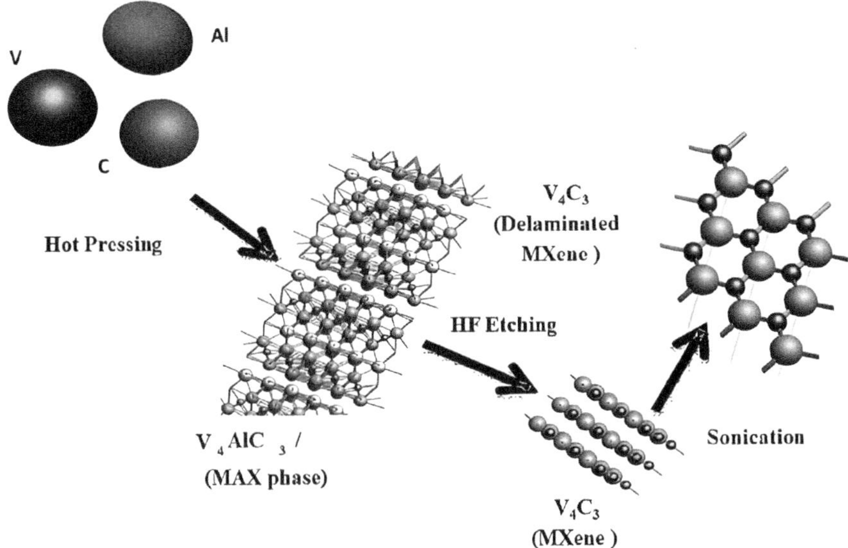

FIGURE 6.53 Synthesis of V_4C_3 MXene.

Source: Reprinted from [220] *Ceramics International*, 46, Ravuri Syamsai and Andrews Nirmala Grace, Synthesis, properties and performance evaluation of vanadium carbide MXene as supercapacitor electrodes, 5323–5330. Copyright 2020, with permission from Elsevier.

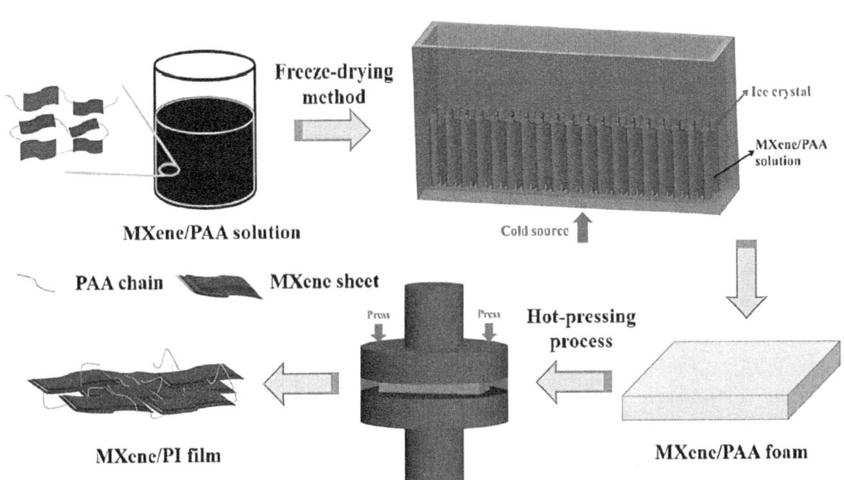

FIGURE 6.54 Schematic illustration of the preparation process of the MXene/PI film.

Source: [221] Yue Zhu, Xingbin Zhao, Qingyu Peng, et al., 2021. Flame-retardant MXene/polyimide film with outstanding thermal and mechanical properties based on the secondary orientation strategy. *Nanoscale Adv.*, 3, 5683. Reproduced with permission from the Royal Society of Chemistry.

heat transfer by reducing thermal resistance. This resulted in MXene/PI films with enhanced fracture strength, sheet friction, and mechanical properties.

In addition to the MXenes, hot pressing technique is widely used to synthesize MAX phases. For example, Nb, Al, and C powders were hot-pressed at 1,700°C with a load equal to 30 MPa (for 1 hour), which resulted in polycrystalline Nb_4AlC_3 [222]. According to XRD analysis, the resultant powders were mostly single-phase Nb_4AlC_3. In another study, a hot-press method was used with solution-processing methods (spin coating and spray coating) to develop $Ti_3C_2T_x$/PVDF films exhibiting high dielectric constant [223]. This process was also utilized for developing Ti_3C_2/UHMWPE composites [224]. When compared to pure UHMWPE, the composites demonstrated improved hardness and yield strength. With the addition of Ti_3C_2, the anti-friction performance of polymer was also drastically reduced. Similarly, surface-modified Ti_3C_2 with polydopamine and amino silane, which was generated by solution blending and hot pressing, enhances the wear resistance and damping capabilities of nitrile butadiene rubber (NBR) composites [225]. The results revealed that surface modification increases the damping, hardness, and tribological parameters of unfilled NBR.

6.3 STRATEGIES TO SYNTHESIZE NITRIDE MXENES

While most of the research on MXene materials is focused on carbides, particularly Ti_3C_2, there are only a few experimental attempts that concentrate on nitride MXenes. Although the limited number of publications on nitride MXenes could be responsible for the delay in their progress, it is not the only factor that affects their development. Nitride MXenes hold great potential in terms of their properties and applications, as anticipated from theoretical and computational work. However, the greatest difficulty in the development of nitride MXenes is the higher formation energy of their MAX precursor and the lower cohesive energy of the M-N bond inside the structure. Another downside to this process is that it also dissolves the MXene, resulting in the breakdown of the structural composition [226]. With perseverance, researchers can overcome the limitations of nitride MXenes and take advantage of their beneficial features. According to DFT calculations, nitrides are predicted to have electrical conductivity, greater active surface area, magnetic moments, Young's modulus, and water stability than carbide MXenes with the same metal and stoichiometry. These improved features directly translate to better performance in nitride MXene applications, such as supercapacitors, CO_2 capture and conversion, HER, OER, ORR, batteries, and sensors [226]. This vast difference in capacitance performance can be seen in Fig. 6.55 [226].

Elements versus areal capacitance and specific capacitance graphs are depicted. Graph a: Vertical axis represents specific capacitance in F/g ranging from 0 to 500 in increments of 100. The data shown in the graph are as follows. M_2C: V, 350; Nb 290; and Mo 150. M_3C_2: Ti 250; Zr, 190; Nb, 150; and Mo 100. M_4C_3: Ti, 250; V, 190; Zr, 200; Nb, 190; and Mo, 100. Graph b: Vertical axis represents specific capacitance in F/g ranging from 0 to 500 in increments of 100. M_2N: Ti, 450; Zr, 350; Nb, 220; Mo, 280. M_3N_2: Ti, 300; Zr, 250; Nb, 150; Nb, 150; and Mo, 200. M_4N_3: Zr, 200; Nb, 100; and Mo, 100. Graph c: Vertical axis represents areal capacitance in mu F/cm² ranging from 0 to 80 in increments of 20. M_2C: V, 55; Nb, 60; Mo, 40. M_3C_2: Ti, 50; Zr, 55;

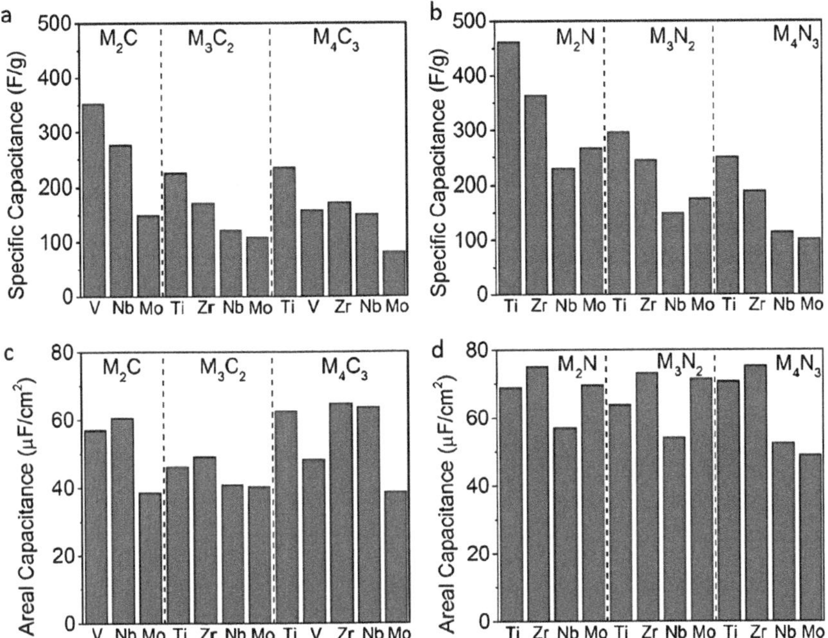

FIGURE 6.55 (a, c) Predicted capacitances of carbide MXenes. (b, d) Predicted capacitances of nitride MXenes.

Source: [226] Johnson D., Qiao Z., Uwadiunor E., Djire A., et al: Holdups in Nitride MXene's Development and Limitations in Advancing the Field of MXene. *Small* 2022, 18, 2106129. Copyright WILEY-VCH Verlag GmbH. Reproduced with permission.

Nb, 40; M_4C_3: Mo, 40; Ti, 60; V, 50; Zr, 65; Nb, 65; and Mo, 40. Graph d: Vertical axis represents areal capacitance in mu F/cm^2 squared ranging from 0 to 80 in increments of 20. M_2N: Ti, 70; Zr, 78; Nb, 59; Mo, 69. M_3N_2: Ti, 63; Zr, 70; Nb, 58; Mo, 70; M_3N_2: Ti, 60; Zr, 70; Nb, 55; Mo, 70; M_4N_3: Ti, 70; Zr, 78; Nb, 55; and Mo, 50.

Recently, it has been observed that in situ/operando approaches are essential in understanding and developing the catalytic characteristics of nitride MXenes. These approaches are essential in uncovering mechanisms and improving the understanding of MXenes' behavior in many applications, including sensing, energy storage, and electronics. So far, these characterization approaches have mainly been used for supercapacitors and batteries [226]. Below, you can find a summary of nitride MXenes, along with a graphic illustration of their synthesis processes. S. Venkateshalu et al. [227] described a novel and simple procedure for synthesizing V_2NT_x MXene and using it as a supercapacitor electrode. Etching Al from its MAX phase V_2AlN yielded the V_2NT_x MXene. The V_2AlN phase was created by ball-milling vanadium and AlN and then cold pressing the resulting pellet. Following that, the pellet was heat treated, which aids in densification and reduces structural defects such as cracks and microvoids. It has been discovered that the pressure used during cold pressing influences the preferred orientation of the grain throughout its growth.

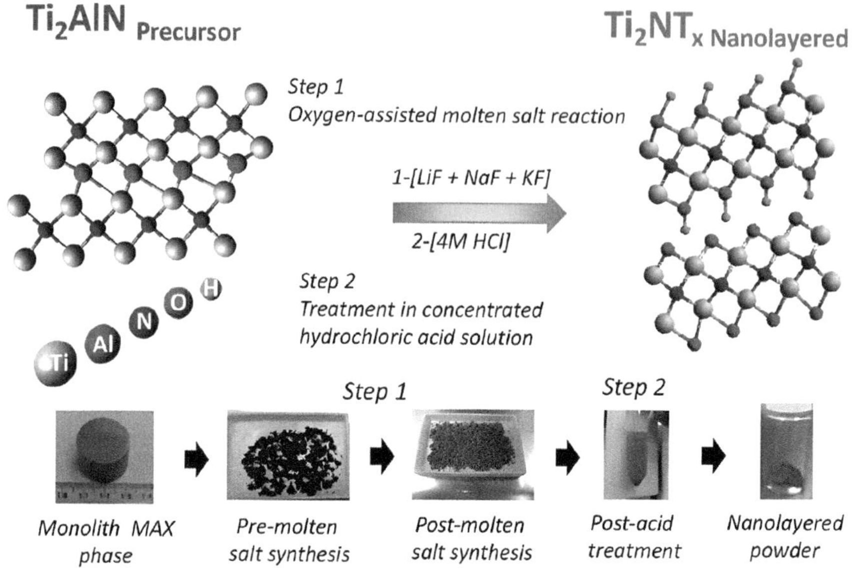

FIGURE 6.56 Synthesis procedure for Ti2NTx MXene via oxygen-assisted molten salt reaction.

Source: Reprinted with permission [228] from Abdoulaye Djire, Andre Bos, Jun Liu, et al. 2019. Pseudocapacitive Storage in Nanolayered Ti_2NT_x MXene Using Mg-Ion Electrolyte. *ACS Appl. Nano Mater.*, 2, 2785–2795. Copyright 2019, American Chemical Society.

A. Djire et al. [228] used an oxygen-assisted molten salt fluoride process to exfoliate Ti_2AlN into the corresponding Ti_2NT_x MXene, as shown in Fig. 6.56 [228]. In a 1:1 mass ratio, Ti_2AlN powder was treated with a combination of fluoride salts (i.e., LiF, NaF, and KF). Under argon flow, the mixture was heated from 25°C to 550°C at a rate of 10°C min^{-1}. The argon flow was converted to 1% O_2 in N_2 gas as soon as the final temperature was reached, and the mixture was kept in these conditions for 1 hour. After 1 hour of 1% O_2 treatment, the sample turned a black brownish color. The powdered MXene and aluminum fluorides that resulted were collected and placed in a sealed container.

B. Soundiraraju et al. [229] synthesized Ti_2N MXene by immersing Ti_2AlN (2 g) in 20 mL of a KF and HCl mixture at ambient temperature for 3 hours. To aid exfoliation and intercalation, the suspension was heated for 1 hour at 40 °C while being bath sonicated. The supernatant was collected, and soluble fluorides were removed by centrifuging the powder mass at 3,500 rpm. After that, the powder was treated with isopropyl alcohol to yield ML-Ti_2NT_x. To delaminate the multilayered structure, DMSO was utilized as the intercalating agent. The resultant MXene was studied for surface-enhanced Raman scattering (SERS) activity by developing SERS substrates from silicon, paper, and glass, as illustrated in Fig. 6.57 [229].

P. Urbankowski et al. [230] synthesized $Ti_4N_3T_x$ by exfoliating its corresponding MAX precursor using molten salts (i.e., 59 wt% KF, 29 wt% LiF, and 12 wt% NaF) in an Ar atmosphere. The resultant nitride MXene displayed a layered hexagonal

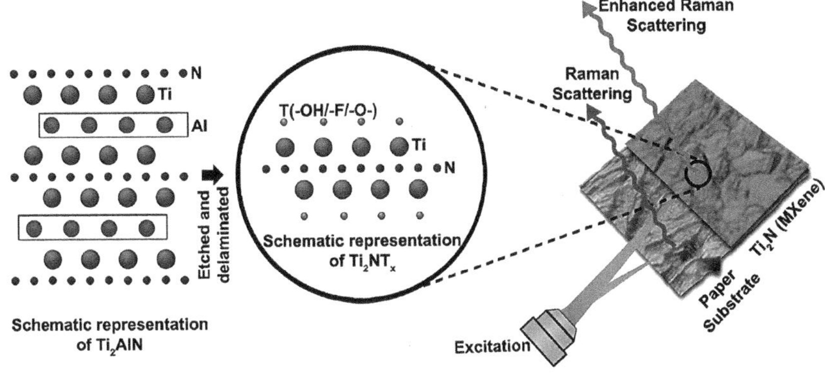

FIGURE 6.57 Schematic representation of the synthesis and SERS activity of the Ti_2NT_x.

Source: Reprinted with permission from [229] Bhuvaneswari Soundiraraju, Benny Kattikkanal George. 2017. Two-Dimensional Titanium Nitride (Ti_2N) MXene: Synthesis, Characterization, and Potential Application as Surface-Enhanced Raman Scattering Substrate. ACS Nano, 11, 8892–8900. Copyright 2017, American Chemical Society.

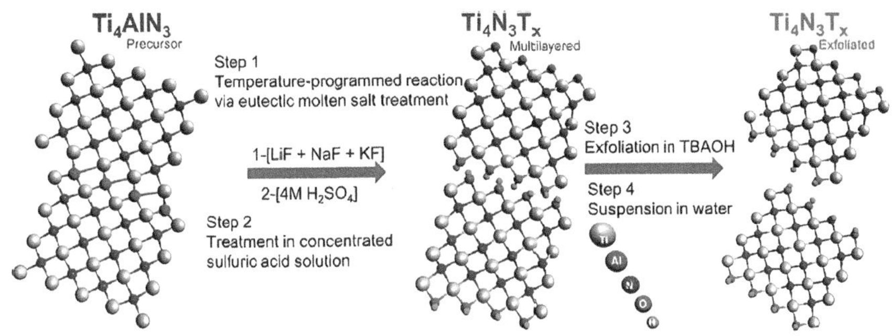

FIGURE 6.58 Step-by-step synthesis process of $Ti_4N_3T_x$.

Source: Reprinted with permission from [231] Abdoulaye Djire, Hanyu Zhang, Jun Liu, et al. 2019. Electrocatalytic and Optoelectronic Characteristics of the Two-Dimensional Titanium Nitride $Ti_4N_3T_x$ MXene. *ACS Appl. Mater. Interfaces*, 11, 11812–11823. Copyright 2019, American Chemical Society.

structure. Similarly, A. Djire et al. [231] reported the temperature-programmed synthesis and characterizations of $Ti_4N_3T_x$, demonstrating that this MXene exhibits both metallic and semiconducting properties. The MAX phase was synthesized by a solid-state reaction between TiH_2, AlN, and TiN at high pressure and temperature, and then the A-site Al^{3+} cation was extracted by separating Ti_4N_3 layers using a reaction with molten fluoride salts (as shown in Fig. 6.58 [231]). They also used absorption and emission spectroscopies to characterize the $Ti_4N_3T_x$ MXene visually, and linear sweep voltammetry to demonstrate its electrocatalytic activity for HER.

G. D. Sun et al. [232] demonstrated a cost-effective approach for developing isolated 2D MoN NSs using Na_2CO_3-assisted topochemical nitridation and NH_3

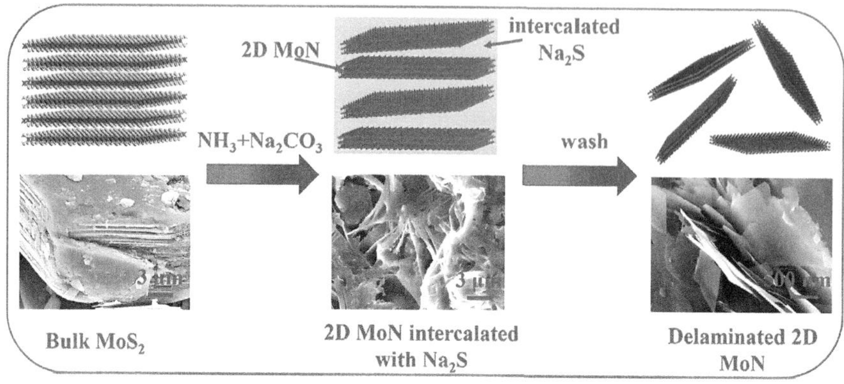

FIGURE 6.59 Synthesis of delaminated 2D MoN MXenes.

Source: Reprinted with permission from [232] Sun G. D., Chang H. Q., et al. 2019. A low-cost and efficient pathway for preparation of 2D MoN nanosheets via Na_2CO_3-assisted nitridation of MoS_2 with NH_3. *Journal of the American Ceramic Society*, 102, 7178–7186. Copyright 2019, the American Ceramic Society.

exfoliation of natural layered $2HMoS_2$ (as illustrated in Fig. 6.59 [232]). First, a MoS_2 and Na_2CO_3 combination with a Na_2CO_3/MoS_2 molar ratio of 2.0–2.5 was treated in an NH_3 environment at 700–800°C. After washing, the removal of intercalated molecules revealed the formation of delaminated ultrathin MoN NSs with a few nanometers of thickness. The results showed that the addition of Na_2CO_3 can significantly increase the nitridation and exfoliation of MoS_2. Through salt-assisted (such as Na_2CO_3) nitridation or carbonization of other transition metal dichalcogenides, this technique has the potential for efficient and large-scale synthesis of other MXenes as well.

A scalable salt-templated technique was carried out to synthesize metallic 2D MoN NSs [233]. The first step was the annealing of Mo-precursor@NaCl powders in an Ar environment at 280°C to obtain the 2D hexagonal MoO_3-coated NaCl (2D h-MoO_3@NaCl). Second, at 650°C the as-developed particles were gently treated in an NH_3 environment. Salt may act as a stabilizer to prevent the changes in morphology during the oxides to nitrides transfer process. P. Urbankowski et al. [234] synthesized multilayered nitride MXenes and delaminated films of Mo_2CT_x and V_2CT_x by heating and cooling at 600°C for 1 hour in an ammonia, NH_3, environment. The reactor's ammonia flow rate was around 300 cm³ min⁻¹. Nitridation happens during the process by replacing C atoms in these MXenes with N atoms. A schematic of the process is shown in Fig. 6.60 [234].

Recently, X. Xiao et al. [235] synthesized ultrathin Mn_3N_2 flakes using a salt-templating technique. In order to synthesize the MXene, manganese chloride was chosen as the precursor, as the synthesis by ammoniation or nitridation always requires high temperatures. Figure 6.61 [235] presents a schematic of the synthesis of 2D Mn_3N_2. To begin, an ethanol-based $MnCl_2$ solution was applied to the surface of the KCl salt template and dried to generate a thin coating of $MnCl_2$ on KCl, which was further treated at 750°C under a steady ammonia flow. $MnCl_2$@KCl was

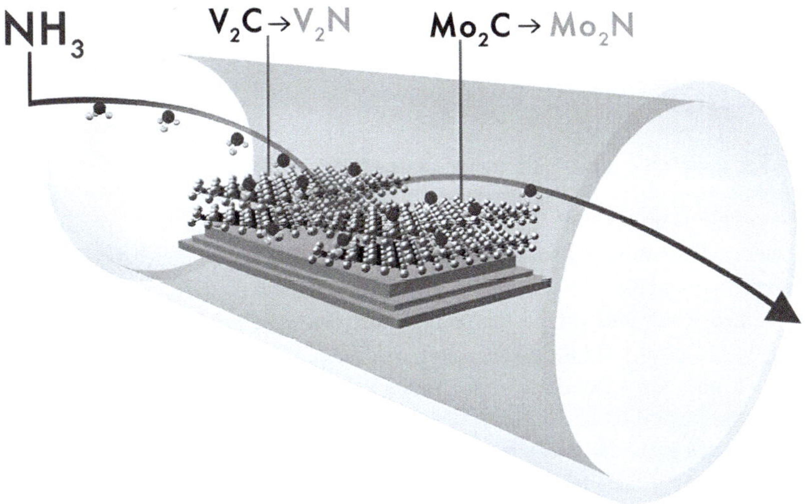

FIGURE 6.60 Synthesis of 2D Mo$_2$CT$_x$ and V$_2$CT$_x$ by ammoniation at elevated temperatures.

Source: [234] Urbankowski P., Anasori B., et al., 2017. 2D molybdenum and vanadium nitrides synthesized by ammoniation of 2D transition metal carbides (MXenes). *Nanoscale*, 9, 17722–17730. Reproduced with permission from the Royal Society of Chemistry.

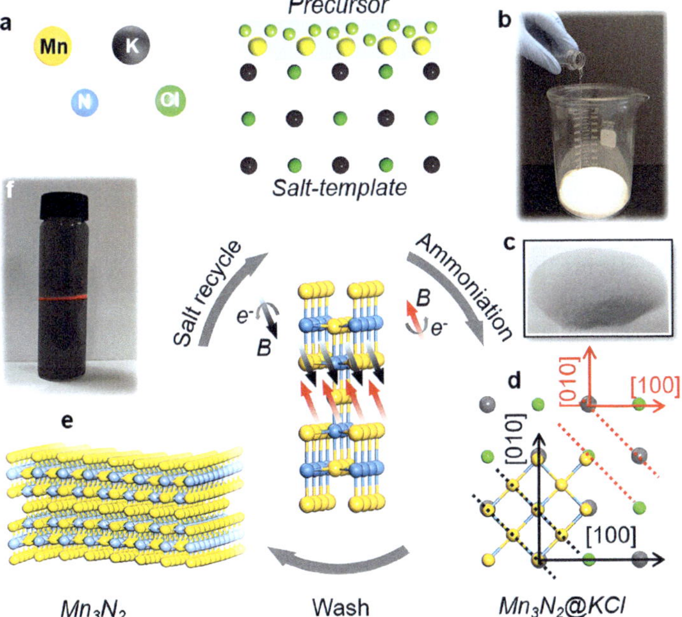

FIGURE 6.61 Step-by-step salt-templated growth mechanism of Mn$_3$N$_2$ flakes.

Source: [235] Xu Xiao, Patrick Urbankowski, et al.: Scalable Synthesis of Ultrathin Mn$_3$N$_2$ Exhibiting Room-Temperature Antiferromagnetism. *Adv. Funct. Mater.* 2019, 1809001. Copyright WILEY-VCH Verlag GmbH & Co. KGaA, Weinheim. Reproduced with permission.

converted into Mn_3N_2@KCl (light grayish powder) during the ammoniation reaction. The KCl salt template was rinsed with de-ionized (DI) water after transformation to obtain 2D Mn_3N_2. The Tyndall effect confirmed that 2D Mn_3N_2 diffused efficiently in DI water to form a black colloidal solution.

6.4 NOVEL STRATEGIES FOR MXENES

J. Li et al. [236] used simple hydrolysis and calcination procedures to produce Mo_2C nanosheets. The ammonia source was ammonium molybdate, the carbon source was 4-Cl-ophenylenediamine, and the ammonium and carbon sources were introduced to deionized water and stirred for 90 minutes. Until the production of amine metal oxide, the pH of this solution was regulated using 1 M HCl. The mixture was then stirred for another 4 hours at 60°C. After centrifuging, washing, and drying, the powder was heated to 850°C in a nitrogen environment for 6 hours. The resultant nanostructures provided Na-ion transport kinetics due to high electron conductivity. This property facilitates electron transmission rate by minimizing the volume expansion. L. Ma et al. [237] reported the synthesis of Mo_2C and Mo_2N using the "urea glass" method. A concentration of 1.45 M ethanol was applied to the solid $MoCl_5$ metal precursor. Mo-orthoesters are formed when $MoCl_5$ interacts with alcohol. To achieve the appropriate molar ratio, various amounts of urea were added to the alcoholic solution. The liquid was then thoroughly mixed until the urea was completely dissolved. It was observed that urea exhibits higher solubility in a metallic precursor than ethanol. The gel-like precursors were calcinated at 800°C with a N_2 gas flow, and silvery-black particles were obtained. Among the non-noble metal catalysts, as-synthesized Mo_2C showed superior HER performance, particularly in the KOH electrolyte.

In a recent study, a UV-induced etching process with milder phosphoric acid was reported to selectively exfoliate the Mo_2Ga_2C precursor [238]. In this process, the double Ga layer was removed from the precursor in the presence of UV light radiation. Similarly, Mo_2C NSs and D-shaped Mo_2C Sas were developed for mode-locked lasers, by employing liquid-phase exfoliation where NSs were deposited on polished fibers [239]. The researchers examined the nonlinear absorption of Mo_2C using a handmade balanced twin-detector measurement setup. The modulation depth was 5.1%, and the saturation intensity was 200 MW cm^{-2}. Highly stable mode-locked fiber lasers at 1 micron and 1.5 micron scales were proven as a typical application in ultrafast photonics. In another study, J. Jeon et al. [240] used thermal annealing to develop epitaxial Mo_2C layers by employing Cu foil as a catalyst under CH_4 and H_2. The transformed Mo_2C area was demonstrated to be dependent on the duration of thermal annealing. Theoretical computations demonstrated that this conversion occurs at the edges of MoS_2 films.

To remove the Al atom from Nb_2AlC, G. Li et al. [241] produced Nb_2C MXene in 30 mL of 40% HF. After heating the HF solution at 40°C, 3 g of Nb_2AlC powder was carefully introduced to a PTFE conical bottle. To avoid overheating the reaction, the addition course was delayed for 510 minutes. The bottle was then wrapped in plastic film with two to three tiny pores. The reaction lasted for 52 hours while being stirred and was kept at 40°C. Following that, the M-Nb$_2$CT$_x$ (multilayered Nb$_2$CT$_x$) powder was created by centrifuging the mixture and rinsing it with ethanol multiple times until the pH was near neutrality; then the mixture was air-dried. The dry M-Nb$_2$CT$_x$ powder was subsequently spread in 20 mL TPAOH and agitated for 3 days at room temperature. After the removal of TPAOH, aqueous dispersion of few-layered Nb$_2$CT$_x$ (F-Nb$_2$CT$_x$) was recovered, and vacuum freeze-drying was used to create F-Nb$_2$CT$_x$ powder. M-Nb$_2$C and F-Nb$_2$C were vacuum-treated at 500°C for 4 hours to remove as many surface-adsorbed groups as feasible during etching and exfoliation.

Q. Xu et al. [242] synthesized the MXene/ZIF-67/CNTs composite for luteolin (LUT) electrochemical detection. As illustrated in Fig. 6.62 [242], ZIF-67/CNTs were first synthesized using an in situ growing technique. Ti$_3$C$_2$-MXene, which has a high specific surface area, was then used as a matrix for the immobilization of ZIF-67/CNTs. The as-synthesized composite has a quick electron transfer ability, abundant active sites, large specific surface area, and strong LUT adsorption capability, displaying superior electrochemical activity toward the LUT redox reaction.

Particles of Co^{2+}, five membered ring molecule (2 Methylimidazole), and cylindrical tubes (CNTs) in situ growth leads ZIF–67 CNTs + Ti$_3$C$_2$. Ultrasound leads to Ti$_3$C$_2$/ZIF-67/CNTs. Drip coating of Ti$_3$C$_2$/ZIF-67/CNTs leads to GCE in the presence of a reaction fused two six-membered ring with carbon, and 1 O as ring atoms with substitutions O, OH and OH, and dihydroxy benzene LUT in the presence electron and removal of proton leads to conversion of one OH in the benzene replaced with O. DPV leads to potential V versus SCE and current (A) graph with bell-shaped peaks.

Recently, a liquid-phase mixing process was used to develop highly flexible MnOx-Ti$_3$C$_2$ films [243]. The process was carried out by simply mixing an Mn(NO$_3$)$_2$

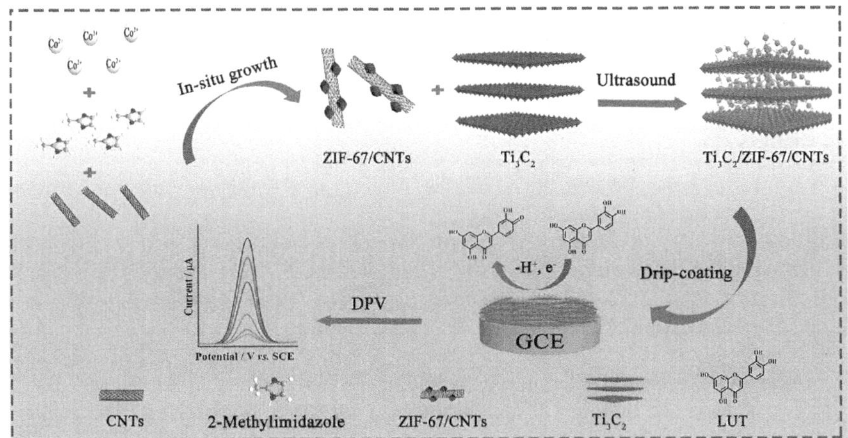

FIGURE 6.62 Schematical representation of the synthesis of MXene/ZIF-67/CNTs composite.

Source: Reprinted from [242] *Journal of Electroanalytical Chemistry*, 880, Quan Xu, Shuxian Chen, Jingkun Xu, et al., Facile synthesis of hierarchical MXene/ZIF-67/CNTs composite for electrochemical sensing of luteolin, 114765. Copyright 2021, with permission from Elsevier.

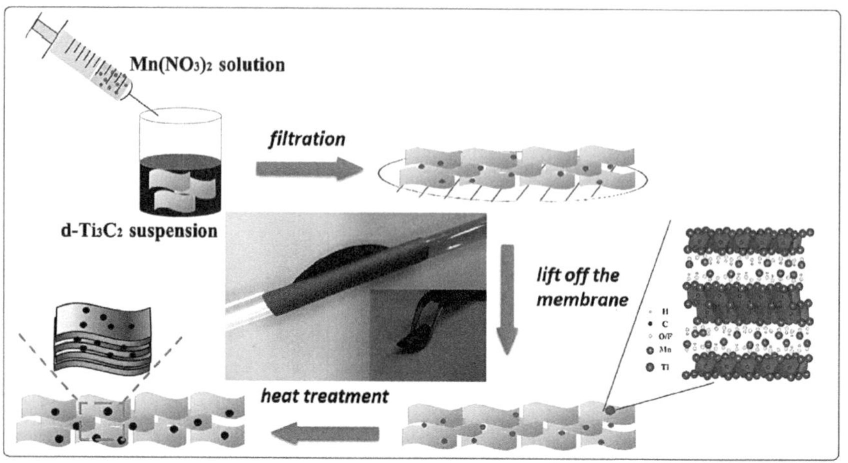

FIGURE 6.63 Schematic illustration of the synthesis of freestanding MnOx-Ti$_3$C$_2$ films.

Source: Reprinted from [243] *Journal of Power Sources*, 359, Yapeng Tian, Chenhui Yang, Wenxiu Que, et al., Flexible and free-standing 2D titanium carbide film decorated with manganese oxide nanoparticles as a high volumetric capacity electrode for supercapacitor, 332–339. Copyright 2017, with permission from Elsevier.

solution with d-Ti$_3$C$_2$ suspension (as shown in Fig. 6.63 [243]). This process resulted into a strong interaction between composite materials, which provided improved electrochemical performances. M. Naguib and colleagues [244] described the synthesis and organic base treatment of V$_2$CT$_x$ and Ti$_3$CNT$_x$ MXenes. In a nutshell, MXene powders were combined with an aqueous solution of 54–56% TBAOH, (C$_4$H$_9$)$_4$NOH in a 1-g MXene to 10-mL base solution ratio. The mixture was then agitated for 2, 4, or 21 hours at room temperature. The mixtures were centrifuged at 2,000 rpm for 0.5 hours after the specified mixing time. After decanting the supernatant liquids, the sediment was employed for further analysis and delamination.

REFERENCES

[1] Ali, I., M. Faraz Ud Din, and Z.-G. Gu, *MXenes thin films: from fabrication to their applications*. Molecules, 2022. **27**(15): p. 4925.

[2] Pogorielov, M., et al., *MXenes—A new class of two-dimensional materials: structure, properties and potential applications*. Nanomaterials, 2021. **11**(12): p. 3412.

[3] Wang, Y., et al., *MXenes: focus on optical and electronic properties and corresponding applications*. Nanophotonics, 2020. **9**(7): p. 1601–1620.

[4] Naguib, M., et al., *25th anniversary article: MXenes: a new family of two-dimensional materials*. Advanced Materials, 2014. **26**(7): p. 992–1005.

[5] Lei, Y.-J., et al., *Tailoring MXene-based materials for sodium-ion storage: synthesis, mechanisms, and applications*. Electrochemical Energy Reviews, 2020. **3**: p. 766–792.

[6] Anasori, B., M.R. Lukatskaya, and Y. Gogotsi, *2D metal carbides and nitrides (MXenes) for energy storage*. Nature Reviews Materials, 2017. **2**(2): p. 1–17.

[7] Fu, B., et al., *MXenes: synthesis, optical properties, and applications in ultrafast photonics*. Small, 2021. **17**(11): p. 2006054.

[8] Wang, B., et al., *Booming development and present advances of two dimensional MXenes for photodetectors.* Chemical Engineering Journal, 2021. **403**: p. 126336.

[9] Mashtalir, O., et al., *Kinetics of aluminum extraction from Ti_3AlC_2 in hydrofluoric acid.* Materials Chemistry and Physics, 2013. **139**(1): p. 147–152.

[10] Chang, F., et al., *Synthesis of a new graphene-like transition metal carbide by de-intercalating Ti_3AlC_2.* Materials Letters, 2013. **109**: p. 295–298.

[11] Hu, M., et al., *Surface functional groups and interlayer water determine the electrochemical capacitance of $Ti_3C_2T_x$ MXene.* ACS Nano, 2018. **12**(4): p. 3578–3586.

[12] Naguib, M., et al., *Two-dimensional transition metal carbides.* ACS Nano, 2012. **6**(2): p. 1322–1331.

[13] Sun, W., et al., *Multiscale and multimodal characterization of 2D titanium carbonitride MXene.* Advanced Materials Interfaces, 2020. **7**(11): p. 1902207.

[14] Iqbal, M., et al., *Co-existence of magnetic phases in two-dimensional MXene.* Materials Today Chemistry, 2020. **16**: p. 100271.

[15] Seh, Z.W., et al., *Two-dimensional molybdenum carbide (MXene) as an efficient electrocatalyst for hydrogen evolution.* ACS Energy Letters, 2016. **1**(3): p. 589–594.

[16] Deeva, E.B., et al., *In situ XANES/XRD study of the structural stability of two-dimensional molybdenum carbide Mo_2CT_x: implications for the catalytic activity in the water—gas shift reaction.* Chemistry of Materials, 2019. **31**(12): p. 4505–4513.

[17] Zhang, J., et al., *Single platinum atoms immobilized on an MXene as an efficient catalyst for the hydrogen evolution reaction.* Nature Catalysis, 2018. **1**(12): p. 985–992.

[18] Halim, J., et al., *XPS of cold pressed multilayered and freestanding delaminated 2D thin films of $Mo_2TiC_2T_z$ and $Mo_2Ti_2C_3T_z$ (MXenes).* Applied Surface Science, 2019. **494**: p. 1138–1147.

[19] Srimuk, P., et al., *Two-dimensional molybdenum carbide (MXene) with divacancy ordering for brackish and seawater desalination via cation and anion intercalation.* ACS Sustainable Chemistry & Engineering, 2018. **6**(3): p. 3739–3747.

[20] Etman, A.S., J. Halim, and J. Rosen, *Mixed MXenes: $Mo_{1.33}CTz$ and $Ti_3C_2T_z$ freestanding composite films for energy storage.* Nano Energy, 2021. **88**: p. 106271.

[21] Halim, J., et al., *Tailored synthesis approach of $(Mo_{2/3}Y_{1/3})_2AlC$ i-MAX and its two-dimensional derivative $Mo_{1.33}CT_z$ MXene: enhancing the yield, quality, and performance in supercapacitor applications.* Nanoscale, 2021. **13**(1): p. 311–319.

[22] Naguib, M., et al., *New two-dimensional niobium and vanadium carbides as promising materials for Li-ion batteries.* Journal of the American Chemical Society, 2013. **135**(43): p. 15966–15969.

[23] Su, T., et al., *One-step synthesis of $Nb_2O_5/C/Nb_2C$ (MXene) composites and their use as photocatalysts for hydrogen evolution.* ChemSusChem, 2018. **11**(4): p. 688–699.

[24] Xiang, H., et al., *Hypoxia-irrelevant photonic thermodynamic cancer nanomedicine.* ACS Nano, 2019. **13**(2): p. 2223–2235.

[25] Gao, L., et al., *Applications of few-layer Nb_2C MXene: narrow-band photodetectors and femtosecond mode-locked fiber lasers.* ACS Nano, 2021. **15**(1): p. 954–965.

[26] Reding, B., et al., *Manipulate intestinal organoids with niobium carbide nanosheets.* Journal of Biomedical Materials Research Part A, 2021. **109**(4): p. 479–487.

[27] Zhao, S., et al., *Li-ion uptake and increase in interlayer spacing of Nb_4C_3 MXene.* Energy Storage Materials, 2017. **8**: p. 42–48.

[28] Liu, Y., et al., *Excellent catalytic activity of a two-dimensional $Nb_4C_3T_x$ (MXene) on hydrogen storage of MgH_2.* Applied Surface Science, 2019. **493**: p. 431–440.

[29] Li, X., et al., *Phase transition induced unusual electrochemical performance of V_2CTX MXene for aqueous zinc hybrid-ion battery.* ACS Nano, 2020. **14**(1): p. 541–551.

[30] Zhou, J., et al., *Synthesis and lithium ion storage performance of two-dimensional V_4C_3 MXene.* Chemical Engineering Journal, 2019. **373**: p. 203–212.

[31] Wang, X., et al., *Two-dimensional V_4C_3 MXene as high performance electrode materials for supercapacitors.* Electrochimica Acta, 2019. **307**: p. 414–421.

[32] Bai, J., et al., *Construction of hierarchical V_4C_3-MXene/MoS$_2$/C nanohybrids for high rate lithium-ion batteries.* Nanoscale, 2020. **12**(2): p. 1144–1154.

[33] Zhou, J., et al., *A two-dimensional zirconium carbide by selective etching of Al_3C_3 from nanolaminated $Zr_3Al_3C_5$.* Angewandte Chemie International Edition, 2016. **55**(16): p. 5008–5013.

[34] Zhou, J., et al., *Synthesis and electrochemical properties of two-dimensional hafnium carbide.* ACS Nano, 2017. **11**(4): p. 3841–3850.

[35] Tang, X., et al., *2D metal carbides and nitrides (MXenes) as high-performance electrode materials for lithium-based batteries.* Advanced Energy Materials, 2018. **8**(33): p. 1801897.

[36] Li, X., Z. Huang, and C. Zhi, *Environmental stability of MXenes as energy storage materials.* Frontiers in Materials, 2019. **6**: p. 312.

[37] Song, P., et al., *MXenes for polymer matrix electromagnetic interference shielding composites: a review.* Composites Communications, 2021. **24**: p. 100653.

[38] Zhang, C., et al., *Two-dimensional transition metal carbides and nitrides (MXenes): synthesis, properties, and electrochemical energy storage applications.* Energy & Environmental Materials, 2020. **3**(1): p. 29–55.

[39] Sang, X., et al., *Atomic defects in monolayer titanium carbide ($Ti_3C_2T_x$) MXene.* ACS Nano, 2016. **10**(10): p. 9193–9200.

[40] Ghidiu, M., et al., *Conductive two-dimensional titanium carbide 'clay' with high volumetric capacitance.* Nature, 2014. **516**(7529): p. 78–81.

[41] Quain, E., et al., *Direct writing of additive-free MXene-in-water ink for electronics and energy storage.* Advanced Materials Technologies, 2019. **4**(1): p. 1800256.

[42] Lipatov, A., et al., *Effect of synthesis on quality, electronic properties and environmental stability of individual monolayer Ti_3C_2 MXene flakes.* Advanced Electronic Materials, 2016. **2**(12): p. 1600255.

[43] Miranda, A., et al., *Electronic properties of freestanding $Ti_3C_2T_x$ MXene monolayers.* Applied Physics Letters, 2016. **108**(3).

[44] Ding, L., et al., *Effective ion sieving with $Ti_3C_2T_x$ MXene membranes for production of drinking water from seawater.* Nature Sustainability, 2020. **3**(4): p. 296–302.

[45] Ding, L., et al., *MXene molecular sieving membranes for highly efficient gas separation.* Nature Communications, 2018. **9**(1): p. 155.

[46] Liu, F., et al., *Preparation of Ti_3C_2 and Ti_2C MXenes by fluoride salts etching and methane adsorptive properties.* Applied Surface Science, 2017. **416**: p. 781–789.

[47] Kajiyama, S., et al., *Enhanced Li-ion accessibility in MXene titanium carbide by steric chloride termination.* Advanced Energy Materials, 2017. **7**(9): p. 1601873.

[48] Yang, Y., et al., *Large-area highly conductive transparent two-dimensional Ti_2CTx film.* The Journal of Physical Chemistry Letters, 2017. **8**(4): p. 859–865.

[49] Hantanasirisakul, K., et al., *Effects of synthesis and processing on optoelectronic properties of titanium carbonitride MXene.* Chemistry of Materials, 2019. **31**(8): p. 2941–2951.

[50] Du, F., et al., *Environmental friendly scalable production of colloidal 2D titanium carbonitride MXene with minimized nanosheets restacking for excellent cycle life lithium-ion batteries.* Electrochimica Acta, 2017. **235**: p. 690–699.

[51] Han, M., et al., *Beyond $Ti_3C_2T_x$: MXenes for electromagnetic interference shielding.* ACS Nano, 2020. **14**(4): p. 5008–5016.

[52] Attanayake, N.H., et al., *Electrocatalytic CO_2 reduction on earth abundant 2D Mo_2C and Ti_3C_2 MXenes.* Chemical Communications, 2021. **57**(13): p. 1675–1678.

[53] Halim, J., et al., *Synthesis and characterization of 2D molybdenum carbide (MXene).* Advanced Functional Materials, 2016. **26**(18): p. 3118–3127.

[54] Feng, W., et al., *Ultrathin molybdenum carbide MXene with fast biodegradability for highly efficient theory-oriented photonic tumor hyperthermia.* Advanced Functional Materials, 2019. **29**(22): p. 1901942.

[55] Maughan, P.A., et al., *Pillared Mo_2TiC_2 MXene for high-power and long-life lithium and sodium-ion batteries.* Nanoscale Advances, 2021. **3**(11): p. 3145–3158.

[56] Guan, Y., et al., *A hydrofluoric acid-free synthesis of 2D vanadium carbide (V_2C) MXene for supercapacitor electrodes.* 2D Materials, 2020. **7**(2): p. 025010.

[57] Ming, F., et al., *Porous MXenes enable high performance potassium ion capacitors.* Nano Energy, 2019. **62**: p. 853–860.

[58] Thörnberg, J., et al., *Synthesis of $(V_{2/3}Sc_{1/3})_2AlC$ i-MAX phase and $V_{2-x}C$ MXene scrolls.* Nanoscale, 2019. **11**(31): p. 14720–14726.

[59] Wang, Z., et al., *VO_2 (p)-V_2C (MXene) grid structure as a lithium polysulfide catalytic host for high-performance Li–S battery.* ACS Applied Materials & Interfaces, 2019. **11**(47): p. 44282–44292.

[60] Xiao, J., et al., *A safe etching route to synthesize highly crystalline Nb_2CT_x MXene for high performance asymmetric supercapacitor applications.* Electrochimica Acta, 2020. **337**: p. 135803.

[61] Ren, X., et al., *Highly catalytic niobium carbide (MXene) promotes hematopoietic recovery after radiation by free radical scavenging.* ACS Nano, 2019. **13**(6): p. 6438–6454.

[62] Zou, X., et al., *A simple approach to synthesis Cr_2CT_x MXene for efficient hydrogen evolution reaction.* Materials Today Energy, 2021. **20**: p. 100668.

[63] Natu, V., et al., *2D $Ti_3C_2T_z$ MXene synthesized by water-free etching of Ti_3AlC_2 in polar organic solvents.* Chem, 2020. **6**(3): p. 616–630.

[64] Feng, A., et al., *Two-dimensional MXene Ti_3C_2 produced by exfoliation of Ti_3AlC_2.* Materials & Design, 2017. **114**: p. 161–166.

[65] Wang, T., et al., *Direct synthesis of hydrogen fluoride-free multilayered Ti_3C_2/TiO_2 composite and its applications in photocatalysis.* Heliyon, 2023. **9**(8).

[66] Halim, J., et al., *Transparent conductive two-dimensional titanium carbide epitaxial thin films.* Chemistry of Materials, 2014. **26**(7): p. 2374–2381.

[67] Feng, A., et al., *Fabrication and thermal stability of NH_4HF_2-etched Ti_3C_2 MXene.* Ceramics International, 2017. **43**(8): p. 6322–6328.

[68] Li, J., et al., *K^+ Intercalation of NH_4HF_2-exfoliated Ti_3C_2 MXene as binder-free electrodes with high electrochemical capacitance.* Physica Status Solidi (a), 2020. **217**(8): p. 1900806.

[69] Wang, C., S. Chen, and L. Song, *Tuning 2D MXenes by surface controlling and interlayer engineering: methods, properties, and synchrotron radiation characterizations.* Advanced Functional Materials, 2020. **30**(47): p. 2000869.

[70] Li, T., et al., *Fluorine-free synthesis of high-purity $Ti_3C_2T_x$ (T = OH, O) via alkali treatment.* Angewandte Chemie International Edition, 2018. **57**(21): p. 6115–6119.

[71] Wei, Z., et al., *Alkali treated $Ti_3C_2T_x$ MXenes and their dye adsorption performance.* Materials Chemistry and Physics, 2018. **206**: p. 270–276.

[72] Li, G., et al., *Highly efficiently delaminated single-layered MXene nanosheets with large lateral size.* Langmuir, 2017. **33**(36): p. 9000–9006.

[73] Carey, M., et al., *Dispersion and stabilization of alkylated 2D MXene in nonpolar solvents and their pseudocapacitive behavior.* Cell Reports Physical Science, 2020. **1**(4).

[74] Li, M., et al., *Element replacement approach by reaction with Lewis acidic molten salts to synthesize nanolaminated MAX phases and MXenes.* Journal of the American Chemical Society, 2019. **141**(11): p. 4730–4737.

[75] Li, Y., et al., *A general Lewis acidic etching route for preparing MXenes with enhanced electrochemical performance in non-aqueous electrolyte.* Nature Materials, 2020. **19**(8): p. 894–899.

[76] Arole, K., et al., *Water-dispersible $Ti_3C_2T_z$ MXene nanosheets by molten salt etching.* iScience, 2021. **24**(12).

[77] Verger, L., et al., *Effect of cationic exchange on the hydration and swelling behavior of $Ti_3C_2T_z$ MXenes.* The Journal of Physical Chemistry C, 2019. **123**(32): p. 20044–20050.

[78] Ghidiu, M., et al., *Ion-exchange and cation solvation reactions in Ti_3C_2 MXene.* Chemistry of Materials, 2016. **28**(10): p. 3507–3514.

[79] Dong, H., et al., *Molten salt derived Nb_2CT_x MXene anode for Li-ion batteries.* ChemElectroChem, 2021. **8**(5): p. 957–962.

[80] Arole, K., et al., *Exfoliation, delamination, and oxidation stability of molten salt etched Nb_2CT_z MXene nanosheets.* Chemical Communications, 2022. **58**(73): p. 10202–10205.

[81] Jin, H., et al., *MXene analogue: a 2D nitridene solid solution for high-rate hydrogen production.* Angewandte Chemie, 2022. **134**(27): p. e202203850.

[82] Chen, J., et al., *Molten salt-shielded synthesis (MS^3) of MXenes in air.* Energy & Environmental Materials, 2023. **6**(2): p. e12328.

[83] Sun, W., et al., *Electrochemical etching of Ti_2AlC to Ti_2CT_x (MXene) in low-concentration hydrochloric acid solution.* Journal of Materials Chemistry A, 2017. **5**(41): p. 21663–21668.

[84] Yang, S., et al., *Fluoride-free synthesis of two-dimensional titanium carbide (MXene) using a binary aqueous system.* Angewandte Chemie, 2018. **130**(47): p. 15717–15721.

[85] Liu, L., et al., *In situ synthesis of MXene with tunable morphology by electrochemical etching of MAX phase prepared in molten salt.* Advanced Energy Materials, 2023. **13**(7): p. 2203805.

[86] Yin, T., et al., *Synthesis of $Ti_3C_2F_x$ MXene with controllable fluorination by electrochemical etching for lithium-ion batteries applications.* Ceramics International, 2021. **47**(20): p. 28642–28649.

[87] Pang, S.-Y., et al., *Universal strategy for HF-free facile and rapid synthesis of two-dimensional MXenes as multifunctional energy materials.* Journal of the American Chemical Society, 2019. **141**(24): p. 9610–9616.

[88] Li, X., et al., *In situ electrochemical synthesis of MXenes without acid/alkali usage in/for an aqueous zinc ion battery.* Advanced Energy Materials, 2020. **10**(36): p. 2001791.

[89] Liu, F., et al., *Preparation of high-purity V_2C MXene and electrochemical properties as Li-ion batteries.* Journal of the Electrochemical Society, 2017. **164**(4): p. A709.

[90] Downes, M., et al., *M_5X_4-a family of MXenes.* 2023.

[91] Anayee, M., et al., *Kinetics of Ti_3AlC_2 etching for $Ti_3C_2T_x$ MXene synthesis.* Chemistry of Materials, 2022. **34**(21): p. 9589–9600.

[92] Ridley, P., et al., *MXene-derived bilayered vanadium oxides with enhanced stability in Li-ion batteries.* ACS Applied Energy Materials, 2020. **3**(11): p. 10892–10901.

[93] Alameda, L.T., et al., *Multi-step topochemical pathway to metastable Mo_2AlB_2 and related two-dimensional nanosheet heterostructures.* Journal of the American Chemical Society, 2019. **141**(27): p. 10852–10861.

[94] Huang, X., et al., *Nanoconfined topochemical conversion from MXene to ultrathin non-layered TiN nanomesh toward superior electrocatalysts for lithium-sulfur batteries.* Small, 2021. **17**(32): p. 2101360.

[95] Alameda, L.T., et al., *Topochemical deintercalation of Al from MoAlB: stepwise etching pathway, layered intergrowth structures, and two-dimensional MBene.* Journal of the American Chemical Society, 2018. **140**(28): p. 8833–8840.

[96] Bhaskar, G., et al., *Topochemical deintercalation of Li from layered LiNiB: toward 2D MBene.* Journal of the American Chemical Society, 2021. **143**(11): p. 4213–4223.

[97] Chen, R., et al., *Topochemical pyrolytic synthesis of quasi-Mxene hybrids via ionic liquid-iron phthalocyanine as a self-template.* Chemical Communications, 2019. **55**(6): p. 771–774.

[98] Kim, K., et al., *Topochemical synthesis of phase-pure Mo$_2$AlB$_2$ through staging mechanism.* Chemical Communications, 2019. **55**(63): p. 9295–9298.

[99] Kesavan, D., et al., *Topochemically synthesized MoS$_2$ nanosheets: a high performance electrode for wide-temperature tolerant aqueous supercapacitors.* Journal of Colloid and Interface Science, 2021. **584**: p. 714–722.

[100] Gong, Y., et al., *Emerging MXenes for functional memories.* Small Science, 2021. **1**(9): p. 2100006.

[101] Rezakazemi, M., et al., *Sustainable MXenes-based membranes for highly energy-efficient separations.* Renewable and Sustainable Energy Reviews, 2021. **143**: p. 110878.

[102] Prakash, N.J. and B. Kandasubramanian, *Nanocomposites of MXene for industrial applications.* Journal of Alloys and Compounds, 2021. **862**: p. 158547.

[103] Geng, D., et al., *Direct synthesis of large-area 2D Mo$_2$C on in situ grown graphene.* Advanced Materials, 2017. **29**(35): p. 1700072.

[104] Xu, C., et al., *Large-area high-quality 2D ultrathin Mo$_2$C superconducting crystals.* Nature Materials, 2015. **14**(11): p. 1135–1141.

[105] Kang, Z., et al., *Controlled growth of an Mo$_2$C—graphene hybrid film as an electrode in self-powered two-sided Mo$_2$C—Graphene/Sb$_2$S$_{0.42}$Se$_{2.58}$/TiO$_2$ photodetectors.* Sensors, 2019. **19**(5): p. 1099.

[106] Sun, W., et al., *Controlled synthesis of 2D Mo$_2$C/graphene heterostructure on liquid Au substrates as enhanced electrocatalytic electrodes.* Nanotechnology, 2019. **30**(38): p. 385601.

[107] Thirumal, V., et al., *Facile single-step synthesis of MXene@CNTs hybrid nanocomposite by CVD method to remove hazardous pollutants.* Chemosphere, 2022. **286**: p. 131733.

[108] Chaitoglou, S., et al., *Insight and control of the chemical vapor deposition growth parameters and morphological characteristics of graphene/Mo$_2$C heterostructures over liquid catalyst.* Journal of Crystal Growth, 2018. **495**: p. 46–53.

[109] Zhang, Z., et al., *Layer-stacking, defects, and robust superconductivity on the Mo-terminated surface of ultrathin Mo$_2$C flakes grown by CVD.* Nano Letters, 2019. **19**(5): p. 3327–3335.

[110] Chaitoglou, S., et al., *Mo$_2$C/graphene heterostructures: low temperature chemical vapor deposition on liquid bimetallic Sn–Cu and hydrogen evolution reaction electrocatalytic properties.* Nanotechnology, 2019. **30**(12): p. 125401.

[111] Li, T., et al., *Probing the domain architecture in 2D α-Mo$_2$C via polarized Raman spectroscopy.* Advanced Materials, 2019. **31**(8): p. 1807160.

[112] Jeon, J., et al., *Transition-metal-carbide (Mo$_2$C) multiperiod gratings for realization of high-sensitivity and broad-spectrum photodetection.* Advanced Functional Materials, 2019. **29**(48): p. 1905384.

[113] Rouhi, S., et al., *Improved coverage of Mo$_2$C Mxenes by copper thin film catalyst using Pvd-Cvd hybrid growth technique.* Available at SSRN 4521749.

[114] Horak, P., et al., *Ion beam sputtering for controlled synthesis of thin max (MXene) phases.* Microscopy and Microanalysis, 2019. **25**(S2): p. 1626–1627.

[115] Pshyk, A., et al., *Low-temperature growth of epitaxial Ti$_2$AlC MAX phase thin films by low-rate layer-by-layer PVD.* Materials Research Letters, 2019. **7**(6): p. 244–250.

[116] Gröner, L., et al., *Microstructural investigations of polycrystalline Ti$_2$AlN prepared by physical vapor deposition of Ti-AlN multilayers.* Surface and Coatings Technology, 2018. **343**: p. 166–171.

[117] Han, F., et al., *Boosting the yield of MXene 2D sheets via a facile hydrothermal-assisted intercalation.* ACS Applied Materials & Interfaces, 2019. **11**(8): p. 8443–8452.

[118] Quyen, V.T., et al., *Advanced synthesis of MXene-derived nanoflower-shaped TiO$_2$@Ti$_3$C$_2$ heterojunction to enhance photocatalytic degradation of Rhodamine B.* Environmental Technology & Innovation, 2021. **21**: p. 101286.

[119] Xue, Q., et al., *Photoluminescent Ti$_3$C$_2$ MXene quantum dots for multicolor cellular imaging.* Advanced Materials, 2017. **29**(15): p. 1604847.

[120] Wang, L., et al., *Synthesis and electrochemical performance of Ti$_3$C$_2$T$_x$ with hydrothermal process.* Electronic Materials Letters, 2016. **12**: p. 702–710.

[121] Xu, M., et al., *Synthesis of MXene-supported layered MoS$_2$ with enhanced electrochemical performance for Mg batteries.* Chinese Chemical Letters, 2018. **29**(8): p. 1313–1316.

[122] Cai, C., et al., *Synthesis of self-assembled phytic acid-MXene nanocomposites via a facile hydrothermal approach with elevated dye adsorption capacities.* Colloids and Surfaces A: Physicochemical and Engineering Aspects, 2020. **589**: p. 124468.

[123] Qiu, F., et al., *Synthesis, characterization and microwave absorption of MXene/NiFe$_2$O$_4$ composites.* Ceramics International, 2021. **47**(17): p. 24713–24720.

[124] Kong, W., et al., *One-step hydrothermal synthesis of fluorescent MXene-like titanium carbonitride quantum dots.* Inorganic Chemistry Communications, 2019. **105**: p. 151–157.

[125] Peng, C., et al., *A hydrothermal etching route to synthesis of 2D MXene (Ti$_3$C$_2$, Nb$_2$C): enhanced exfoliation and improved adsorption performance.* Ceramics International, 2018. **44**(15): p. 18886–18893.

[126] Huang, D., et al., *Demonstration of a white laser with V$_2$C MXene-based quantum dots.* Advanced Materials, 2019. **31**(24): p. 1901117.

[127] Wang, X., et al., *Heterostructures of Ni–Co–Al layered double hydroxide assembled on V$_4$C$_3$ MXene for high-energy hybrid supercapacitors.* Journal of Materials Chemistry A, 2019. **7**(5): p. 2291–2300.

[128] Yan, F., et al., *Solvothermal synthesis of nitrogen-doped MXene quantum dots for the detection of alizarin red based on inner filter effect.* Dyes and Pigments, 2021. **195**: p. 109720.

[129] Xu, G., et al., *Preparation of Ti$_3$C$_2$T$_x$ MXene-derived quantum dots with white/blue-emitting photoluminescence and electrochemiluminescence.* Advanced Optical Materials, 2018. **6**(24): p. 1800951.

[130] Cui, G., et al., *Synthesis of CuS nanoparticles decorated Ti$_3$C$_2$T$_x$ MXene with enhanced microwave absorption performance.* Progress in Natural Science: Materials International, 2020. **30**(3): p. 343–351.

[131] Ding, M., et al., *Evidencing interfacial charge transfer in 2D CdS/2D MXene Schottky heterojunctions toward high-efficiency photocatalytic hydrogen production.* Solar RRL, 2021. **5**(2): p. 2000414.

[132] Luo, S., et al., *Preparation and dye degradation performances of self-assembled MXene-Co$_3$O$_4$ nanocomposites synthesized via solvothermal approach.* ACS Omega, 2019. **4**(2): p. 3946–3953.

[133] Gao, X.T., et al., *Ultrathin MXene nanosheets decorated with TiO$_2$ quantum dots as an efficient sulfur host toward fast and stable Li–S batteries.* Small, 2018. **14**(41): p. 1802443.

[134] Chen, J., et al., *Morphology and photocatalytic activity of TiO$_2$/MXene composites by in-situ solvothermal method.* Ceramics International, 2020. **46**(12): p. 20088–20096.

[135] Sun, S., et al., *W$_{18}$O$_{49}$/Ti$_3$C$_2$T$_x$ Mxene nanocomposites for highly sensitive acetone gas sensor with low detection limit.* Sensors and Actuators B: Chemical, 2020. **304**: p. 127274.

[136] Liu, X., et al., *Facile solvothermal synthesis of ZnO/Ti$_3$C$_2$T$_x$ MXene nanocomposites for NO$_2$ detection at low working temperature.* Sensors and Actuators B: Chemical, 2022. **367**: p. 132025.

[137] Wu, N., et al., *Molybdenum carbide MXene embedded with nickel sulfide clusters as an efficient electrocatalyst for hydrogen evolution reaction.* International Journal of Hydrogen Energy, 2023. **48**(46): p. 17526–17535.

[138] Huang, J., et al., *In situ construction of 1D CdS/2D Nb_2CT_x MXene Schottky heterojunction for enhanced photocatalytic hydrogen production activity.* Applied Surface Science, 2022. **573**: p. 151491.

[139] Xu, F., et al., *Synthesis and photocatalytic H_2-production activity of plasma-treated $Ti_3C_2T_x$ MXene modified graphitic carbon nitride.* Journal of the American Ceramic Society, 2020. **103**(2): p. 849–858.

[140] Wang, J., et al., *Plasma oxidized $Ti_3C_2T_x$ MXene as electron transport layer for efficient perovskite solar cells.* ACS Applied Materials & Interfaces, 2021. **13**(27): p. 32495–32502.

[141] Zhang, F., et al., *Plasma-enhanced pulsed-laser deposition of single-crystalline Mo_2C ultrathin superconducting films.* Physical Review Materials, 2017. **1**(3): p. 034002.

[142] Zhang, Z., et al., *Substrate orientation-induced epitaxial growth of face centered cubic Mo_2C superconductive thin film.* Journal of Materials Chemistry C, 2017. **5**(41): p. 10822–10827.

[143] Stevens, M., et al., *Pulsed laser deposition of epitaxial Cr_2AlC MAX phase thin films on MgO (111) and Al_2O_3 (0001).* Materials Research Letters, 2021. **9**(8): p. 343–349.

[144] Zhang, S. and J. Xu, *PDMS/Ag/Mxene/Polyurethane conductive yarn as a highly reliable and stretchable strain sensor for human motion monitoring.* Polymers, 2022. **14**(24): p. 5401.

[145] Halim, J., et al., *Electronic and optical characterization of 2D Ti_2C and Nb_2C (MXene) thin films.* Journal of Physics: Condensed Matter, 2019. **31**(16): p. 165301.

[146] Li, G., et al., *Third-order nonlinear optical response of few-layer MXene Nb_2C and applications for square-wave laser pulse generation.* Advanced Materials Interfaces, 2021. **8**(6): p. 2001805.

[147] Meshkian, R., et al., *Synthesis of two-dimensional molybdenum carbide, Mo_2C, from the gallium based atomic laminate Mo_2Ga_2C.* Scripta Materialia, 2015. **108**: p. 147–150.

[148] Chen, Q., et al., *Optical properties of two-dimensional semi-conductive MXene Sc_2CO_x produced by sputtering.* Optik, 2020. **219**: p. 165046.

[149] Tarasov, A.S., et al., *Growth process, structure and electronic properties of Cr_2GeC and Cr_2-xMnxGeC thin films prepared by magnetron sputtering.* Processes, 2023. **11**(8): p. 2236.

[150] Zhang, Q., et al., *Fabrication of 3D interconnected porous MXene-based PtNPs as highly efficient electrocatalysts for methanol oxidation.* Journal of Electroanalytical Chemistry, 2021. **894**: p. 115338.

[151] Guo, M., et al., *3D hollow MXene (Ti_3C_2)/reduced graphene oxide hybrid nanospheres for high-performance Li-ion storage.* Journal of Materials Chemistry A, 2021. **9**(42): p. 23841–23849.

[152] Li, C., et al., *A self-sacrifice template strategy to synthesize Co-LDH/MXene for lithium-ion batteries.* Chemical Communications, 2021. **57**(86): p. 11378–11381.

[153] Hu, Z., et al., *Hierarchical $Ti_3C_2T_x$ MXene/carbon nanotubes for low overpotential and long-life Li-CO_2 batteries.* ACS Nano, 2021. **15**(5): p. 8407–8417.

[154] Sun, B., et al., *Room-temperature gas sensors based on three-dimensional Co_3O_4/ Al_2O_3 @ $Ti_3C_2T_x$ MXene nanocomposite for highly sensitive NOx detection.* Sensors and Actuators B: Chemical, 2022. **368**: p. 132206.

[155] Wu, J., et al., *Scalable synthesis of 2D Mo_2C and thickness-dependent hydrogen evolution on its basal plane and edges.* Advanced Materials, 2023: p. 2209954.

[156] Abdolhosseinzadeh, S., et al., *Perspectives on solution processing of two-dimensional MXenes.* Materials Today, 2021. **48**: p. 214–240.

[157] Sun, S., et al., *Hybrid energy storage mechanisms for sulfur-decorated Ti_3C_2 MXene anode material for high-rate and long-life sodium-ion batteries.* Chemical Engineering Journal, 2019. **366**: p. 460–467.

[158] Ding, Y., et al., *Facile synthesis of FePS$_3$ nanosheets@MXene composite as a high-performance anode material for sodium storage.* Nano-Micro Letters, 2020. **12**: p. 1–12.

[159] Luo, J., et al., *Pillared structure design of MXene with ultralarge interlayer spacing for high-performance lithium-ion capacitors.* ACS Nano, 2017. **11**(3): p. 2459–2469.

[160] Cheng, Y., et al., *Bioinspired microspines for a high-performance spray Ti$_3$C$_2$T$_x$ MXene-based piezoresistive sensor.* ACS Nano, 2020. **14**(2): p. 2145–2155.

[161] Sarycheva, A., et al., *2D titanium carbide (MXene) for wireless communication.* Science Advances, 2018. **4**(9): p. eaau0920.

[162] Chen, W., et al., *Flexible, transparent, and conductive Ti$_3$C$_2$T$_x$ MXene—silver nanowire films with smart acoustic sensitivity for high-performance electromagnetic interference shielding.* ACS Nano, 2020. **14**(12): p. 16643–16653.

[163] Zheng, X., et al., *Multifunctional RGO/Ti$_3$C$_2$T$_x$ MXene fabrics for electrochemical energy storage, electromagnetic interference shielding, electrothermal and human motion detection.* Materials & Design, 2021. **200**: p. 109442.

[164] Wang, Z., H. Kim, and H.N. Alshareef, *OXIDE thin-film electronics using all-MXene electrical contacts.* Advanced Materials, 2018. **30**(15): p. 1706656.

[165] Zhao, M.Q., et al., *Scalable manufacturing of large and flexible sheets of MXene/graphene heterostructures.* Advanced Materials Technologies, 2019. **4**(5): p. 1800639.

[166] Salles, P., et al., *Automated scalpel patterning of solution processed thin films for fabrication of transparent MXene microsupercapacitors.* Small, 2018. **14**(44): p. 1802864.

[167] Kumar, K.A., et al., *Dip-coating of MXene and transition metal dichalcogenides on 3D-printed nanocarbon electrodes for the hydrogen evolution reaction.* Electrochemistry Communications, 2021. **122**: p. 106890.

[168] Salles, P., et al., *Electrochromic effect in titanium carbide MXene thin films produced by dip-coating.* Advanced Functional Materials, 2019. **29**(17): p. 1809223.

[169] Li, L., et al., *New application of MXene in polymer composites toward remarkable anti-dripping performance for flame retardancy.* Composites Part A: Applied Science and Manufacturing, 2019. **127**: p. 105649.

[170] Yue, Y., et al., *3D hybrid porous Mxene-sponge network and its application in piezoresistive sensor.* Nano Energy, 2018. **50**: p. 79–87.

[171] Hu, D., et al., *Flexible and durable cellulose/MXene nanocomposite paper for efficient electromagnetic interference shielding.* Composites Science and Technology, 2020. **188**: p. 107995.

[172] Wang, Y., et al., *MXene-coated conductive composite film with ultrathin, flexible, self-cleaning for high-performance electromagnetic interference shielding.* Chemical Engineering Journal, 2021. **412**: p. 128681.

[173] Cheng, H., et al., *Ni flower/MXene-melamine foam derived 3D magnetic/conductive networks for ultra-efficient microwave absorption and infrared stealth.* Nano-Micro Letters, 2022. **14**(1): p. 63.

[174] Yao, S., et al., *Effects of dip-coating processing parameters on the functional performances of Ti$_3$C$_2$T$_x$ MXene-silk yarns.* Journal of Fiber Bioengineering and Informatics, 2023. **16**(1): p. 15–30.

[175] Yang, L., et al., *Surface-modified metallic Ti$_3$C$_2$T$_x$ MXene as electron transport layer for planar heterojunction perovskite solar cells.* Advanced Functional Materials, 2019. **29**(46): p. 1905694.

[176] Ying, G., et al., *Conductive transparent V$_2$CT$_x$ (MXene) films.* FlatChem, 2018. **8**: p. 25–30.

[177] Montazeri, K., et al., *Beyond gold: spin-coated Ti$_3$C$_2$-based MXene photodetectors.* Advanced Materials, 2019. **31**(43): p. 1903271.

[178] Ying, G., et al., *Transparent, conductive solution processed spincast 2D Ti$_2$CT$_x$ (mxene) films.* Materials Research Letters, 2017. **5**(6): p. 391–398.

[179] Yang, L., et al., *SnO$_2$–Ti$_3$C$_2$ MXene electron transport layers for perovskite solar cells.* Journal of Materials Chemistry A, 2019. **7**(10): p. 5635–5642.

[180] Huang, H., et al., *Scalable, and low-cost treating-cutting-coating manufacture platform for MXene-based on-chip micro-supercapacitors.* Nano Energy, 2020. **69**: p. 104431.

[181] Li, L., et al., *Hydrophobic and stable MXene–polymer pressure sensors for wearable electronics.* ACS Applied Materials & Interfaces, 2020. **12**(13): p. 15362–15369.

[182] Dillon, A.D., et al., *Highly conductive optical quality solution-processed films of 2D titanium carbide.* Advanced Functional Materials, 2016. **26**(23): p. 4162–4168.

[183] Zhang, M., et al., *Formation of new MXene film using spinning coating method with DMSO solution and its application in advanced memristive device.* Ceramics International, 2019. **45**(15): p. 19467–19472.

[184] Jiang, Q., et al., *All pseudocapacitive MXene-RuO$_2$ asymmetric supercapacitors.* Advanced Energy Materials, 2018. **8**(13): p. 1703043.

[185] Chen, W., et al., *MXene loaded onto clean wiper by a dot-matrix drop-casting method as a free-standing electrode for stretchable and flexible supercapacitors.* Chemical Engineering Journal, 2021. **423**: p. 130242.

[186] Lee, E., et al., *Two-dimensional vanadium carbide MXene for gas sensors with ultra-high sensitivity toward nonpolar gases.* ACS Sensors, 2019. **4**(6): p. 1603–1611.

[187] Singh, M.K., S. Krishnan, and D.K. Rai, *Rational design of Ti$_3$C$_2$T$_x$ MXene coupled with hierarchical CoS for a flexible supercapattery.* Electrochimica Acta, 2023. **441**: p. 141825.

[188] Wu, X., et al., *Polymer- Ti$_3$C$_2$T$_x$ composite membranes to overcome the trade-off in solvent resistant nanofiltration for alcohol-based system.* Journal of Membrane Science, 2016. **515**: p. 175–188.

[189] Tan, K., et al., *Optical and conductivity studies of polyvinyl alcohol-MXene (PVA-MXene) nanocomposite thin films for electronic applications.* Optics & Laser Technology, 2021. **136**: p. 106772.

[190] Mashtalir, O., et al., *Amine-assisted delamination of Nb$_2$C MXene for Li-ion energy storage devices.* Advanced Materials, 2015. **27**(23): p. 3501–3506.

[191] Wang, C., et al., *Delaminating vanadium carbides for zinc-ion storage: hydrate precipitation and H$^+$/Zn^{2+} Co-action mechanism.* Small Methods, 2019. **3**(12): p. 1900495.

[192] Luo, J., et al., *Sn^{4+} ion decorated highly conductive Ti$_3$C$_2$ MXene: promising lithium-ion anodes with enhanced volumetric capacity and cyclic performance.* ACS Nano, 2016. **10**(2): p. 2491–2499.

[193] Lu, M., et al., *Tent-pitching-inspired high-valence period 3-cation pre-intercalation excels for anode of 2D titanium carbide (MXene) with high Li storage capacity.* Energy Storage Materials, 2019. **16**: p. 163–168.

[194] Peng, M., et al., *Manipulating the interlayer spacing of 3D MXenes with improved stability and zinc-ion storage capability.* Advanced Functional Materials, 2022. **32**(7): p. 2109524.

[195] Cho, S., D.Y. Kim, and Y. Seo, *Binder-free high-performance MXene supercapacitors fabricated by a simple electrospray deposition technique.* Advanced Materials Interfaces, 2020. **7**(22): p. 2000750.

[196] Sun, B., et al., *Ti$_3$C$_2$ MXene-bridged Ag/Ag$_3$PO$_4$ hybrids toward enhanced visible-light-driven photocatalytic activity.* Applied Surface Science, 2021. **535**: p. 147354.

[197] Zhang, B., et al., *Transforming Ti$_3$C$_2$T$_x$ MXene's intrinsic hydrophilicity into superhydrophobicity for efficient photothermal membrane desalination.* Nature Communications, 2022. **13**(1): p. 3315.

[198] Iqbal, A., et al., *Anomalous absorption of electromagnetic waves by 2D transition metal carbonitride Ti$_3$CNT$_x$ (MXene).* Science, 2020. **369**(6502): p. 446–450.

[199] Cao, W.-T., et al., *Binary strengthening and toughening of MXene/cellulose nanofiber composite paper with nacre-inspired structure and superior electromagnetic interference shielding properties.* ACS Nano, 2018. **12**(5): p. 4583–4593.

[200] Ren, C.E., et al., *Charge-and size-selective ion sieving through $Ti_3C_2T_x$ MXene membranes.* The Journal of Physical Chemistry Letters, 2015. **6**(20): p. 4026–4031.

[201] Xin, W., et al., *Lightweight and flexible MXene/CNF/silver composite membranes with a brick-like structure and high-performance electromagnetic-interference shielding.* RSC Advances, 2019. **9**(51): p. 29636–29644.

[202] Liu, W., et al., *Molecularly stacking manganese dioxide/titanium carbide sheets to produce highly flexible and conductive film electrodes with improved pseudocapacitive performances.* Advanced Energy Materials, 2017. **7**(22): p. 1602834.

[203] Iqbal, J., et al., *A hydrofluoric acid-free green synthesis of magnetic M. Ti_2CT_x nanostructures for the sequestration of cesium and strontium radionuclide.* Nanomaterials, 2022. **12**(18): p. 3253.

[204] Rozmysłowska-Wojciechowska, A., et al., *A simple, low-cost and green method for controlling the cytotoxicity of MXenes.* Materials Science and Engineering: C, 2020. **111**: p. 110790.

[205] Xiu, L.-Y., Z.-Y. Wang, and J.-S. Qiu, *General synthesis of MXene by green etching chemistry of fluoride-free Lewis acidic melts.* Rare Metals, 2020. **39**(11): p. 1237–1238.

[206] Ramírez, R., et al., *Green, HF-free synthesis of MXene quantum dots and their photocatalytic activity for hydrogen evolution.* Small Methods, 2023: p. 2300063.

[207] Limbu, T.B., et al., *Green synthesis of reduced $Ti_3C_2T_x$ MXene nanosheets with enhanced conductivity, oxidation stability, and SERS activity.* Journal of Materials Chemistry C, 2020. **8**(14): p. 4722–4731.

[208] Huang, Y., et al., *A safe and fast-charging lithium-ion battery anode using MXene supported Li_3VO_4.* Journal of Materials Chemistry A, 2019. **7**(18): p. 11250–11256.

[209] Wang, J., et al., *Bioinspired, high-strength, and flexible MXene/aramid fiber for electromagnetic interference shielding papers with joule heating performance.* ACS Nano, 2022. **16**(4): p. 6700–6711.

[210] Li, Y. and X. Zhang, *Electrically conductive, optically responsive, and highly orientated $Ti_3C_2T_x$ MXene aerogel fibers.* Advanced Functional Materials, 2022. **32**(4): p. 2107767.

[211] Siebert, J.P., et al., *Sol-gel based synthesis and enhanced processability of MAX phase Cr_2GaC.* Journal of Materials Chemistry C, 2019. **7**(20): p. 6034–6040.

[212] Sinclair, J., et al., *Sol-gel-based synthesis of the phosphorus-containing MAX phase V_2PC.* Inorganic Chemistry, 2022. **61**(43): p. 16976–16980.

[213] Zhang, H., et al., *MXene-rGO aerogel assisted $Na_{3.5}MnTi(PO_4)_3$ cathode for high-performance sodium-ion batteries.* Chemical Engineering Journal, 2023. **466**: p. 143132.

[214] Sergiienko, S.A., et al., *Photocatalytic removal of benzene over $Ti_3C_2T_x$ MXene and TiO_2—MXene composite materials under solar and NIR irradiation.* Journal of Materials Chemistry C, 2022. **10**(2): p. 626–639.

[215] Wu, Z.-H., et al., *Silane modified MXene/polybenzazole nanocomposite aerogels with exceptional surface hydrophobicity, flame retardance and thermal insulation.* Composites Communications, 2023. **37**: p. 101402.

[216] Zhou, J., et al., *Hierarchical porous and three-dimensional MXene/SiO_2 hybrid aerogel through a sol-gel approach for lithium–sulfur batteries.* Molecules, 2022. **27**(20): p. 7073.

[217] Tian, R., et al., *Functionalized Pt nanoparticles between α/γ-Fe_2O_3 and MXene for superior acetone sensing.* Sensors and Actuators B: Chemical, 2023. **383**: p. 133584.

[218] Mohammed, I., et al., *Synthesis, characterization, and electromagnetic properties of $Ba_{1.8}Sr_{0.2}Co_2Fe_{11.9}Pr_{0.1}O_{22}/Ti_3C_2T_x$-MXene composites.* Physica Status Solidi (b), 2023. **260**(9): p. 2300149.

[219] Chen, F., et al., *Preparation of polyacrylamide/MXene hydrogels as highly-efficient electro-adsorbents for methylene blue removal*. Polymer-Plastics Technology and Materials, 2021. **60**(14): p. 1568–1584.

[220] Syamsai, R. and A.N. Grace, *Synthesis, properties and performance evaluation of vanadium carbide MXene as supercapacitor electrodes*. Ceramics International, 2020. **46**(4): p. 5323–5330.

[221] Zhu, Y., et al., *Flame-retardant MXene/polyimide film with outstanding thermal and mechanical properties based on the secondary orientation strategy*. Nanoscale Advances, 2021. **3**(19): p. 5683–5693.

[222] Ghidiu, M., et al., *Synthesis and characterization of two-dimensional Nb_4C_3 (MXene)*. Chemical Communications, 2014. **50**(67): p. 9517–9520.

[223] Li, W., et al., *Multilayer-structured transparent MXene/PVDF film with excellent dielectric and energy storage performance*. Journal of Materials Chemistry C, 2019. **7**(33): p. 10371–10378.

[224] Zhang, H., et al., *Preparation, mechanical and anti-friction performance of MXene/polymer composites*. Materials & Design, 2016. **92**: p. 682–689.

[225] Qu, C., et al., *Surface modification of Ti_3C_2-MXene with polydopamine and amino silane for high performance nitrile butadiene rubber composites*. Tribology International, 2021. **163**: p. 107150.

[226] Johnson, D., et al., *Holdups in nitride MXene's development and limitations in advancing the field of MXene*. Small, 2022. **18**(17): p. 2106129.

[227] Venkateshalu, S., et al., *New method for the synthesis of 2D vanadium nitride (MXene) and its application as a supercapacitor electrode*. ACS Omega, 2020. **5**(29): p. 17983–17992.

[228] Djire, A., et al., *Pseudocapacitive storage in nanolayered Ti_2NT_x MXene using Mg-ion electrolyte*. ACS Applied Nano Materials, 2019. **2**(5): p. 2785–2795.

[229] Soundiraraju, B. and B.K. George, *Two-dimensional titanium nitride (Ti_2N) MXene: synthesis, characterization, and potential application as surface-enhanced Raman scattering substrate*. ACS Nano, 2017. **11**(9): p. 8892–8900.

[230] Urbankowski, P., et al., *Synthesis of two-dimensional titanium nitride Ti_4N_3 (MXene)*. Nanoscale, 2016. **8**(22): p. 11385–11391.

[231] Djire, A., et al., *Electrocatalytic and optoelectronic characteristics of the two-dimensional titanium nitride $Ti_4N_3T_x$ MXene*. ACS Applied Materials & Interfaces, 2019. **11**(12): p. 11812–11823.

[232] Sun, G.D., et al., *A low-cost and efficient pathway for preparation of 2D MoN nanosheets via Na_2CO_3-assisted nitridation of MoS_2 with NH_3*. Journal of the American Ceramic Society, 2019. **102**(12): p. 7178–7186.

[233] Xiao, X., et al., *Salt-templated synthesis of 2D metallic MoN and other nitrides*. ACS Nano, 2017. **11**(2): p. 2180–2186.

[234] Urbankowski, P., et al., *2D molybdenum and vanadium nitrides synthesized by ammoniation of 2D transition metal carbides (MXenes)*. Nanoscale, 2017. **9**(45): p. 17722–17730.

[235] Xiao, X., et al., *Scalable synthesis of ultrathin Mn_3N_2 exhibiting room-temperature antiferromagnetism*. Advanced Functional Materials, 2019. **29**(17): p. 1809001.

[236] Li, J., et al., *Design of lamellar Mo_2C nanosheets assembled by Mo_2C nanoparticles as an anode material toward excellent sodium-ion capacitors*. ACS Sustainable Chemistry & Engineering, 2019. **7**(22): p. 18375–18383.

[237] Ma, L., et al., *Efficient hydrogen evolution reaction catalyzed by molybdenum carbide and molybdenum nitride nanocatalysts synthesized via the urea glass route*. Journal of Materials Chemistry A, 2015. **3**(16): p. 8361–8368.

[238] Mei, J., et al., *Two-dimensional fluorine-free mesoporous Mo_2C MXene via UV-induced selective etching of Mo_2Ga_2C for energy storage*. Sustainable Materials and Technologies, 2020. **25**: p. e00156.

[239] Liu, S., et al., *Ultrafast photonics applications based on evanescent field interactions with 2D molybdenum carbide (Mo₂C)*. Journal of Materials Chemistry C, 2021. **9**(19): p. 6187–6192.

[240] Jeon, J., et al., *Epitaxial synthesis of molybdenum carbide and formation of a Mo₂C/MoS₂ hybrid structure via chemical conversion of molybdenum disulfide*. ACS Nano, 2018. **12**(1): p. 338–346.

[241] Li, G., et al., *Highly efficient Nb₂C MXene cathode catalyst with uniform O-terminated surface for lithium—oxygen batteries*. Advanced Energy Materials, 2021. **11**(1): p. 2002721.

[242] Xu, Q., et al., *Facile synthesis of hierarchical MXene/ZIF-67/CNTs composite for electrochemical sensing of luteolin*. Journal of Electroanalytical Chemistry, 2021. **880**: p. 114765.

[243] Tian, Y., et al., *Flexible and free-standing 2D titanium carbide film decorated with manganese oxide nanoparticles as a high volumetric capacity electrode for supercapacitor*. Journal of Power Sources, 2017. **359**: p. 332–339.

[244] Naguib, M., et al., *Large-scale delamination of multi-layers transition metal carbides and carbonitrides "MXenes"*. Dalton Transactions, 2015. **44**(20): p. 9353–9358.

7 Properties of MXenes

7.1 STRUCTURAL PROPERTIES

Similar to their corresponding MAX phases, MXenes are made up of a hexagonal layered structure, with densely packed M atoms and an octahedral alignment of the X atoms. In MXenes, van der Waals interactions keep the structural layers together, similar to those found in other 2D materials. The majority of the methods that are used to synthesize these materials involve the aqueous solutions, which results in the production of MXenes with surface functionalization. The possible sites in MXenes for these surface groups can be directly on top of the M atom, between the X atoms, or between the metal atoms. Despite extensive theoretical simulations to investigate the effect of surface groups, fabrication of MXenes without these groups is still in its initial stages. $-F$ and $-OH$ groups require one electron to achieve stability compared to the $-O$ group, which requires two electrons. Therefore, the vacant sites between two metal atoms are more likely to be occupied by $-O$ termination groups [1].

A few physical properties of MXenes, that is, surface area, change in chemical composition, and layered structure, are similar to those of other 2D materials. MXenes also exhibit similar properties to 2D graphene, such as hydrophilicity, structural stability, and electronic conductivity. What distinguishes MXenes from other 2D materials are their ceramic-like properties, which make these materials chemically and mechanically stable; tunable interlayer spacing; and thicknesses, which can be extended to several units. Furthermore, surface functionalization also makes these materials extraordinary among others. These functional groups have the ability to modify the electronic structure, thereby producing a combination of high values of electronegativity and work function, which play an important role in different applications. For example, there is lower resistance for charge transfer in terminated MXenes, which facilitates electron migration and enhances electrocatalytic activity [2].

Many theoretical and experimental investigations have been undertaken to understand and predict the properties of MXenes. For example, following the first-principle calculations, researchers defined the electronic properties of $Hf_3C_2F_2$ by calculating the density of the state and band structure. At room temperature, the structural stability of $Hf_3C_2F_2$ was determined using ab initio molecular dynamics (AIMD) [3]. According to density functional theory (DFT) calculations, the required cohesive energy (E_c) to form $Hf_3C_2F_2$ monolayer is -51.25 eV, which determines its mechanical stability. They also calculated the Hf-C bond length, which illustrated a strong Hf–F interaction. H. Zhang et al. [4] used DFT calculations to study the optical, electrical, and structural properties of three different compositions of surface-functionalized MXenes, that is, M_2CT_2-I, M_2CT_2-II, and M_2CT_2-III (where Zr, Ti, or Hf were used as M). The kinetic stabilities of these compositions were observed by the phonon spectra, where no negative frequency was detected, except for M_2CO_2-II and

DOI: 10.1201/9781003465768-7

M_2CO_2-III. The most stable composition was observed to be M_2CT_2-I, as it possessed highest values of cohesive energies. The geometries of bare and O-functionalized Ti_2C MXenes are illustrated in Fig. 7.1 [4]. Similarly, the atomic structures of DTM MXenes are illustrated in Fig. 7.2 [5], where high-symmetry adsorption sites (for −O and −OH groups) on the outermost layer are represented by I, II, and III sites.

Studies have shown that electrical conductivity, structural stability, and electrochemical behavior of functionalized $Mo_2TiC_2T_x$ can be greatly influenced by structural defects. The formation energies of different structural defects were calculated,

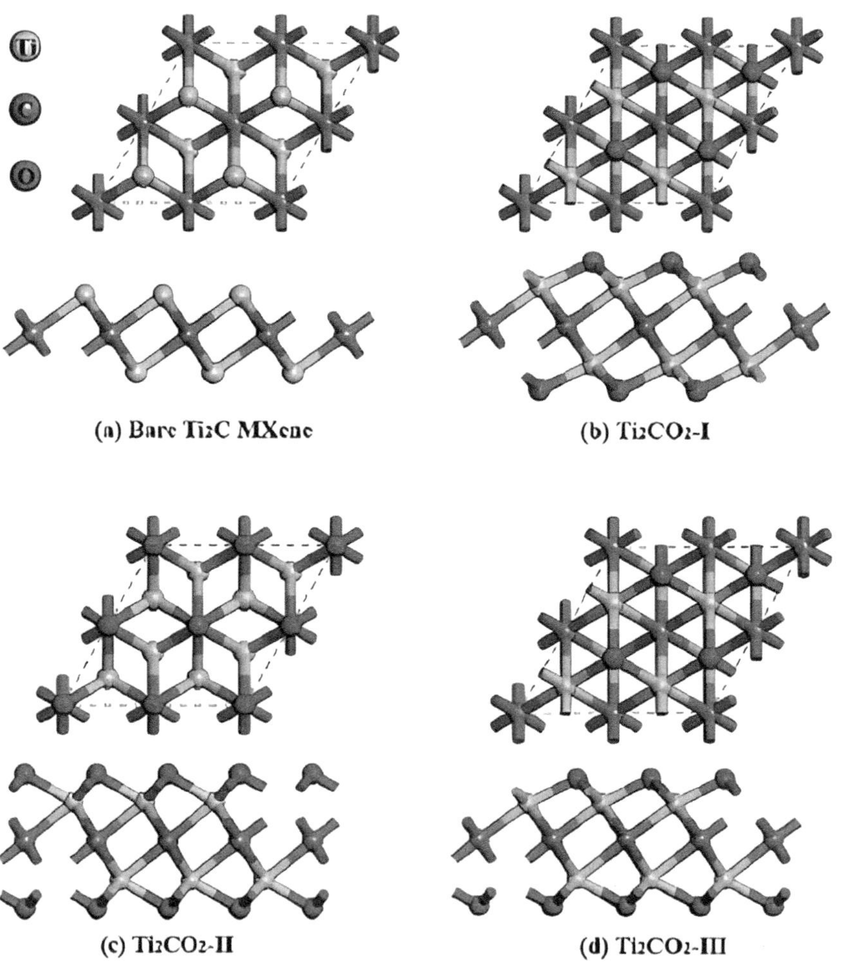

FIGURE 7.1 Geometries of (a) bare Ti_2C MXene and (b-d) O-functionalized Ti_2C MXene.

Source: [4] Zhang H., Yang G., Zuo X. et al. 2016. Computational studies on the structural, electronic and optical properties of graphene-like MXenes (M_2CT_2, M = Ti, Zr, Hf; T = O, F, OH) and their potential applications as visible-light driven photocatalysts. *J. Mater. Chem. A*, 4, 12913–12920. Reproduced with permission from the Royal Society of Chemistry.

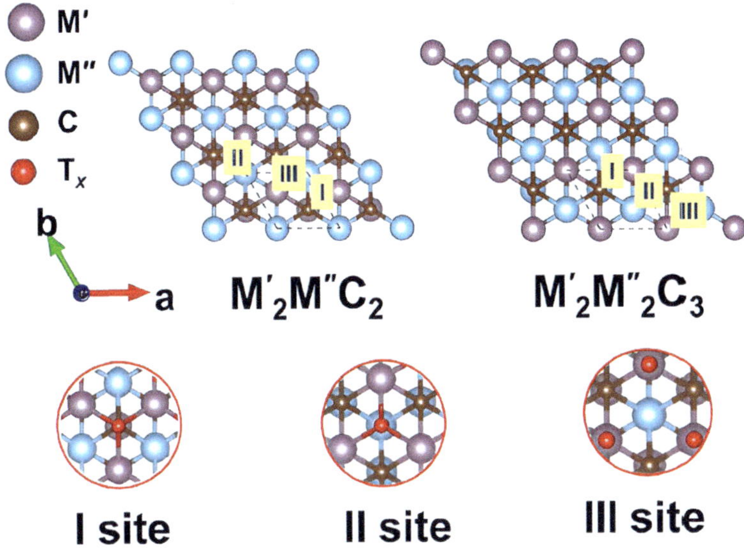

FIGURE 7.2 Atomic structures of $M'_2M''C_2$ and $M'_2M''_2C_3$ MXenes.

Source: Reprinted with permission from [5] Di Jin, Luke R. Johnson, Abhinav S. Raman, et al. 2020. Computational Screening of 2D Ordered Double-Transition Metal Carbides (MXenes) as Electrocatalysts for Hydrogen Evolution Reaction. *J. Phys. Chem. C*, 124, 10584–10592. Copyright 2020, American Chemical Society.

and the results showed that the formation of defects is surface termination-dependent, with OH- and F-functionalized MXenes requiring less energy than O-functionalized ones. It was also discovered that the outer Mo layers are more likely to produce defects than the inner Ti layers [6]. Figure 7.3a [6] illustrates the top and side views of MXene without structural defects (or perfect MXene). The top views of different types of structural defects are shown in Fig. 7.3b [6]. These defects include different types of single or two adjacent vacancies created on the same or different sublayers of the structure, such as V_{Tx}, V_{Tx2}, V_{Mo}, V_{Mo2}, V_{MoTi}, V_{Mo2Ti}, and V_{MoMo}.

7.2 MECHANICAL STRENGTH AND PROPERTIES

Mechanical strength and properties of MXenes are getting more attention of researchers due to the presence of stronger M-N and M-C bonding. Although MXenes have been observed to possess lower elastic properties than graphene, they exhibit highest bending characteristics of 1,050 GPa, which facilitates their ability to produce heterostructures. Therefore, in composite applications, the interactions of MXenes with a polymeric matrix are much better than that of graphene. Further research revealed that increasing the layer number decreases the Young's modulus of both carbide- and nitride-based MXenes. Furthermore, nitride-based components have the highest values compared to carbide MXenes. The immobility of terminations significantly reduces c_{11} parameter values while increasing essential deformations. Despite the

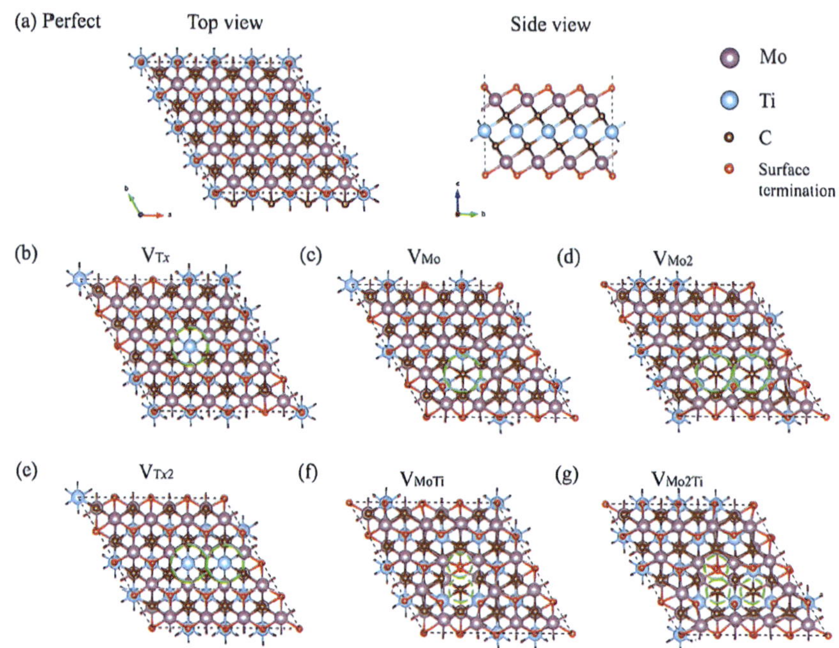

FIGURE 7.3 (a) Top and side views of the perfect MXene. Top views of structural defects of (b) V_{Tx}, (c) V_{Mo}, (d) V_{Mo2}, (e) V_{Tx2}, (f) V_{MoTi}, and (g) V_{Mo2Ti}.

Source: [6] Khaledialidusti R., Mishra A. K., Barnoush A. 2020. Atomic defects in monolayer ordered double-transition metals carbide ($Mo_2TiC_2T_x$) MXene and CO_2 Activation. *J. Mater. Chem. C*, 8, 4771–4779. Reproduced with permission from the Royal Society of Chemistry.

availability of different methodologies to characterize the bulk materials, evaluation of the mechanical properties of 2D materials is still in its early stages [7, 8].

Majority of the characterizations are being carried out by using the nanoindentation method, which involves an antiferromagnetic (AFM) tip, that applies force at the center point of the material. By using this approach, a Young's modulus of 333 ± 20 was obtained for $Ti_3C_2T_x$ monolayer, which is slightly higher than that of $Ti_3C_2O_2$ (i.e., 386 GPa) but lower than that of MoS_2 and GO. Due to the deficiency of measurement techniques, there are several limitations that effect the mechanical evaluation of MXenes, that is, the difficulty in surface control, interfaces among weak composites, and formation of surface defects. As a result, new studies should concentrate on developing the synthesis approaches in order to determine its defect composition and separate various functional groups of bare/pristine components. Based on theory and experiments, feasibly a complete analysis of mechanical traits and noticeable functional groups has yet to be developed [7–9].

A study employing classical molecular dynamics was carried out to evaluate the mechanical properties of 2D $Ti_{n+1}C_n$. In order to compute Young's modulus, tensile deformation of samples produced a linear component of stress/strain curves. All $Ti_{n+1}C_n$ samples showed strain-rate effects. According to the radial distribution function, the studied sample structures, during the process of deformation, remain unchanged. The calculated elastic constant values correlate well with known DFT data [10]. For all three $Ti_{n+1}C_n$, the stress/strain simulations were performed to confirm these results. These simulations employed three values of strain rates, that is, 0.0002, 0.0004, and 0.001 p s^{-1}. According to the results of elastic deformation, a linear region was obtained for all the curves, with a slope independent of strain rate.

N. Zhang et al. [11] theoretically studied the mechanical and electronic properties of pristine and functionalized $Ti_{n+1}C_n/Ti_{n+1}N_n$. As a result, higher monolayer thickness and lattice constants were observed for carbide-based MXenes. On the other hand, $Ti_{n+1}N_n$ exhibited higher in-plane Young's moduli compared to $Ti_{n+1}C_n$. However, as the monolayer thickness increases, these values were observed to decrease in both MXenes. Therefore, MXenes with high monolayer thickness were reported to be structurally stable. In another DFT study, the mechanical properties of M_2CT_2 were observed. The results showed that the properties of these materials are influenced by M atoms and the surface groups. Generally, MXenes terminated with the O-group were observed to possess better mechanical strength and smaller lattice parameters, compared to OH- and F-functionalized ones. Sc_2CO_2 was observed to have the least interlayer thickness and the greatest mechanical strength [12].

The elastic constant (c_{11}) of graphene was calculated to be 1,035 Gpa, which agrees with previous investigations. The MXene elastic constants were then examined in detail, and the c_{11} values are shown in Fig. 7.4 [12]. Except for Cr_2CO_2, oxygen-functionalized MXenes have greater elastic constants than comparable MXenes functionalized by fluorine and hydroxyl groups, as seen in the figure. Hf_2CO_2 c_{11}, for example, is discovered to be as high as 488.5 GPa, whereas the values for

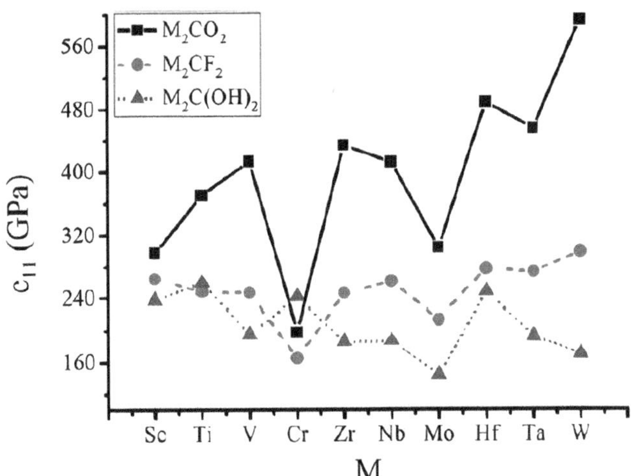

FIGURE 7.4 Elastic constants of c11 for M_2CT_2 MXenes (color online).

Source: Reprinted with permission from [12] Xian-Hu Zha, Kan Luo, Qiuwu Li, et al. 2015. Role of the surface effect on the structural, electronic and mechanical properties of the carbide MXenes. *Europhysics Letters*, 111, 26007. Copyright 2015, EPLA.

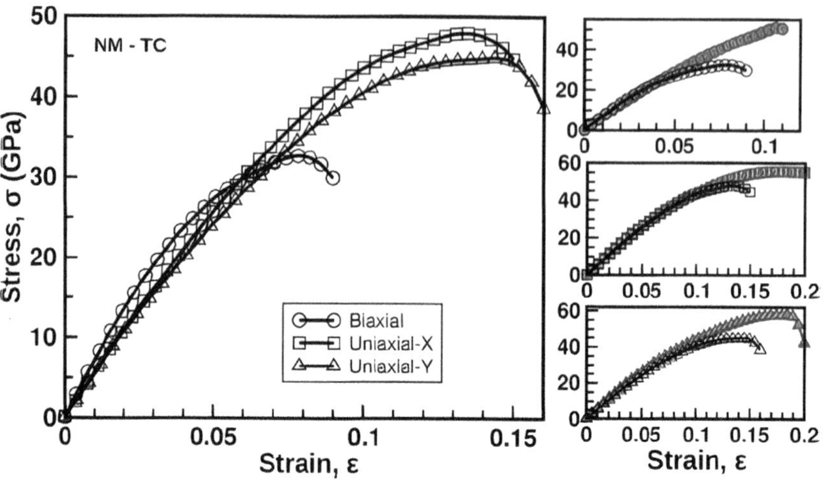

FIGURE 7.5 Stress–strain curve of the pristine Ti_2C under biaxial and uniaxial tensile strains along the zigzag (uniaxial-X) and armchair (uniaxial-Y) directions.

Source: Reprinted with permission from [14] Poulami Chakraborty et al. 2017. Manipulating the mechanical properties of Ti_2C MXene: Effect of substitutional doping. *Phys. Rev. B* 95, 184106. Copyright 2017 by the American Physical Society.

$Hf_2C(OH)_2$ and Hf_2CF_2 are only 247.8 and 277.1 GPa, respectively. Among all of the observed MXenes, W_2CO_2 has the maximum mechanical strength, with a c_{11} value of 592.7 GPa. Cr_2CO_2 has the lowest c_{11} value, which is just 33.4% of that of W_2CO_2. $Mo_2C(OH)_2$ has the lowest c_{11} value of all the MXenes studied, which is just 24.3% that of W_2CO_2.

Using first-principle calculations, the elastic properties of bare and functionalized Ti_2CT_2 and $Ti_3C_2T_2$ monolayers were studied. This study showed that the crystal and electronic structure, optical and elastic characteristics, and chemical bonding can be influenced by the surface groups. Surface termination caused the in-plane lattice constant to drop in Ti_2CT_2 and expand in $Ti_3C_2T_2$, with the largest change occurring in the O-functionalized ones [13]. P. Chakraborty et al. [14] investigated B and V substitutional doping at the Ti and C sites of Ti_2C. Through first-principle calculations of variables such as Young's modulus, in-plane stiffness, and strain, it was determined that B doping is particularly successful in increasing elastic characteristics. Figure 7.5 [14] depicts the computed stress–strain curves for pristine Ti_2C. The results of nonmagnetic (NM) Ti_2C are presented in the left panel, while a comparison between the curves of NM and A-AFM Ti_2C is illustrated in the right panels. Similarly, Fig. 7.6 [14] shows the comparison of curves between pristine and B/V-doped Ti_2C.

7.3 ELECTRONIC PROPERTIES

Pristine MXenes are analogous to MAX phases in terms of their metallic nature. Some MXenes can acquire semiconducting properties through the addition of functional termination groups onto their surfaces. The bandgaps of these materials can

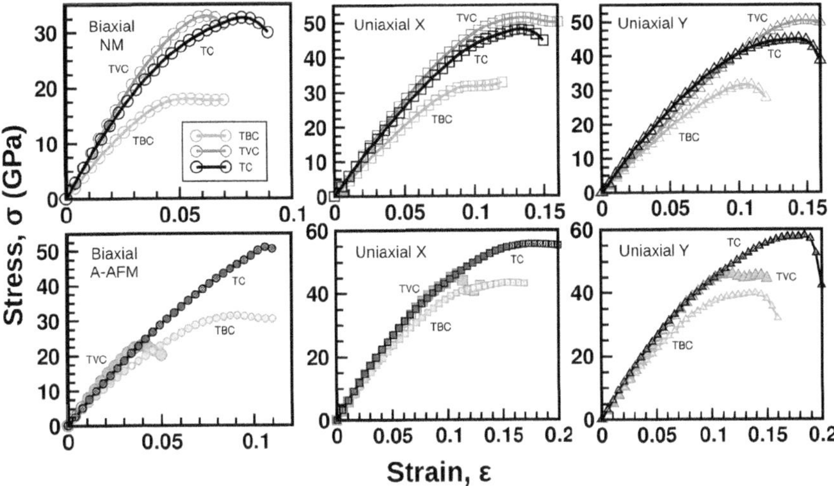

FIGURE 7.6 Stress–strain curve of Ti_2C, $Ti_2(C_{0.5}, B_{0.5})$, and $(Ti,V)C$ under biaxial and uniaxial tensile strains along the X and Y directions.

Source: Reprinted with permission from [14] Poulami Chakraborty et al. 2017. Manipulating the mechanical properties of Ti_2C MXene: Effect of substitutional doping. *Phys. Rev. B* 95, 184106. Copyright 2017 by the American Physical Society.

be aligned in the optoelectronic heterostructures, due to the similarity of these bandgaps with the solar spectrum. MXenes can be classified as topologically trivial or nontrivial based on their spin-orbital coupling (SOC). Furthermore, MXenes can be classified as metallic, semimetallic, or semiconducting based on their electrical conductivity. Due to the surface groups, Sc_2CT_2 (T = F, O, or OH), Zr_2CO_2, Ti_2CO_2, and Hf_2CO_2 are semiconductors. Bandgaps for these materials were measured using general gradient approximation, and except for $Sc_2C(OH)_2$, indirect bandgaps were observed for all the other semiconductors. Due to their same oxidation states, F and OH groups have similar impacts on the electrical structures of Zr-, Ti-, and Hf-based MXenes. However, O groups exhibit various effects in an equilibrium state as they accept two electrons [7, 15].

Electronic structures of 4d and 5d metal-based MXenes have been observed to be significantly influenced by relativistic SOC. When coupling is not taken into account, $M'_2M''C_2O_2$ (where $M' = W$ or Mo and $M'' = Ti$, Zr, or W), W_2CO_2, and Mo_2CO_2, are semimetallic (with zero bandgaps). However, when SOC is considered, the degeneracy of the conduction band minima and valence band maxima can be observed due to the formation of bandgap. As a result, these MXenes become semiconductors with indirect bandgaps, such as 0.472 and 0.401 eV for W_2CO_2 and $W_2HfC_2O_2$, respectively. $M'_2M''_2C_3O_2$, on the other hand, remains semimetallic. The bandgaps can also be tuned by applying external electric fields or by altering the mechanical properties. For example, with a 2% strain, the bandgap of Sc_2CO_2 changes from indirect to direct. Within a 14% strain, other materials, that is, Zr_2CO_2, Hf_2CO_2, and Ti_2CO_2, also modulate their bandgaps [7, 15].

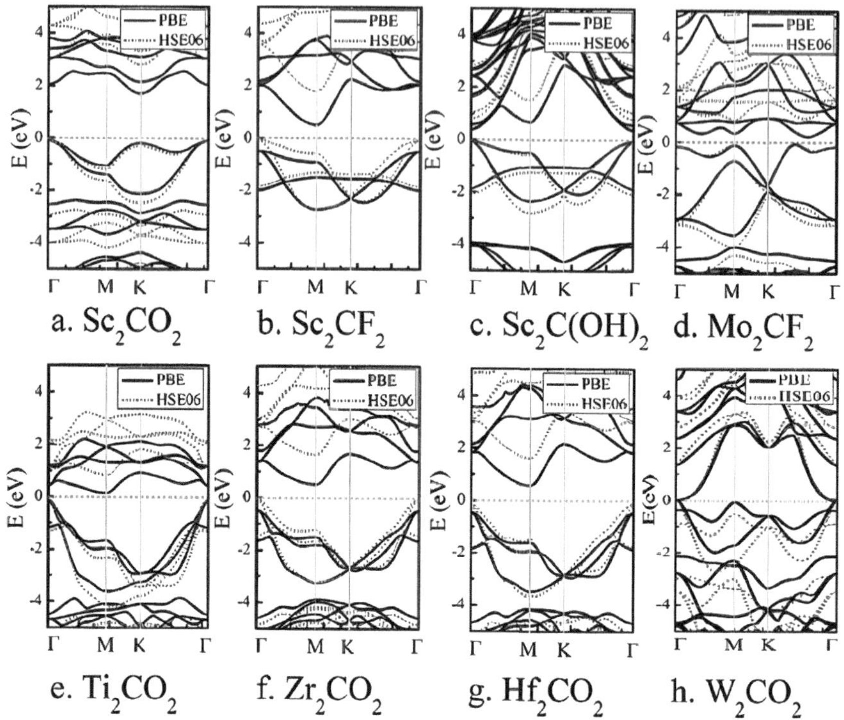

FIGURE 7.7 Electronic band structures of semiconducting M_2CT_2 MXenes (with Fermi level at zero) (color online).

Source: Reprinted with permission from [12] Xian-Hu Zha, Kan Luo, Qiuwu Li, et al. 2015. Role of the surface effect on the structural, electronic and mechanical properties of the carbide MXenes. *Europhysics Letters*, 111, 26007. Copyright 2015, EPLA.

Using the DFT calculations, researchers investigated the electronic properties of pristine and functionalized $Ti_{n+1}C_n/Ti_{n+1}N_n$ [11]. Adsorption calculations revealed that $Ti_{n+1}N_n$ preferentially adheres to terminal groups, thereby implying more active sites. More notably, outside the surfaces of $Ti_{n+1}N_n(OH)_2$, free electron states providing ideal electron transmission channels were revealed. The overall electrical conductivity of $Ti_{n+1}N_n$ was found to be higher than that of $Ti_{n+1}C_n$. Similarly, a theoretical study on the electronic properties of M_2CT_2 revealed that except for $Sc_2C(OH)_2$ (with a direct bandgap), Sc_2CO_2, Ti_2CO_2, Hf_2CO_2, Zr_2CO_2, Mo_2CF_2, Sc_2CF_2, and $Sc_2C(OH)_2$ all exhibit semiconducting properties [12]. Figure 7.7 [12] depicts the equivalent electronic energy bands. The HSE06 functional, as seen in this illustration, enhances the bandgap values from GGA data.

K. D. Fredrickson et al. [16] also used DFT to examine the effects of surface groups, applied potential, and water intercalation on the electrical properties of Ti_2C and Mo_2C MXenes. The lattice parameter (c) was observed to be changed with the surface group and to increase significantly during water intercalation. At zero applied potential, both MXenes were discovered to be O-functionalized, while

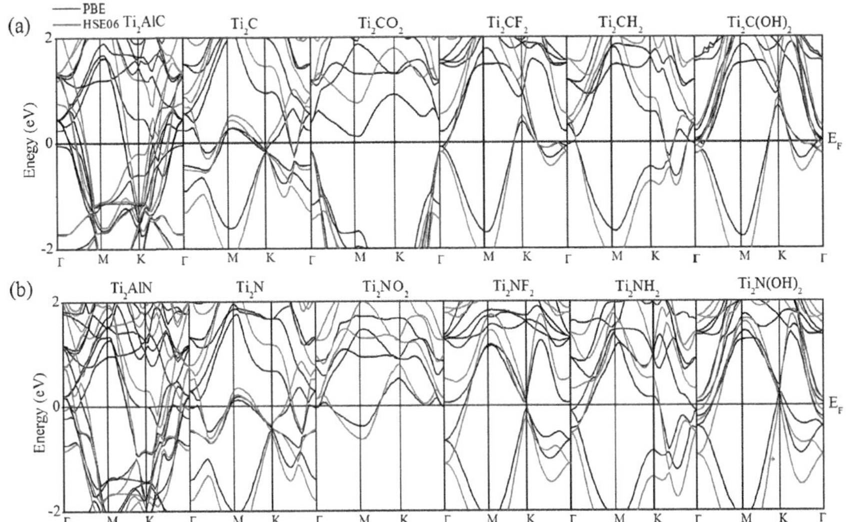

FIGURE 7.8 (a) Band structures of Ti_2CT_2 and (b) Ti_2NT_2, and related MAX phases. The HSE06 and PBE produce identical results.

Source: Reprinted with permission from [17] Yu Xie and P. R. C. Kent. 2013. Hybrid density functional study of structural and electronic properties of functionalized $Ti_{n+1}X_n$ (X = C, N) monolayers. *Phys. Rev. B* 87, 235441. Copyright 2013 by American Physical Society.

pristine materials were observed to be unstable (regardless of the potential). In another study, PBE and HSE06 simulations yielded identical results for the electrical structure analysis of functionalized $Ti_{n+1}N_n$ and $Ti_{n+1}C_n$ (n = 1–9). Except for Ti_2CO_2, which is believed to be semiconducting, all of the materials investigated were metallic. For thicker MXenes, the computed density of states (DOS) at the Fermi level was significantly larger than that of thin MXenes, which indicates their difference in surface chemistry and electronic conductivity [17]. They started by investigating the electrical characteristics of the thinnest Ti_2XT_2 monolayers. Figure 7.8 [17] shows and compares the computed PBE and HSE06 results of these monolayers to those of pure MXenes and parent MAX phases. With the significant exception of Ti_2CO_2, all other materials were observed to be metallic (as illustrated in Fig. 7.8).

Figure 7.9 [17] illustrates the partial DOS of various materials to provide a clearer description of their electrical characteristics. As we can see for MAX phases, Ti 3d orbitals dominate the DOS at the Fermi level [N(E$_F$)]. Below this level, the valence states can be separated into bands A and B, generated by Ti 3d-Al 3p and hybridized Ti 3d-C 2p and Ti 3d-Al 3s orbitals, respectively. The Ti-C and Ti-Al bands are represented by these combined states. The extraction of Al atoms narrows both bands and creates a gap between them. The N(E$_F$) of carbides and nitrides increases from 1.88 to 3.15 and from 2.77 to 4.84, respectively, due to the breakdown of Ti-Al bonds. Because of the new energy states between Ti and terminations, the N(E$_F$) lowers after functionalization. The hybridization of Ti with surface groups results in the formation of band C. In nitrides, unlike carbides, functional groups contribute

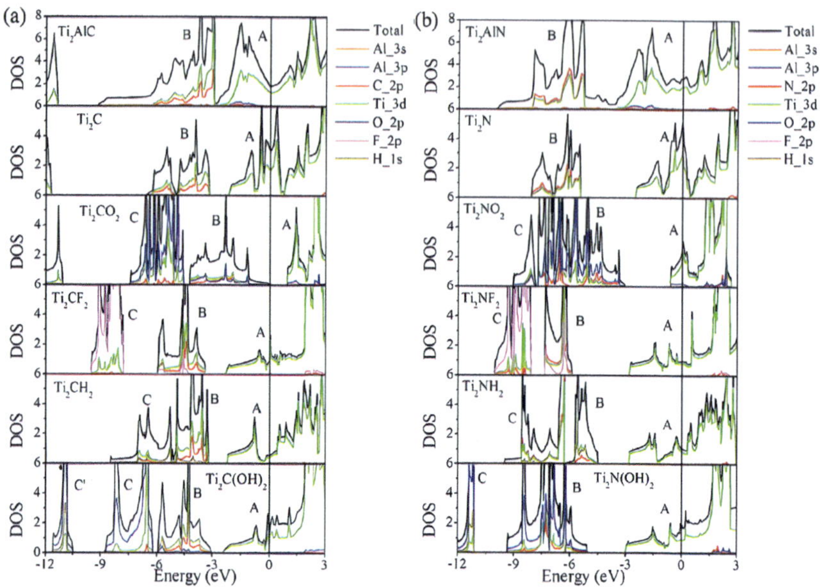

FIGURE 7.9 (a) Partial DOS of Ti_2CT_2 and (b) Ti_2NT_2, and MAX phases, computed using the HSE06 functional.

Source: Reprinted with permission from [17] Yu Xie and P. R. C. Kent. 2013. Hybrid density functional study of structural and electronic properties of functionalized $Ti_{n+1}X_n$ (X = C, N) monolayers. Phys. Rev. B 87, 235441. Copyright 2013 by American Physical Society.

significantly to band B, with the exception of the newly formed band C. When we compare Ti_2CO_2 to other materials, we find that the Ti 3d, C 2p, and O 2p orbitals all contribute similarly to band B.

Among the electronic characteristics of bare and functionalized Ti_2CT_2 and $Ti_3C_2T_2$, the optical bandgap was found only in Ti_2CO_2, owing to its semiconductor-like structure, which makes it suitable for use in optical and electronic devices [13]. In a work by B. Anasori et al., the electrical structures of $Mo_2TiC_2T_x$ were compared to that of their Ti_3C_2 counterparts [18]. It was discovered that the decrease in temperature causes the resistance to be increased, which results in changing the metallic nature of MXenes. The results of the DFT study revealed that $Mo_2TiC_2T_x$ is a semiconductor, whereas $Ti_3C_2T_x$ is a metal. C. Si et al. [19] projected that $Mo_2MC_2O_2$ (where M = Zr, Ti, or Hf) are robust quantum spin Hall (QSH) insulators based on first-principle calculations. To characterize the QSH states in $Mo_2MC_2O_2$, a tight-binding model based on different orbital bases in a triangular lattice was also developed. It demonstrated that the topological gap is contributed by the SOC strength of M. As a result, with varied M atoms, $Mo_2MC_2O_2$ displayed substantial gaps ranging from 0.1 to 0.2 eV, which were large enough to realize room-temperature QSH effects. Figure 7.10 [19] displays the band configurations (GGA and GGA + SOC) of $Mo_2MC_2O_2$. In the absence of SOC, these three $Mo_2MC_2O_2$ structures have zero bandgap, with degenerating conduction and valence bands.

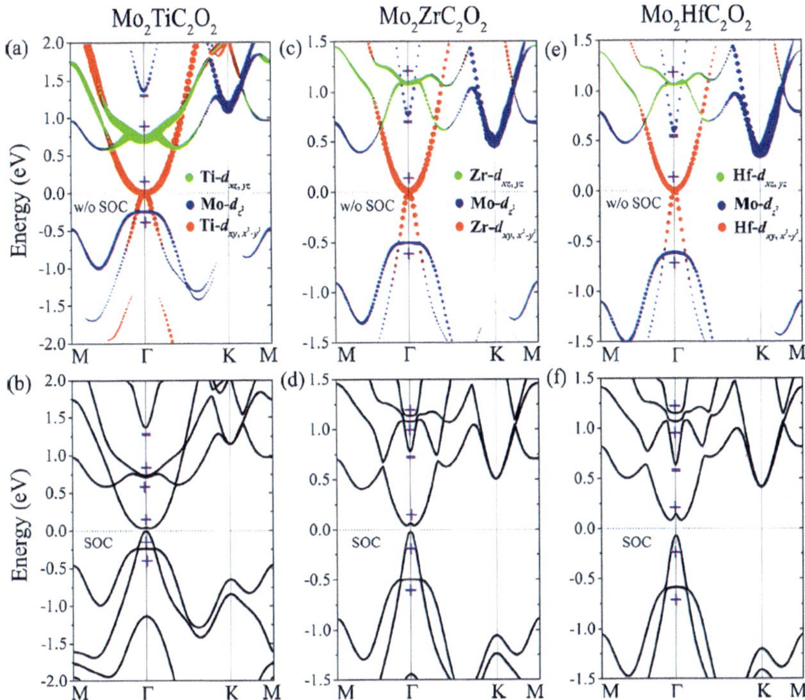

FIGURE 7.10 Calculated GGA band structures for (a, b) $Mo_2TiC_2O_2$, (c, d) $Mo_2ZrC_2O_2$, and (e, f) $Mo_2HfC_2O_2$, with and without SOC.

Source: Reprinted with permission from [19] Chen Si, Kyung-Hwan Jin, Jian Zhou, et al. 2016. Large-Gap Quantum Spin Hall State in MXenes: d-Band Topological Order in a Triangular Lattice. *Nano Lett.*, 16, 6584–6591. Copyright 2016, American Chemical Society.

2D functionalized Sc_2CT_2 structures are widely recognized to be good semiconductors. Another work revealed that MAX precursors of these materials can be doped by different transition metals (TMs), that is, V, Mn, Ti, or Cr, under ScAl-rich circumstances. The resulting $TM-Sc_2C(OH)_2$ and $TM-ScCO_2$ structures were observed to be n- and p-type semiconductors, respectively. The disparity is due to O having higher empty p orbitals than OH. Similarly, Mn- and Cr-doped Sc_2CT_2 structures were observed to exhibit magnetic properties as well [20]. Similarly, DFT was used to analyze the electrical structures of Si, $Ge-Sc_2C(OH)_2$ monolayers. Band inversion was seen in the doped systems, which were discovered to have topological invariants. When SOC was considered, bandgaps were observed to be formed, which resulted in the topological insulating nature of the doped materials [21].

7.4 OPTICAL PROPERTIES

MXenes possess a range of linear and nonlinear optical properties, such as photoluminescence, absorption, and nonlinear refractive index, that are closely related to their electronic structures. The optical properties are determined by the dispersion of refractive indices or dielectric functions. According to the theoretical studies, there are some important parameters that influence optical properties of MXenes. For example, the bandgap of M_2C (i.e., 0.92–1.75 eV) corresponds to the visible light absorption range based on the composition of M. By modifying the surface termination, the bandgap can also be tuned in the UV region. Thus, a wide range of spectrum (UV to NIR) and absorption capabilities enable MXenes to be used in photothermal therapy and photoacoustic imaging. Because of the excellent metal conductivity and significant local surface plasmon resonance effect of MXenes, they are potential substrates for surface-enhanced Raman spectroscopy. Furthermore, the hydrophilic and flexible surface of the Raman label can give stability and tight contact [1, 9, 22, 23].

Both of these characteristics help improve the Raman scattering signal. Light-emitting MQDs (MXene quantum dots) have recently been discovered by producing very small dot-phase MXene, which is another important feature, as opposed to standard non-radiative MXene. Under specific excitation wavelength (primarily UV blue), some emissions can be generated in these dot-phase MXenes via defect-induced luminescence or dimensionally induced quantum confinement. Excitation-related emission behavior was observed in Ti_3C_2 MQD colloids (i.e., hydrothermal graphene or carbon QDs). This effect could be induced by various optical size selections or surface imperfections during sample capture. The fluorescence peak shifted from 460 to 580 nm when the excitation wavelength increased from 340 to 500 nm. These characteristics are advantageous for optical-based biosensing and biological imaging [1, 9, 22, 23].

In a theoretical study, researchers used DFT calculations to explore the optical properties of three different functionalized compositions, that is, M_2CT_2-I, M_2CT_2-II, and M_2CT_2-III (where Zr, Ti, and Hf were used as M). The kinetic stabilities of these compositions were confirmed by phonon spectra, where no negative frequency was observed (except for M_2CO_2-II and M_2CO_2-III) [4]. Similarly, the optical properties of Si, Ge, Sn, F, B, N, or S-doped Ti_2CF_2 were also examined using DFT analysis [24]. The majority of the examined Ti_2CF_2 MXenes were found to be promising dielectric materials, with the greatest absorption coefficient in the infrared region at around 34.3 nm. It was observed that as the wavelength increases, the values of absorption coefficients for pure and B-doped Ti_2CF_2 reach almost zero; however, these values do not decrease for other doped MXenes. For all of these MXenes, the relation of dielectric function w.r.t photon energy is illustrated in Fig. 7.11 [24].

Yan, F. et al. [25] produced N-MQDs by using Ti_3C_2 MXenes and analyzed the morphology, element content, and optical properties of as-synthesized material by using different characterization techniques, that is, X-ray photoelectron spectroscopy (XPS), TEM, etc. These N-MQDs, composed of oxygen, nitrogen, and carbon, were observed to be a scattered quasi-spherical exhibiting sufficient quantum field (5.42%) and yellow fluorescence (580 nm). Furthermore, UV-Vis absorption and fluorescence spectroscopy was employed to study the optical characteristics, as shown in Fig. 7.12 [25]. The fluorescence intensity was also measured under different conditions, to examine the optical stabilities of N-MQDs.

Halim, J. et al. [26] used physical vapor deposition to fabricate epitaxial thin films of two different MAX phases, that is, Ti_2AlC and Nb_2AlC on sapphire substrates. Below 50 K, metallic conductivity (with weak localization) was observed

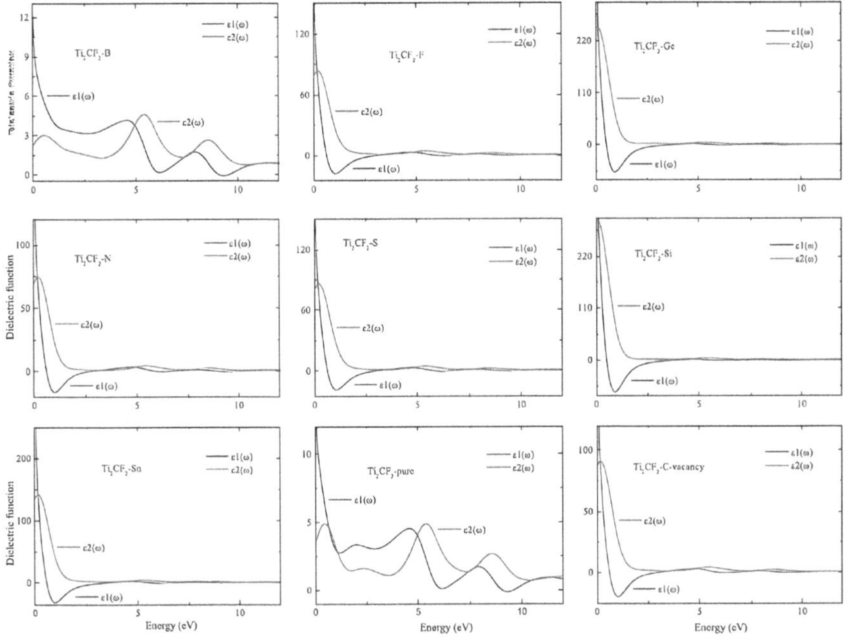

FIGURE 7.11 Dielectric function of the studied pure and doped Ti_2CF_2.

Source: Reprinted from [24] *Physica B: Condensed Matter*, 561, Rui-Zhou Zhang, Hong-Ling Cui, Xiao-Hong Li, First-principles study of structural, electronic and optical properties of doped Ti_2CF_2 MXenes, 90–96, Copyright 2019, with permission from Elsevier.

for Ti-based films. However, as the temperature decreases from room temperature to 40 K, the resistivity of Nb-based films was observed to be increased. They used spectroscopic ellipsometry (0.75–3.50 eV) to evaluate the optical characteristics of both films. The metallic behavior was confirmed by the results for Ti_2CT_z films. The dielectric functions of these films are illustrated in Figs. 7.13(a) and (b) [26], which show 0.73 eV–3.34 eV spectral region for Ti- and Nb-based films, respectively. Straight lines indicate the outcomes of a model dielectric functions (MDF)-based analysis with numerous parametrized contributions (as indicated by dashed lines). Q. Chen et al. [27] used magnetron sputtering to produce semi-conductive MXene Sc_2CO_x. The bandgap of the Sc_2CO_x sample was evaluated using UV–Vis absorption and photoluminescence spectroscopy.

7.5 ELECTRICAL PROPERTIES

MXenes have a wide range of conductivities. They can be metallic, semiconducting, or insulating, depending on their surface groups, lateral size, formation and concentration of defects, interlayer d-spacing, and delamination yield. In general, high conductivity of MXenes can be achieved when the defect concentration is minimal between large flakes. Similarly, compared to cold-pressed films, spin-cast $Ti_3C_2T_x$

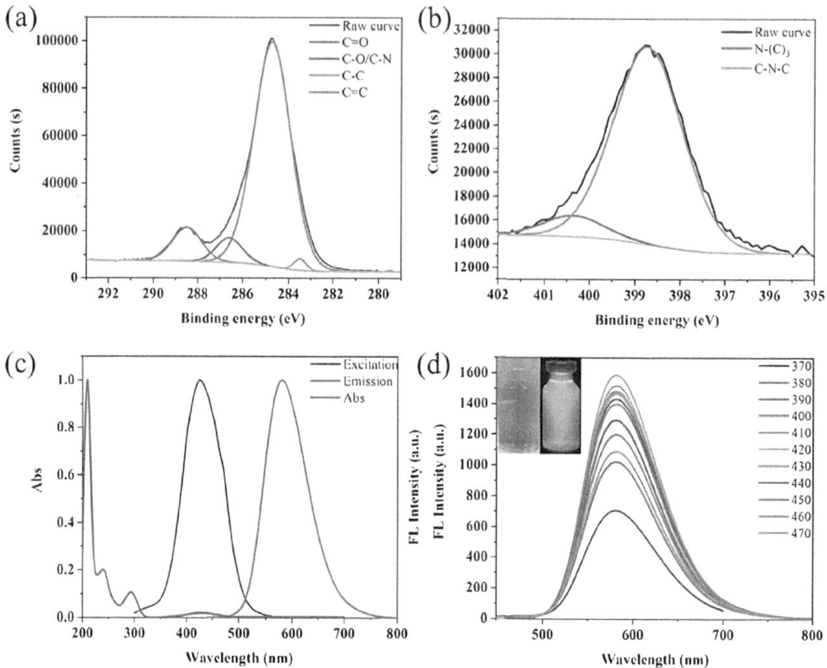

FIGURE 7.12 (a) C1s, (b) N1s spectrum, (c) UV–Vis absorption and fluorescence spectra of N-MQDs, and (d) fluorescence emission spectra at different excitation wavelengths.

Source: Reprinted from [25] *Dyes and Pigments*, 195, Fanyong Yan, Jingru Sun, Yueyan Zang, et al., Solvothermal synthesis of nitrogen-doped MXene quantum dots for the detection of alizarin red based on inner filter effect, 109720. Copyright 2021, with permission from Elsevier.

MXene films, for example, have a greater conductivity (about 9,880 S cm^{-1}). High conductivities of MXenes enable them to be used in various electronic and energy-related applications [2, 7, 9, 28]. G. Gao et al. [29] examined different MXene compositions and presented a detailed study on the variation in the electronic structures of these MXenes depending on different functional groups (as illustrated in Fig. 7.14 [29]). It was demonstrated that, under standard conditions, V_2C, Ti_2C, and Ti_3C_2 can be functionalized by a mixture of −OH and −O groups, while Nb-based MXenes (i.e., Nb_2C and $Nb_4C_3O_2$) are only O-functionalized. Figure 7.15 [29] shows the electronic band structures of pristine and functionalized MXenes. The effects of surface groups can be observed by the corresponding difference in band structures.

Z. Huang et al. [30] used first-principle calculations to determine the effects of a wide range of surface terminations (i.e., H, O, OH, F, Br, I, or Cl) on the electronic structures and conductivities of $M'_2M''C_2$ (where M' = Ta, Nb, or V and M'' = Zr, Hf, or Ti). The topological phase with semimetallic band structures was seen in the majority of the systems. However, $M'_2M''C_2F_2$-fluorinated MXenes were observed to be topological insulators. The hybrid functional computations revealed large non-trivial bandgaps (from 34 to 318 meV), which can be estimated for quantum spin

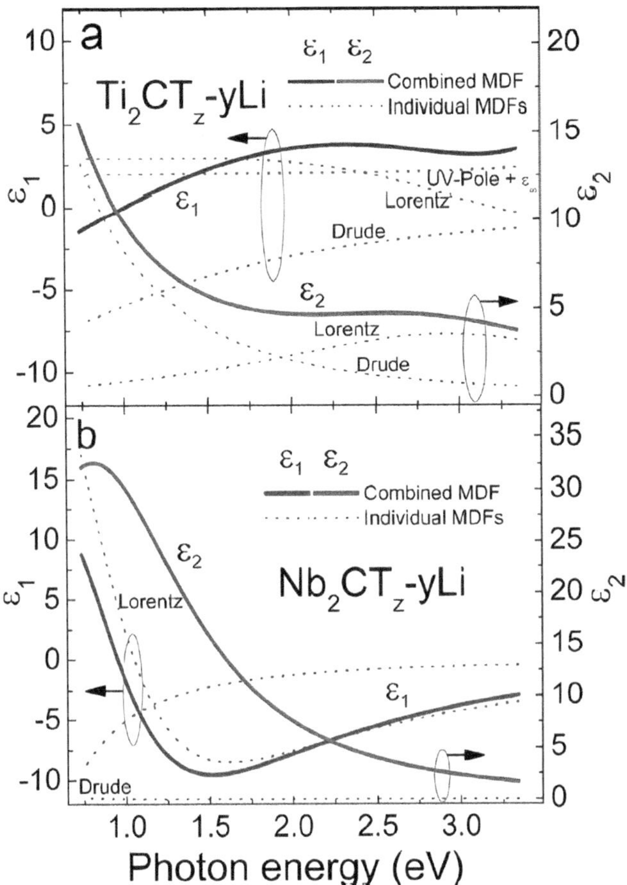

FIGURE 7.13 Dielectric functions for (a) Ti$_2$CT$_{z-y}$ Li and (b) Nb$_2$CT$_{z-y}$ Li thin films in the 0.73 eV –3.34 eV spectral range.

Source: Reprinted from [26] Joseph Halim et al. 2019. Electronic and optical characterization of 2D Ti$_2$C and Nb$_2$C (MXene) thin films. *J. Phys.: Condens. Matter* 31, 165301. With permission from IOP Science.

Hall effects (even at room temperature). Similarly, thermoelectric properties of vacuum-assisted filtered freestanding Mo$_2$CT$_x$, Mo$_2$TiC$_2$T$_x$, and Mo$_2$Ti$_2$C$_3$T$_x$ sheets were evaluated [31]. After heating at 800 K, n-type Seebeck coefficient and high value of conductivity were displayed by these sheets. At 803 K, the thermoelectric power of the Mo$_2$TiC$_2$T$_x$ MXene was 3.09 × 10^{-4} Wm^{-1} K^{-2}. While MXenes do not have the best thermoelectric characteristics, they do outperform their ternary and quaternary carbides.

Using SCAN-rVV10+U, Z. Jing et al. [32] demonstrated that pristine and functionalized Cr$_2$TiC$_2$ MXenes are AFM semiconductors with intermediate bandgap. All studied MXene structures displayed high Seebeck coefficients (>400 V K^{-1}), particularly Cr$_2$TiC$_2$ and Cr$_2$TiC$_2$F$_2$, that exhibited >800 and >700 V K^{-1}, respectively. The

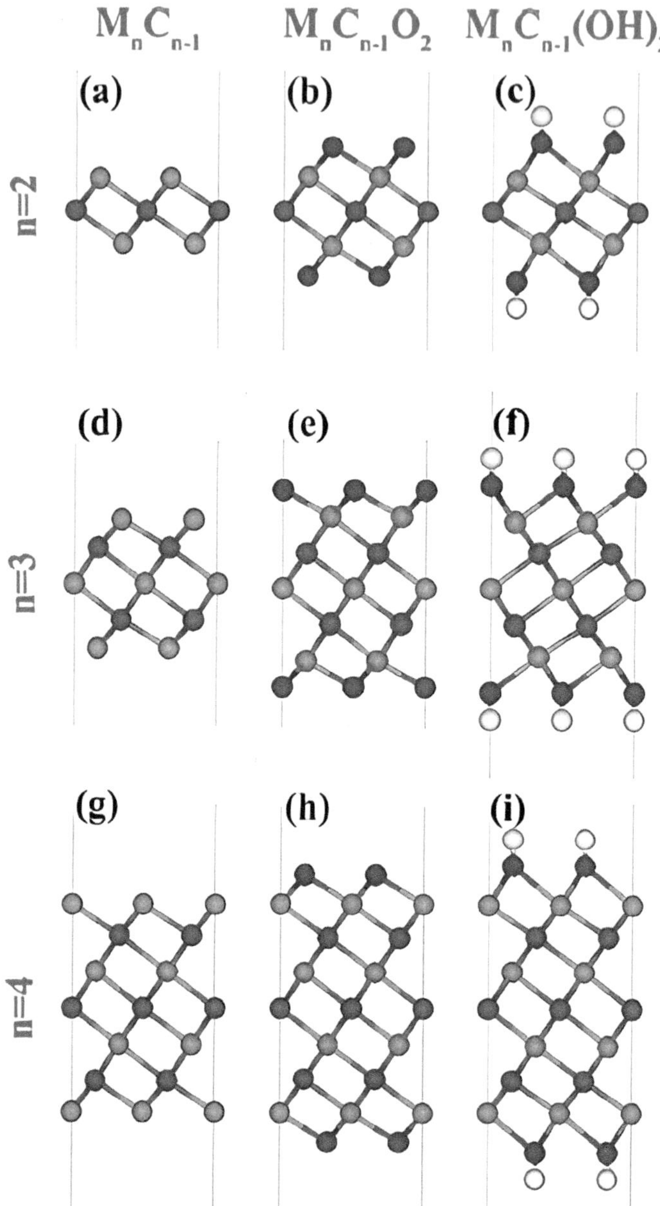

FIGURE 7.14 Models of pristine (M_nC_{n-1}), O-terminated ($M_nC_{n-1}O_2$), and OH-terminated [$M_nC_{n-1}(OH)_2$] MXenes (color code: cyan, gray, red, and white for metal, carbon, oxygen, and hydrogen, respectively).

Source: Reprinted with permission from [29] Guoping Gao, Anthony P. O'Mullane, Aijun Du. 2017. 2D MXenes: A New Family of Promising Catalysts for the Hydrogen Evolution Reaction. ACS Catal., 7, 494–500. Copyright 2017, American Chemical Society.

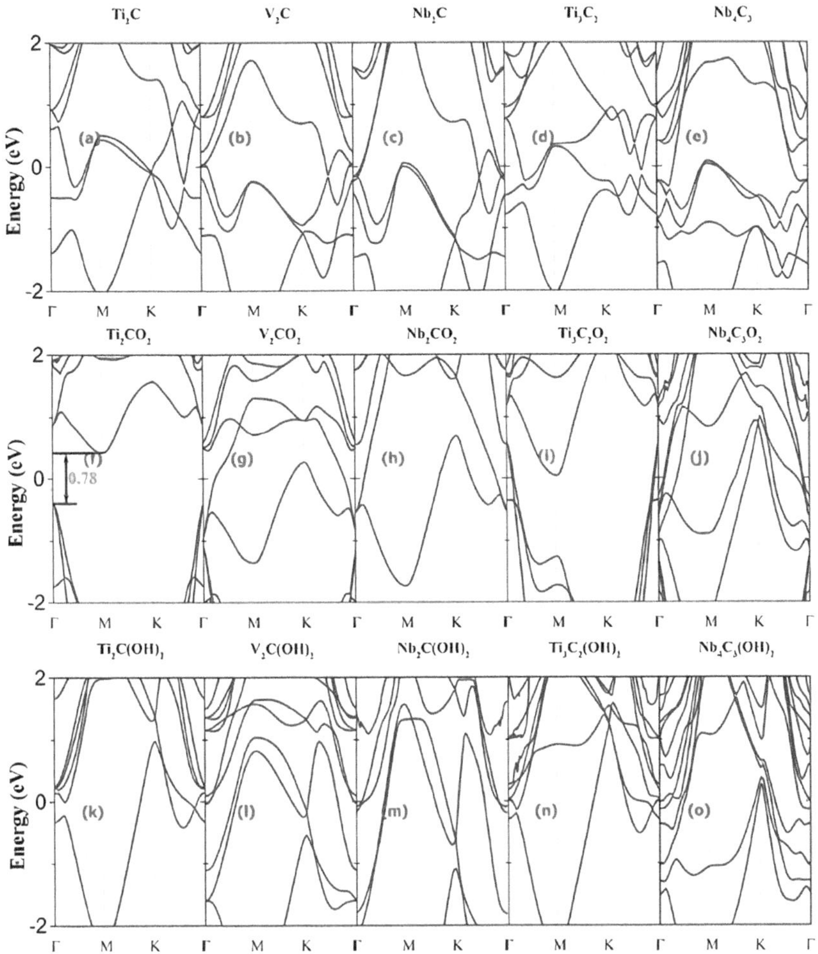

FIGURE 7.15 Band structure of (a–e) pristine, (f–j) O-terminated, and (k–o) OH-terminated MXenes. The Femi level is set to zero.

Source: Reprinted with permission from [29] Guoping Gao, Anthony P. O'Mullane, Aijun Du. 2017. 2D MXenes: A New Family of Promising Catalysts for the Hydrogen Evolution Reaction. *ACS Catal.*, 7, 494–500. Copyright 2017, American Chemical Society.

hole relaxation time of p-type $Cr_2TiC_2(OH)_2$ is determined to be 8 ps, indicating that it has superior electron transport capabilities when compared to other MXenes. The thermal conduction of phonon was also observed to be reduced by surface groups; therefore, lowest lattice thermal conductivity (~6.5 Wm K^{-1}) was examined for $Cr_2TiC_2(OH)_2$. The magnitude of the Seebeck coefficients of Cr_2TiC_2 and $Cr_2TiC_2T_2$ for both p- and n-type dopants with high carrier concentration (1019 cm^{-3}) is shown in Fig. 7.16a [32]. Furthermore, Fig. 7.16b [32] shows that there is a decrease in relaxation time (τ) for the given carrier concentration. The relation between electrical conductivity and temperature is demonstrated in Fig. 7.16c [32] using the electronic transport

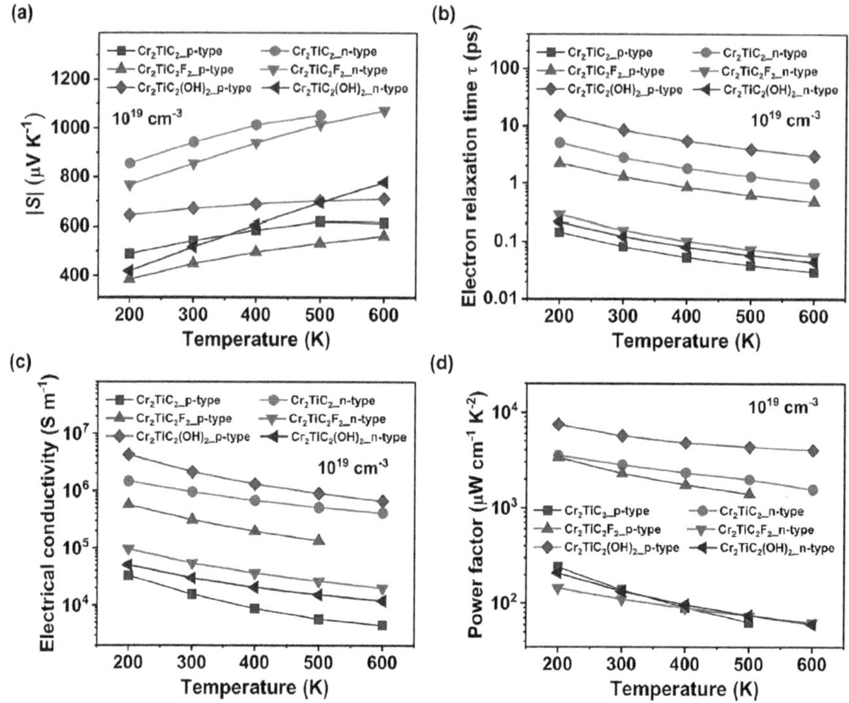

FIGURE 7.16 Thermoelectric and electron transport properties of pristine and functionalized Cr_2TiC_2 for both p- and n-type carriers: (a) Seebeck coefficient, (b) electron relaxation time, (c) electrical conductivity, and (d) thermoelectric power factor.

Source: Reprinted with permission from [32] Ziang Jing, Hangyu Wang, Xianghui Feng, et al. 2019. Superior Thermoelectric Performance of Ordered Double-Transition Metal MXenes: $Cr_2TiC_2T_2$ (T = −OH or −F). *J. Phys. Chem. Lett.*, 10, 5721–5728. Copyright 2019, American Chemical Society.

parameters acquired from the BoltzTraP code and τ computed by the single parabolic band (SPB) model. In general, the higher the electrical conductivity, the longer the relaxation time. The power factors ($S2\sigma$) are shown in Fig. 7.16d [32].

In another study, the semiconducting properties of OH- and F-functionalized $ScYCT_2$ were also investigated by using first-principle calculations [33]. Similarly, the findings of Boltzmann transport theory and the Slack model presented the excellent thermoelectric properties of these MXenes. The findings revealed that the examined MXene structures have excellent thermoelectric characteristics. At 900 K, the highest power factor of 0.072 Wm K^{-2} was observed for the n-type $ScYC(OH)_2$ along with a maximum ($\approx$3) ZT value.

7.6 CHEMICAL STABILITY AND STRUCTURE

Multilayered MXene derivatives, like other 2D materials, can be combined with polymers and organic compounds to create heterostructures, which improves their stability and surface functionality. The colloidal solutions of MXenes are quite important as they are becoming a top priority to be used in various applications. Furthermore, single or few-layer nanosheets (NSs) are being employed to produce MXene-based materials. The MXene-based aqueous solutions can be synthesized by sonication or shaking in water, which can then be used to fabricate MXene-based thin films, reinforce polymer-based composites, or manufacture thick membranes. During the last decade, it has been discovered that both single and multilayered flakes (most commonly examined $Ti_3C_2T_x$ MXene) deteriorate slowly in either open air or water. Therefore, it is essential to explore the stabilities of MXene-base aqueous solutions, as many industrial applications rely on the solution rather than a precipitate dispensation [1, 2, 34].

Oxygen-functionalized M_2CO_2 MXenes are emerging as challenging materials, with intriguing device applications. Researchers presented a theoretical study to examine the stability of these MXenes, using first-principle calculations. It was observed that depending on the position of the oxygen atom, there are two structural phases for M_2CO_2. The O atom usually occupies a position on one side of the MXene that is directly on top of M. However, for M_2CO_2, the O atom was observed to occupy two structural phases, that is, the BB′ phase (on top of the M atom) and CB phase (on top of the C atom). They discovered that the CB phase is stable for M = Y and Sc, whereas the BB′ phase is stable for other metal atoms (i.e., M = V, Ti, Hf, Zr, Ta, or Nb). The stability of these two phases was explained by atom-projected DOS, charge transfer, electron localization function, and Bader charge analyses. The negative formation energies of these phases for all M_2C MXenes (as shown in Fig. 7.17 [35]) suggested the feasibility of their synthesis [35].

Elements versus formation energy in E^f (eV) graph depicts values in BB′ MXene and CB MXene as follows: Sc_2CO_2, −9.4 and −10 respectively; Y_2CO_2 −9 and −9.5 respectively; Ti_2CO_2, −10.8 and −10 respectively; Zr_2CO_2, −11.3 and −10.5 respectively; Hf_2CO_2, −12 and −11 respectively; V_2CO_2, −7.8 and −7.5 respectively; Nb_2CO_2, −8.3 and −8.2 respectively; and Ta_2CO_2, −9, −8.8 respectively.

The phase stability study of Mo_2ScAlC_2 precursor was carried out both theoretically and experimentally [36]. Compared to the unstable Sc_3AlC_2 and Mo_3AlC_2, the studied alloy was observed to be stable and exhibited a least value formation enthalpy (−24 meV atom^{-1}). E. B. Deeva et al. [37] reported that in a reducing environment, 2D Mo_2CT_x maintains its stability without displaying any sintering up to 550–600°C. At higher temperatures, the surface groups defunctionalize the material, causing a transition to bulk β-Mo_2C that is accomplished at around 730°C. An in situ temperature programmed reduction-thermogravimetric analysis (TGA-TPR) experiment was carried out to investigate the reducibility of surface groups in Mo_2CT_x. At temperatures ranging from 50 to 800°C, 10% H_2 in N_2 was used across Mo_2CT_x, and a steady weight loss was observed (Fig. 7.18a [37]). During the TPR experiment, XRD patterns and XANES spectra were also developed at different temperatures (Fig. 7.18b[37]). As the c cell parameter decreases, the edge position of Mo spectra shifts to lower energies, indicating a steady decline in the oxidation state of Mo (Fig. 7.18c [37]). SEM analysis of Mo_2CTx-500 indicated no apparent variations from the original Mo_2CT_x, with platelet morphology remaining intact (Fig. 7.18d [37]).

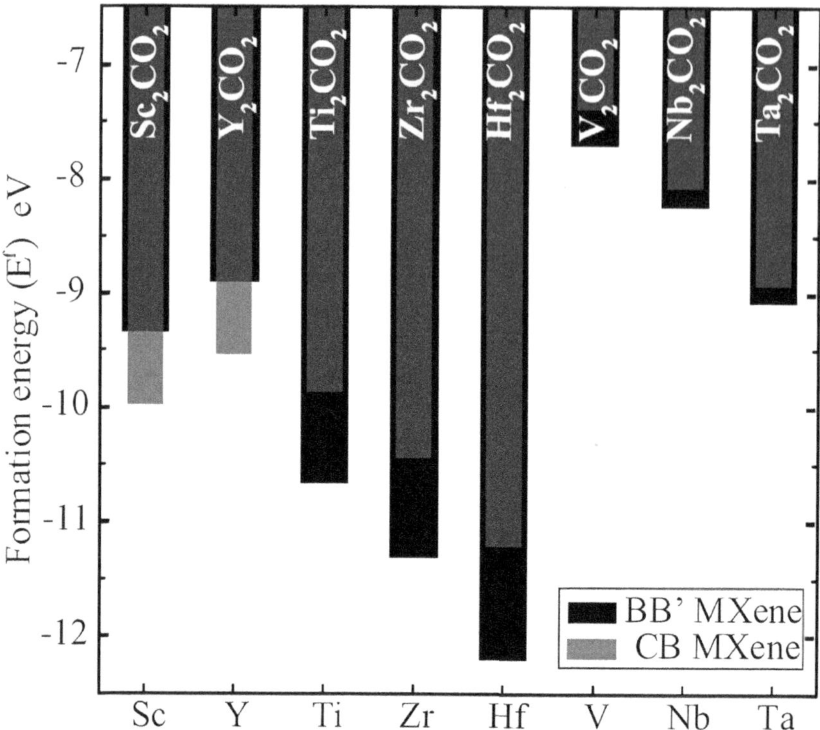

FIGURE 7.17 Formation energies of CB and BB′ phases of O-functionalized M$_2$CO$_2$.

Source: Reprinted with permission from [35] Avanish Mishra, Pooja Srivastava, Abel Carreras, et al. 2017. Atomistic Origin of Phase Stability in Oxygen-Functionalized MXene: A Comparative Study. *J. Phys. Chem. C*, 121, 34, 18947–18953. Copyright 2017, American Chemical Society.

7.7 HYDROPHILICITY

In addition to electronic conductivity, MXenes have outstanding hydrophilicity, comparable to graphene oxide. During the synthesis process of MXenes by selective etching, various hydrophilic functional groups such as OH, O, and F can be added to the surface of the material. With contact angles ranging from 21.5° to 35°, this surface functionalization provides hydrophilic characteristics to the MXene. In general, the hydrophilicity of MXenes changes depending on the substance. However, there are not much explanations for these modifications. The stability of delaminated MXenes, as their different concentrations are dispersed in aqueous colloidal solutions, is assured by highly negative surface charges and hydrophilicity. It was observed that MXene-based non-Pt hydrophilic materials are efficient oxygen reduction reaction (ORR) catalysts. This is due to the adsorption and molecular oxygen interactions, which are produced due to an increase in water-active sites on the material's surface. However, these active sites and hydrophilicity influence the ORR activity of non-Pt catalysts, that requires more research [2]. For example, H. Ang et al. [38] reported the synthesis of NS-doped Mo$_2$C NSs with 1.0 nm thickness and

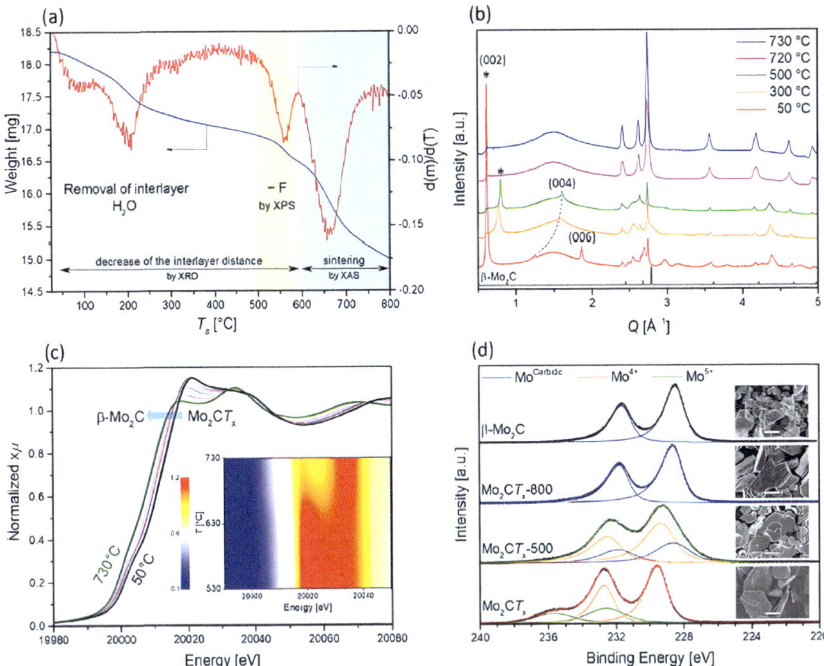

FIGURE 7.18 In situ structural stability study of Mo_2CT_x followed by (a) TGA, (b, c) XANES/XRD, and (d) ex situ XPS data of the Mo 3d core levels of partially reduced Mo_2CT_x and reference materials.

Source: Reprinted with permission from [37] Evgeniya B. Deeva, Alexey Kurlov, Paula M. Abdala, et al. 2019. In Situ XANES/XRD Study of the Structural Stability of Two-Dimensional Molybdenum Carbide Mo_2CT_x: Implications for the Catalytic Activity in the Water–Gas Shift Reaction. *Chem. Mater.*, 31, 4505–4513. Copyright 2019, American Chemical Society.

a large surface area (139 m^2 g^{-1}). Furthermore, they demonstrated that introducing these dopants into Mo_2C considerably improves its hydrophilicity, which accelerates ion diffusion and increases the rate of hydrogen evolution reaction (HER).

7.8 THERMAL STABILITY

According to the mass spectrometry and thermogravimetric analysis, the thermal stability of MXenes was discovered to be substantially dependent on their environment and chemical composition. According to a recent study, $Ti_3C_2T_x$ retains its stability even at 800°C in the Ar environment. The thermogravimetric analysis showed that beyond 800°C, $Ti_3C_2T_x$ converts to TiC by losing a significant composition. However, when this material is annealed in an O atmosphere, it partially oxidizes into anatase TiO_2 nanocrystals at 200°C and completely transforms into rutile TiO_2 at 1000°C. This shows that by modifying annealing temperatures, oxidation duration, and heating rate, $Ti_3C_2T_x$ can be transformed into TiO_2 with diverse morphologies, resulting in a variety of MXene-based hybrids or derivatives. On the other hand, MXenes with

M located on the surface are usually thermodynamically metastable, which exhibit high surface energies and oxidize spontaneously in the air. For pristine Ti_2C, unsaturated titanium interact significantly with O_2 molecules, which results in efficient O_2 dissociation for Ti_2C MXene. As a result, the adsorbed O on Ti_2C compromises the latter's thermodynamic stability. Furthermore, MXene's high thermal conductivity is advantageous for electronic devices [39].

M. Ashton et al. [40] employed first-principle simulations to compare the binding energies of functionalized MXenes, to estimate the thermodynamic stability of these compounds as a function of their chemical composition. When MXenes were exposed to solutions with low µH (i.e., hydrogen chemical potential) value, all surfaces were observed to be saturated with oxygen. Furthermore, solutions with greater µH value can also produce F-functionalized Sc-based MXenes. This study examined the thermodynamic stability of 54 O-functionalized MXene compositions. The results revealed that the formation energies of 38 structures were expected to be less than 200 meV atom^{-1}, while the remaining six were expected to exhibit energy values less than 100 meV atom^{-1} (as illustrated in Fig. 7.19 [40]).

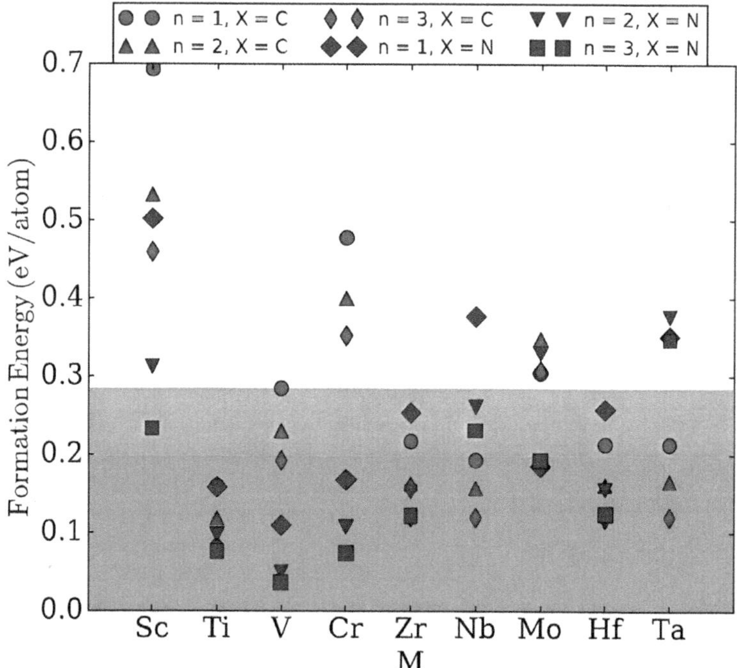

FIGURE 7.19 Formation energies for $M_{n+1}X_nO_2$, relative to the lowest energy mixture of competing bulk phases.

Source: Reprinted with permission from [40] Michael Ashton, Kiran Mathew, Richard G. Hennig, et al. 2016. Predicted Surface Composition and Thermodynamic Stability of MXenes in Solution. *J. Phys. Chem. C*, 120, 6, 3550–3556. Copyright 2016, American Chemical Society.

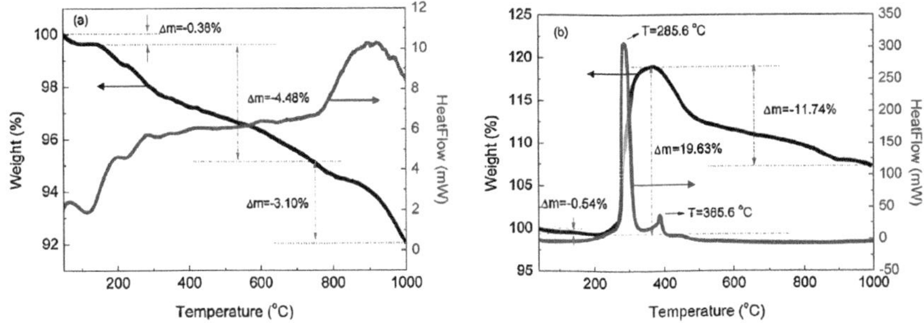

FIGURE 7.20 TG and DSC curves of MXenes from room temperature to 1,000°C in (a) argon and (b) oxygen atmosphere.

Source: Reprinted from [41] *Materials Science and Engineering B*, 191, Z. Li, L. Wang, et al., Synthesis and thermal stability of two-dimensional carbide MXene Ti_3C_2, 33–40. Copyright 2015, with permission from Elsevier.

In another study, Ti_3C_2 synthesized from a pressure-less technique was discovered to be highly stable when compared to Ti_3C_2 produced from hot-pressed Ti_3AlC_2 [41]. Intercalation with urea, DMSO, or ammonia could further exfoliate this material. Thermogravimetry (TG) and differential scanning calorimetry (DSC) revealed that in an Ar atmosphere, OH- or F-terminated Ti_3C_2 remains stable even at 800°C. However, when these materials were annealed in an O atmosphere, they partially oxidized into anatase nanocrystals at 200°C and completely transformed into rutile at 1,000°C. Figure 7.20 [41] illustrates the TG and DSC curves of MXenes in argon, from room temperature to 1,000°C (Fig. 7.20a) or oxygen atmospheres (Fig. 7.20b). The structural changes are presented by three different stages as a function of temperature, which represent 0.38%, 4.48%, and 2.47% weight loss, for the first, second and third stage, respectively.

Figure 7.21 [41] illustrates the SEM and XRD results of MXenes treated at high temperatures. The SEM image of the sample treated at 1,000°C in an Ar atmosphere is shown in Fig. 7.21a. The sample retained MXene's flexible quasi-2D structure. Furthermore, SEM images of samples treated at 200°C or 1,000°C in an oxygen atmosphere are shown in Figs. 7.21b and c, respectively. At 1,000°C, the crystals expanded and joined to form a net structure, as seen in Fig. 7.21c. The MXene's two-dimensional structure was barely visible. The XRD patterns of these samples are depicted in Fig. 7.21d. J. Yang et al. [42] used the DFT approach to analyze 20 different sulfur-terminated MXenes (M_2XS_2). V_2CS_2, Nb_2CS_2, Ta_2CS_2, Hf_2CS_2, and Cr_2NS_2 were observed to be dynamically stable at 1,000 K. According to formation enthalpy studies, the most stable structures are the easiest ones to synthesize. The phonon frequencies for these MXenes were observed to be positive, as shown in Fig. 7.22 [42], showing that the projected structures are dynamically stable. The phonon frequencies for Sc_2XS_2, Ti_2XS_2, Mo_2XS_2, W_2XS_2, Hf_2NS_2, V_2NS_2, Nb_2XS_2, and Ta_2NS_2 are negative, indicating the instability of these structures. Compounds of group VB (V, Nb, and Ta) or neighboring transition metals (Cr and Hf) dominate the dynamically stable structures. This is compatible with the experimental results obtained after synthesizing Ta_2CS_2 and Nb_2CS_2.

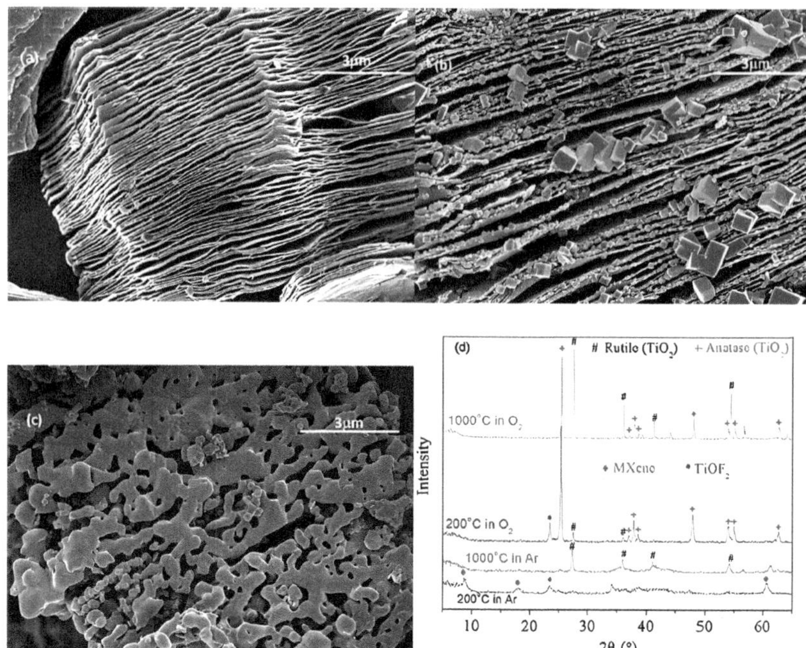

FIGURE 7.21 Results of the thermal analysis of MXenes. (a) SEM image of analysis at 1,000°C in Ar, (b) SEM image of analysis at 200°C in O_2, (c) SEM image of analysis at 1,000°C in O_2, and (d) XRD patterns of analysis in O_2 or Ar.

Source: Reprinted from [41] *Materials Science and Engineering B*, 191, Z. Li, L. Wang, et al., Synthesis and thermal stability of two-dimensional carbide MXene Ti_3C_2, 33–40. Copyright 2015, with permission from Elsevier.

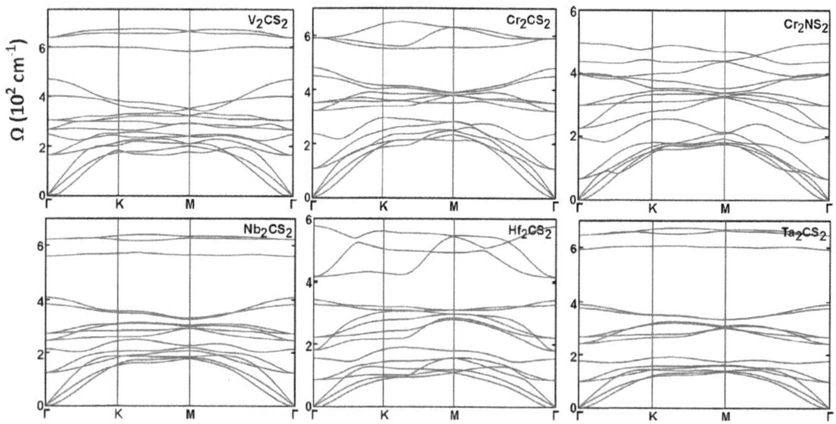

FIGURE 7.22 Calculated phonon branches of two-dimensional single-layer M_2XS_2 structures.

Source: Reprinted from [42] *Computational Materials Science*, 153, J. Yang, A. Wang, et al., Stability and electronic properties of sulfur terminated two-dimensional early transition metal carbides and nitrides (MXene), 303–308. Copyright 2018, with permission from Elsevier.

7.9   HYDROGEN STORAGE FEATURE

The development of materials for hydrogen storage with maximal storage capacity as well as exceptional desorption and adsorption properties that can be used under atmospheric circumstances is a critical undertaking that many chemists and material engineers must do. Currently available hydrogen storage materials, that is, CNTs, MOFs, graphene, COFs, etc., are not sufficient enough as at lower temperatures, they conserve hydrogen surrounded by nitrogen, due to the effect of van der Waals forces on hydrogen bonding. However, in case of metallic hybrids, the conserved hydrogen's desorption process allows the choice of higher temperatures, due to the chemical adsorption mechanism of hydrogen. MXene components have been shown in recent investigations to expand hydrogen storage capabilities. In MXenes, H bonds are bound using three mechanisms, that is, chemisorption, Kubas-type particle contact, and physisorption. Because certain physicochemical properties of MXenes have not been assessed by numerous studies, first principle and DFT calculations are advised as appropriate methods for predicting them [9].

For example, the hydrogen storage capability of HF-etched $Ti_3C_2T_x$ was determined along with the bonding mechanism of hydrogen molecules. It was demonstrated that the pores created during the HF etching will be occupied by the molecular hydrogen. It was also reported that the formation of carbon vacancies can increase the hydrogen storage mechanism [9]. In another study, S. Zhengyang et al. [43] synthesized a novel $(Ti_{0.5}V_{0.5})_3C_2$ from its corresponding MAX precursor, and the influence of this catalyst on the Mg hydrogen storage process was comprehensively examined. The dehydrogenation temperature of MgH_2 was observed to be significantly reduced (from 266 to 196°C) by the addition of 10 wt% MXene. Within 20 minutes, about 5.0 wt% H_2 was released (at 250°C) from the MXene containing MgH_2 (as shown in Fig. 7.23 [43]). At 120°C, this sample rapidly absorbed 4.8 wt% H_2; these hydrogenation

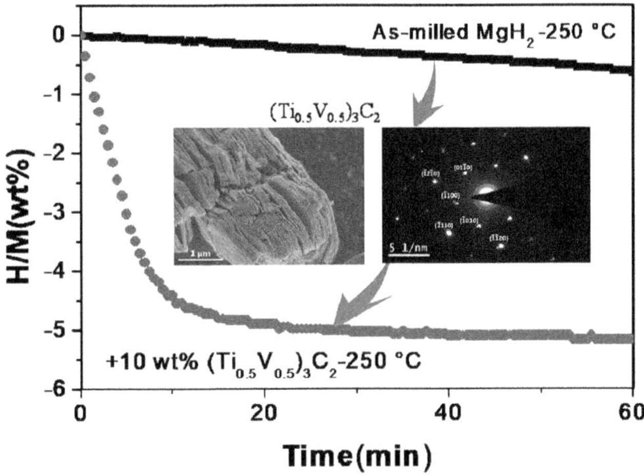

FIGURE 7.23 Graphical representation of the catalytic effects of $(Ti_{0.5}V_{0.5})_3AlC_2$ on the hydrogen storage reaction of Mg.

Source: Reprinted from [43] *Materialia*, 1, Zhengyang Shen, Zeyi Wang, Min Zhang, et al., A novel solid-solution MXene $(Ti_{0.5}V_{0.5})_3C_2$ with high catalytic activity for hydrogen storage in MgH_2, 114–120. Copyright 2018, with permission from Elsevier.

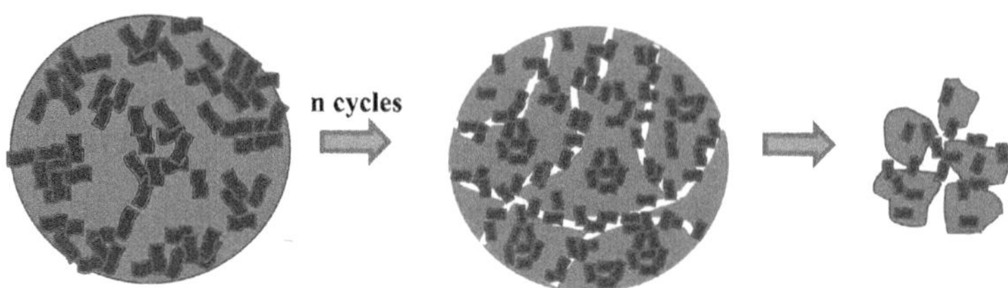

FIGURE 7.24 Hydrogen storage mechanism of the as-milled MgH_2–5 wt%$Nb_4C_3T_x$ composite.

Source: Reprinted from [44] *Applied Surface Science*, 493, Yana Liu, Haiguang Gao, Yunfeng Zhu, et al., Excellent catalytic activity of a two-dimensional $Nb_4C_3T_x$ (MXene) on hydrogen storage of MgH_2, 431–440. Copyright 2019, with permission from Elsevier.

kinetics outperform even the well-studied Nb_2O_5 catalyst. The activation energy of the MgH_2–10 wt% $(Ti_{0.5}V_{0.5})_3C_2$ was determined to be 77.3 kJ mol^{-1}.

In another investigation, chemically exfoliated $Nb_4C_3T_x$ was ball-milled into MgH_2. The resultant MgH_2–5 wt% $Nb_4C_3T_x$ composite was observed to be an excellent hydrogen storage material [44]. The onset temperature of activated composite was decreased from 296.5°C to 150.6°C. It was observed that by following a rate constant of 1.09164 wt% min^{-1} at 250°C, H_2 would be released completely within 800 s. Furthermore, a high activation energy (81.2 kJ mol^{-1} H_2) for dehydrogenation of the activated composite was observed, which was significantly lower than the apparent activation energy for pure ball-milled Mg (i.e., 153.8 kJ mol^{-1} H_2). At 50°C within 2 hours, the sample absorbed 3.50 wt% H_2 after dehydrogenation, and the activation energy was dropped to 27.8 kJ mol^{-1} H_2. The hydrogen storage method is depicted in Fig. 7.24 [44]. Figures 7.25a and c [44] show low- and high-magnification SEM images of the balled MgH_2–5 wt% $Nb_4C_3T_x$ composite. It is illustrated that the morphology of the composite balled for 40 hours comprises micrometric particles with a nonuniform particle size distribution. There are more small particles less than 0.3 m and clustered particles larger than 2 m.

MgH_2 particles have many defects after four hydrogenation/dehydrogenation cycles, as seen in Figs. 7.25b and d [44], and are smaller and more evenly sized as a result of the expansion in volume during the process. It is generally known that lowering the particle size improves hydrogen sorption kinetics due to the shorter diffusion distance for H and higher surface-to-volume ratio. Furthermore, structural imperfections can act as nucleation sites, and increasing the density of structural defects can increase the nucleation rate. It was observed that catalyst dispersion is one of the possible factors responsible for the better hydrogen kinetics and lowered E_a of the composite. The catalytic efficacy of a catalyst was determined not only by its intrinsic activity but also by its distribution state. TEM was used to examine the distribution of $Nb_4C_3T_x$ in the MgH_2 matrix. In Fig. 7.25e [44], 20–30 nm $Nb_4C_3T_x$ particles are scattered in a MgH_2 matrix. $Nb_4C_3T_x$ particles, on the other hand, become monolayer, smaller, and more uniformly dispersed after de-/rehydrogenation cycles, as seen in Fig. 7.25f [44]. Thus, increasing the catalyst–matrix contact may contribute to the activated MgH_2–5 wt%$Nb_4C_3T_x$ composite's increased hydrogenation or dehydrogenation capabilities.

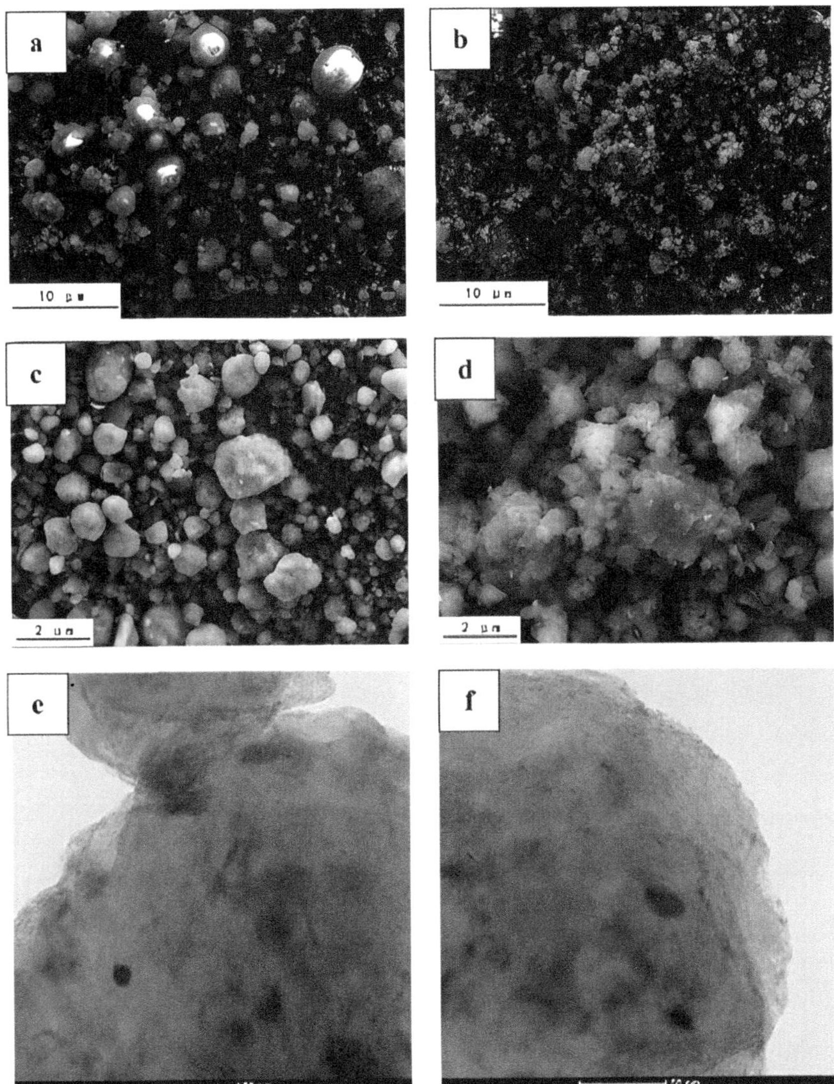

FIGURE 7.25 SEM and TEM images of (a, c, e) the as-milled MgH_2–5 wt%$Nb_4C_3T_x$ composite and (b, d, f) after 4 de-/rehydrogenation cycles.

Source: Reprinted from [44] *Applied Surface Science*, 493, Yana Liu, Haiguang Gao, Yunfeng Zhu, et al., Excellent catalytic activity of a two-dimensional $Nb_4C_3T_x$ (MXene) on hydrogen storage of MgH_2, 431–440. Copyright 2019, with permission from Elsevier.

7.10 MAGNETIC PROPERTIES

For various MAX phases, both experimental and theoretical studies have identified a significant magnetic ground state. Most commonly, two metal elements (Cr and/or Mn) have been observed in the composition of almost all magnetic MAX phases. According to these studies, different MAX phases, that is, Cr_2AlC, Cr_2AlN, Cr_2GeC, Cr_2GaN, Mn_2AlC, Mn_2GeC, Cr_4AlN_3, and $(Cr_2Ti)AlC_2$, are all observed to be magnetic in nature. Furthermore, $(M_{2/3}Sc_{1/3})_2AlC$ (where M = Mn or Cr), $(Mo_{2/3}RE_{1/3})_2AlC$ (where RE stands for the rare earth metals), and $(Mo_{2/3}RE_{1/3})_2GaC$ phases are the recent additions to this family, which exhibit tunable magnetic properties. The magnetic characteristics of these materials can be modified by varying the chemical ordering and ratio of M elements, which will be useful in future applications. However, these properties have been thoroughly investigated using first-principle calculations only, and except for a few, none of the magnetic phases have been exfoliated into MXenes yet. The magnetic ground state of MXenes was determined by using spin-polarized calculations, which include a nonmagnetic (NM), a ferromagnetic (FM), and few AFM configurations [7].

Although most pristine MXenes have NM configurations, a few of them were observed to exhibit a FM ground state (i.e., Cr_2C, Ti_2C, Mn_2C, Ti_2N, and Mn_2N), whereas Cr_2N and V_2C exhibited AFM configurations (as illustrated in Fig. 7.26

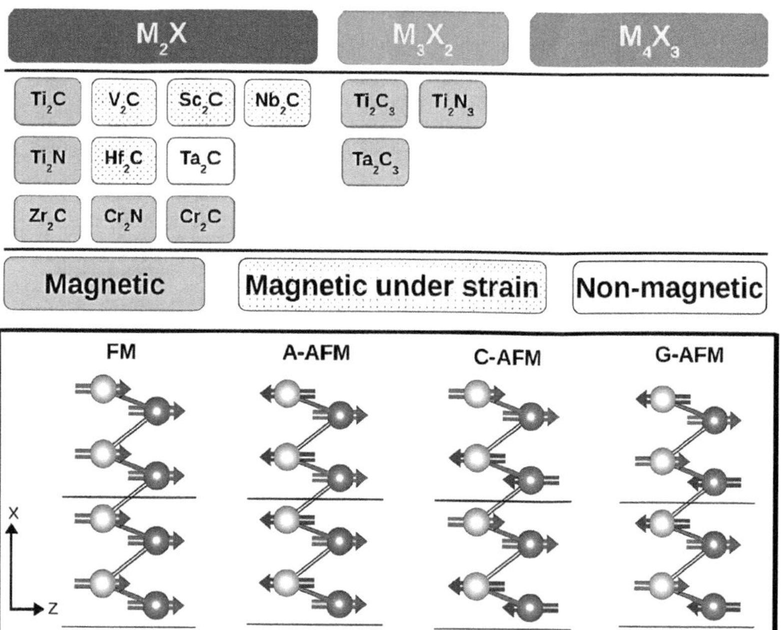

FIGURE 7.26 Various un-passivated magnetic MXenes belonging to M_2X, M_3X_2, and M_4X_3 families, as reported in the literature.

Source: Reprinted from [45] *Elsevier Books*, Poulami Chakraborty, Tilak Das, Tanusri Saha-Dasgupta, CompreShensive Nanoscience and Nanotechnology, 319–330. Copyright 2019, with permission from Elsevier.

[45]). Despite these characteristics, the synthesis of these pristine MXenes is still in its early stages due to the insertion of surface groups during experiments, which is difficult to be avoided. Interestingly, some functionalized MXenes, that is, Cr_2NO_2, Ti_2NO_2, Mn_2NT_2, and Mn_2CT_2, have been projected to retain a magnetic moment regardless of these surface groups. These structures can maintain their magnetism close to room temperature, as they have been predicted to exhibit values of magnetic moments up to 3 μB. Furthermore, doping of MXenes with Cr or Mn and modification in mechanical properties can produce significant magnetic moments [1, 7].

Half-metallicity in magnetic MXenes has recently been explored. In half-metallicity, electronic conductivity is provided by a spin-charge carrier channel, in which one spin channel is insulating and the other one is metallic. Researchers predicted half-metallicity in Cr_2C for the first time using HSE, with a spin gap of 2.85 eV. As a result, near-half-metallicity in Ti_2C and Ti_2N systems has been predicted, although actual half-metallicity is only found in biaxial strains. This property, like FM, vanishes following functionalization in all three systems. This characteristic was expected in Mn_2CF_2 and Mn_2NT_2 as well. These observations imply that these structures hold significant promise to be used in spintronic devices, including magnetic sensors, spin injectors, and spin filters [7].

Using first-principle calculations, FM configurations were observed for Cr_2NO_2, Ti_2NO_2, and functionalized M_2NT_x [46]. They reported excellent magnetic properties for these MXenes, including high Curie temperatures (1,877–566 K), robust FM configurations, half-metallicity, and magnetic moments up to 9 μB, which indicate their efficiency for spintronic applications. To theoretically predict the magnetic properties of these MXenes, the results of crystal field theory provided different structural models based on various M groups, as illustrated in Figs. 7.27a and b [46]. The electrical structures are observed to be greatly influenced by the number of d-electrons and coordination of M groups. Figure 7.27c represents an arrangement of nonbonding d-orbitals between σ (bonding) and σ* (nonbonding) states of MXenes. A wide range of magnetic properties were observed depending on the arrangement of these orbitals and availability of electrons. Furthermore, the positions of localized electrons on the centers of M atoms are also presented in Fig. 7.27d.

The most desirable phase for magnetic materials is the FM ground state. In their quest for FM MXenes, researchers discovered five distinct nitride MXenes with these configurations, including Ti_2NO_2, Cr_2NO_2, Mn_2NO_2, Mn_2NF_2, and $Mn_2N(OH)_2$. The ground state energy of the NM phase (7.1 eV) was observed to be greater than that of the FM phase, indicating the stability of magnetically ordered Mn_2NF_2. The charge density distribution for different configurations is presented in Fig. 7.28 [46].

In another study, the magnetic properties of six functionalized nitride-based MXenes, that is, V_2NO_2, Mn_2NO_2, Mo_2NO_2, Cr_2NF_2, V_2NF_2, and Mo_2NF_2, were calculated with intrinsic bandgaps [47]. These MXenes displayed orbital ordering, which in certain conditions produces magnetoelastic or magnetoelectric coupling. In V_2NO_2, Cr_2NF_2, and Mo_2NF_2, the ferroelectric phase existed as an excited state, with the magnetic order associated with polar displacements via orbital ordering. Similarly, in Janus MXenes (i.e., $Mo_8N_4F_7O$, $Cr_8N_4F_7O$, and $V_8N_4O_7F$), these magnetic effects were also observed. Using DFT study, J. Yang et al. [48] studied the magnetic properties of $Cr_2M_2C_3T_2$ (where M = V, Ti, Ta, or Nb), and also observed

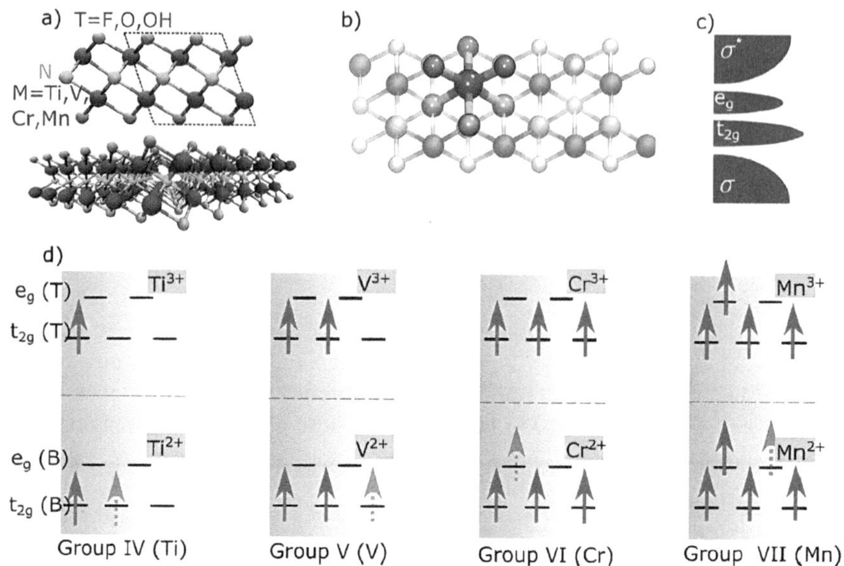

FIGURE 7.27 (a) Side views of the crystal structure. (b) Top view of a monolayer MXene. (c) Simplified DOS. (d) Arrangement of localized electrons on the center of M atoms.

Source: Reprinted with permission from [46] Hemant Kumar, Nathan C. Frey, Liang Dong, et al. 2017. Tunable Magnetism and Transport Properties in Nitride MXenes. *ACS Nano*, 11, 7648–7655. Copyright 2017, American Chemical Society.

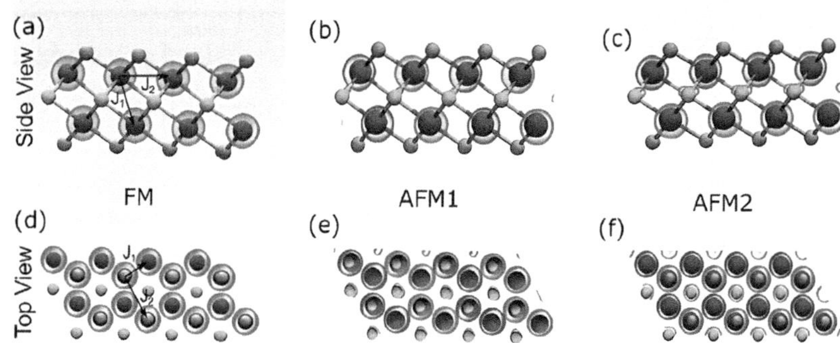

FIGURE 7.28 Spin-polarized charge density distribution for (a, d) FM, (b, e) AFM-1, and (c, f) AFM-2.

Source: Reprinted with permission from [46] Hemant Kumar, Nathan C. Frey, Liang Dong, et al. 2017. Tunable Magnetism and Transport Properties in Nitride MXenes. *ACS Nano*, 11, 7648–7655. Copyright 2017, American Chemical Society.

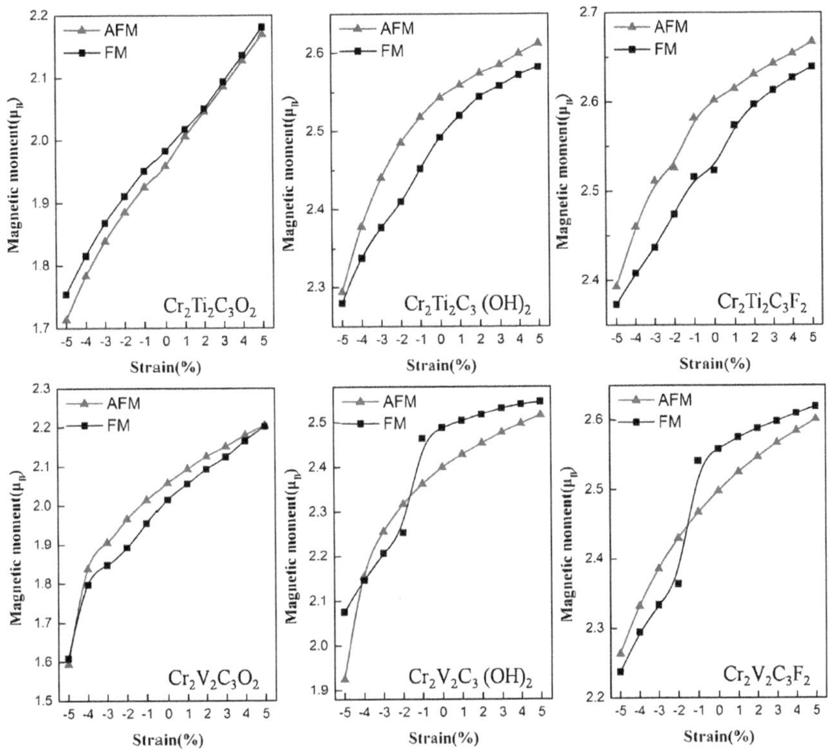

FIGURE 7.29 Magnetic moment of Cr atoms with extensile strain.

Source: Reprinted from [48] *Computational Materials Science*, 139, Jianhui Yang, Xuepiao Luo, Xumeng Zhou, et al., Tuning magnetic properties of $Cr_2M_2C_3T_2$ (M=Ti and V) using extensile strain, 313–319. Copyright 2017, with permission from Elsevier.

the influence of extensile strain on these properties. According to the calculated results, $Cr_2V_2C_3O_2$ and $Cr_2Ti_2C_3O_2$ (in their strain-free state) exhibit FM configuration. Using the Heisenberg model with mean-field approximation, Curie temperatures (720.6 K and 246.8 K) were also observed. The variations in the magnetic moment of these structures are illustrated in Fig. 7.29 [48]. For $Cr_2Nb_2C_3T_2$ and $Cr_2Nb_2C_3T_2$ with extensile stresses ranging from −5% to 5%, the AFM configuration was observed to be energetically favorable. The projected density of states revealed that as the strain grows, Cr-3d orbitals become more localized, and the DOS around the fermi level declines, potentially reducing conductivity.

Another work used first-principle simulations to analyze the magnetic characteristics of $Cr_2M'C_2T_2$ (where M′ = V or Ti). According to the results, depending on the bond coupling interactions and the selection of metal atoms and surface groups, these structures can exhibit FM, NM, or even AFM configurations [49]. Similarly, based on theoretical studies, FM configurations were observed for $Hf_2VC_2O_2$ and $Hf_2MnC_2O_2$ monolayers [50]. These MXenes outperformed other 2D FM materials by showing better Curie temperatures (495 K–1,133 K) and magnetic moments (i.e., 3–4 µB/unit cell). Furthermore, in the presence of less tensile strain (0–3%), phase transitions from FM to AFM and semimetal to semiconductor were also observed. In another spin-polarized DFT study, 50 DTM pristine or functionalized MCr_2CT_x compositions were examined [51]. They observed FM or AFM half-metallicity and AFM semiconduction for these MXenes. It was reported that FM half-metallic $YCr_2C_2H_2$ and pristine/functionalized $ScCr_2C_2$-based materials exhibit high Curie temperatures along with wide bandgaps.

Furthermore, moderate bandgaps and high Néel temperatures were expected for AFM semiconductors, that is, $ZrCr_2C_2$, $TiCr_2C_2$, and $ZrCr_2C_2(OH)_2$. Another first-principle calculation revealed that Sc_2AlC can be doped with different M elements under ScAl-rich conditions [20]. The resulting $TM-Sc_2C(OH)_2$ structures were observed to be n-type, while $TM-ScCO_2$ structures were p-type semiconductors. The disparity is due to O having higher empty p orbitals than OH. Magnetic properties were discovered in all Cr- and Mn-doped Sc_2CT_2 structures. Figure 7.30 [20] shows that Cr- and Mn-doped structures exhibit high magnetic moments, compared to the Ti- or V-doped structures. To be more specific, the values of magnetic moments for $Ti-Sc_2CO_2$ and $V-Sc_2CF_2$ were 0.53 and 0.43 µB, respectively. In the Mn-doped $ScCO_2$, the magnetic moment can reach 3.2 µB. As a result, Mn and Cr are promising doping elements for magnetically doped Sc_2CT_2 systems.

Another study reported the bipolar AFM semiconducting behavior of Cr_2TiC_2FCl with vanishing magnetism. Surprisingly, gate voltage was observed to exhibit the ability to control their spin orientation, which resulted in phase transition into half-metallic antiferromagnets [52]. J. He et al. [53] predicted the magnetic properties of asymmetrically functionalized Janus Cr_2C MXenes (represented as Cr_2CXX'). These materials exhibited bipolar magnetic semiconducting behavior with vanishing

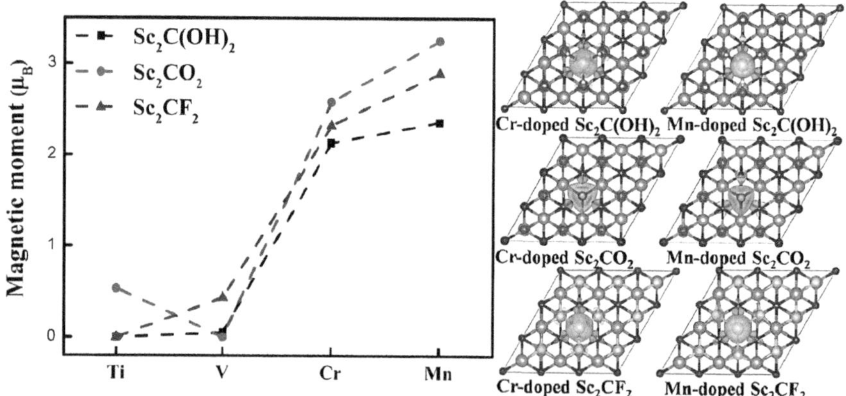

FIGURE 7.30 Magnetic moments of Ti-, V-, Cr-, and Mn-doped Sc_2CT_2 (T = O, F, or OH) systems.

Source: [20] Yang Jianhui, Luo Xuepiao, et al. 2016. Investigation of magnetic and electronic properties of transition metal doped Sc_2CT_2 (T = O, OH or F) using a first principles study. *Phys. Chem. Chem. Phys.*,18, 12914–12919. Reproduced with permission from the Royal Society of Chemistry.

magnetism, as their conduction and valence bands comprised opposite spin channels. Cr_2CClBr, Cr_2CFCl, Cr_2CHF, Cr_2CHCl, and Cr_2CFOH materials have been discovered to have Néel temperatures as high as 400 K. Surprisingly, electron or hole doping may easily control spin carrier direction and the resulting phase to half-metallic AFMs. Similarly, DFT calculations were carried out to determine the effects of TM doping and structural defects on the magnetic properties of Ti_4N_3 NSs [54]. Based on the predicted binding and formation energies, the surface Ti and sub-surface N vacancies were observed to be stable. Furthermore, the former appears to be simpler than the latter. Both of them were observed to have the ability to increase the magnetism of NSs. X. Xiao et al. [55] used lattice-matching synthesis of ultrathin Mn_3N_2. Even at 300 K, Mn_3N_2 flakes demonstrated intrinsic magnetic behavior, allowing for possible room-temperature applications.

7.11 OPTOELECTRONIC PROPERTIES

Compared to other 2D materials (such as graphene), the optoelectronic properties of MXenes indicate their significant promise in transparent energy storage devices, photothermal conversion, and transparent conductive coatings. The fabrication approaches that are used to produce MXene thin films are of significant importance, as optoelectronic performance of spin-coated $Ti_3C_2T_x$ films outperforms that of sputtered or spray-coated transparent films. Percolation issues in very transparent conductive coatings are another consideration. Typically, films will experience a considerable increase in sheet resistance, as thickness of a film is reduced to a diffusion threshold. Percolation is an unavoidable and unwanted phenomenon in practice. This type of behavior is mostly common in films with no such apparent issues. Because of their superior optoelectronic capabilities, MXene-based transparent films can sufficiently be thin without influencing their electrical structure and conductivity. As a result, MXenes can function as transparent supercapacitors without the need for extra current collectors [39].

Another study developed a MXene electrochromic device by combining titanium carbide's optoelectronic and pseudocapacitive properties. When cycled in acidic electrolytes (i.e., H_3PO_4 and H_2SO_4) at < 1 V, the device displayed a significant reversible transmittance change (12%) along with a visible absorption peak shift (from 770 to 670 nm), compared to the neutral electrolytes (e.g., $MgSO_4$). The XRD and Raman spectroscopy results revealed that protonation or deprotonation of oxide-like surface groups is responsible for the reversible shift in the absorption peak. In this device, a dip-coated $Ti_3C_2T_x$ acted as an active material and a transparent conductive coating [56]. L. Yang et al. [57] used $Ti_3C_2T_x$ NSs as a novel electron transport layer (ETL) in perovskite solar cells, manufactured at low temperatures. Interestingly, the surface Ti-O bonds were increased by a simple UV-ozone treatment, which increased its suitability as an ETL. After 30 minutes of treatment, the power conversion efficiency was increased from 5% to 17.17%, due to reduced recombination and electron transfer at the ETL–perovskite interface.

A.D. Dillon et al. [58] demonstrated that optically active 2D Ti_3C_2 thin films can be designed from aqueous solutions, which transmit visible light (>97%) per nanometer thickness. As a result, the conductivity of the spin-cast films on a macroscopic scale

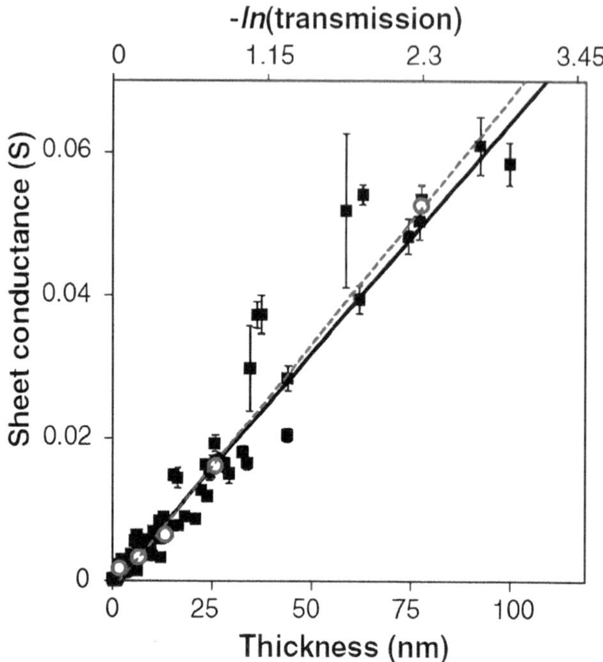

FIGURE 7.31 Electrical and optical properties of spin-casted $Ti_3C_2T_x$ films. Sheet conductance versus film thickness (lower *x*-axis) and ln(transmission) (upper *x*-axis).

Source: [58] Dillon Andrew D., Ghidiu Michael J. et al.: Highly Conductive Optical Quality Solution-Processed Films of 2D Titanium Carbide. *Adv. Funct. Mater.* 2016, *26*, 4162–4168. Copyright WILEY-VCH Verlag GmbH & Co. KGaA, Weinheim. Reproduced with permission.

was comparable to that of the corresponding 2D sheets. Therefore, Ti_3C_2 is considered as the first example of optoelectronic MXenes and a building block for near-infrared plasmonic applications. The relationship between thickness of films and measured sheet conductance is illustrated in Fig. 7.31 [58]. The sheet conductance of a collection of films was measured and plotted against the thickness measured by AFM to better understand the transport properties of $Ti_3C_2T_x$ (Fig. 7.31). A linear fit to these data yields a bulk conductivity of 6600 S cm^{-1} for the spin-casted films, which was observed to be higher than that of previously reported vacuum filtered films.

K. H. Tan et al. [59] used a drop-casting approach to develop thin films of $Ti_3C_2T_x$-PVA (polyvinyl alcohol) nanocomposites. The thickness of this structure ranges from 7.20 to 7.88 μm. When compared to pure PVA (1 × 10^{-13} S m^{-1}), the electrical conductivity of the PVA was observed to be improved by at least 3,000 times. The highest value of 7.25 × 10^{-3} S m^{-1} was reported for this nanocomposite along with an optical absorption coefficient of 4,000–5,000 cm^{-1}. Similarly, the optical and electric characteristics of $Ti_4N_3T_x$ were observed, and both semiconducting and metallic behavior were reported [60]. The optical measurements revealed the visible light absorption (at ~2.0 eV) and structural defects dependent on the excitation wavelength (at ~2.9 eV) for MXenes combined with the surface oxide layer.

7.12 BIOCOMPATIBILITY AND STABILITY

Biocompatibility and stabilities of MXenes have also been studied by researchers. For example, a thermodynamic cancer therapeutic method was presented by H. Xiang et al. [61] that uses oxygen-irrelevant free radicals. They used Nb_2C NSs to develop a free radical nanogenerator, in which the mesopores served as initiator reservoirs and the MXene core as the photonic thermal trigger at near-infrared (NIR)-II bio-window. The formation of free radicals was facilitated by the photothermal conversion effect of Nb_2C, which enabled the speedy disintegration and fast release of the encapsulated initiators 2,2'-azobis[2-(2-imidazolin-2-yl)propane]-dihydrochloride (AIPH). Figure 7.32 [61] illustrates a broad therapeutic technique based on creating the $Nb_2C@mSiO_2$ structure. This nanocomposite behaved as an

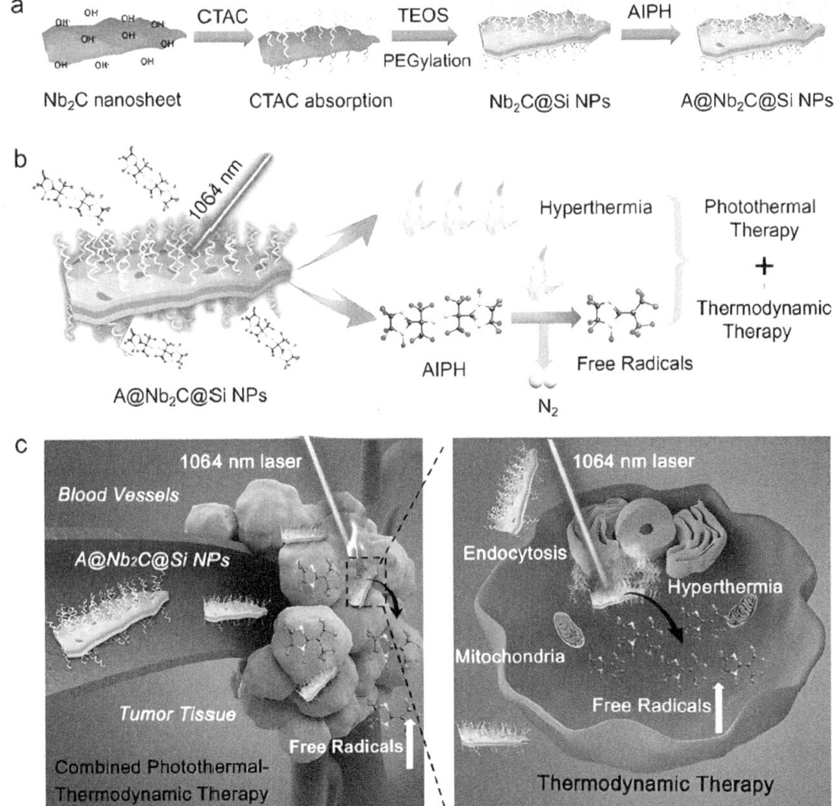

FIGURE 7.32 Schematic illustration of (a) the fabrication of AIPH@Nb$_2$C@mSiO2 NPs, (b) thermal decomposition for radical production, and (c) systematic delivery of NPs for photothermal cancer hyperthermia and photonic thermodynamic cancer therapy.

Source: Reprinted with permission from [61] Huijing Xiang, Han Lin, Luodan Yu, et al. 2019. Hypoxia-Irrelevant Photonic Thermodynamic Cancer Nanomedicine. *ACS Nano*, 13, 2223–2235. Copyright 2019, American Chemical Society.

efficient carrier of free radicals due to the hydrophilicity and biocompatibility of the MXene. Furthermore, their well-defined structural properties, which include hollow interiors and a wide surface area, provided a high loading of initiators and regulated free-radical release. AIPH molecules were loaded into the mesopores of $Nb_2C@mSiO_2$ NPs as the free radical source to create the $AIPH@Nb_2C@mSiO_2$ structure.

W. Feng et al. [62] developed Mo_2C by exfoliating bulk Mo_2Ga_2C and utilized it for photonic tumor hyperthermia. The photonic behavior of the MXene was predicted by using computational simulations particularly, which indicated that Mo_2C exhibits intense NIR absorption, covering NIR-I and II. Mo_2C-PVA nanoflakes displayed rapid degradability and efficient biocompatibility. Importantly, Mo_2C-PVA nanoflakes exhibited an unusually broad absorption band along with 24.5% and 43.3% photothermal conversion efficiency for NIR-I and NIR-II, respectively. Similarly, ultrathin Nb_2C MXene was also reported to be utilized as a radioprotectant [63]. The antioxidant capabilities of the 2D Nb_2C MXene were fascinating, thereby successfully removing hydrogen peroxide, hydroxyl, and superoxide radicals. This MXene was also pretreated with polyvinylpyrrolidone (PVP) to enhance biocompatibility, and the resultant composite was used as free radical against IR. It was also discovered that adsorption and desorption of collagen on the MXene's surface can easily be managed by combining zeta potential and dynamic light scattering measurements [64].

Reaction (a) of Nb_2C nanosheet and OH with CTAC leading to CTAC absorption. Further with TEOS and PEGylation gives Nb_2C at Si NPs, which reacts with AIPH leading to A at Nb_2C at Si NPs. Reaction (b) of A at Nb_2C at Si NPs with 1064 nm leading to hyperthermia and AIPH gives N_2 and free radicals leading to photothermal therapy plus thermodynamic therapy. (C) Combined photothermal thermodynamic therapy. Blood vessels with A at Nb_2C at Si NPs enters into tumor tissue on which 1064 nm laser and free radicals passes. Thermodynamic therapy. 1064 nm laser, and free radicals pass toward the cel with mitochondria, endocytosis, and hyperthermia occurs.

7.13 PLASMONIC PROPERTIES

MXenes have strong plasmonic capabilities that can be used in plasmonic photodetectors and surface-enhanced Raman scattering (SERS). MXenes have plasmonic features, which means that when light resonantly illuminates them, the electrons in the MXene move collectively and in one direction, resulting in an electric dipole moment. These characteristics of MXenes are classified as bulk and surface plasmon. Higher-solution electron energy-loss spectroscopy (EELS) observations of most $Ti_3C_2T_x$ MXenes assigned unique characteristics at 0.44, 0.66, 0.8, 0.95, 1.7, and 3.5 eV, as illustrated in Fig. 7.33a [22]. The multipolar longitudinal modes of the surface plasmon were attributed to the first four modes below 1 eV. Transversal surface plasmon mode was assigned to the 1.7-eV response to 780 nm absorption. The inter-band transition was responsible for the comparatively mild 3.5–6 eV broad peak. Furthermore, by modifying surface functionalization, these modes could be adjusted to the range of blue shift. [22].

In addition, the bulk plasmonic properties were observed to be not dependent on layers, which distinguished them from other plasmonic materials. In other words,

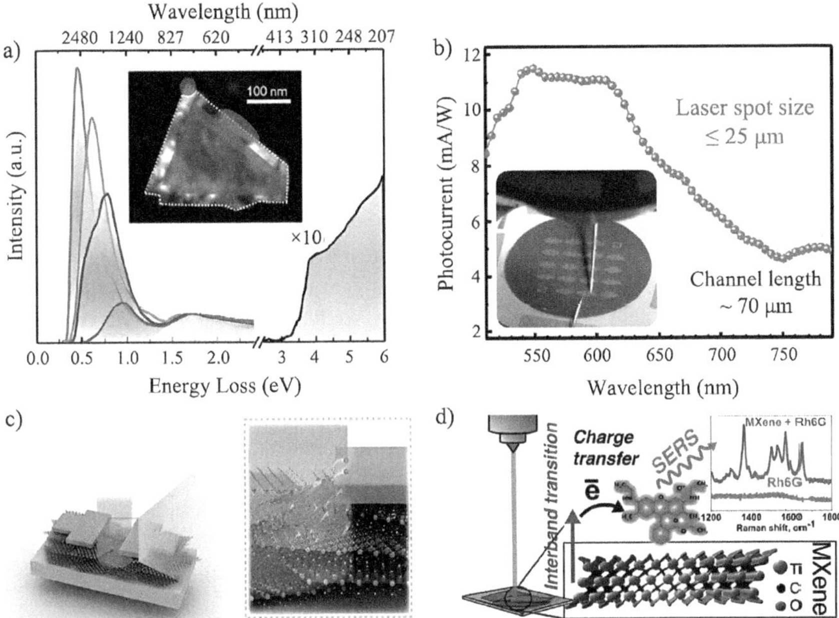

FIGURE 7.33 Plasmonic properties and applications of MXenes. (a) Small-sized $Ti_3C_2T_x$ nanoflake EELS. (b) Photo-response of a Mo_2CT_x thin-film photodetector as a function of the wavelength. (c) Schematic representation of the photodetector under light illumination. d) Schematic of SERS enhancement of Rh6G on $Ti_3C_2T_x$ substrates.

Source: [22] Fu Bo, Sun Jingxuan, et al.: MXenes: Synthesis, Optical Properties, and Applications in Ultrafast Photonics. *Small* 2021, 17, 2006054. Copyright WILEY-VCH Verlag GmbH & Co. KGaA, Weinheim. Reproduced with permission. The sources of the material [Reprinted with permission from Jehad K. El-Demellawi, et al. 2018. Tunable Multipolar Surface Plasmons in 2D $Ti_3C_2T_x$ MXene Flakes. *ACS Nano*, 12, 8, 8485–8493. Copyright 2018, American Chemical Society. Reprinted with permission from Asia Sarycheva, Taron Makaryan, et al. 2017. Two-Dimensional Titanium Carbide (MXene) as Surface-Enhanced Raman Scattering Substrate. *J. Phys. Chem. C*, 121, 36, 19983–19988. Copyright 2017, American Chemical Society, and Velusamy D. B. et al.: MXenes for Plasmonic Photodetection. *Advanced materials*, 31, 2019, 1807658. Copyright WILEY-VCH Verlag GmbH & Co. KGaA, Weinheim. Reproduced with permission] are also acknowledged.

MXenes do not have to be single layers to use light through plasmonic characteristics. Because of these properties, MXenes are capable of being utilized in photon detectors. High-conductivity materials are undesirable for plasmonic photodetector applications due to the generation of high dark current. In the visible range, Mo_2CT_x with lower electronic conductivity has recently been reported as a potential plasmonic photodetector (as shown in Fig. 7.33b). These NSs have the potential to reach 5×10^{11} J detectivity and 9 AW^{-1} responsivity. These values were observed to be three to four orders of magnitude greater than that of MoS_2 and graphene-based photodetectors. Furthermore, Fig. 7.33c schematically represents the photodetector under light illumination.

MXenes are sensitive SERS substrates due to their plasmonic characteristics in the NIR spectrum. Furthermore, the negatively charged MXene surface interacted strongly with oppositely charged molecules, which facilitates the charge transfer to dye. The chemical mechanism is the dominant mechanism for MXene SERS action. This is not the case with a gold flat surface, which benefits from both electromagnetic and chemical mechanisms. Due to the charge transfer mechanism, as shown in Fig. 7.33d, $Ti_3C_2T_x$ nanoflakes can increase the positively charged dye molecules (by 10^6). MXenes have excellent SERS sensitivity and low detection limits due to the charge transfer process. Further research with LiF-etched $Ti_3C_2T_x$ flakes revealed a considerably improved effect, with the lowest measured Raman signals reaching 10^{-8} M. At 532 nm, the enhancement for Rh6G was increased by 10^{12}, due to the change in substrate to Ti_2NT_x. MXenes with nanocomposites thus serve as effective SERS detection substrates in a variety of applications, such as high-sensitivity environmental analysis, food safety monitoring, and biological imaging [22].

Tej B. Limbu et al. [65] presented an environmentally friendly synthesis of reduced $Ti_3C_2T_x$ MXene and observed its oxidation stability, electrical conductivity, and SERS activity. It was demonstrated that r-$Ti_3C_2T_x$ exhibits better stabilities and SERS activity compared to the conventional $Ti_3C_2T_x$. Surface plasmon-related events in nitride and carbonitride MXenes are projected to be more apparent than in bulk TM nitrides. To investigate the plasmonic and photonic characteristics of MXenes further, collaborative efforts from both modeling and experimentation are required [66].

7.14 VIBRATIONAL MODES

The interlayer interactions and covalent bonds determine the vibrational modes in MXene structures, which facilitate the analysis of their surface group and structural stability. These modes can be studied by using IR or Raman spectroscopy. The quality and layer number of graphene have widely been determined by using Raman spectroscopy. In 2D modes, a multi-peak fitting technique is used to extract the information on layer number. The vibrational spectra of MoS_2 and phosphorene also exhibit these modes that depend on the layer number. Corresponding to the out-of-plane and in-plane modes, MXenes exhibit A_{1g} and E_g modes, respectively. Figures 7.34a and b [15] illustrate the Raman spectra of the MXene for different modes of vibrations. For 2D layered materials, Raman spectroscopy has been observed to be insensitive to the surface groups. For example, the peak of rGO is initially observed at 1350 cm^{-1} (D mode), but it varies with any change in levels of functionalization. However, IR spectroscopy is a useful complement to Raman spectroscopy due to its sensitivity toward surface groups. IR light is employed in this technique to excite the molecules, as a result information on the covalent bonding is revealed through vibrational modes. For example, this technique can identify the surface groups developed on the surface of GO during the synthesis. Because of the abundance of functional groups supplied to MXene surfaces during production, IR spectroscopy is suitable for their analysis. The molecular vibrations produced by IR light are classified as E_u and A_u modes [15].

For the surface terminations of MXenes, three characteristic peaks for the C=O stretching mode, C-H deformation mode, and O-H stretching mode at 1700, 1460, and 3500 cm^{-1}, respectively, are identified. Different external species can also be intercalated by post-modification, beyond these intrinsic functional groups. For example, the intercalation of SO_3H can be detected between 1020 and 1160 cm^{-1} (Fig. 7.34c).

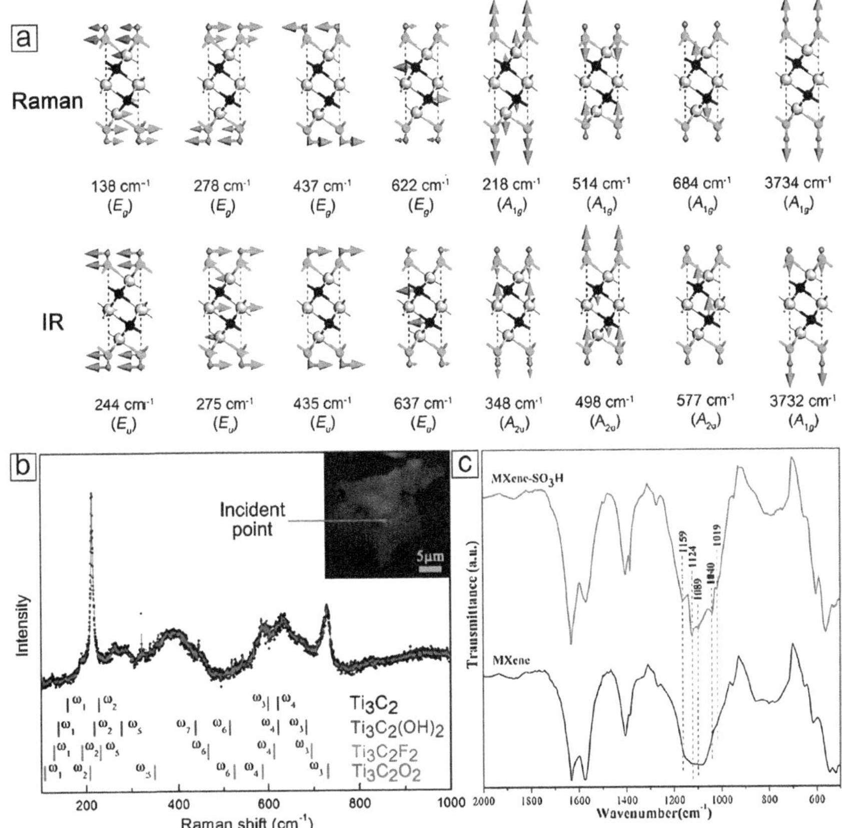

FIGURE 7.34 (a) Atomic bond vibrations for different wavenumbers in Raman and IR active modes. (b) A typical Raman spectrum of the MXene Ti_3C_2. (c) FTIR spectra of bare and functionalized MXene.

Source: [15] Bachmatiuk A., Gemming T. et al. 2019. Applications of 2D MXenes in energy conversion and storage systems. *Chem. Soc. Rev.*, 48, 72–133. Reproduced with permission from the Royal Society of Chemistry. The sources of the material (Hu T., Wang J. et al. 2015. Vibrational properties of Ti_3C_2 and $Ti_3C_2T_2$ (T = O, F, OH) monosheets by first-principles calculations: a comparative study. *Phys. Chem. Chem. Phys.*, 17, 9997–10003. Reproduced with permission from the Royal Society of Chemistry) and (Reprinted from *Applied Surface Science*, 384, Hongbing Wang, Jianfeng Zhang, et al., Surface modified MXene Ti_3C_2 multilayers by aryl diazonium salts leading to large-scale delamination, 287–293. Copyright 2016, with permission from Elsevier) are also acknowledged.

This shows that in addition to the traditional groups, newly introduced surface groups can also be investigated by this technique [15]. Similarly, the phonon spectra of pristine and functionalized M_2XT_2 were also computed with the four termination configurations, confirming some unique stabilities. Many properties are shared by all M_2C systems because of their comparable crystal structure and chemical interaction and are also shared by higher-order M_3C_2 and M_4C_3 MXenes. Three acoustic modes were discovered in the phonon dispersions. Two of them were observed to exhibit a linear dispersion corresponding to in-plane vibrations, while the third mode exhibited a quadratic dispersion corresponding to out-of-plane vibration. A large phonon band-gap was also discovered, which separated the low- and high-frequency vibrations of M and C elements, respectively [7].

7.15 TRANSPARENCY AND CONDUCTIVE PROPERTIES

MXenes are potential candidates for transparent conductors due to their high electrical conductivity, 2D shape, processability, and outstanding mechanical characteristics. The first transparent MXene film was created by exfoliating the magnetron-sputtered Ti_3AlC_2 phase on a sapphire substrate. The resultant films have transmittances ranging from 14 to 85% and conductivities ranging from ≈ 2000 to 5000 S cm^{-1}. In visible light, the MAX phase was also observed to be transparent, and it absorbs more light than its corresponding MXene. As the interlayer spacing increases, the lattice parameter also expands. Therefore, MXenes with larger interlayer spacing have shown to be less conductive and more transparent, compared to those that exhibit smaller interlayer spacing. Different solution-processing methods have been used to deposit MXene films of various thicknesses on flexible or glass substrates [66]. During the spin-casting method, a centrifugal shear force is created due to the high-speed spinning, which results in developing high-quality and uniform MXene flakes. The $Ti_3C_2T_x$ sheets exhibited high conductivity of 6600 S cm^{-1}, with 97% transparency. Notably, these films also display substantial IR absorption with a wavelength comparable to the negative $\varepsilon 1$ in permittivity measurements. These findings are consistent with the DFT expectations. The conductivity of the film increased thrice following storage in a dry N_2 environment. Furthermore, vacuum-annealing was also observed to greatly increase conductivity while retaining transparency [66].

Using first-principle computations, X. Zha et al. [67] examined the electrical and thermal characteristics of functionalized Sc_2CT_2 semiconductors. At ambient temperature, anisotropic electron mobilities of Sc_2CF_2 were observed, with values of 1.07×10^3 and 5.03×10^3 cm^2 V^{-1} s^{-1} in the armchair and zigzag directions, respectively. This value of electron mobility in the zigzag direction was observed to be roughly four times that of phosphorene in the armchair direction. The first investigation on the optoelectronic characteristics of flexible and transparent V_2CT_x films was published by G. Ying et al. [68]. Films were created by dispersing aqueous colloidal suspensions with tetrabutylammonium hydroxide. Throughout the visible range, the optical absorption spectrum was found to be devoid of noticeable transitions. The absorption coefficient at 550 nm was calculated to be $\approx 1.22 \pm 0.05 \times 10^5$ cm^{-1}. Figure 7.35 [68] illustrates AFM profiles of films spin-casted from 6 and 12 mg ml^{-1} colloidal solution at spin rates ranging from 500 to 2,000 rpm. The thicknesses of these coatings ranged from 17 to 112 nm. The typical surface roughness is between 2 and 8 nm.

In another study, transparent conductive Ti_2CT_x films were developed using a spin-coating technique. These films exhibit an figures of merit (FOM) of around 5, which is equivalent to that of spin-cast $Ti_3C_2T_x$ films. The values of absorptivity of these films were comparable, with $\alpha\ 550 \approx 2.7 \times 10^5$ cm^{-1}. This does not indicate that they are technologically similar. The Ti_2CT_x films contain one fewer Ti atom

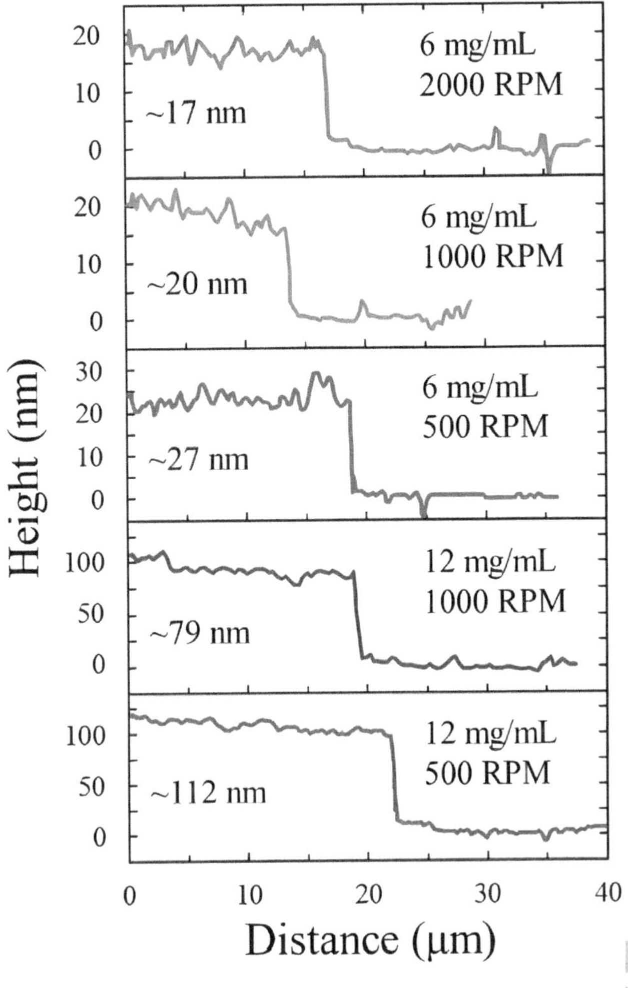

FIGURE 7.35 AFM profiles of spin-casted films. Dependence of thicknesses on the initial concentration of colloidal suspension and spinning speed.

Source: Reprinted from [68] *FlatChem*, 8, Guobing Ying, Sankalp Kota, et al., Conductive transparent V_2CT_x (MXene) films, 25–30. Copyright 2018, with permission from Elsevier.

per formula unit [69]. Figure 7.36a [69] shows the film thickness, h, as a function of the reciprocal of the spinning speed, and Fig. 7.36b displays the effects of concentration, c, on thickness at different spin speeds. Figure 7.36c depicts the dependence of τ (transmittance) on ω (spincasting speed) and c at 550 nm. Not unexpectedly, quicker rotation speeds and smaller solid loadings result in thinner, more transparent films. Varying the spinning speed from 500 to 4,000 rpm results in a variation of τ in the 43–65% range.

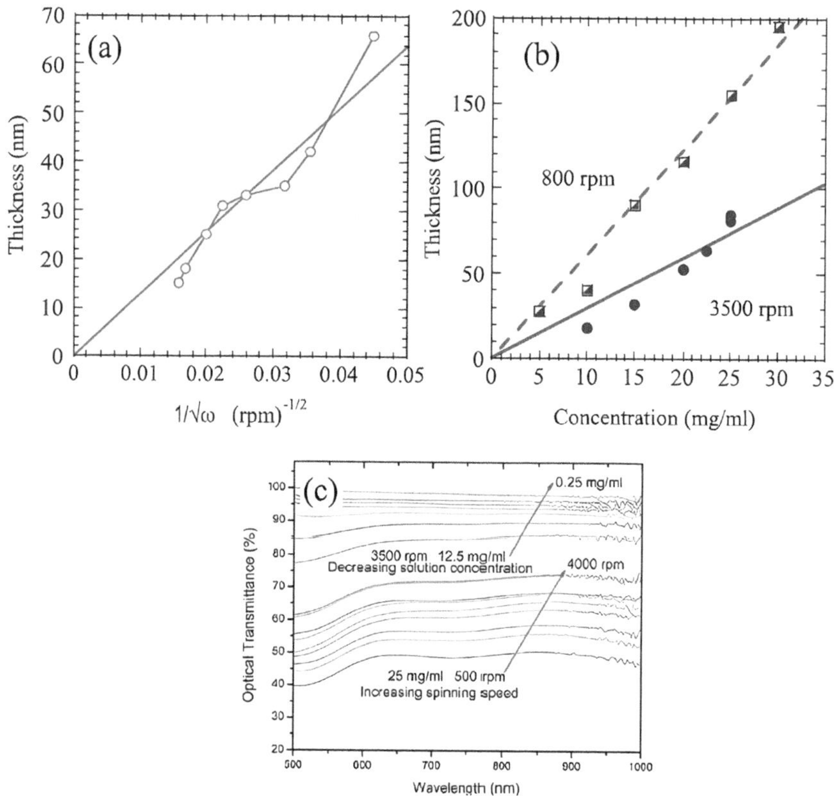

FIGURE 7.36 (a) Dependencies of film thicknesses, h, on (a) $1/\sqrt{\omega}$ for 25 mg ml^{-1}, (b) concentration, c, at different spin speeds, and (c) optical transmittance, τ, depending on ω (lower cluster) and c (top cluster).

Source: Reproduced from [69] Ying G., Dillon A. D. et al. 2017. Transparent, conductive solution processed spin cast 2D Ti$_2$CT$_x$ (MXene) films. *MATER. RES. LETT.*, 5, 391–398. With permission from Taylor & Francis Group.

7.16 NONLINEAR OPTICS AND OPTICAL DETECTION

Nonlinear optics corresponds with the interaction of light and matter in the range in which materials behave nonlinearly to electromagnetic fields. This phenomenon is important for optical communication, photonic devices, and laser optics. The nonlinear optical characteristics of MXenes have also recently been discovered. Ti$_3$C$_2$T$_x$ shows the nonlinear absorption of light as measured by Z-scan, which means that illuminating intensity and transmission have a nonlinear relation. Typically, illuminating intensity increases with a decrease in nonlinear absorption coefficient, until a threshold value (represented as I$_{th}$) is reached. For Ti$_3$C$_2$T$_x$, the value of the nonlinear absorption coefficient was observed to be -10^{-21} m^2 V^{-2}, indicating possible usage in optical switching applications. Saturable absorption (SA) was also detected at various illumination wavelengths (ranging from 800 to 1,800 nm), which implies that more than one mechanism contributes to nonlinear absorption. This capability of MXenes allows them to be employed in ultrafast laser applications. MXenes are used as a spectral filter to generate mode-locked laser pulses in this application. In femtosecond mode-locked lasers, metallic $Ti_3C_2T_x$ and Ti_3CNT_x were employed. $Ti_3C_2T_x$ also provided similar outcomes. The broadband wavelength of $Ti_3C_2T_x$ ranging from 1,550 to 1,620 nm covers the critical telecommunication spectrum (C band).

MXenes have nonlinear optical properties that are comparable, if not superior, to those of other 2D materials such as transition metal dichalcogenides, graphene, and black phosphorus. $Ti_3C_2T_x$ was also shown to be capable of creating an optical diode when linked with a reverse saturable material. The device could be used to filter high-rectification-ratio optical signals. When $Ti_3C_2T_x$ diffused in the rhodamine 101 solution, it produced random lasing, generating a metamaterial. Although research into these properties is still in its early stages, it has already shown promising benefits in various applications. However, the interaction of MXenes with light requires further thorough research. The photonic, optoelectronic, and plasmonic properties of various MXenes need to be investigated further before they can be used in practical applications [66].

To address these concerns, Y. Wang et al. [70] created a 2D Nb_2C/MoS_2 heterostructure that surpasses pristine Nb_2C in both nonlinear and linear optical performance. The Nb_2C/MoS_2 inherited the predominance of Nb_2C and MoS_2 in absorption, resulting in an increased optical absorption capability. They achieved enhanced NIR nonlinear optical modulation in addition to linear optical modulation, with a nonlinear absorption coefficient of Nb_2C/MoS_2 that is more than double that of pure Nb_2C. The optical absorption (OA) Z-scan approach was used to evaluate the nonlinear optical absorption (NLO) response of Nb_2C/MoS_2 and Nb_2C in the NIR region. The OA Z-scan findings are presented in Figs. 7.37a and b [70]. Both Nb_2C/MoS_2 and Nb_2C display typical SA, which means that as the sample approaches the point, the normalized transmittance increases.

In another study, the nonlinear optical response of magnetron sputtered 2D Nb_2C NSs was studied using Z-scan measurements, which exhibited outstanding light modulation capabilities [71]. The SA process of this MXene produced an effective nonlinear absorption coefficient ($\beta eff \approx -10^5$ cm GW^{-1}), which was observed to be two times greater than that produced by reversible SA process. D. Huang et al. [72] used V_2C MQDs to develop a nonlinear broadband scattering system with increased gain to show a white laser. Similarly, S. Li et al. [73] used a liquid-phase exfoliation process to synthesize Mo_2C NSs and investigated their nonlinear optical characteristics. They observed 5.1% saturation intensity along with a high value of modulation depth, that is, 200 MW cm^{-2}. Furthermore, using evanescent field interactions, they also developed a mode-locker based on a D-shaped fiber Mo_2C, which displayed high value of pulse energy and peak power.

7.17 PHOTOTHERMAL EFFECTS AND CONVERSION

When light is absorbed in MXene NSs, the photon energy is converted into phonon within the structure, which considerably elevates their temperature. Studies have shown that $Ti_3C_2T_x$ displays excellent photothermal conversion efficiency after

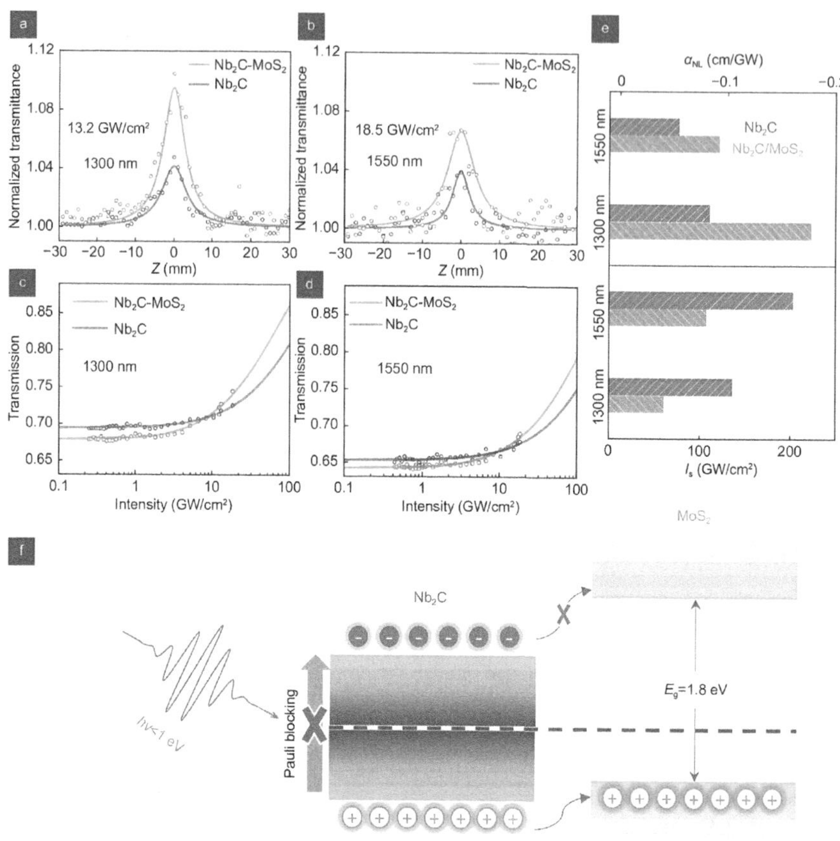

FIGURE 7.37 OA Z-scan results of Nb_2C and Nb_2C/MoS_2 with an excitation wavelength of (a) 1,300 nm and (b) 1,550 nm. (c, d) Corresponding nonlinear transmittance curves. (e) Histogram of Is and αNL. (f) Schematic illustration of the transfer process.

Source: Reproduced from [70] Wang YD, Wang YW, Dong YL, et al. 2023. 2D Nb_2CT_x MXene/MoS_2 heterostructure construction for nonlinear optical absorption modulation. *Opto-Electron Adv* 6, 220162. Copyright 2023, with permission.

absorbing NIR light. Other investigations have shown that light irradiation can cause oxygen stress on the surface of flakes. The photothermal property of MXenes is another interesting use in mechanical light-driven actuators. All optical modulators based on MXenes, due to their broadband operation, ease of integration, and low loss, have been utilized for information processing and optical communications. These properties also make them capable for optoelectronic and photonic devices [22, 66]. To increase the stability and rate of freshwater production, B. Zhang et al. [74] developed a $Ti_3C_2T_x$-engineered membrane, which exhibited strong hydrophobicity along with an efficient photothermal effect. It was observed that in order to maintain a high temperature differential, a photothermal membrane with a low thermal conductivity that successfully produces photothermal effects allows for localized surface heating while preventing unexpected transmission of heat.

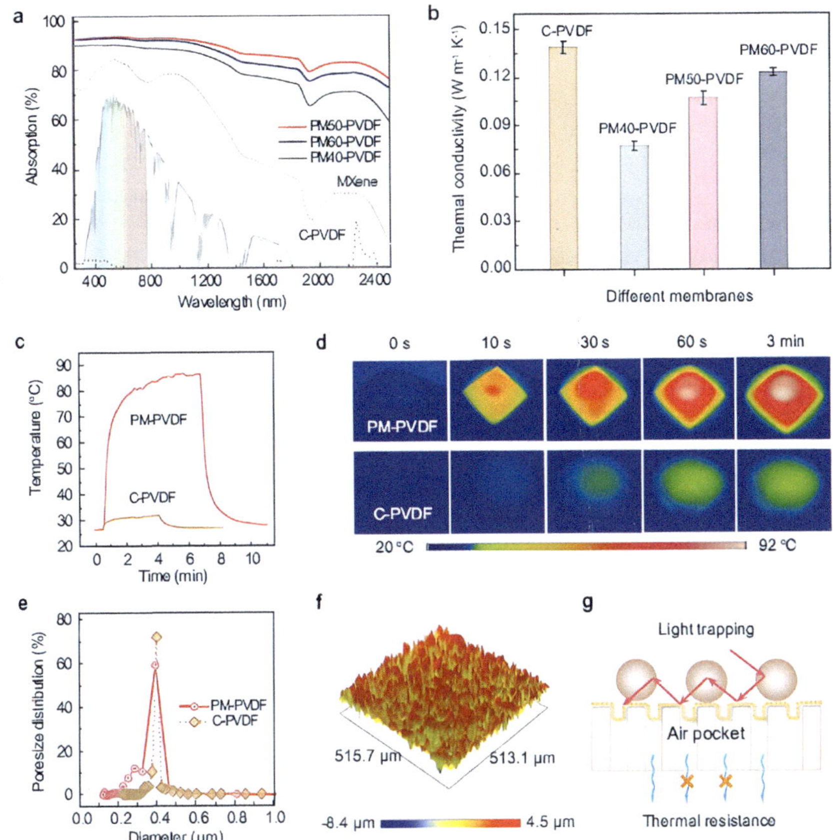

FIGURE 7.38 (a) UV-Vis-NIR absorption spectra. (b) Corresponding thermal conductivities. (c) Surface temperature profiles as a function of time and (d) IR thermal images. (e) Pore size distribution. (f) Typical optical surface profile. (g) Schematic illustration with favorable light trapping and thermal resistance.

Source: Reprinted with permission from [74] Springer Nature. *Nature Communications*, Baoping Zhang et al. Transforming $Ti_3C_2T_x$ MXene's intrinsic hydrophilicity into super-hydrophobicity for efficient photothermal membrane desalination. 13: 3315, copyright 2022.

The UV-visible spectra of the membranes developed with different chemical compositions are shown in Fig. 7.38a [74]. Significant absorption of $Ti_3C_2T_x$ MXene was revealed by two strong absorption peaks (at 610 and 1,148 nm) of PM-PVDF membranes with variable thicknesses of MXene layers, while a significantly lower light absorption capability was demonstrated by C-PVDF membrane. Despite being infused with a highly thermal conductive MXene, the PM-PVDF membranes displayed lower thermal conductivity ($0.08 \sim 0.12$ Wm^{-1} K^{-1}) than the C-PVDF membrane (0.14 Wm^{-1} K^{-1}) (Fig. 7.38b). As a result, when subjected to one-sun illumination, the surface temperature of the PM50-PVDF membrane increases to 84.8°C in 3 minutes (Fig. 7.38c), exhibiting its exceptional surface self-heating behavior. This membrane delivered an increase of 60.5°C, which is about 10 times that of the C-PVDF membrane (6.2°C) under identical conditions, as confirmed by the IR thermal pictures in Fig. 7.38d. The porous structure was confirmed by the formation of trapped air pockets (Fig. 7.38e) and optical surface profiles (Fig. 7.38f) of the PM-PVDF membrane. Furthermore, the combined functioning of MXene layers is schematically presented in Fig. 7.38g. The thermal resistance was provided by the porous structure exhibiting trapped air pockets, which facilitates efficient utilization of light along with the thermal conduction [74].

Another work discovered a simple method for producing in situ $N\text{-}Ti_3C_2$ MQDs via amine-assisted solvothermal tailoring. MXene with a few layers was synthesized by exfoliating pristine MXene. These N-MQDs responded differently from several metal cations in terms of fluorescence quenching, which facilitates the detection of Cu^{2+} [75]. J. Jeon et al. [76] presented a hybrid phototransistor based on Mo_2C and MoS_2. A low Schottky barrier height (≈70 meV) was produced due to the combination of both Mo-based materials. This resulted in the transfer of hot carriers from the Mo_2C to MoS_2 channel due to the plasmonic resonance produced by metallic Mo_2C grating. It was also discovered that photo-response of light can be varied from visible to NIR by modifying the grating period of Mo_2C (400–1,000 nm).

7.18 ULTRAFAST DYNAMICS

After a detailed analysis of optical, optoelectronic, and photothermal properties of MXenes, it has been observed that a complete understanding of physical processes that drive responses of MXenes toward light stimuli is essential to be carried out for all photo-response applications, that is, photodetection, SERS, photothermal ablation, and photo-actuator. Light stimulates electron creation inside NSs, electron energy translates to lattice movements, and thermal energy diffuses out of these structures. An ultrafast spectroscopy is used to observe these energetic processes that take place on the ultrafast time scale, that is, fs and ps. The THz results revealed that the primary effects are centered on the carrier-free dynamics of $Ti_3C_2T_x$. The results also demonstrated that ultrafast pump lights can modify $Ti_3C_2T_x$ conductivity and THz transmission. It was also reported that in MXenes, a temporary broadband THz transparency can also be caused by ultrafast laser pulses of visible and NIR wavelengths, with a nanosecond lifetime. Some investigations have aimed at using transient absorption spectroscopy to understand the optical applications caused by light–MXene interactions.

As previously stated, the novel photodetector material Mo_2CT_x has a lower electrical density than typical plasmonic metals. As a result, the experimental electron–electron scattering relaxation period is considerably longer. According to the ultrafast transient absorption of Mo_2CT_x, the synthesis of electrons and holes take 200 fs, while their corresponding recombination takes 53 and 13 ps, respectively. This delayed relaxation matches Raman detection signals well when compared to plasmonic gold nanostructures, demonstrating that the photo-response coincides with that of surface plasmon-assisted heat carriers. The ultrafast analysis also aided in understanding the spatial self-phase modulation (SSPM) of $Ti_3C_2T_x$. The SSPM pattern was thus generated using the diffraction ring, which is produced by modifying the refractive index.

In recent research, the transient spectra of MXene ($Ti_3C_2T_x$) were collected using a two-color pump–probe configuration [22].

Figure 7.39a [22] illustrates a nonequilibrium excitation condition caused by the illumination of $Ti_3C_2T_x$ by different powers and wavelengths. The results showed that the absorption spectra altered when the pump light was perturbed. The physical image of this interaction was initially explained by combining the dynamic signals with transient spectra (450–1,100 nm). Two temporary absorption zones (in the visible and NIR ranges) were generated during excitation by light (in the 260–800 nm range), as illustrated in Fig. 7.39b. From the dynamic view in Fig. 7.39c, photon stimuli respond in three phases, that is, early-stage electron–electron interaction, electron–phonon coupling, and thermal diffusion. The $Ti_3C_2T_x$ nanoflakes have already been thermalized after 2 ps. The transient spectrum signal in this final step is linearly related to the absorption efficiency and excitation fluence. A recent ultrafast study investigated the dynamic processes of $Ti_3C_2T_x$ under various interfacial circumstances, including ethanol and water solution along with polymer combination. The heat dissipation channels might split into a hydrogen bond-dominated (fast) channel and a lattice motion-mediated (slow) channel, as shown in Fig. 7.39d. The findings also revealed that $Ti_3C_2T_x$ flash energy migration might heat the bilayer molecules [22].

7.19   RECENT ADVANCEMENTS IN THE CHARACTERISTICS OF MXENES

This significant and fast-expanding family of MXenes has been shown to have appealing electrical, optical, electrochemical, and mechanical properties, leading to various applications. The characteristics and compositions of MXenes can be modified in many ways due to the presence of metals and surface groups. MXenes are ideal for solution processing, as they can easily be dispersed in polar organic solvents or water due to negative zeta potential and hydrophilicity. Applications such as optical communication, ultrafast lasers, metamaterial broadband absorbers, SERS substrates, and light-to-heat conversion have been reported to utilize their optical and plasmonic capabilities [66]. Different forms of magnetic ordering in MXenes with enormous SOC promise unique quantum anomalous Hall systems. The sensitivity of MXene-based devices to external electric fields, mechanical strain, and photo-irradiation is critical for customizing novel phenomena and applications. Exploring the influence of gate voltage on magnetic order control in MXenes is an intriguing field of research. Furthermore, barely anything is known about superconducting MXenes [77].

Recently, an increasing interest has been observed in the plasmonic features of MXenes, which is efficient for several electrical devices. Although research into the nonlinear optical characteristics of MXenes is still in its early stages, it has revealed a bright potential for MXenes in a variety of applications. More research into MXenes' interaction with light as well as their optoelectronic, plasmonic, and photonic properties is needed [78]. Furthermore, $Ti_3C_2T_z$ thin films exhibit transparency with 97% transmission of visible light (per nanometer thickness), which is comparable to 97.7% transmission of a graphene monolayer. In comparison, due to the high electrical conductivity of M_2CT_z thin films (with M = V or Ti) and high transparency, these

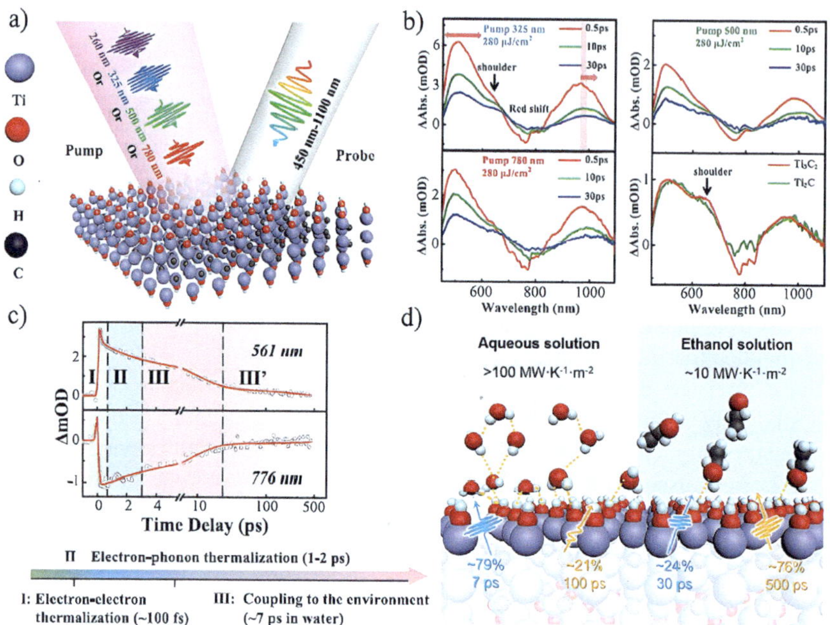

FIGURE 7.39 Ultrafast spectroscopy studies of MXenes. a) Schematic of two-color pump–probe spectroscopy. b) Transient spectra under different pump excitation wavelengths. c) Dynamics of MXenes in water solutions for three different stages. d) MXene–solvent interfacial thermal energy migration pathways.

Source: [22] Fu Bo, Sun Jingxuan, et al.: MXenes: Synthesis, Optical Properties, and Applications in Ultrafast Photonics. *Small* 2021, 17, 2006054. Copyright WILEY-VCH Verlag GmbH & Co. KGaA, Weinheim. Reproduced with permission. The sources of the material (Reprinted with permission from Qi Zhang, Li Yan, Mingsheng Yang, et al. 2020. Ultrafast Transient Spectra and Dynamics of MXene (Ti$_3$C$_2$T$_x$) in Response to Light Excitations of Various Wavelengths. *J. Phys. Chem. C*, 124, 11, 6441–6447. Copyright 2020, American Chemical Society and Reprinted with permission from Jiebo Li, Zhen Chi, Ruzhan Qin, et al. 2020. Hydrogen Bond Interaction Promotes Flash Energy Transport at MXene-Solvent Interface. *J. Phys. Chem. C*, 124, 19, 10306–10314. Copyright 2020, American Chemical Society) are also acknowledged.

materials have become a priority to be utilized as transparent conductive electrodes. Similarly, the optical properties of MXenes are functionalization-dependent and can be modified by employing electrochemical ion intercalation. MXenes' high electromagnetic interference (EMI) protection is another intriguing physical characteristic. MXenes are thus attractive 2D materials for EMI electrical devices due to their high flexibility and lightweight. MXenes have also demonstrated acceptable affinity for other gases such as CO_2, ethanol, and ammonia. Furthermore, first-principle simulations have predicted that Ti$_2$C has an efficient physisorption capability with regard to various gas molecules [7].

The structural defects in MXene NSs give rise to the oxidation mechanism. The internal electric fields are a significant reason for this mechanism, which produces vacancies and metal oxides by promoting the relocation of electrons and Ti cations.

Thus, oxidation can be reduced by minimizing the production of internal electric fields caused by surrounding structural defects. However, during the exfoliation processes, structural defects cannot be avoided. The stability of MXene solutions is also primarily significant, as it plays an important role in developing the MXene-based composites and hetero-structures for various applications. Moreover, MXenes are hydrophilic in nature, and their stability in aqueous solutions needs further research. Furthermore, oxidation can be reduced by optimizing MXenes with different surface functionalizations (such as inert oxides). The size optimization of MXene NSs in solutions is also very important. Separation of MXene NSs with larger lateral sizes from those with smaller lateral sizes not only increases storage time but also increases MXene activity for various applications [79, 80].

Another essential technique to protect MXenes from oxidation is to improve their storage conditions, that is, in a refrigerating or a deoxygenated atmosphere. Storing these materials in organic solvents (without moisture) will improve their stability, which facilitates the formation of MXene-based polymer composites for several solution-based applications [79, 80]. Furthermore, most of the theoretical studies are based on uniformly functionalized or pristine MXenes. This means that theoretical data are predicting the properties of those materials that are difficult to be synthesized experimentally, as the formation of surface groups during the synthesis processes cannot be avoided. Therefore, theoretical simulations should also evaluate the properties of those MXenes that exhibit structural compositions with heterogeneous terminations [7].

REFERENCES

[1] Meng, W., et al., *Advances and challenges in 2D MXenes: from structures to energy storage and conversions.* Nano Today, 2021. **40**: p. 101273.

[2] Ahmad Junaidi, N.H., et al., *A comprehensive review of MXenes as catalyst supports for the oxygen reduction reaction in fuel cells.* International Journal of Energy Research, 2021. **45**(11): p. 15760–15782.

[3] Jadav, R.P., et al., *Structural stability and electronic properties of 2D MXene $Hf_3C_2F_2$ monolayer by density functional theory approach.* Biointerface Research in Applied Chemistry, 2022. **13**(2): p. 152.

[4] Zhang, H., et al., *Computational studies on the structural, electronic and optical properties of graphene-like MXenes (M_2CT_2, M = Ti, Zr, Hf; T = O, F, OH) and their potential applications as visible-light driven photocatalysts.* Journal of Materials Chemistry A, 2016. **4**(33): p. 12913–12920.

[5] Jin, D., et al., *Computational screening of 2D ordered double transition-metal carbides (MXenes) as electrocatalysts for hydrogen evolution reaction.* The Journal of Physical Chemistry C, 2020. **124**(19): p. 10584–10592.

[6] Khaledialidusti, R., A.K. Mishra, and A. Barnoush, *Atomic defects in monolayer ordered double transition metal carbide ($Mo_2TiC_2T_x$) MXene and CO_2 adsorption.* Journal of Materials Chemistry C, 2020. **8**(14): p. 4771–4779.

[7] Champagne, A. and J.-C. Charlier, *Physical properties of 2D MXenes: from a theoretical perspective.* Journal of Physics: Materials, 2020. **3**(3): p. 032006.

[8] Ronchi, R.M., J.T. Arantes, and S.F. Santos, *Synthesis, structure, properties and applications of MXenes: current status and perspectives.* Ceramics International, 2019. **45**(15): p. 18167–18188.

[9] Kumar, J.A., et al., *Methods of synthesis, characteristics, and environmental applications of MXene: a comprehensive review*. Chemosphere, 2022. **286**: p. 131607.

[10] Borysiuk, V.N., V.N. Mochalin, and Y. Gogotsi, *Molecular dynamic study of the mechanical properties of two-dimensional titanium carbides $Ti_{n+1}Cn$ (MXenes)*. Nanotechnology, 2015. **26**(26): p. 265705.

[11] Zhang, N., et al., *Superior structural, elastic and electronic properties of 2D titanium nitride MXenes over carbide MXenes: a comprehensive first principles study*. 2D Materials, 2018. **5**(4): p. 045004.

[12] Zha, X.-H., et al., *Role of the surface effect on the structural, electronic and mechanical properties of the carbide MXenes*. Europhysics Letters, 2015. **111**(2): p. 26007.

[13] Bai, Y., et al., *Dependence of elastic and optical properties on surface terminated groups in two-dimensional MXene monolayers: a first-principles study*. RSC Advances, 2016. **6**(42): p. 35731–35739.

[14] Chakraborty, P., et al., *Manipulating the mechanical properties of Ti_2C MXene: effect of substitutional doping*. Physical Review B, 2017. **95**(18): p. 184106.

[15] Pang, J., et al., *Applications of 2D MXenes in energy conversion and storage systems*. Chemical Society Reviews, 2019. **48**(1): p. 72–133.

[16] Fredrickson, K.D., et al., *Effects of applied potential and water intercalation on the surface chemistry of Ti_2C and Mo_2C MXenes*. The Journal of Physical Chemistry C, 2016. **120**(50): p. 28432–28440.

[17] Xie, Y. and P. Kent, *Hybrid density functional study of structural and electronic properties of functionalized $Ti_{n+1}X_n$ (X = C, N) monolayers*. Physical Review B, 2013. **87**(23): p. 235441.

[18] Anasori, B., et al., *Control of electronic properties of 2D carbides (MXenes) by manipulating their transition metal layers*. Nanoscale Horizons, 2016. **1**(3): p. 227–234.

[19] Si, C., et al., *Large-gap quantum spin Hall state in MXenes: d-band topological order in a triangular lattice*. Nano Letters, 2016. **16**(10): p. 6584–6591.

[20] Yang, J., et al., *Investigation of magnetic and electronic properties of transition metal doped Sc_2CT_2 (T = O, OH or F) using a first principles study*. Physical Chemistry Chemical Physics, 2016. **18**(18): p. 12914–12919.

[21] Balcı, E., Ü.Ö. Akkuş, and S. Berber, *Doped $Sc_2C(OH)_2$ MXene: new type s-pd band inversion topological insulator*. Journal of Physics: Condensed Matter, 2018. **30**(15): p. 155501.

[22] Fu, B., et al., *MXenes: synthesis, optical properties, and applications in ultrafast photonics*. Small, 2021. **17**(11): p. 2006054.

[23] Wang, B., et al., *Booming development and present advances of two dimensional MXenes for photodetectors*. Chemical Engineering Journal, 2021. **403**: p. 126336.

[24] Zhang, R.-Z., H.-L. Cui, and X.-H. Li, *First-principles study of structural, electronic and optical properties of doped Ti_2CF_2 MXenes*. Physica B: Condensed Matter, 2019. **561**: p. 90–96.

[25] Yan, F., et al., *Solvothermal synthesis of nitrogen-doped MXene quantum dots for the detection of alizarin red based on inner filter effect*. Dyes and Pigments, 2021. **195**: p. 109720.

[26] Halim, J., et al., *Electronic and optical characterization of 2D Ti_2C and Nb_2C (MXene) thin films*. Journal of Physics: Condensed Matter, 2019. **31**(16): p. 165301.

[27] Chen, Q., et al., *Optical properties of two-dimensional semi-conductive MXene Sc_2COx produced by sputtering*. Optik, 2020. **219**: p. 165046.

[28] Tang, M., et al., *Surface terminations of MXene: synthesis, characterization, and properties*. Symmetry, 2022. **14**(11): p. 2232.

[29] Gao, G., A.P. O'Mullane, and A. Du, *2D MXenes: a new family of promising catalysts for the hydrogen evolution reaction*. ACS Catalysis, 2017. **7**(1): p. 494–500.

[30] Huang, Z.-Q., et al., *Large-gap topological insulators in functionalized ordered double transition metal carbide MXenes.* Physical Review B, 2020. **102**(7): p. 075306.

[31] Kim, H., et al., *Thermoelectric properties of two-dimensional molybdenum-based MXenes.* Chemistry of Materials, 2017. **29**(15): p. 6472–6479.

[32] Jing, Z., et al., *Superior thermoelectric performance of ordered double transition metal MXenes: $Cr_2TiC_2T_2$ (T = −OH or −F).* The Journal of Physical Chemistry Letters, 2019. **10**(19): p. 5721–5728.

[33] Chang, W.-L., et al., *Thermoelectric properties of two-dimensional double transition metal MXenes: $ScYCT_2$ (T = F, OH).* Journal of Physics and Chemistry of Solids, 2023. **176**: p. 111210.

[34] Abbasi, N.M., et al., *Recent advancement for the synthesis of MXene derivatives and their sensing protocol.* Advanced Materials Technologies, 2021. **6**(10): p. 2001197.

[35] Mishra, A., et al., *Atomistic origin of phase stability in oxygen-functionalized MXene: a comparative study.* The Journal of Physical Chemistry C, 2017. **121**(34): p. 18947–18953.

[36] Meshkian, R., et al., *Theoretical stability and materials synthesis of a chemically ordered MAX phase, Mo_2ScAlC_2, and its two-dimensional derivate Mo_2ScC_2 MXene.* Acta Materialia, 2017. **125**: p. 476–480.

[37] Deeva, E.B., et al., *In situ XANES/XRD study of the structural stability of two-dimensional molybdenum carbide Mo_2CT_x: implications for the catalytic activity in the water–gas shift reaction.* Chemistry of Materials, 2019. **31**(12): p. 4505–4513.

[38] Ang, H., et al., *Hydrophilic nitrogen and sulfur Co-doped molybdenum carbide nanosheets for electrochemical hydrogen evolution.* Small, 2015. **11**(47): p. 6278–6284.

[39] Zhang, C., et al., *Two-dimensional transition metal carbides and nitrides (MXenes): synthesis, properties, and electrochemical energy storage applications.* Energy & Environmental Materials, 2020. **3**(1): p. 29–55.

[40] Ashton, M., et al., *Predicted surface composition and thermodynamic stability of MXenes in solution.* The Journal of Physical Chemistry C, 2016. **120**(6): p. 3550–3556.

[41] Li, Z., et al., *Synthesis and thermal stability of two-dimensional carbide MXene Ti_3C_2.* Materials Science and Engineering: B, 2015. **191**: p. 33–40.

[42] Yang, J., et al., *Stability and electronic properties of sulfur terminated two-dimensional early transition metal carbides and nitrides (MXene).* Computational Materials Science, 2018. **153**: p. 303–308.

[43] Shen, Z., et al., *A novel solid-solution MXene $(Ti_{0.5}V_{0.5})_3C_2$ with high catalytic activity for hydrogen storage in MgH_2.* Materialia, 2018. **1**: p. 114–120.

[44] Liu, Y., et al., *Excellent catalytic activity of a two-dimensional Nb_4C_3Tx (MXene) on hydrogen storage of MgH_2.* Applied Surface Science, 2019. **493**: p. 431–440.

[45] Chakraborty, P., T. Das, and T. Saha-Dasgupta, *MXene: a new trend in 2D materials science.* Comprehensive Nanoscience and Nanotechnology, 2019. **319**.

[46] Kumar, H., et al., *Tunable magnetism and transport properties in nitride MXenes.* ACS Nano, 2017. **11**(8): p. 7648–7655.

[47] Teh, S. and H.-T. Jeng, *Magnetoelastic and magnetoelectric coupling in two-dimensional nitride MXene: a density functional theory study.* Nanomaterials, 2023. **13**(19): p. 2644.

[48] Yang, J., et al., *Tuning magnetic properties of $Cr_2M_2C_3T_2$ (M = Ti and V) using extensile strain.* Computational Materials Science, 2017. **139**: p. 313–319.

[49] Yang, J., et al., *Tunable electronic and magnetic properties of $Cr_2M'C_2T_2$ (M' = Ti or V; T = O, OH or F).* Applied Physics Letters, 2016. **109**(20).

[50] Dong, L., et al., *Rational design of two-dimensional metallic and semiconducting spintronic materials based on ordered double-transition-metal MXenes.* The Journal of Physical Chemistry Letters, 2017. **8**(2): p. 422–428.

[51] Zhang, Y., et al., *Computational design of double transition metal MXenes with intrinsic magnetic properties.* Nanoscale Horizons, 2022. **7**(3): p. 276–287.

[52] He, J., et al., *Cr_2TiC_2-based double MXenes: novel 2D bipolar antiferromagnetic semiconductor with gate-controllable spin orientation toward antiferromagnetic spintronics.* Nanoscale, 2019. **11**(1): p. 356–364.

[53] He, J., et al., *High temperature spin-polarized semiconductivity with zero magnetization in two-dimensional Janus MXenes.* Journal of Materials Chemistry C, 2016. **4**(27): p. 6500–6509.

[54] Zhou, T., et al., *Atomic vacancy defect, Frenkel defect and transition metals (Sc, V, Zr) doping in Ti_4N_3 MXene nanosheet: a first-principles investigation.* Applied Sciences, 2020. **10**(7): p. 2450.

[55] Xiao, X., et al., *Scalable synthesis of ultrathin Mn_3N_2 exhibiting room-temperature antiferromagnetism.* Advanced Functional Materials, 2019. **29**(17): p. 1809001.

[56] Salles, P., et al., *Electrochromic effect in titanium carbide MXene thin films produced by dip-coating.* Advanced Functional Materials, 2019. **29**(17): p. 1809223.

[57] Yang, L., et al., *Surface-modified metallic $Ti_3C_2T_x$ MXene as electron transport layer for planar heterojunction perovskite solar cells.* Advanced Functional Materials, 2019. **29**(46): p. 1905694.

[58] Dillon, A.D., et al., *Highly conductive optical quality solution-processed films of 2D titanium carbide.* Advanced Functional Materials, 2016. **26**(23): p. 4162–4168.

[59] Tan, K., et al., *Optical and conductivity studies of polyvinyl alcohol-MXene (PVA-MXene) nanocomposite thin films for electronic applications.* Optics & Laser Technology, 2021. **136**: p. 106772.

[60] Djire, A., et al., *Electrocatalytic and optoelectronic characteristics of the two-dimensional titanium nitride Ti_4N_3Tx MXene.* ACS Applied Materials & Interfaces, 2019. **11**(12): p. 11812–11823.

[61] Xiang, H., et al., *Hypoxia-irrelevant photonic thermodynamic cancer nanomedicine.* ACS Nano, 2019. **13**(2): p. 2223–2235.

[62] Feng, W., et al., *Ultrathin molybdenum carbide MXene with fast biodegradability for highly efficient theory-oriented photonic tumor hyperthermia.* Advanced Functional Materials, 2019. **29**(22): p. 1901942.

[63] Ren, X., et al., *Highly catalytic niobium carbide (MXene) promotes hematopoietic recovery after radiation by free radical scavenging.* ACS Nano, 2019. **13**(6): p. 6438–6454.

[64] Rozmysłowska-Wojciechowska, A., et al., *A simple, low-cost and green method for controlling the cytotoxicity of MXenes.* Materials Science and Engineering C, 2020. **111**: p. 110790.

[65] Limbu, T.B., et al., *Green synthesis of reduced Ti_3C_2Tx MXene nanosheets with enhanced conductivity, oxidation stability, and SERS activity.* Journal of Materials Chemistry C, 2020. **8**(14): p. 4722–4731.

[66] Hantanasirisakul, K. and Y. Gogotsi, *Electronic and optical properties of 2D transition metal carbides and nitrides (MXenes).* Advanced Materials, 2018. **30**(52): p. 1804779.

[67] Zha, X.-H., et al., *Promising electron mobility and high thermal conductivity in Sc_2CT_2 (T = F, OH) MXenes.* Nanoscale, 2016. **8**(11): p. 6110–6117.

[68] Ying, G., et al., *Conductive transparent V_2CT_x (MXene) films.* FlatChem, 2018. **8**: p. 25–30.

[69] Ying, G., et al., *Transparent, conductive solution processed spincast 2D Ti_2CT_x (MXene) films.* Materials Research Letters, 2017. **5**(6): p. 391–398.

[70] Wang, Y., et al., *2D Nb_2CT_x MXene/MoS_2 heterostructure construction for nonlinear optical absorption modulation.* Opto-Electronic Advances, 2023: p. 220162-1-220162-11.

[71] Li, G., et al., *Third-order nonlinear optical response of few-layer MXene Nb_2C and applications for square-wave laser pulse generation.* Advanced Materials Interfaces, 2021. **8**(6): p. 2001805.

[72] Huang, D., et al., *Demonstration of a white laser with V$_2$C MXene-based quantum dots.* Advanced Materials, 2019. **31**(24): p. 1901117.

[73] Liu, S., et al., *Ultrafast photonics applications based on evanescent field interactions with 2D molybdenum carbide (Mo$_2$C).* Journal of Materials Chemistry C, 2021. **9**(19): p. 6187–6192.

[74] Zhang, B., et al., *Transforming Ti$_3$C$_2$T$_x$ MXene's intrinsic hydrophilicity into superhydrophobicity for efficient photothermal membrane desalination.* Nature Communications, 2022. **13**(1): p. 3315.

[75] Feng, Y., et al., *Solvothermal synthesis of in situ nitrogen-doped Ti$_3$C$_2$ MXene fluorescent quantum dots for selective Cu^{2+} detection.* Ceramics International, 2020. **46**(6): p. 8320–8327.

[76] Jeon, J., et al., *Transition-metal-carbide (Mo$_2$C) multiperiod gratings for realization of high-sensitivity and broad-spectrum photodetection.* Advanced Functional Materials, 2019. **29**(48): p. 1905384.

[77] Khazaei, M., et al., *Recent advances in MXenes: from fundamentals to applications.* Current Opinion in Solid State and Materials Science, 2019. **23**(3): p. 164–178.

[78] Wang, Y., et al., *MXenes: focus on optical and electronic properties and corresponding applications.* Nanophotonics, 2020. **9**(7): p. 1601–1620.

[79] Cao, F., et al., *Recent advances in oxidation stable chemistry of 2D MXenes.* Advanced Materials, 2022. **34**(13): p. 2107554.

[80] Razium A.S. and F. Baomin, *Progression in the oxidation stability of MXenes.* 2023, Springer.

8 MXenes for Catalysis Applications

8.1 CATALYSTS FOR ENERGY CONVERSION

Electrocatalysis, an important branch of electrochemistry, is driven by concerns and regulations about environmental sustainability. One of the primary benefits of electrocatalysis is the possibility of producing cleaner energy and reducing greenhouse gas emissions. In these processes, certain redox reactions take place on the surface of electrodes. However, the efficiency for massive overpotentials and low selectivity severely limits the promise of electrocatalytic reactions. For this purpose, different catalysts are inserted at the surface of electrodes, to reduce their limitation of overpotential. Therefore, discovering new materials is essential to improve the efficiency of electrocatalysis [1]. MXenes offer a significant advantage over other 2D materials since their surface can be functionalized with different termination groups and can also be used to modify their bandgap configuration. MXenes also exhibit enhanced polarity, which enables them to be combined with other semiconductors to make heterostructures. Furthermore, these materials possess large surface area, high metallic conductivity, hydrophilicity, fast charge transfer, facile functionalization, and stability in aqueous solution, making them ideal co-catalysts [2].

As a result, MXenes are being identified as excellent catalysts for energy conversion processes, such as carbon dioxide (CO_2) hydrogenation, synthetic reactions, CO_2 reduction, nitrogen reduction reaction (NRR), oxygen evolution reaction (OER), hydrogen evolution reaction (HER), and oxygen reduction reaction (ORR). This section presents a detailed discussion of latest researches of pristine and doped MXenes in catalysis.

8.2 CATALYSTS FOR CARBON DIOXIDE HYDROGENATION

When CO_2 is transformed into hydrocarbon fuel, it does not release any greenhouse gases upon combustion. The conventional catalysts for CO_2 conversion include Cu, TiO_2, Cu_2O, and other 2D materials, such as $g\text{-}C_3N_4$ and functionalized GO. However, there is still a lack of complete understanding of these developments, and more research is required. The CO_2 conversion process involves several electroreduction steps, which includes electron and proton pairs. CO_2 fixation requires pressure and external energy for chemisorption, while a negative reduction potential is essential for hydrogenation. Hydrocarbon fuels (i.e., H_2CO, CO, CH_4, and CH_3OH) are classified into different types depending on the proton and electron pairs transported following the completion of the reaction. A complete study of the reaction pathway is required when creating innovative catalysts for specific product targeting. Various MXenes (Ti_2CO_2, V_2CO_2, and $Ti_3C_2O_2$) were examined in a computational analysis.

DOI: 10.1201/9781003465768-8

The results of these analysis showed that H_2 and CO create oxygen vacancies on the material, facilitating the catalytic reduction process. Among all of the studied MXenes, Ti_2CO_2 exhibited better catalytic activity.

In the first step, CO_2 and proton chemisorb in an oxygen vacancy (OV) and a nearby site, respectively. To migrate toward the C site, a proton has to overcome an energy barrier (i.e., 0.45 eV), which results in an HCOO intermediate. In the second step, HCOOH (0.53 eV) is created due to the movement of another proton into the O site. The hydrogenation of HCOO is considered to be exothermic (0.51 eV) in contrast to the desorption of HCOOH (0.59 eV), which requires lower energy barrier. Ti_2CO_2 MXene is thus a suitable photocatalyst for CO_2 reduction due to its effective visible light absorption and bandgap (0.91 eV). DFT simulations were used to find the possible reaction route for the production for CH_4 from CO_2, for which the catalysts were observed to be Cr_3C_2 and Mo_3C_2. The active lowest energy indicated the chemisorption of methane across the surface of Mo_3C_2. By including four electron–hole pairs on the MXene surface, chemisorbed CH_4 was generated during a six-step hydrogenation reaction. The total energy barrier for this reaction is 1.31 eV. To convert CO_2 to methane, the minimum input required by Cr_3C_2 is 1.05 eV. The predicted energy barriers for O- and OH-terminated MXene (Mo_3C_2) are considerably lower, which shows that a humid environment could be effective [1, 3].

Another factor that influences the electrochemical properties of functionalized MXenes is structural defects. For functionalized $Mo_2TiC_2T_x$, the formation energies of these defects were observed to be dependent on the surface groups. The computed defects showed that MXenes functionalized with hydroxyl and fluorine require less energy than oxygen-functionalized ones. Researchers discovered that these defects are more likely to be produced in the outer layers of molybdenum than in the inner layers (containing titanium) [4]. According to the computed results, CO_2 utilizes a nonspontaneous reaction energy (0.21 eV) to be physisorbed on the surface of $Mo_2TiC_2O_2$ (as shown in Fig. 8.1a [4]). Furthermore, for $Mo_3C_2O_2$, the physisorption process was observed to exhibit a Gibbs free energy of 0.23 eV. The ability of defective $Mo_2TiC_2O_2$ MXenes to adsorb CO_2 was further investigated, and the results showed that these Mo-based materials are weakly reactive to capture CO_2. A qualitative understanding of the behavior of electrons over the surface along with electron localization is illustrated in Fig. 8.2 [4].

C. Cheng et al. [5] used first-principle simulations to explore the catalytic oxidation of pure and defective Cu_3-doped monolayers of Mo_2CO_2. The stability of catalyst was comprehensively demonstrated using geometry distortion, molecular dynamics simulations (at finite temperature), and energy analysis. To determine the catalytic ability, adsorption energy differences along with oxidation of both materials were examined under O_2 and CO atmospheres. They discovered that Cu_3/d-Mo_2CO_2 could be a superior catalyst for CO oxidation due to its high stability and reactivity. In another study, DFT calculations were used to analyze the catalytic characteristics of a single Pd atom with OV combined with pristine and defective Mo_2CO_2 monolayers. It was discovered that OV can stabilize the dopant (i.e., Pd atom), making the Pd/OV-Mo_2CO_2 combination an effective monodisperse atomic catalyst. The dopant served as the active center, causing the intermediates to react favorably around it. To investigate the catalytic activity, three reaction pathways were

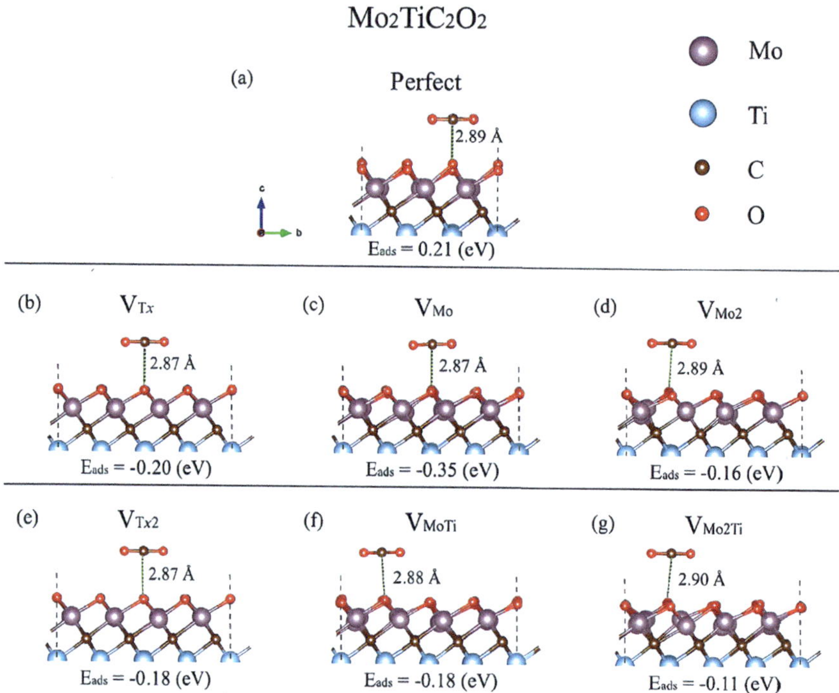

FIGURE 8.1 Relaxed adsorption configurations of a CO_2 molecule on the perfect and defected $Mo_2TiC_2O_2$ MXene. (a) Perfect MXene, (b) MXene–V_{Tx}, (c) MXene–V_{Mo}, (d) MXene–V_{MO2}, (e) MXene–V_{Tx2}, (f) MXene–V_{MoTi}, and (g) MXene–V_{Mo2Ti}.

Source: [4] Khaledialidusti R., Mishra A. K., Barnoush A. 2020. Atomic defects in monolayer ordered double-transition metal carbide ($Mo_2TiC_2T_x$) MXene and CO_2 Activation. *J. Mater. Chem. C*, 8, 4771–4779. Reproduced with permission from the Royal Society of Chemistry.

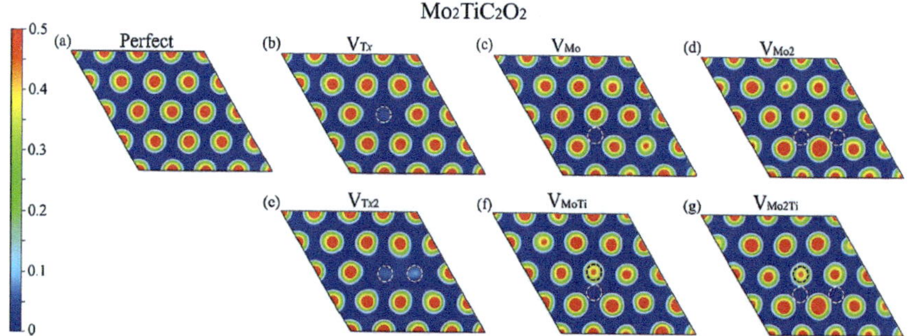

FIGURE 8.2 Electron localization function on the plane perpendicular to the c-axis at a close distance on top of surface terminations of the defected $Mo_2TiC_2O_2$.

Source: [4] Khaledialidusti R., Mishra A. K., Barnoush A. 2020. Atomic defects in monolayer ordered double-transition metal carbide ($Mo_2TiC_2T_x$) MXene and CO_2 Activation. J. Mater. Chem. C, 8, 4771–4779. Reproduced with permission from the Royal Society of Chemistry.

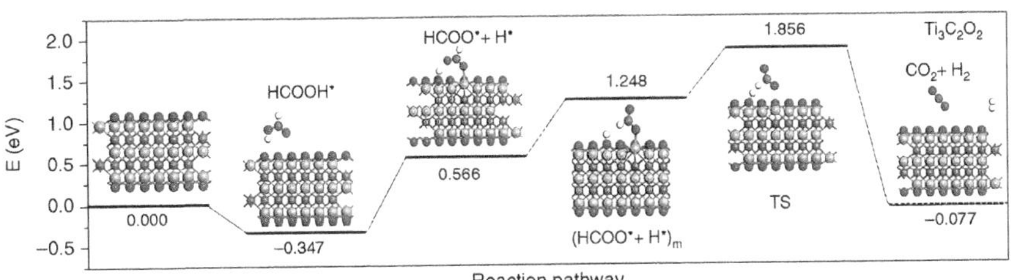

FIGURE 8.3 Reaction paths for the dehydrogenation of HCOOH over $Ti_3C_2O_2$. (* and subscript m represent the adsorbed species and meta-stable state, respectively).

Source: Reprinted with permission from [7] Springer Nature. *Nature Communications*. Tingting Hou et al. 2020. Modulating oxygen coverage of $Ti_3C_2T_x$ MXenes to boost catalytic activity for HCOOH dehydrogenation. 11: 4251, copyright 2020.

examined for CO oxidation, which showed better activity with a rate-limiting energy barrier (0.49 eV) [6].

Formic acid (HCOOH) is a promising hydrogen transporter as it is safe and renewable. In HCOOH dehydrogenation, better catalytic activity is reported by noble metal-based materials. However, formation of high-efficiency non-noble metal-based catalysts exhibits a significant difficulty. T. Hou et al. [7] increased the catalytic activity of $Ti_3C_2T_x$ MXenes by controlling the oxygen atoms on their surfaces. Without having any effect on the crystalline structure, MXenes treated with air at 250°C significantly increased the number of oxygen atoms on the surface. The resultant catalyst displayed 100% selectivity for H_2 (at 80°C) and a mass activity of 365 mmol g^{-1} h^{-1}, which is more than two times greater than that of Pd/C or Pt/C. In comparison to $Ti_3C_2O_2$, HCOOH displayed an adsorption energy of 0.35 eV along with dissociation of a hydrogen atom to generate a hydroxyl group (Fig. 8.3 [7]). Following that, HCOO* changed the adsorption arrangement to produce a metastable state with an enhanced energy (0.68 eV). The dissociation energy (E_a) for HCOO* over $Ti_3C_2O_2$ via a meta-stable state was only 0.61 eV, which was substantially lower than the E_a for Ti_3C_2 and $Ti_3C_2O_2$-V.

8.3 CATALYSTS FOR SYNTHETIC REACTIONS

The agricultural fertilizer ammonia (NH_3, as a fuel) has been reinstated as a possible alternative to petroleum. During combustion, the molecular energy bond between nitrogen and hydrogen will be released, producing only nitrogen and water (zero carbon emission). For traditional NH_3 synthesis, a Haber–Bosch process, which transforms H_2 and N_2 at high pressures and temperatures, is required. MXenes exhibit potential catalytic nitrogen reduction and capture activities, determined by their transition metals (i.e., d2: Zr, Hf, or Ti; d3: Ta, V, or Nb; or d4: Mo or Cr). The first phase of this process is the interaction between N_2 and the catalytic surface (i.e., N_2 capture), similar to CO_2 and H_2 capture. According to the theoretical calculations, N_2 utilizes a negative change in Gibbs energy (without any physisorption) to change its state from a gas phase to a chemisorbed one. The binding energies of V_3C_2 (−2.17 eV) and Nb_3C_2 (−2.44 eV) are considerable, which indicates a strong reactivity with N_2. In both theoretical and experimental investigations, the dissociation of nitrogen was observed to be the rate-limiting step in this process. When MXenes are used as catalysts, weakening of the N≡N bond and activation occur spontaneously [3].

This phenomenon was also confirmed by another theoretical study [8]. According to the DFT results, even more than H_2O and CO_2 capture, MXenes are potential materials for N_2 capture. Under mild circumstances, weakening of the N≡N triple bond on surface of the material occurs, which promotes the catalytic conversion into ammonia. When N_2 is chemisorbed on V_3C_2 and Nb_3C_2 NSs, they can activate spontaneously with extremely low overpotentials (i.e., 0.64 V and 0.90 V) vs. standard hydrogen electrode (SHE). According to the comparison of transition states and reaction energies, both these MXenes were observed to be most promising materials for nitrogen conversion applications. NH_3 was produced from the intermediate phases after the gain of sixth electron–hole pair and the highest reaction energies for Nb_3C_2 and V_3C_2 and materials were observed to be only 0.39 and 0.32 eV vs. SHE, respectively, as shown in Fig. 8.4 [8]. Cr_3C_2 and, to a lesser extent, Mo_3C_2 also exhibit good catalytic potential. It is important to note that V_3C_2 also displays a uniform series of

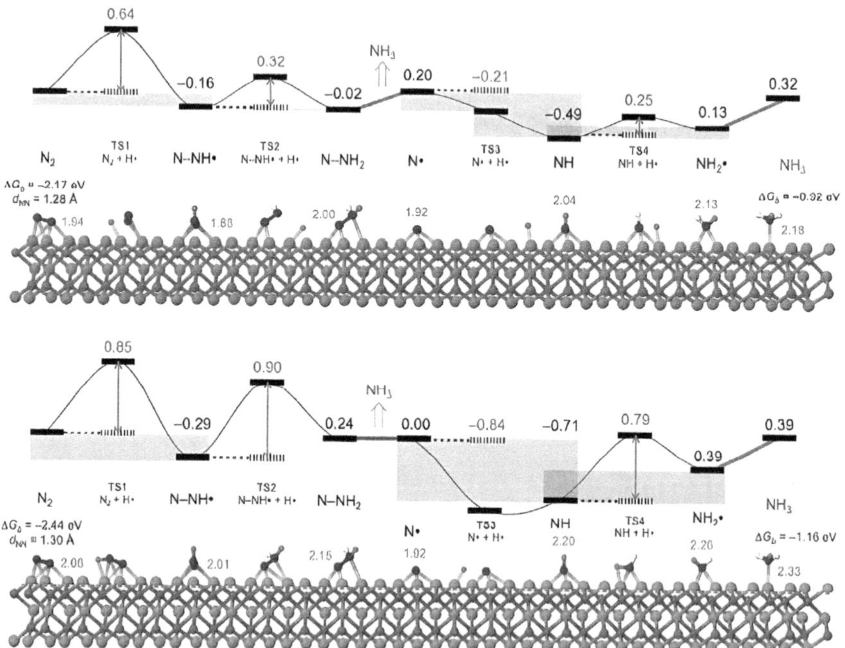

FIGURE 8.4 Minimum energy path for N_2 conversion into NH_3 catalyzed by V_3C_2 (top) and Nb_3C_2 (bottom) MXenes, calculated at the DFT + D3 computational level.

Source: [8] Azofra L. M., Li N., MacFarlane D. R., Sun, C. 2016. Promising prospects for 2D d²–d⁴ M_3C_2 transition metal carbides (MXenes) in N_2 capture and conversion into ammonia. *Energy Environ. Sci.*,9, 2545–2549. Reproduced with permission from the Royal Society of Chemistry.

reactions throughout the conversion process along with superior catalytic activity. For example, the transition states that lead to TS_2 and TS_4 have activation barriers of 0.32 and 0.25 eV, whereas for Nb_3C_2, these barriers are 0.90 and 0.79 eV.

8.4 CATALYSTS FOR THE REDUCTION OF CO_2

Anthropogenic CO_2 emissions produced by the unsustainable depletion of fossil fuels have disturbed the carbon balance of nature, resulting in environmental challenges such as the rise in sea level, global warming, and species extinction. These challenges can be overcome by recent developments in the utilization and conversion of CO_2. One of these environmentally friendly processes is CO_2 electroreduction reaction (CRR), which converts CO_2 in useful fuels and chemicals. MXenes have been explored for CRR using computer-aided screening or theoretical studies. Low-coordinated metals that are present on the surface of MXenes (i.e., three coordinated terminal metals) provide the active sites for the chemisorption process. For the conversion of CO_2 to CH_4, two MXenes, that is, Cr_3C_2 and Mo_3C_2 (with the structural composition M_3X_2), were observed to be the best possibility. The energy input for CRR over $Cr_3C_2T_x$ and $Mo_3C_2T_x$ could be further lowered with the inclusion of surface terminations such as $-O$ and $-OH$ compared to bare MXenes. However, the CRR over MXene-based catalysts is still in its initial phases when compared to that over other 2D nanomaterials [2, 9, 10].

N. Li et al. [11] also used well-resolved density functional theory (DFT) simulations to examine the composition M_3C_2 for CO_2 conversion. For the conversion process, of all the observed compositional structures, Cr_3C_2 and Mo_3C_2 were reported to be the most promising ones. During the early steps of hydrogenation, spontaneous reactions resulted in the production of radical species, that is, OCHO· and HOCO·. This results in atomic-level understandings of the catalysts and comprehension of the reduction of CO_2 into hydrocarbon fuels, which is critical for elucidating the underlying stages for CO_2 fixation. Figures 8.5 and 8.6 [11] provide the conversion pathway of CO_2 catalyzed by functionalized $Mo_3C_2T_x$. The Gibbs free energy exhibited by physisorption on $Mo_3C_2(OH)_2$ (0.35 eV) or $Mo_3C_2O_2$ (0.23 eV) is much greater than that on the bare MXene. It shows that CO_2 capture depends on the functionalization. For $Mo_3C_2(OH)_2$, a lower reaction energy (−0.92 eV) is measured in the first hydrogenation step, compared to the bare MXene (−0.64 eV). As a result, OH-functionalized material will behave as an efficient CO_2 conversion catalyst due to comparatively easy CO_2 hydrogenation. Meanwhile, to create a radical specie in $Mo_3C_2O_2$ (i.e., *COOH), CO_2 is captured with an absorption energy of 0.49 eV (Fig. 8.6 [11]).

In the CO_2 reduction process, in addition to producing and separating electron–hole pairs during photoexcitation, the anatase surface also helps in the adsorption of CO_2, thereby allowing the transfer of protons and electrons from the surface to CO_2. Either CO or HCOOH is produced due to the reduction of CO_2 at 2e. The reaction paths to HCOOH use initial 1e and 2e reduction reactions. In a pathway involving 1e reduction, an electron is transmitted to the CO_2 adsorbate from the surface of TiO_2, resulting in the activated CO_2^- anion radical exhibiting an activation barrier (0.87 eV). As a result, it appears that this is the rate-limiting step in the pathway to HCOOH [12].

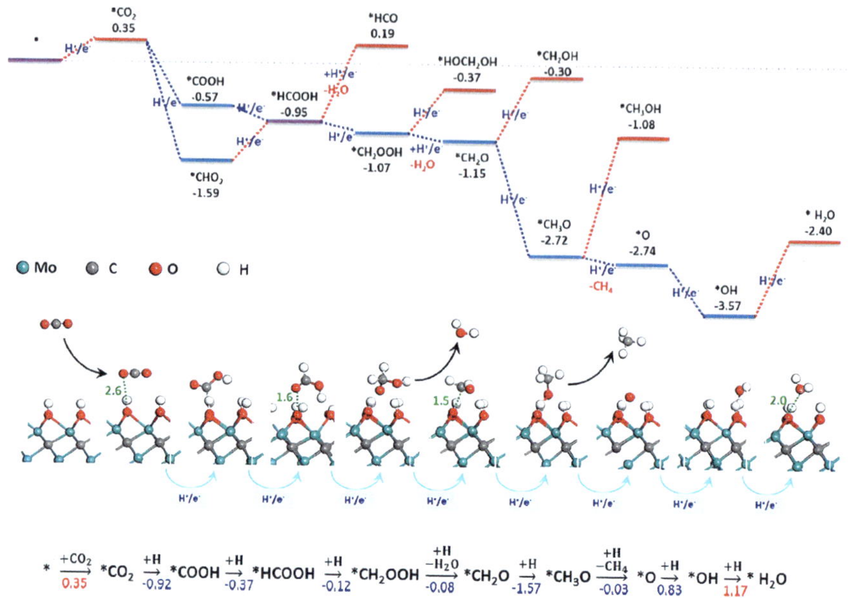

FIGURE 8.5 Minimum energy pathway for CO_2 conversion into CH_4 and H_2O over $Mo_3C_2(OH)_2$. (Top) Gibbs free energies along the pathway. (Middle) Side view and some selected distances are shown in Å. (Bottom) Corresponding formula with Gibbs free energy.

Source: Reprinted with permission from [11] Neng Li, Xingzhu Chen, Wee-Jun Ong, et al. 2017. Understanding of Electrochemical Mechanisms for CO_2 Capture and Conversion into Hydrocarbon Fuels in Transition-Metal Carbides (MXenes). *ACS Nano*, 11, 11, 10825–10833. Copyright 2017, American Chemical Society.

A.D. Handoko et al. [13] theoretically introduced a new CO_2 reduction route with −H coordination for O-terminated MXene-based catalysts due to weaker *CO binding. They developed new scaling relations depending on the alternating coordination of −C and −H intermediates along the minimal energy pathway. For C-bonded intermediates, the binding energies were discovered to be significantly associated, in such a way that the E_b of *COOH is linearly related to that of *CHO, *CH₂OH, and *CH₃, as shown in Fig. 8.7a [13]. Similarly, the E_b of *HCOOH was also observed to be linearly related to that of *H₂CO and *HOCH₃ in the case of H-bonded intermediates (Fig. 8.7b). Furthermore, Fig. 8.7c shows that the binding energies of *COOH and *HCOOH are decoupled, since they are attached to the surface of MXenes.

X. Zhang et al. [14] studied CO_2 reduction at OVs on V_2CO_2, Ti_2CO_2, and $Ti_3C_2O_2$ monolayers. The chemical route that exhibited the energy barrier of 0.53 eV was observed to be the most favorable one. The energy barriers of other single-carbon organic compounds' reaction pathways were found to be substantially greater, suggesting exceptional selectivity of HCOOH. Four compounds were considered: HCHO, CH_3OH, HCOOH, and CH_4. They also proposed that CO and H_2 can add enough OVs to the MXenes' surface. Among all of the observed MXenes, Ti_2CO_2 demonstrated the best catalytic performance.

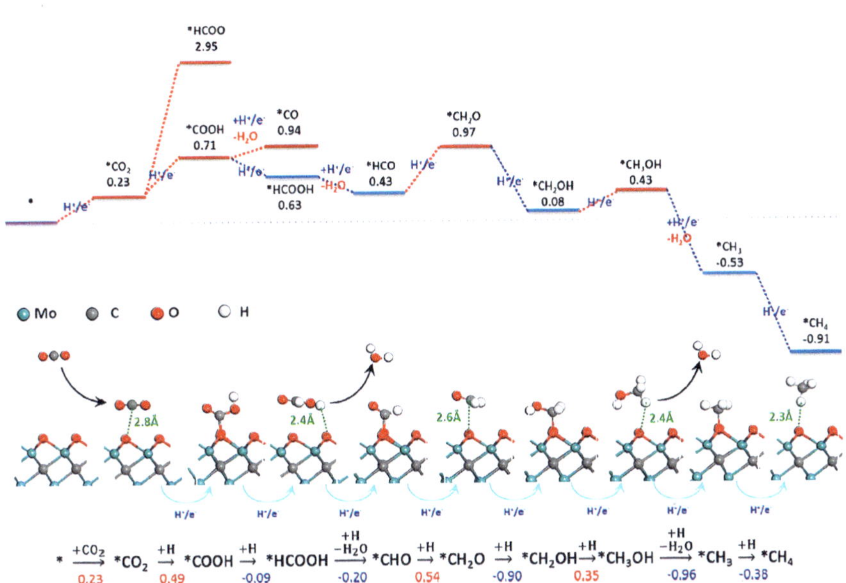

FIGURE 8.6 Minimum energy pathway for CO_2 conversion into CH_4 and H_2O over $Mo_3C_2O_2$. (Top) Gibbs free energies along the pathway. (Middle) Side view and some selected distances are shown in Å. (Bottom) Corresponding formula with Gibbs free energy change is shown in eV.

Source: Reprinted with permission from [11] Neng Li, Xingzhu Chen, Wee-Jun Ong, et al. 2017. Understanding of Electrochemical Mechanisms for CO_2 Capture and Conversion into Hydrocarbon Fuels in Transition-Metal Carbides (MXenes). *ACS Nano*, 11, 11, 10825–10833. Copyright 2017, American Chemical Society.

8.5 CATALYSTS FOR NRR

Ammonia is a type of chemical material that is used in industrial and agricultural production, as well as in energy conversion and storage. In industries, ammonia is being synthesized using the conventional Haber–Bosch process, which requires high pressures and temperatures. This results in the emission of greenhouse gases along with the consumption of a lot of energy. In comparison to this approach, ammonia can also be produced by using the NRR. This method allows the synthesis of ammonia under mild conditions, as there are abundant sources of nitrogen and water. In recent years, NRR has received a lot of attention by researchers. NRR involves lower reaction rates and coulomb efficiency; therefore, selection of the catalyst should be done carefully to produce more ammonia. MXenes are observed to be the efficient NRR catalysts due to their high electronic conductivity. The surface terminations of MXenes play an important role in the conversion of nitrogen to ammonia [1, 2, 15].

There are a number of researches that display the NRR electrocatalytic performance of MXenes. For example, L. Li et al. [16] used DFT calculations to demonstrate the electrocatalytic behavior of transition metal (TM)-doped $Mo_2TiC_2O_2$. The results showed that doping of TMs (i.e., Ta, Zr, W, Hf, Re, etc.) can greatly accelerate

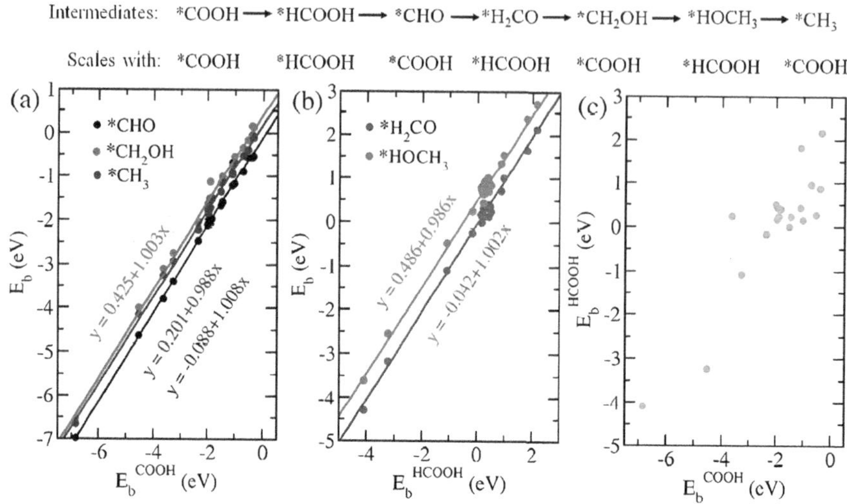

FIGURE 8.7 Binding energies of CRR intermediates on M_2XO_2 MXene surfaces: (a) *CHO, *CH$_2$OH, and *CH$_3$ plotted as a function of *COOH. (b) *H$_2$CO and *HOCH$_3$ plotted as a function of *HCOOH. (c) *HCOOH plotted as a function of *COOH.

Source: [13] Handoko A. D., Jin H. et al. 2018. Establishing New Scaling Relations on Two-Dimensional MXenes for CO$_2$ Electroreduction. *J. Mater. Chem. A*, 6, 21885–21890. Reproduced with permission from the Royal Society of Chemistry.

the performance of MXenes for the NRR process. Among them, the Mo$_2$TiC$_2$O$_2$-ZrSA exhibited the lowest potential-determining step barrier (0.15 eV) and good selectivity over HER competition. H. Luo et al. [17] theoretically investigated the feasibility of employing P-, N-, S-, F-, or Cl-doped Fe/Ti$_3$C$_2$O$_2$ or Fe/Ti$_3$C$_2$O$_{2-x}$ as a NRR catalyst. Gibbs free energy and adsorption energies were determined to explain the limiting potential reduction and activation of N$_2$.

Figure 8.8 [17] illustrates the most likely reaction pathways for several catalysts. Furthermore, nonmetal doping has a significant effect on NRR. When compared to the non-doped structure, E$_{ads}$ is lower for N$_2$ adsorption, which may explain why NH$_3$ desorption is easier in the final step. Doping makes the hydrogenation process phases easier or more difficult. N, P, and Cl doping makes it difficult for *N$_2$ to generate *NNH, whereas F and S doping facilitates the synthesis of the *NNH phase. Meanwhile, the doping of P causes the conversion of *NNH to *NNH$_2$ to occur spontaneously. Due to the exothermic nature, nonmetal doping has no effect on *NNH$_2$ → *NHNH$_2$ → *NH$_2$NH$_2$ → *NH$_2$NH$_2$ → *NH$_2$NH$_2$ → *NH$_2$NH$_3$. Of all the structures, Fe/S-Ti$_3$C$_2$O$_{2-x}$ and Fe/F-Ti$_3$C$_2$O$_{2-x}$ were observed to be stable catalysts.

Similarly, another DFT study demonstrated that B-doped W$_2$CO$_2$ and Mo$_2$CO$_2$ are excellent NRR catalysts due to their limiting potentials of 0.24 and 0.20 V, respectively [18]. It was discovered that the strong B-N bonding can accelerate the hydrogenation of nitrogen along with the conversion of *NH$_2$ to *NH$_3$. Y. Luo et al. [19] reported that a metal host combined with Ti$_3$C$_2$T$_x$ NSs results in 5.78%

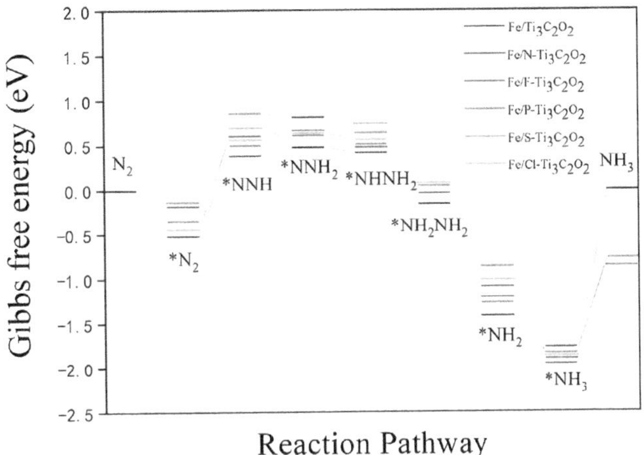

FIGURE 8.8 NRR pathways for all structures.

Source: Reprinted from [17] H. Luo, X. Wang, C. Wan, et al.: A Theoretical Study of Fe Adsorbed on Pure and Nonmetal (N, F, P, S, Cl)-Doped $Ti_3C_2O_2$ for Electrocatalytic Nitrogen Reduction. *Nanomaterials* 2022, 12(7), 1081. Open access.

faradic efficiency at an ultralow voltage for the conversion of NH_3, under ambient settings. This study demonstrated that the catalytic activity of MXenes depends on their active sites, which shows that adjusting the active sites and delaying hydrogen evolution activity can considerably increase NH_3 electrosynthesis selectivity. Five potential mechanisms were investigated, as shown in Fig. 8.9a [19], with two feasible low-energy paths (solid arrows). Similarly, the computations and experimental results support the existence of a low-energy channel on the edge plane (Fig. 8.9b). A comparison of the reaction energy in the routes reveals that the NRR is most active in the center Ti on the edge plane. For the first TS, particularly [M-Ti]*N_2 phase necessitates an activation barrier (0.64 eV) to create [M-Ti]*NNH. According to the least energy route of (1), the rate-determining step of the entire process was observed to be the addition of H to nitrogen.

NRR is both fuel-efficient and environmentally benign; however, the majority of the NRR catalysts described are oxides or noble metals, which are inefficient and expensive. MXene-based quantum dots (QDs) that are being synthesized exhibit better NRR catalytic activity [20]. Similarly, functionalized $Ti_3C_2T_x$ were reported as efficient catalysts for nitrogen reduction and production of ammonia [21]. The catalysts achieved structural and electrochemical stability in 0.1 M HCl, along with the 9.3% faradic efficiency at extremely low overpotentials (i.e., −0.4 V). According to the DFT studies, the distal NRR mechanism was observed to be more favorable, and the rate-limiting phase was the conversion of ammonia (i.e., the final step). The reaction pathways are shown in Fig. 8.10 [21]. The entire reaction was partitioned into two half reactions at the distal mechanism. The first half reaction is exothermic with free energy transfer (2.68 eV) from N_2 to *N.

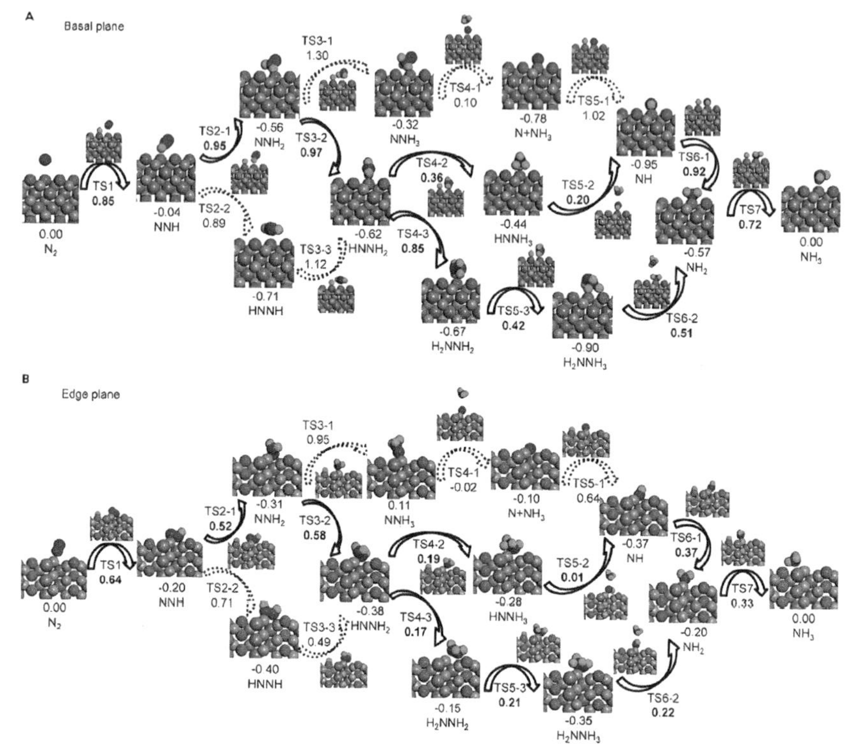

FIGURE 8.9 Comparison of the reaction pathways on the basal plane (a) and the edge plane (b).

Source: Reprinted from [19] *Joule*, 3, Y. Luo et al., Efficient Electrocatalytic N_2 Fixation with MXene under Ambient Conditions, 279–289, Copyright 2019, with permission from Elsevier.

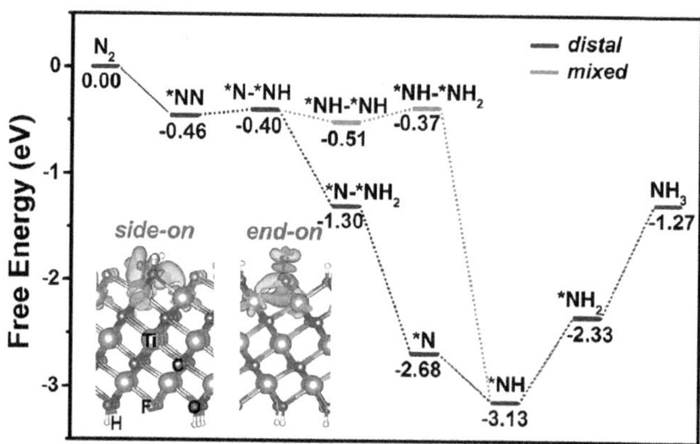

FIGURE 8.10 DFT-calculated energy profile for electrocatalytic NRR on $Ti_3C_2T_x$.

Source: [21] Zhao J., Zhang L. et al. 2018. $Ti_3C_2T_x$ (T = F, OH) MXene nanosheets: conductive 2D catalysts for ambient electrohydrogenation of N_2 to NH_3. *J. Mater. Chem. A*, 6, 24031–24035. Reproduced with permission from the Royal Society of Chemistry.

8.6 CATALYSTS FOR HYDROGEN EVOLUTION REACTION

Electrocatalytic water splitting is a cost-effective method that utilizes renewable energy sources. In this process, H is evolved at the electrode when a bias is provided between two electrodes. Usually, Pt is considered an optimum HER catalyst as it exhibits low Gibbs free energy (ΔG_H^*) for adsorption of hydrogen. According to the Bronsted–Evans–Polanyi equation, this energy is directly related to HER activation; thus, it is considered as a rate-determining step. However, Pt has certain limitations regarding the cost and lack of availability. As a result, different researches were carried out to find the ideal substitute for Pt, having a hydrogen adsorption energy close to zero. Theoretical and experimental studies reveal that MXenes are potential HER catalysts [1–3, 22]. In contrast to other 2D materials, these structures exhibit a high density of active catalytic sites. To increase the HER activity of MXenes, multiple strategies were used, including structural engineering, altering the surface terminal groups, doping, and combining with other active materials. Among these methods, doping is the most feasible option because of its capacity to naturally optimize the electrical structure. Doping with sulfur and nitrogen can greatly improve the efficiency of Mo_2C nanosheets (NSs), according to research [1].

TM modification is a powerful technique for improving the catalytic capabilities of MXenes in HER. To prevent the atomic aggregation associated with typical TM modification, researchers implanted TM atoms into MXenes. DFT simulations were used to investigate the catalytic characteristics of TM-doped M_2CO_2 (where M = Zr, Ti, Ta, Hf, and V). The results demonstrated that doping can modify the Gibbs free energies and the conductivities of MXenes. Some MXenes containing TM atoms could attain good HER catalytic activity and metallic conductivity at the same time [23]. Another work used well-defined DFT calculations to investigate the effect of heteroatom (N, P, S, or B) doping on the HER performance of bare or O-functionalized M_2C (where M = Mo or Ti). The X-doped functionalized M_2CT_2 outperformed the HER activity of X-doped pure M_2C. Furthermore, computed ΔG_H showed that N-doped Ti_2CO_2 has better electrocatalytic behavior [24]. In another study, it was discovered that Fe-, Ni-, or Co-doped oxidized V_2CO_2 exhibits extraordinary HER catalytic activity [25]. The first-principle calculations revealed that the addition of one of these metals results in significantly weakening of the strong hydrogen–oxygen bond, thereby attaining the optimal value (0 eV) of hydrogen adsorption-free energy. This is done by selecting the suitable type and exposure of the active sites along with the promoters. Figure 8.11 [25] depicts the HER performance of TM-absorbed V_2CO_2.

In order to study the structural stability and HER electrocatalytic performance, Y. Cheng et al. [26] used the DFT simulations for carbon vacancy engineering and variation in the TMs of Cr_2CO_2. According to these findings, the electrical conductivity of pure and functionalized Cr_2C makes it an efficient HER catalyst. Furthermore, the adsorption-free energy can be tuned to an ideal value (i.e., 0 eV) by an adequate TM coverage modification. X. Zou et al. [27] suggested that Cr_2CTx synthesized via etching of the corresponding MAX phase would exhibit minimum Al_2O_3, and it results in better HER performance than Pt-based catalysts (at high current density). Furthermore, the remarkable stability for more than 160 hours (without attenuation)

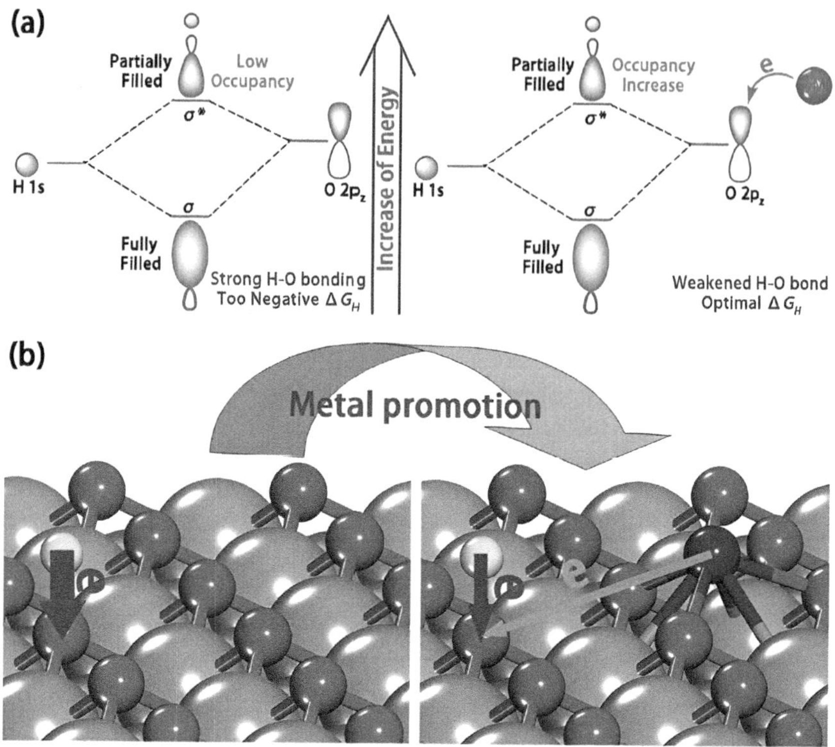

FIGURE 8.11 HER performance of TM-absorbed V_2CO_2. a) The combination of H 1s orbital and O 2pz orbital forms a fully filled bonding orbital and a partially filled antibonding orbital. b) Charge transfer from H to O will occur when H adsorbs on O.

Source: [25] J. Wang, Q. Chen, et al.: Transition Metal-Promoted V_2CO_2 (MXenes): A New and Highly Active Catalyst for Hydrogen Evolution Reaction. *Adv. Sci.* 2016, 1600180. Copyright WILEY-VCH Verlag GmbH & Co. KGaA, Weinheim. Reproduced with permission.

was found to be superior to any other MXene-based catalyst performance. G. Gao et al. [28] revealed that under typical conditions, a few MXenes contain a mixture of −OH and −O groups (i.e., V_2C, Ti_2C, and Ti_3C_2), while Nb_2C and $Nb_4C_3O_2$ are fully O-functionalized.

Furthermore, in typical circumstances, all of these MXenes were conductive, allowing for significant charge transfer kinetics. The hydrogen adsorption free energies (ΔG_{H*}^{0}) of $M_nC_{n-1}O_2$ were calculated at different coverages, as shown in Fig. 8.12 [28]. The active sites for HER were observed to be provided by two surface O* layers. There are eight O* atoms in a $M_nC_{n-1}O_2$ supercell, and simulations of varied H can be carried out by considering an increment of 1/8. The larger −ve value of ΔG_{H*}^{0} shows that the oxygen and H* interactions are strong at lesser coverages (1/8 and 2/8), preventing further hydrogen release from the surface of the catalyst. Furthermore, ΔG_H^{*0} approaches zero as these coverages extend to 3/8 and 4/8, as seen in Figs. 8.12b and d. In contrast, Nb_2CO_2 and $Nb_4C_3O_2$ (as illustrated in Figs. 8.12c and e) exhibit better catalytic activity at relatively low coverages (i.e., 1/8 and 2/8, respectively).

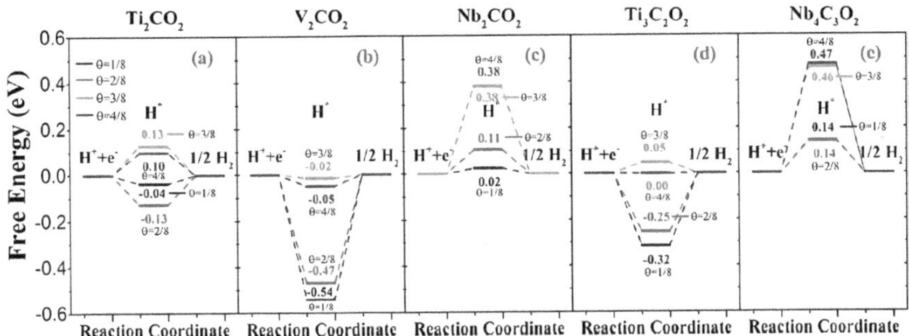

FIGURE 8.12 Free energy diagram of HER processing on Ti_2CO_2, V_2CO_2, Nb_2CO_2, $Ti_3C_2O_2$, and $Nb_4C_3O_2$ under standard conditions.

Source: Reprinted with permission from [28] G. Gao, A. P. O'Mullane, A. Du. 2017. 2D MXenes: A New Family of Promising Catalysts for the Hydrogen Evolution Reaction. *ACS Catal.*, 7, 494–500. Copyright 2017, American Chemical Society.

Another study reported a hybrid $Nb_2O_5/C/Nb_2C$ as an effective HER catalyst for water splitting. The photogenerated charge carrier's separation at the hybrid's interface and interaction between Nb_2O_5 and MXene resulted in improved HER performance [29]. The favorable effect of cobalt doping on the redox characteristics of Mo_2CT_x: Co was also observed, resulting in significantly increased HER activity compared to that of pure catalyst (without doping) [30]. Researchers studied the catalytic activity of ten mono-metal carbides, and among all the materials, W_2CO_2 and Ti_2CO_2 were discovered to be active HER catalysts [31]. Similarly, due to electrocatalytic activity and large surface area, B, S-$Ti_3C_2T_x$ NSs were also discovered to be efficient electrocatalysts for HER. These catalysts exhibited a low Tafel slope of ~54 mV dec^{-1} along with a low overpotential for the HER (−110 mV) vs. reversible hydrogen electrode (RHE) [32].

Z. W. Seh et al. [33] conducted a computational screening investigation and predicted the catalytic activity of bare and functionalized Mo_2CT_x. The drop-casting method was used to develop Ti_2CT_x and Mo_2CT_x onto glassy carbon electrodes with Nafion as a binder to test the electrocatalytic activity for HER. These materials were then tested in 0.5 M H_2SO_4(aq) in a three-electrode electrochemical cell with a rotating disc electrode device at 1,600 rpm. The linear sweep voltammograms (LSVs) for carbon and MXenes are shown in Fig. 8.13a [33]. Within the desired potential window, there is no detectable action in the bare glassy carbon. Ti_2CT_x has some HER activity; however, it requires high values of overpotential (609 mV) that make it a weak catalyst.

Structural investigations on N-Mo_2C NSs revealed that the surfaces of these NSs are surrounded by apical Mo atoms, providing a suitable catalyst prototype to investigate the role of these atoms during HER catalysis [34]. HER activity was found in N-Mo_2C NSs with a Tafel slope (i.e., 44.5 mV dec^{-1}) and an extremely low overpotential (99 mV vs. RHE) at 10 mA cm^{-2} with exceptional long-term stability. Similarly, P-doped materials also possess better performance as HER electrocatalysts [35]. The P-Mo_2C@C NWs with controlled 2.9% P-doping delivered a low overpotential (i.e.,

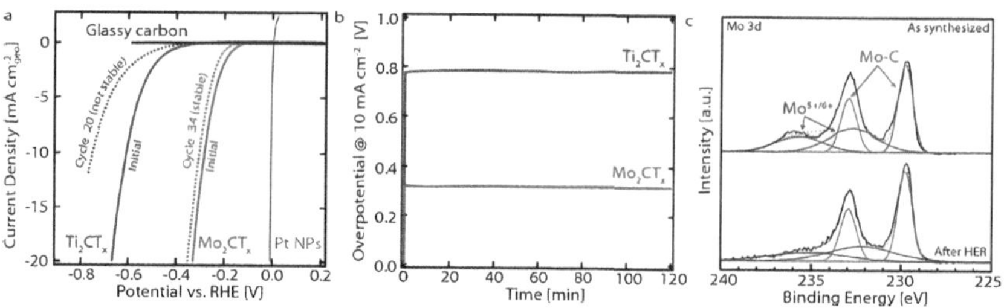

FIGURE 8.13 (a) Anodic-going iR-corrected LSVs of Mo_2CT_x and Ti_2CT_x, with bare glassy carbon. (b) Evolution of the potential required to maintain 10 mA cm^{-2} geo. (c) Mo 3d XPS spectra of Mo_2CT_x.

Source: Reprinted with permission from [33] Zhi Wei Seh, Kurt D. Fredrickson, Babak Anasori, et al. 2016. Two-Dimensional Molybdenum Carbide (MXene) as an Efficient Electrocatalyst for Hydrogen Evolution. *ACS Energy Lett.*, 1, 589–594. Copyright 2016, American Chemical Society.

89 mV at ~10 mA cm^{-2}) in acidic electrolytes. Another work demonstrated the formation of single-crystalline Mo_2C NSs with thicknesses ranging from 1.0 to 1.2 nm and a large specific surface area (122–153 m^2 g^{-1}). When compared to bulk-layered Mo_2C, this unique NS structure provided a shorter electron transport channel during HER, resulting in a 15-mV reduction in overpotential [36]. Similarly, P-Mo_2C/Ti_3C_2@NC can likewise function as a high-performance electrocatalyst [37].

2D $Ti_4N_3T_x$ MXene was discovered to exhibit semiconductor and metallic characteristics along with the excellent HER activity, in terms of a Tafel slope of ~190 mV dec^{-1} and low overpotential (~300 mV at −10 mA cm^{-2}) [38]. Recently, DFT calculations were used to screen 24 double transition metal (DTM) carbides with compositions $M_2'M''C_2T_x$ and $M_2'M_2''C_3T_x$ to determine the suitable HER catalyst. $Mo_2NbC_2O_2$ had the lowest overpotential among the 18 carbides projected to be active electrocatalyst candidates [39]. Figure 8.14a [39] depicts the HER activity volcano plot. Fully O-functionalized surfaces of MXenes with Mo and Ti as the outermost metals have been observed to exhibit considerable hydrogen adsorption, and the majority of them are active around the plots. MXenes with Cr and V as the outermost metals and that are O-functionalized have strong H and MXene interaction, while MXenes with a mixed OH and O termination have a reduced HER potential. $Nb_2Ta_2C_3$ has insufficient adsorption energetics, indicating that it is unsuitable for use as a HER catalyst. Figure 8.14b depicts an extensive view of the volcano plots, displaying 18 MXenes with overpotentials of less than 0.2 V, including those indicated as prospective HER candidates.

In another study, the theoretical HER mechanism of different O-functionalized $M_2'M''C_2O_2$ (where M′ = Mo or Cr; M″ = V, Ti, Ta, or Nb) was investigated [40]. DFT computations were used to objectively compare the related monometallic $M_3'C_2O_2$. Based on the theoretical results, creating a "sandwich-like" ordered DTM arrangement can significantly increase the HER performance of M-MXenes. Furthermore, the catalytic activity of Mo-based materials (i.e., $Mo_2M''C_2O_2$) outperforms that of $Cr_2M''C_2O_2$, which indicates the superior catalytic performance of $Mo_2NbC_2O_2$ and $Mo_2VC_2O_2$, compared to that of Pt(111). Z. Zeng et al. [41] used well-defined DFT

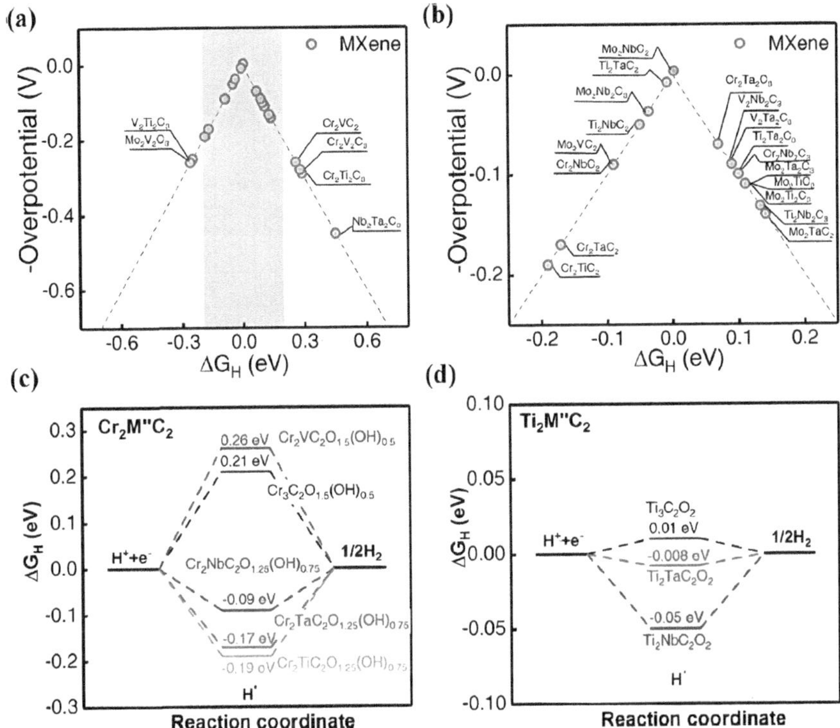

FIGURE 8.14 (a) HER activity volcano plot with theoretically calculated overpotentials for MXenes. (b) Zoomed-in portion of the top of the volcano with 18 compositions. (c) Calculated free energy diagram for HER of Cr-based materials. (d) Calculated free energy diagram for HER of Ti-based materials.

Source: Reprinted with permission from [39] Di Jin, Luke R. Johnson, Abhinav S. Raman, et al. 2020. Computational Screening of 2D Ordered Double Transition-Metal Carbides (MXenes) as Electrocatalysts for Hydrogen Evolution Reaction. *J. Phys. Chem. C*, 124, 10584–10592. Copyright 2020, American Chemical Society.

calculations to examine the HER activity of all 64 O-functionalized ordered DTM carbonitrides with composition $M'_2M''CNO_2$ (where M' and $M'' = $ V, Ti, Cr, Nb, Zr, Hf, Mo, or Ta). Figure 8.15a [41] illustrates the H coverage of Ti_2ZrCNO_2 (i.e., 1/4, 1/2, 3/4, and 1). Figure 8.15b depicts the link between Gibbs' free energy and H coverage for the 11 $M'_2M''CNO_2$ MXenes. With increased H coverage, GH shows an overall rising trend. The majority of the tested $M'_2M''CNO_2$ MXenes exhibited electrocatalytic activity only when the H coverage is 1/4.

8.7 CATALYSTS FOR OXYGEN EVOLUTION REACTION

OER is an important electrocatalytic energy conversion process with comparatively slow kinetics, thereby demanding the use of an effective electrocatalyst, having the ability to lower the energy barrier. The use of noble metal-based oxides, which exhibit

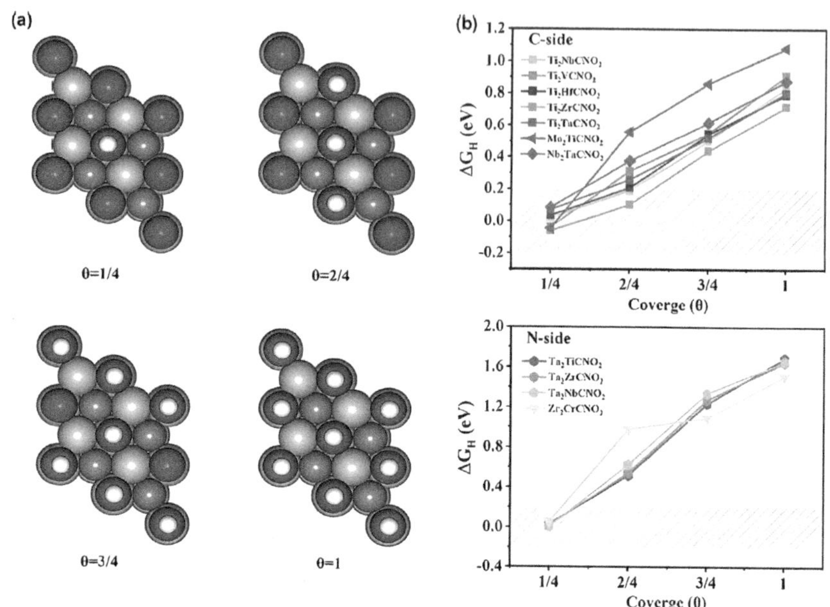

FIGURE 8.15 Top view of the H coverage and the ΔGH with different H coverages for 11 M′₂M″CNO₂ MXenes. (a) Top view of the H coverage for the C-side of Ti₂ZrCNO₂. (b) Calculated ΔGH with different H coverages for the C- or N-side of 11 M′₂M″CNO₂ MXenes.

Source: Reprinted by permission from [41] Springer Nature. *npj Computational Materials*. Zhoulan Zeng et al. 2021. Computational screening study of double transition metal carbonitrides M′₂M″CNO₂-MXene as catalysts for hydrogen evolution reaction. 80, copyright 2021.

good OER performance, is limited by their lack of resources and high cost. As a result, it is critical to design highly active catalysts capable of replacing these compounds. At the moment, MXene-based materials are a center of discussion due to their possible applications in OER [1, 2, 22]. Through DFT calculations, Y. Cheng et al. [42] reported the bifunctional electrocatalytic behavior of a single transition metal atom combined with the surface of Cr_2CO_2. The results demonstrated that Ni/Cr_2CO_2 MXenes exhibit better HER and OER catalytic activity with low overpotentials, that is, 0.16 V and 0.46 V, respectively (Fig. 8.16 [42]). On the surface of MXenes, the binding strength was enhanced by the transfer of a large numbers of electrons from Ni. Furthermore, ab initio molecular dynamics (AIMD) simulations indicated that the aggregation to form clusters can be prevented by the stable immobilization of Ni atoms on the MXene substrate. Figure 8.17 [42] illustrates the results of the OER performances of TM/Cr_2CO_2.

Y. Tang et al. [43] described a simple and controllable method for producing $Ti_3C_{1.6}N_{0.4}$ and $Ti_3C_{1.8}N_{0.2}$ flakes, by using an in situ nitrogen solid solution along with an etching procedure, and observed their corresponding OER electrocatalytic performance. Similarly, in another study, the OER activities of $Ti_3C_2T_x$ and $Ti_3C_2T_x$-N_y in 1 M KOH electrolyte were investigated [44]. One-step NH_3/Ar plasma treatment

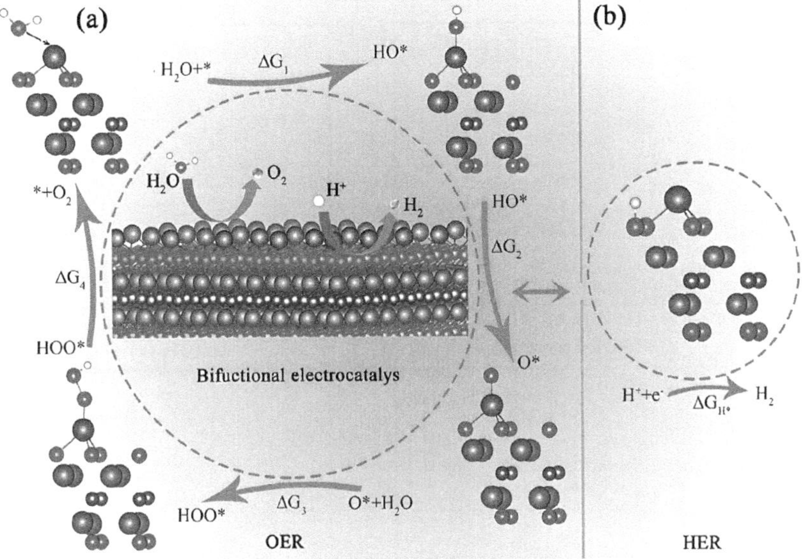

FIGURE 8.16 Schematic of TM/Cr$_2$CO$_2$ surface as bifunctional catalyst. Elementary reaction structures of (a) OER and (b) HER.

Source: Reprinted with permission from [42] Yuwen Cheng, Jianhong Dai, Yan Song, et al. 2019. Nanostructure of Cr$_2$CO$_2$ MXene Supported Single Metal Atom as an Efficient Bifunctional Electrocatalyst for Overall Water Splitting. *ACS Appl. Energy Mater.*, 2, 6851–6859. Copyright 2019, American Chemical Society.

was used to synthesize these materials, which allows modification of the nitrogen content and structural properties of materials by changing the volume ratio of Ar and NH$_3$ in the plasma. Figure 8.18a [44] depicts the polarization curves of all samples, while Fig. 8.18b depicts the OER overpotential necessary for J = 100 mA cm^{-2} in Fig. 8.18a. The overpotential of all Ti$_3$C$_2$T$_x$-N$_y$ was observed to be lower than that of Ti$_3$C$_2$T$_x$, as shown in Fig. 8.18b, which indicates that the electrocatalytic activity is improved by N doping. Ti$_3$C$_2$T$_x$-N$_6$ also exhibited a Tafel slope of 76.68 mV dec^{-1}, indicating the quickest reaction kinetics (as shown in Fig. 8.18c). According to the time-dependent potential (Fig. 8.18d), the relative current can sustain around 73% of its starting value after long-term stability measurement, and Ti$_3$C$_2$T$_x$-N$_6$ has demonstrated greater cycling stability than Ti$_3$C$_2$T$_x$.

J. Liu et al. [45] described a fast chemical reaction method for producing hierarchical Co-Bi/Ti$_3$C$_2$T$_x$ at room temperature. During the reaction, hierarchical structure of the hybrid material assists in shortening the charge transfer pathways and producing more active sites. For quick redox processes, the strong interaction of the MXene with Co-Bi enhances the electrostatic attraction of more anionic intermediates. As a result, this hybrid material demonstrated exceptional OER catalytic activity, with a 250-mV overpotential delivering a current density of 10 mA cm^{-2} and a Tafel slope of roughly 53 mV dec^{-1}. In another study, a highly effective oxygen electrode was created using a hybrid film (TCCN) comprising g-C$_3$N$_4$ combined with MXene NSs.

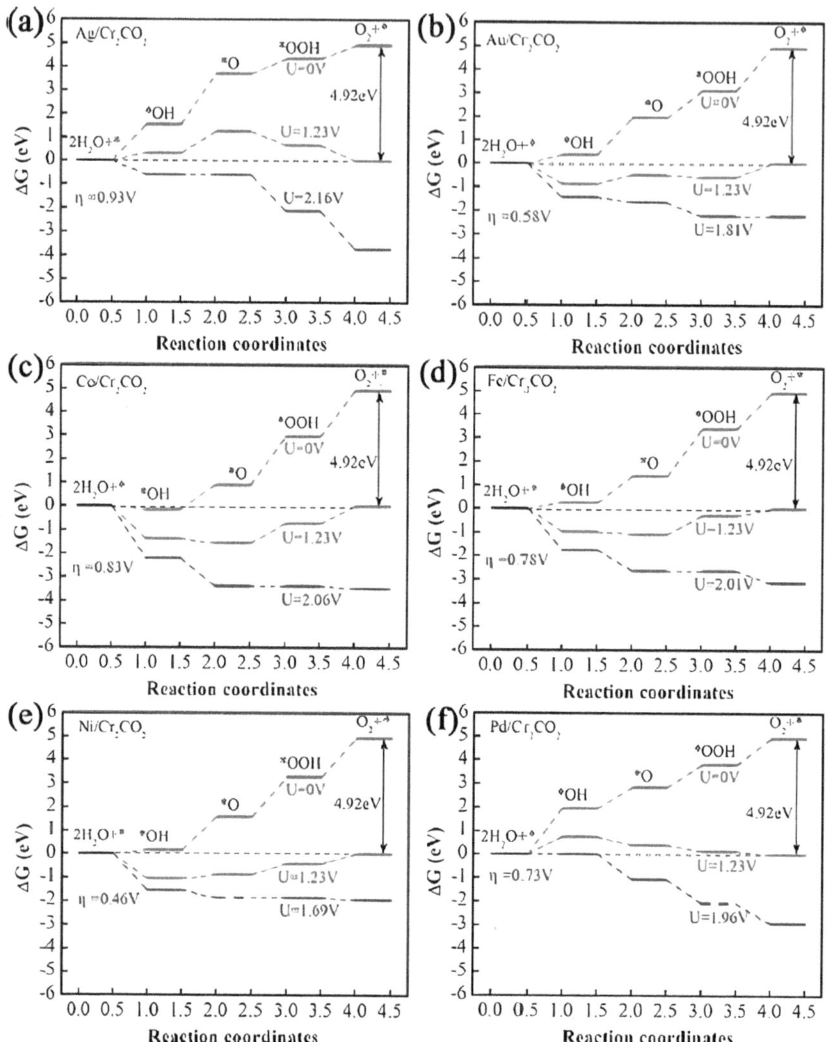

FIGURE 8.17 Standard free energy diagrams for OER under different electrode potentials: (a) Ag/Cr$_2$CO$_2$, (b) Au/Cr$_2$CO$_2$, (c) Co/Cr$_2$CO$_2$, (d) Fe/Cr$_2$CO$_2$, (e) Ni/Cr$_2$CO$_2$, and (f) Pd/Cr$_2$CO$_2$.

Source: Reprinted with permission from [42] Yuwen Cheng, Jianhong Dai, Yan Song, et al. 2019. Nanostructure of Cr$_2$CO$_2$ MXene Supported Single Metal Atom as an Efficient Bifunctional Electrocatalyst for Overall Water Splitting. *ACS Appl. Energy Mater.*, 2, 6851–6859. Copyright 2019, American Chemical Society.

Ti$_3$C$_2$ is connected with g-C$_3$N$_4$ via Ti–Nx interaction, resulting in a hydrophilic and porous freestanding film that shows outstanding OER [46].

L. Zhao et al. [47] developed a hybrid material containing 2D cobalt 1,4-benzene-dicarboxylate with Ti$_3$C$_2$T$_x$ NSs, by using an in situ interdiffusion reaction-assisted technique. In the OER, the resultant hybrid material exhibited a Tafel slope of 48.2 mV dec^{-1} (in 0.1 M KOH) at 1.64 V vs. RHE. Another study used a

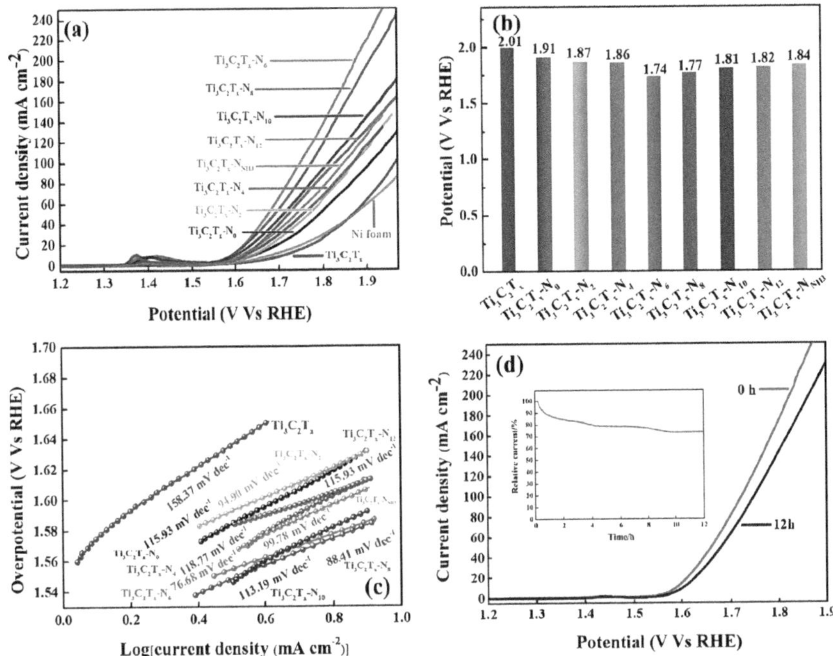

FIGURE 8.18 (a) LSV curves at a scan rate of 5 mV s^{-1}. (b) OER overpotential required for J = 100 mA cm^{-2}. (c) Corresponding Tafel plots. (d) Comparison of OER overpotential and initial value after the long-term stability test.

Source: Reprinted from [44] *Chemical Engineering Journal*, 420, X. Chen et al., Tunable nitrogen-doped delaminated 2D MXene obtained by NH$_3$/Ar plasma treatment as highly efficient hydrogen and oxygen evolution reaction electrocatalyst, 129832, Copyright 2021, with permission from Elsevier.

metalorganic framework to combine a novel NiCoS on Ti$_3$C$_2$T$_x$. The hybrid exhibited an increased surface area with high charge transfer conductivity and hence superior activity towards OERs [48]. Y. Wen et al. [49] created a CoNi-ZIF-67@Ti$_3$C$_2$T$_x$ hybrid by using a coprecipitation method, in which bimetallic CoNi-ZIF-67 rhombic dodecahedrons were grown on the Ti$_3$C$_2$T$_x$ matrix. The presence of the MXene increases the average oxidation of Co/Ni elements, making the hybrid material an outstanding OER electrocatalyst.

8.8 CATALYSTS FOR OXYGEN REDUCTION REACTION

ORR has a direct impact on the conversion efficiency of fuel cell. Among several catalysts, Pt-based materials offer the highest catalytic activity, but they are not cost-effective and have poor stability. As a result, finding low-cost, high-efficiency alternatives to replace these materials is critical. MXenes have been extensively explored in terms of ORR catalysis due to their high electrical conductivity and larger surface area [2]. MXenes are durable under oxidative, mechanical, and acidic environments. These materials exhibit conductive channels for electrons with an efficiency to avoid corrosion, in contrast to noble metal/carbon catalysts. In an alkaline electrolyte, for example, Ag exhibits a relatively strong ORR activity; however, their reported performances are insufficient. Preparing a more conductive material paired with bimetallic catalysts is one interesting strategy for improving catalytic performance [22].

D. Kan et al. [50] proposed an efficient method for constructing bifunctional catalysts by using Pt/Pd single atoms to control the electronic structures of functionalized Nb_2CT_2. Nb_2CF_2-VF-Pt exhibited low overpotentials for ORR (i.e., 0.40 eV), which was observed to be better than the Pt(111) and IrO_2(110) catalysts. The catalytic nature of this hybrid structure was reported to be dependent on F-terminated groups, Pt, and the electron donor capacity. In another study, researchers used first-principle calculations to simulate Pt/v-$Ti_{n+1}C_nT_x$ surfaces (Fig. 8.19 [51]). They concentrated on the MXene's termination effects, which may be a crucial element in improving ORR performance. Extensive computational investigations of the geometries, charges, and electronic structures of various surfaces explained their features [51].

Different energy corrections were computed to characterize the overall ORR electrochemical reaction profile, as shown in the free energy diagram at U = 0 and U = 1.23 V on different material surfaces (Fig. 8.20 [51]). F-terminated surfaces were observed to have lower free energy than O-terminated ones. It is important to note that a lower free energy does not always imply greater ORR performance because the overpotential must be chosen by the highest free energy difference throughout the four electrochemical phases.

In another study, M-N_4-Gr/V_2C heterostructures were studied as a potential ORR catalysts by using eight different TMs [52]. According to the calculations, Zn-N_4-Gr/V_2C

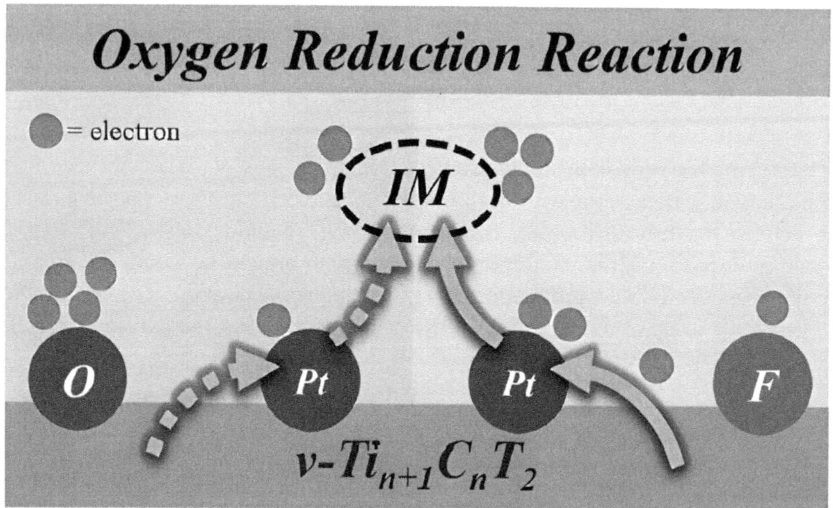

FIGURE 8.19 Schematic electron transfer on different Pt/v-$Ti_{n+1}C_nT_2$ surfaces with ORR intermediates (IMs).

Source: Reprinted with permission from [51] Chi-You Liu, Elise Y. Li. 2019. Termination Effects of Pt/v-$Ti_{n+1}C_nT_2$ MXene Surfaces for Oxygen Reduction Reaction Catalysis. *ACS Appl. Mater. Interfaces*, 11, 1, 1638–1644. Copyright 2019, American Chemical Society.

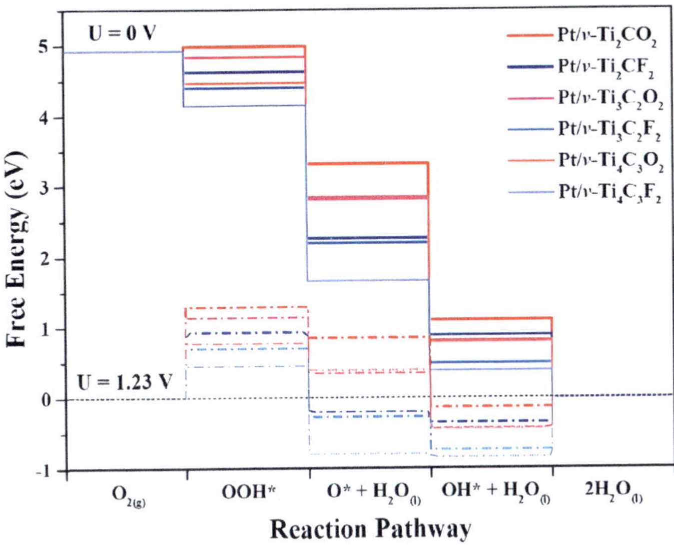

FIGURE 8.20 Free energy diagram of ORR IMs on Pt/v-Ti$_{n+1}$C$_n$T$_2$ surfaces.

Source: Reprinted with permission from [51] Chi-You Liu, Elise Y. Li. 2019. Termination Effects of Pt/v-Ti$_{n+1}$C$_n$T$_2$ MXene Surfaces for Oxygen Reduction Reaction Catalysis. *ACS Appl. Mater. Interfaces*, 11, 1, 1638–1644. Copyright 2019, American Chemical Society.

heterostructures were thermodynamically stable, while Ni-N$_4$-Gr/V$_2$C and Co-N$_4$-Gr/V$_2$C exhibited low overpotentials (i.e., 0.32 and 0.45 V), thereby demonstrating higher ORR activity. Another study summarized the synthesis, structure, and current progress in the possible ORR MXene-based catalysts [53]. According to first-principle calculations, the graphitic sheet on Mo$_2$C and V$_2$C is an active ORR catalyst, with low kinetic barriers and overpotentials (0.36 V). The electronic connections between the MXene and doping sheet were computed to be related to the band center of C-atoms located at the surface and the work function of the heterostructure [54]. They also investigated the reaction barriers for oxygen to dissociate on the active sites of doped materials, to further evaluate the catalytic performance. These reactions take place by following two reaction pathways: it can either combine with a water molecule to create an OOH* group or dissociate directly into two O* species, as shown in Fig. 8.21 [54]. Both reaction routes are exothermic and kinetically easily occur due to the strong binding capabilities of these heterostructures.

Other efficient TM (Ni$_1$/Ni$_2$ and Fe$_1$/Ni$_2$)-doped MXene-based bifunctional (OER/ORR) catalysts were reported, which exhibited better performance with low overpotential than other materials, such as Pt(111) and IrO$_2$(110). The electron-grabbing capabilities and interactions between metal adsorbents and the MXene substrate are attributable to the superior catalytic activity in the MXene-based double atomic catalysts (DACs), according to electronic structure analyses. Their findings showed not only intriguing options for these catalysts but also fresh theoretical insights into developing noble-metal-free bifunctional DACs rationally [55]. Figure 8.22a [55] depicts the ORR and OER kinetic mechanism route diagrams of the activated

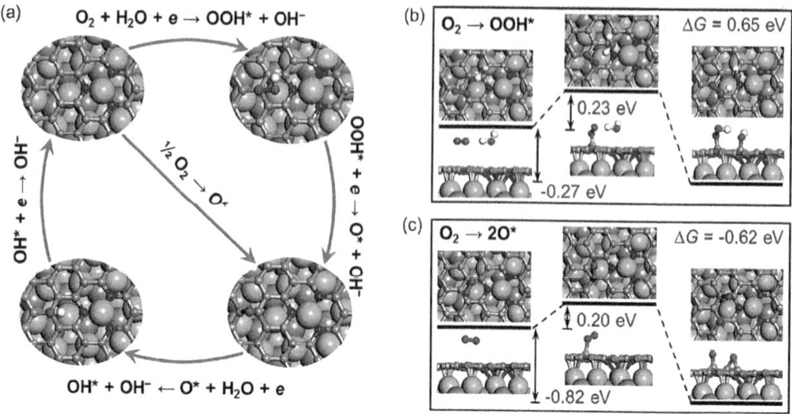

FIGURE 8.21 (a) Dual reaction pathways for ORR on N-doped graphene on monolayer of V_2C. (b, c) Kinetic barriers and transition states (middle panel) for O_2 dissociation via two reaction pathways.

Source: [54] Zhou S. Yang X. et al. 2018. Heterostructures of MXenes and N-doped graphene as highly active bifunctional electrocatalysts. *Nanoscale*, 10, 10876–10883. Reproduced with permission from the Royal Society of Chemistry.

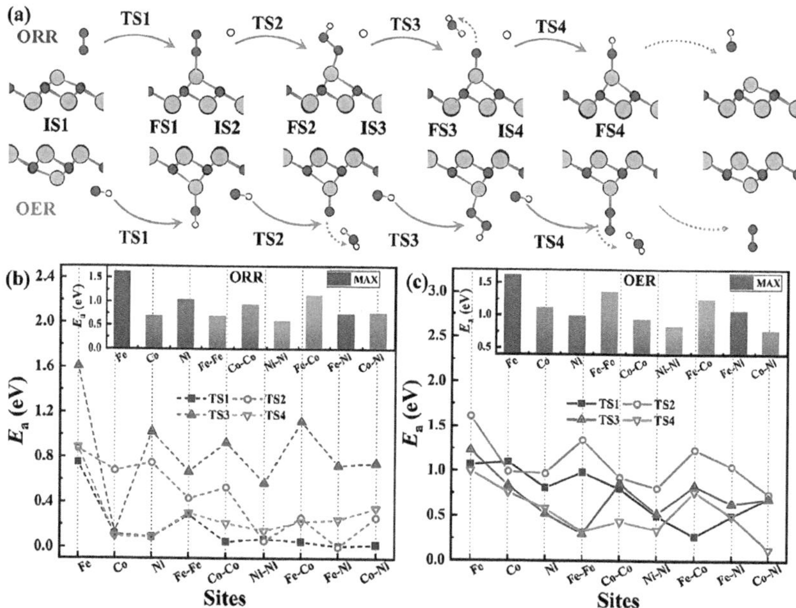

FIGURE 8.22 (a) Schematic diagram of the kinetic reaction process for ORR and OER. (b, c) Calculated energy barriers for ORR (b) and OER (d) for nine ACs.

Source: [55] Wei B.; Fu Z., Legut D. et al.: Rational Design of Highly Stable and Active MXene-Based Bifunctional ORR/OER Double-Atom Catalysts. *Adv. Mater*. 2021, 2102595. Copyright WILEY-VCH Verlag GmbH & Co. KGaA, Weinheim. Reproduced with permission.

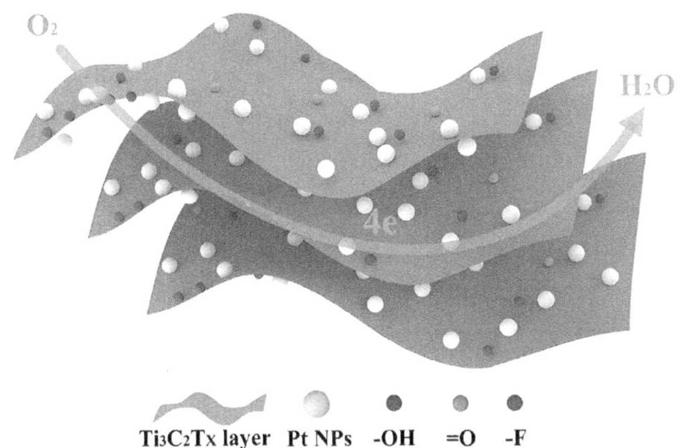

FIGURE 8.23 Depiction of the electrocatalytic ORR mechanisms on the Pt/Ti$_3$C$_2$T$_x$ catalyst.

Source: Reprinted from [57] *Journal of Electroanalytical Chemistry*, 865, Chuang Zhang, Ben Ma, Yingke Zhou, Cheng Wang, Highly active and durable Pt/MXene nanocatalysts for ORR in both alkaline and acidic conditions, 114142, Copyright 2020, with permission from Elsevier.

carbons (ACs). Figures 8.22b and c depict the E$_a$ for each OER and ORR phase on nine Fe/Co/Ni-ACs. The bar charts illustrate the largest values of the energy barriers for the nine ACs.

Y. Zhang et al. [56] developed a new Co/N-CNTs@Ti$_3$C$_2$T$_x$ as a bifunctional electrocatalyst toward ORR and OER. In another investigation, the Pt/MXene catalyst outlasted and outperformed the Pt/C (commercial) catalyst for ORR in both alkaline and acid environments (Fig. 8.23 [57]). The reaction was dominated by a four-electron oxygen reduction mechanism. The simple, green, and cost-effective technique of producing nanocomposite catalysts with three-phase reaction regions to improve fuel cell performance [57] looked promising. Similarly, Y. Lei et al. [58] concentrated on the synthesis of ORR electrocatalysts composed of porous N-rich carbon/MXene. With onset and half-wave potentials of 0.85 and 0.71 V, respectively, the electrocatalyst demonstrated good electrocatalytic activity and stability.

REFERENCES

[1] Dey, A., et al., *Doped MXenes—a new paradigm in 2D systems: synthesis, properties and applications.* Progress in Materials Science, 2023: p. 101166.

[2] Meng, W., et al., *Advances and challenges in 2D MXenes: from structures to energy storage and conversions.* Nano Today, 2021. **40**: p. 101273.

[3] Pang, J., et al., *Applications of 2D MXenes in energy conversion and storage systems.* Chemical Society Reviews, 2019. **48**(1): p. 72–133.

[4] Khaledialidusti, R., A.K. Mishra, and A. Barnoush, *Atomic defects in monolayer ordered double transition metal carbide (Mo$_2$TiC$_2$T$_x$) MXene and CO$_2$ adsorption.* Journal of Materials Chemistry C, 2020. **8**(14): p. 4771–4779.

[5] Cheng, C., et al., *Cu3-cluster-doped monolayer Mo$_2$CO$_2$ (MXene) as an electron reservoir for catalyzing a CO oxidation reaction.* ACS Applied Materials & Interfaces, 2018. **10**(38): p. 32903–32912.

[6] Cheng, C., et al., *Single Pd atomic catalyst on Mo_2CO_2 monolayer (MXene): unusual activity for CO oxidation by trimolecular Eley–Rideal mechanism*. Physical Chemistry Chemical Physics, 2018. **20**(5): p. 3504–3513.

[7] Hou, T., et al., *Modulating oxygen coverage of $Ti_3C_2T_x$ MXenes to boost catalytic activity for HCOOH dehydrogenation*. Nature Communications, 2020. **11**(1): p. 4251.

[8] Azofra, L.M., et al., *Promising prospects for 2D d^2–$d^4M_3C_2$ transition metal carbides (MXenes) in N_2 capture and conversion into ammonia*. Energy & Environmental Science, 2016. **9**(8): p. 2545–2549.

[9] Li, Z. and Y. Wu, *2D early transition metal carbides (MXenes) for catalysis*. Small, 2019. **15**(29): p. 1804736.

[10] Liu, J., et al., *2D MXene-based materials for electrocatalysis*. Transactions of Tianjin University, 2020. **26**: p. 149–171.

[11] Li, N., et al., *Understanding of electrochemical mechanisms for CO_2 capture and conversion into hydrocarbon fuels in transition-metal carbides (MXenes)*. ACS Nano, 2017. **11**(11): p. 10825–10833.

[12] He, H., P. Zapol, and L.A. Curtiss, *Computational screening of dopants for photocatalytic two-electron reduction of CO_2 on anatase (101) surfaces*. Energy & Environmental Science, 2012. **5**(3): p. 6196–6205.

[13] Handoko, A.D., et al., *Establishing new scaling relations on two-dimensional MXenes for CO_2 electroreduction*. Journal of Materials Chemistry A, 2018. **6**(44): p. 21885–21890.

[14] Zhang, X., et al., *Ti_2CO_2 MXene: a highly active and selective photocatalyst for CO_2 reduction*. Journal of Materials Chemistry A, 2017. **5**(25): p. 12899–12903.

[15] Sun, J., et al., *Recent advances of MXene as promising catalysts for electrochemical nitrogen reduction reaction*. Chinese Chemical Letters, 2020. **31**(4): p. 953–960.

[16] Li, L., et al., *Theoretical screening of single transition metal atoms embedded in MXene defects as superior electrocatalyst of nitrogen reduction reaction*. Small Methods, 2019. **3**(11): p. 1900337.

[17] Luo, H., et al., *A theoretical study of Fe adsorbed on pure and nonmetal (N, F, P, S, Cl)-doped $Ti_3C_2O_2$ for electrocatalytic nitrogen reduction*. Nanomaterials, 2022. **12**(7): p. 1081.

[18] Zheng, S., et al., *Electrochemical nitrogen reduction reaction performance of single-boron catalysts tuned by MXene substrates*. The Journal of Physical Chemistry Letters, 2019. **10**(22): p. 6984–6989.

[19] Luo, Y., et al., *Efficient electrocatalytic N_2 fixation with MXene under ambient conditions*. Joule, 2019. **3**(1): p. 279–289.

[20] Li, T., et al., *Fluorine-free $Ti_3C_2T_x$ (T=O, OH) nanosheets (~50–100 nm) for nitrogen fixation under ambient conditions*. Journal of Materials Chemistry A, 2019. **7**(24): p. 14462–14465.

[21] Zhao, J., et al., *$Ti_3C_2T_x$ (T=F, OH) MXene nanosheets: conductive 2D catalysts for ambient electrohydrogenation of N_2 to NH_3*. Journal of Materials Chemistry A, 2018. **6**(47): p. 24031–24035.

22] Zhu, J., et al., *Recent advance in MXenes: a promising 2D material for catalysis, sensor and chemical adsorption*. Coordination Chemistry Reviews, 2017. **352**: p. 306–327.

[23] Chen, Z., et al., *Transition metal atoms implanted into MXenes (M_2CO_2) for enhanced electrocatalytic hydrogen evolution reaction*. Applied Surface Science, 2020. **509**: p. 145319.

[24] Ding, B., et al., *Uncovering the electrochemical mechanisms for hydrogen evolution reaction of heteroatom doped M_2C MXene (M=Ti, Mo)*. Applied Surface Science, 2020. **500**: p. 143987.

[25] Ling, C., et al., *Transition metal-promoted V_2CO_2 (MXenes): a new and highly active catalyst for hydrogen evolution reaction*. Advanced Science, 2016. **3**(11): p. 1600180.

[26] Cheng, Y.-W., et al., *Transition metal modification and carbon vacancy promoted Cr_2CO_2 (MXenes): a new opportunity for a highly active catalyst for the hydrogen evolution reaction.* Journal of Materials Chemistry A, 2018. **6**(42): p. 20956–20965.

[27] Zou, X., et al., *A simple approach to synthesis Cr_2CT_x MXene for efficient hydrogen evolution reaction.* Materials Today Energy, 2021. **20**: p. 100668.

[28] Gao, G., A.P. O'Mullane, and A. Du, *2D MXenes: a new family of promising catalysts for the hydrogen evolution reaction.* ACS Catalysis, 2017. **7**(1): p. 494–500.

[29] Su, T., et al., *One-step synthesis of $Nb_2O_5/C/Nb_2C$ (MXene) composites and their use as photocatalysts for hydrogen evolution.* ChemSusChem, 2018. **11**(4): p. 688–699.

[30] Kuznetsov, D.A., et al., *Single site cobalt substitution in 2D molybdenum carbide (MXene) enhances catalytic activity in the hydrogen evolution reaction.* Journal of the American Chemical Society, 2019. **141**(44): p. 17809–17816.

[31] Ling, C., et al., *Searching for highly active catalysts for hydrogen evolution reaction based on O-terminated MXenes through a simple descriptor.* Chemistry of Materials, 2016. **28**(24): p. 9026–9032.

[32] Tekalgne, M.A., et al., *Efficient electrocatalysts for hydrogen evolution reaction using heteroatom-doped MXene nanosheet.* International Journal of Energy Research, 2023. **2023**.

[33] Seh, Z.W., et al., *Two-dimensional molybdenum carbide (MXene) as an efficient electrocatalyst for hydrogen evolution.* ACS Energy Letters, 2016. **1**(3): p. 589–594.

[34] Jia, J., et al., *Ultrathin N-doped Mo_2C nanosheets with exposed active sites as efficient electrocatalyst for hydrogen evolution reactions.* ACS Nano, 2017. **11**(12): p. 12509–12518.

[35] Shi, Z., et al., *Phosphorus-Mo_2 C@ carbon nanowires toward efficient electrochemical hydrogen evolution: composition, structural and electronic regulation.* Energy & Environmental Science, 2017. **10**(5): p. 1262–1271.

[36] Ang, H., et al., *Hydrophilic nitrogen and sulfur Co-doped molybdenum carbide nanosheets for electrochemical hydrogen evolution.* Small, 2015. **11**(47): p. 6278–6284.

[37] Tang, Y., et al., *Synergistically coupling phosphorus-doped molybdenum carbide with MXene as a highly efficient and stable electrocatalyst for hydrogen evolution reaction.* ACS Sustainable Chemistry & Engineering, 2020. **8**(34): p. 12990–12998.

[38] Djire, A., et al., *Electrocatalytic and optoelectronic characteristics of the two-dimensional titanium nitride $Ti_4N_3T_x$ MXene.* ACS Applied Materials & Interfaces, 2019. **11**(12): p. 11812–11823.

[39] Jin, D., et al., *Computational screening of 2D ordered double transition-metal carbides (MXenes) as electrocatalysts for hydrogen evolution reaction.* The Journal of Physical Chemistry C, 2020. **124**(19): p. 10584–10592.

[40] Li, N., et al., *Double transition metal carbides MXenes (D-MXenes) as promising electrocatalysts for hydrogen reduction reaction: Ab initio calculations.* ACS Omega, 2021. **6**(37): p. 23676–23682.

[41] Zeng, Z., et al., *Computational screening study of double transition metal carbonitrides $M'_2M''CNO_2$-MXene as catalysts for hydrogen evolution reaction.* npj Computational Materials, 2021. **7**(1): p. 80.

[42] Cheng, Y., et al., *Nanostructure of Cr_2CO_2 MXene supported single metal atom as an efficient bifunctional electrocatalyst for overall water splitting.* ACS Applied Energy Materials, 2019. **2**(9): p. 6851–6859.

[43] Tang, Y., et al., *The effect of in situ nitrogen doping on the oxygen evolution reaction of MXenes.* Nanoscale Advances, 2020. **2**(3): p. 1187–1194.

[44] Chen, X., et al., *Tunable nitrogen-doped delaminated 2D MXene obtained by NH_3/Ar plasma treatment as highly efficient hydrogen and oxygen evolution reaction electrocatalyst.* Chemical Engineering Journal, 2021. **420**: p. 129832.

[45] Liu, J., et al., *Hierarchical cobalt borate/MXenes hybrid with extraordinary electrocatalytic performance in oxygen evolution reaction.* ChemSusChem, 2018. **11**(21): p. 3758–3765.

[46] Ma, T.Y., et al., *Interacting carbon nitride and titanium carbide nanosheets for high-performance oxygen evolution.* Angewandte Chemie International Edition, 2016. **55**(3): p. 1138–1142.

[47] Zhao, L., et al., *Interdiffusion reaction-assisted hybridization of two-dimensional metal—organic frameworks and $Ti_3C_2T_x$ nanosheets for electrocatalytic oxygen evolution.* ACS Nano, 2017. **11**(6): p. 5800–5807.

[48] Zou, H., et al., *Metal—organic framework-derived nickel—cobalt sulfide on ultrathin MXene nanosheets for electrocatalytic oxygen evolution.* ACS Applied Materials & Interfaces, 2018. **10**(26): p. 22311–22319.

[49] Wen, Y., et al., *MXene boosted CoNi-ZIF-67 as highly efficient electrocatalysts for oxygen evolution.* Nanomaterials, 2019. **9**(5): p. 775.

[50] Kan, D., et al., *Rational design of bifunctional ORR/OER catalysts based on Pt/Pd-doped Nb_2CT_2 MXene by first-principles calculations.* Journal of Materials Chemistry A, 2020. **8**(6): p. 3097–3108.

[51] Liu, C.-Y. and E.Y. Li, *Termination effects of Pt/v-$Ti_{n+1}C_nT_2$ MXene surfaces for oxygen reduction reaction catalysis.* ACS Applied Materials & Interfaces, 2018. **11**(1): p. 1638–1644.

[52] Chen, Y., et al., *Rational design of $M–N_4–Gr/V_2C$ heterostructures as highly active ORR catalysts: a density functional theory study.* RSC Advances, 2022. **12**(23): p. 14368–14376.

[53] Zhang, Y., et al., *Recent progress of MXenes as the support of catalysts for the CO oxidation and oxygen reduction reaction.* Chinese Chemical Letters, 2020. **31**(4): p. 931–936.

[54] Zhou, S., et al., *Heterostructures of MXenes and N-doped graphene as highly active bifunctional electrocatalysts.* Nanoscale, 2018. **10**(23): p. 10876–10883.

[55] Wei, B., et al., *Rational design of highly stable and active MXene-based bifunctional ORR/OER double-atom catalysts.* Advanced Materials, 2021. **33**(40): p. 2102595.

[56] Zhang, Y., et al., *In situ growth of cobalt nanoparticles encapsulated nitrogen-doped carbon nanotubes among $Ti_3C_2T_x$ (MXene) matrix for oxygen reduction and evolution.* Advanced Materials Interfaces, 2018. **5**(16): p. 1800392.

[57] Zhang, C., et al., *Highly active and durable Pt/MXene nanocatalysts for ORR in both alkaline and acidic conditions.* Journal of Electroanalytical Chemistry, 2020. **865**: p. 114142.

[58] Lei, Y., et al., *Synthesis of porous N-rich carbon/MXene from MXene@polypyrrole hybrid nanosheets as oxygen reduction reaction electrocatalysts.* Journal of the Electrochemical Society, 2020. **167**(11): p. 116503.

9 MXenes for Supercapacitor Applications

9.1 ELECTROCHEMICAL PROPERTIES

In supercapacitors, energy is stored by developing a double layer of charges at the interfaces of electrodes and electrolytes. In comparison to the dielectric capacitors, supercapacitors exhibit higher power densities. Furthermore, their long cycling life and high rates of charging or discharging rates make them suitable for usage in cars and backup power stations. Depending on the charge storage mechanisms, these devices can be classified as pseudocapacitors, electrochemical double-layer capacitors (EDLCs), and hybrid capacitors. For asymmetric hybrid capacitors, both electrodes store charges differently, with one as pseudocapacitive and the other one as an EDLC. Recently, symmetric hybrid capacitors have also been evolved, which consist of two identical electrodes. Electrical energy is stored in EDLCs by fast ion adsorption at multiple pores via a physical electro-sorption method. In contrast, a faradaic charge transfer occurs at the electrodes of pseudocapacitors, and a combination of these two mechanisms takes place in hybrid capacitors [1–5].

This area has made progress with the development of a wide range of devices, from simply understanding the energy storage mechanisms of typical supercapacitors exhibiting high capacitances to ultra-flexible micro-supercapacitors (MSCs), which hold the future of microscale electronics. The performance of supercapacitors is greatly influenced by the properties of the electrode material; therefore, selection of an active material depends on the application it would be used for. For example, EDLCs can be developed by using a material like graphene, which exhibits a large surface area. But this quality is insufficient for future microelectronics, which require significant energy production. 2D materials other than graphene, that is, transition metal (oxy)hydroxides (TMOs) and transition metal dichalcogenides, have also been introduced in this field to improve the energy density limitation. However, they do not possess high electrical conductivities, and their rate performance needs to be improved by the addition of current collectors or carbon-based binders [6].

2D MXenes have been presented as a solution to these problems, as they possess a tunable layered structure, high electronic conductivities, structural stability, mechanical flexibility, and chemical stability. These materials exhibit better electrochemical properties due to charge transfer feasibility, produced by the layered structure and oxidation states of transition metals. Different metallic, organic, or inorganic ions can also be intercalated between MXene layers. A wide range of cations, including as Li^+, Na^+, K^+, Mg^+, and NH^{4+}, can be chemically and electrochemically intercalated into MXenes, resulting in high volume capacity. Strong electronic conductivity of MXenes allows for rapid charge transfer, potentially improving the performance of composites. An in situ investigation employing X-ray absorption spectroscopy

DOI: 10.1201/9781003465768-9

(XAS) revealed that MXenes combined with oxide exhibit a pseudocapacitive behavior. Adsorption of ions also depends on the surface terminations of MXenes, which results in an improved volume capacity [5, 7].

Areal and volumetric capacitance are two critical electrochemical features of supercapacitor electrodes that must be taken into account. Areal capacitance (mF cm^{-2}) is the combination of mass loading (mg cm^{-2}) and specific capacitance (F g^{-1}). Improving these two factors have been shown to be effective methods for enhancing the areal capacitance of MXene-based supercapacitors. For example, an electrode based on $Ti_3C_2T_x$ showed an areal capacitance of 579 mF cm^{-2}, when it had a high value of mass loading (7.6 mg cm^{-2}). As the value of mass loading was reduced to 1.8 mg cm^{-2}, areal capacitance significantly reduced to 211 mF cm^{-2}. Another essential parameter to consider is volumetric capacitance (F cm^{-3}), which is a combination of electrode tap density (g cm^{-3}) and specific capacitance (F g^{-1}). MXene sheets have a significantly higher tap density than carbon electrodes, resulting in better volumetric performance [4].

Researchers have emphasized the need to define the scan rate when reporting capacitance as it has a significant impact on ion diffusion kinetics and consequently active site utilization. On the other hand, specific capacitance can also be improved by altering the morphology of the electrode, extending the interlayer gap, modifying the surface groups, increasing the redox active sites, introducing heteroatoms, and combining with electro-active materials. Fabrication of hybrid materials, that is, MXene/conducting polymer or MXene/metal oxide hybrids, is another method for introducing pseudocapacitive sites. 2D MXene-doped conductive polymer-conjugated films with greatly increased areal capacitance may be easily made using electrochemical polymerization in MXene colloidal solutions. Aside from the strategies outlined above, a better design of the device has also been shown to be an efficient way to increase areal capacitance. MSCs, for example, make good use of the elaborate design's advantages, such as a quicker ion diffusion route and co-planar structure, without the need for a separator. High areal capacitance is typically obtained while fabricating MXene-based MSCs with high mass loading [4, 8].

9.2 ENERGY STORAGE MECHANISMS

As we have discussed before, energy is stored in EDLCs by fast ion adsorption at multiple pores via a physical electro-sorption method. In contrast, a faradaic charge transfer occurs at the electrodes of pseudocapacitors. Pseudocapacitance is classified into three types based on their electrochemical capacitive properties: redox pseudocapacitance, intercalation, and underpotential deposition. Understanding these mechanisms of MXenes is critical for developing MXene-based electrodes and explaining their superior capacitive properties. Because MXenes are readily available in functional groups, it is hypothesized that the active groups between the layers, as well as between the interface and the surface, are the key sites of energy storage. While energy is commonly stored in rechargeable batteries via an adsorption–desorption process, supercapacitors take advantage of the electrical double layer on the surface of electrodes. Research on these mechanisms is being conducted to integrate materials with types of storage, which is essential in gaining stability and high energy density [9].

The structure and properties of electrodes are important factors that determine the capacitance of a supercapacitor. Generally, for high value of capacitance, an electrode material should exhibit a large surface area and high electronic conductivity. Therefore, MXenes are efficient electrodes as they possess these essential features among others. They can be utilized as a current collector (to transfer electrons) and an active material (to store charge). This shows that, in contrast to other 2D materials, MXenes do not require any separate current collector [4]. Intercalation of MXenes is another important factor that influences the energy storage mechanisms. Various cations, for example, Li^+, K^+, Na^+, Cs^+, TEA^+, Ca^{2+}, Mg^{2+}, Al^{3+}, and NH^{4+}, can chemically or electrochemically intercalate the MXene layers, occupy the active sites, and facilitate the storage of energy. Initially, ions are accommodated on the edges or the shallow adsorption sites, but later on, they get adsorbed on the material's core or deep adsorption sites. These processes ensure quick ion adsorption and superior rate performance of MXenes [9, 10]. Therefore, it is essential to understand these driving characteristics in order to develop a supercapacitor with the required performance.

9.2.1 Influence of Morphology on Energy Storage

Different characteristics of MXenes give rise to corresponding energy storage mechanisms or electrochemical behaviors. For example, EDL capacitance is due to the large surface area, intercalation mechanism results in pseudocapacitance, and redox-active surface with an acidic electrolyte gives rise to high volumetric capacitance (nearly $900–1500\,F\,cm^{-3}$). Figure 9.1 [6] schematically illustrates the features of MXene-based electrodes. It is important to note that the materials on which intercalation and redox processes rely exhibit slow-rate performance, which limits their use in high-rate applications (e.g., AC-line filtering). However, combining MXenes with conductive polymers and the formation of hybrid structure will solve the problem, and it will lead to mesopore retention and rapid ion transfer [6]. To generalize, we can say that design of MXene-based electrodes is essential to achieve high-performance supercapacitors. Particularly, porous structure, surface groups, and interlayer spacing play a significant role in determining the electrochemical behavior [5].

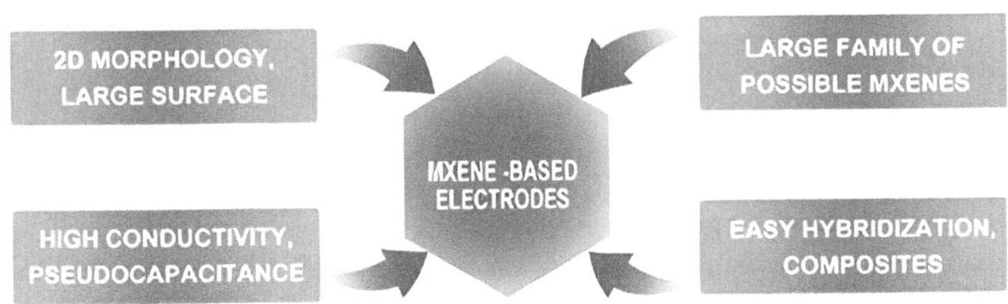

FIGURE 9.1 Properties of MXene as an active material for supercapacitive electrodes.

Source: Reprinted from [6] Pratteek Das and Zhong-Shuai Wu. 2020. MXene for energy storage: present status and future perspectives. *J. Phys. Energy*, 2, 032004. With permission from IOP Science.

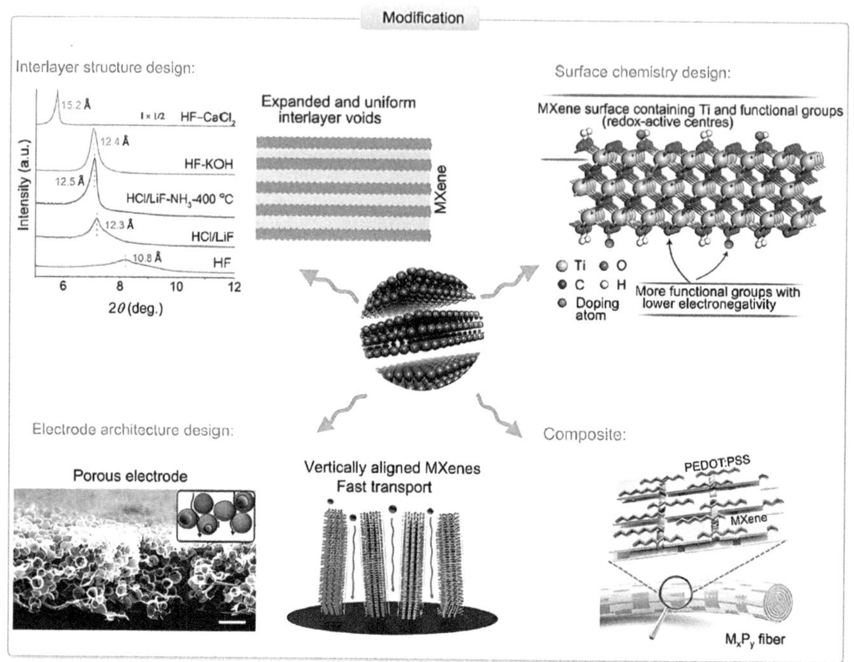

FIGURE 9.2 Modification design strategies of MXene, including interlayer structure design, surface chemistry design, electrode architecture design, and composite.

Source: [9] Hu M., Zhang H., Hu T. et al. 2020. Emerging 2D MXenes for supercapacitors: status, challenges and prospects. *Chem. Soc. Rev.*, 49, 6666–6693. Reproduced with permission from the Royal Society of Chemistry.

This section will focus on presenting an overview of the influence of MXene-based electrodes on energy storage from four perspectives, that is, interlayer structure design, electrode architecture design, composites, and surface chemistry design (as shown in Fig. 9.2 [9]). Intercalation pseudocapacitance occurs in MXenes by redox reactions, with rapid addition of ions in acidic electrolytes. Active surface exposure and faster ion intercalation will be facilitated by a large interlayer spacing, and it results in the storage of charge with high density [9].

The synthesis processes always have an effect on the MXene structure, which results in varying the performance of the electrode. MXenes etched with low-concentration hydrofluoric acid (HF) exhibit high value of capacitance, as they possess high-mobility H_2O molecules and an increased interlayer spacing. However, the temperature and duration of etching have no significant influence on electrochemical properties. For example, capacitance only changes from 100 to 120 F g⁻¹ if there is an increase in the etching time by more than 100 times. But due to the replacement of the etchant, that is, HF with HCl/LiF solution, the value of capacitance was observed to increase significantly to 900 F cm⁻³ in H_2SO_4 solution, likely due to the addition of water molecules and Li⁺ ions, which inhibits restacking of MXene sheets.

Furthermore, the presence of −Cl terminations (with greater ionic radius) formed on the edges of these sheets might widen the interlayer spacing, resulting in rapid ion diffusion. Furthermore, partial removal of Al may eliminate the problems of restacking and low conductivity. Post-modification, such as intercalation and heat treatment, can also be used to tune the interlayer structure.

When multilayered MXenes are delaminated by sonication, the interlayer becomes broader, allowing for easier ion access and increasing capacitance. KOH or CH_3COOK treatment increased the capacitance by expanding the interlayer gaps by cation intercalation. K^+ intercalation also results in a more uniform layered structure, providing a rapid ion diffusion. Apart from this, interlayer spacing can also be enhanced by annealing with ammonia. Interestingly, an approach involving oxidation of anodes can introduce pores into the layers without affecting the electrochemically active sites, resulting in improved high-rate electrochemical performance. MXenes exhibit size-dependent electrochemical characteristics in addition to the interlayer structural impact. MXenes with smaller sizes decrease the ion diffusion path and facilitate electrolyte access to active sites, resulting in improved electrochemical performance. Electrode architecture optimization is also significant in increasing capacity. The volumetric capacitance of hydrogel $Ti_3C_2T_x$ electrodes was observed to be ~1,500 F cm^{-3} [6, 9].

9.2.2 Influence of Electrolytes on Energy Storage

Electrolytes or various solvents used in supercapacitors have many features along with some limitations, as summarized in Fig. 9.3 [6]. In acidic/salt solutions, there is a restriction of the lower upper limit on the voltage window (0.6 V). This limitation

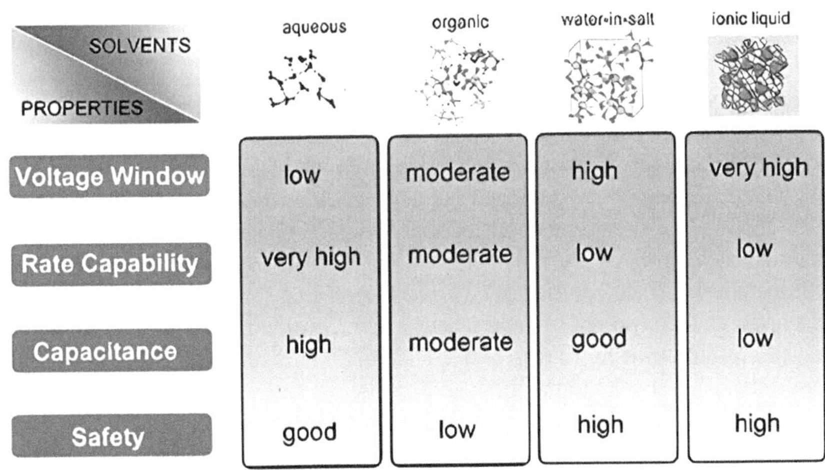

FIGURE 9.3 Effects of different types of solvents on the performance and safety of MXene supercapacitors.

Source: Reprinted from [6] Pratteek Das and Zhong-Shuai Wu. 2020. MXene for energy storage: present status and future perspectives. *J. Phys. Energy*, 2, 032004. With permission from IOP Science.

can be overcome to enhance the energy density, by extending the potential window. According to researchers, the voltage window can be expanded to ~2.4 V in organic solvent-based electrolytes. Another important way is using electrolytes that do not decompose during electrode interaction. Furthermore, various organic solvents result in variations in intercalation, resulting in a significant variance in the total charge storage. Thus, research into novel electrolytes has helped the growth of MXene-based supercapacitors [6].

In aqueous electrolytes, the ion insertion process results in the variation of c value during cycling. Unlike in carbon materials, the deformation patterns in MXenes differ depending on the ions used. In general, highly charged smaller-sized cations contract and slightly charged larger-sized cations expand the interlayer spaces. This shows that expansion due to cation intercalation and electrostatic attraction between MXene sheets (negatively charged) and cations is in an inverse relationship. Furthermore, deformation behavior of MXenes is also influenced by the number of intercalated cations. Various intercalated H^+ ions can reverse the shrinkage of the interlayer gaps caused by the intercalation of lesser H^+ ions. The structural changes that occur during these processes are highly reversible, followed by a change in the mechanical properties of MXene electrodes, thereby influencing the cycling lifetime [9]. MXenes exhibit outstanding cycle stability; however, to minimize energy loss, it is important to discover an electrolyte in which there is no change in volume of MXene while charging/discharging. Figure 9.4 [9] illustrates that in the H_2SO_4 electrolyte, $Ti_3C_2T_x$ is an intercalation pseudocapacitive material.

Due to their extended voltage window, nonaqueous electrolytes (i.e., organic and ionic-liquid gel) have also been studied in addition to aqueous electrolytes, which produce higher energy densities. Nonaqueous electrolytes have been observed to exhibit electrochemical performance quite different from that of aqueous electrolytes. For example, a set of broad peaks (at -0.2 V and 0.6 V vs. Ag) were observed for the CV curves of $Ti_3C_2T_x$ MXene in acetonitrile electrolyte. These current peaks are due to the intercalation of EMI^+ cations and $TFSI^-$ anions between MXene layers. Similarly, for $Ti_3C_2T_x$ in the $LiPF_6$/EC-DMC (ethylene or dimethyl carbonate) electrolyte, etched with HCl/LiF or HF, first there is partial removal of Li^+ at the interface of the electrolyte and MXene, which is followed by the redox reactions to store charge. This process is carried out across a wide range from 0.05 V to 3 V vs. Li^+/Li. However, it should be observed that removal of Li^+ does not takes place in all nonaqueous electrolytes [9].

In a study by H. Shao et al. [11], it was found that the presence of water molecules between the layers of MXene is crucial for the pseudocapacitive behavior. This allows for proton transfer to trigger the Ti atoms' redox reaction. Another study [12] showed that $Ti_3C_2T_x$-DC (dipicolinic acid) achieved an excellent capacitance in KCl electrolyte. This DC-intercalated material was observed to exhibit high capacitance (353 F g^{-1} at 0.5 A g^{-1}) in an aqueous electrolyte. This value of capacitance was reported to be higher than that of pure MXene (i.e., 127 F g^{-1} in KCl) or the one in an acidic electrolyte (i.e., 320 F g^{-1}). Furthermore, the intercalated MXene demonstrated an excellent capacitance retention of 100% at 1.0 A g^{-1} (after 10,000 cycles). X. Wang et al. [13] revealed that using the carbonate solvent for an electrolyte solution can enhance the pseudocapacitive charge storage of Ti_3C_2 by two times.

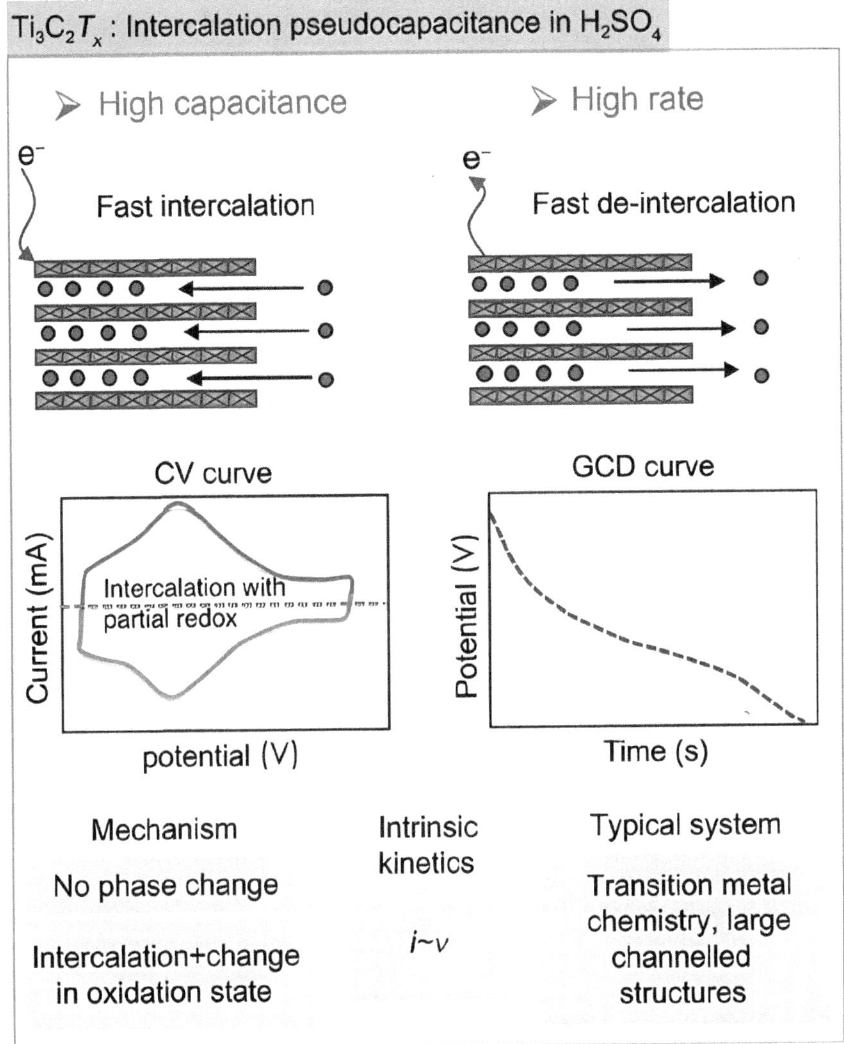

FIGURE 9.4 Schematic of the charge storage mechanism, cyclic voltammetry (CV) curve, galvanostatic charge–discharge (GCD) profile, key mechanism descriptions, intrinsic kinetics, and typical systems for intercalation pseudocapacitors.

Source: [9] Hu M., Zhang H., Hu T. et al. 2020. Emerging 2D MXenes for supercapacitors: status, challenges and prospects. *Chem. Soc. Rev.*, 49, 6666–6693. Reproduced with permission from the Royal Society of Chemistry.

9.3 MXENE-BASED SUPERCAPACITOR ELECTRODES

MXene-based composites can be developed by combining the MXenes with electro-active elements such as metal compounds, conducting polymers, and carbonaceous materials. This results in a synergistic effect, where MXenes exhibit exceptional electronic conductivity, increased accessible surface area, and stabilized structure. These materials also facilitate an increase in interlayer separation and prevent MXene nanosheet restacking. Therefore, supercapacitor electrodes made up of these hybrid structures exhibit high energy and power densities [8]. Some examples of MXene-based electrodes are presented below. R. Meshkian et al. [14] used free-standing $W_{1.33}C$ films with up to 10% polymer (PEDOT:PSS), which displayed a high capacitance (600 F cm^{-3}) in 1M H_2SO_4 along with better capacitance retention after 10,000 cycles. In another study, using 1 M Na_2SO_4 electrolyte, V_2C displayed good cycling stability, specific capacitance of 164 F g^{-1} (at 2 mV s^{-1}), and ~90% specific capacitance retention after 10,000 cycles [15]. In another study, V_2C demonstrated electric double-layer capacitance with high capacitance (i.e., 223.5 F g^{-1}) in 1 M Na_2SO_4 and 94.7% cycling stability (after 5,000 cycles) [16]. In three aqueous electrolytes, V_2C displayed specific capacitances of 487, 225, and 184 F g^{-1} in 1 M H_2SO_4, $MgSO_4$, and KOH, respectively [17].

Another study reported that the V_4C_3 MXenes exhibit more electrochemically active sites, due to the penetration of electrolyte ions between the layers. In 1M H_2SO_4 electrolyte, the electrode material displayed a high capacitance of 330 F g^{-1} (at 5 mV s^{-1}), with 90% retention rate even after 3,000 cycles [18]. Y. Xin and colleagues [19] demonstrated that quantum capacitance limitation of graphene-based electrodes can be overcome by bare and terminated niobium carbides. Although this value of capacitance is significantly reduced by surface groups at positive electrodes, except for the Nb_2C, no specific effect was observed at negative electrodes. In 1 M H_2SO_4, KOH, and $MgSO_4$, $Nb_4C_3T_x$ freestanding films (at 5 mV s^{-1}) displayed notable capacitances of 1075, 687, and 506 F cm^{-3}, respectively [20]. For 2D tantalum carbide, a capacitance of 481 F g^{-1} was observed in a 0.1M H_2SO_4 electrolyte, with 89% cyclic retention (even after 2,000 cycles) [21]. Furthermore, the single-layer $Ti_2V_{0.9}Cr_{0.1}C_2T_x$ electrode demonstrated a high capacitance of 553.27 F g^{-1} (at 2 mV s^{-1}) [22]. In the following sections, we will discuss the most recent progress in MXene-based supercapacitors, including symmetric, asymmetric, hybrid, and micro-supercapacitors.

9.3.1 SYMMETRIC SUPERCAPACITORS

In aqueous electrolytes, MXenes provide exceptional specific capacitance in aqueous solutions; however, the narrow potential window limits the energy density. On the other hand, ionic liquid electrolytes have large-sized ions, which make the intercalation difficult, and there is slow ion transport, even though they exhibit higher energy along with a wide potential window. Therefore, it would be preferable to investigate MXenes with larger interlayer spacing, as it would facilitate the intercalation process. D. Gandla et al. [23] reported the first acetonitrile electrolyte application of the freestanding $Mo_2Ti_2C_3$ electrode. The resultant f-$Mo_2Ti_2C_3$ outperformed the typical f-$Ti_3C_2T_x$ structure in terms of electrochemical characteristics, charge

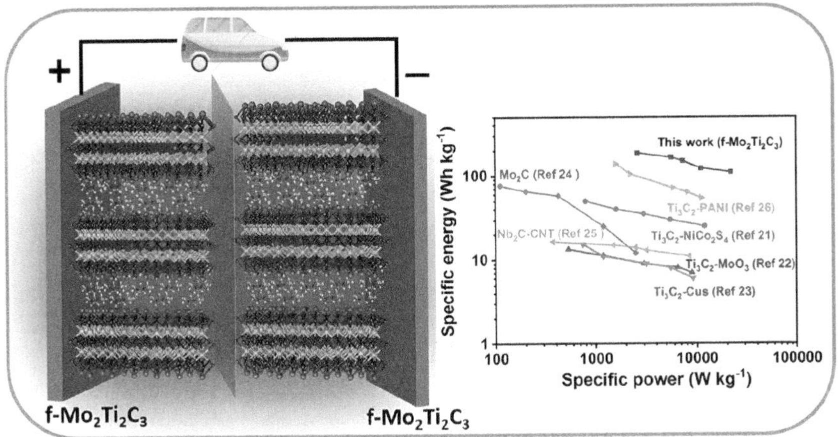

FIGURE 9.5 Schematic illustration of symmetric supercapacitors and Ragone plot showing specific energy vs. specific power of the f-Mo$_2$Ti$_2$C$_3$ symmetric device in comparison to present energy storage devices.

Source: Reprinted with permission from [23] D. Gandla, F. Zhang, and D. Q. Tan. 2022. Advantage of Larger Interlayer Spacing of a Mo$_2$Ti$_2$C$_3$ MXene Free-Standing Film Electrode toward an Excellent Performance Supercapacitor in a Binary Ionic Liquid–Organic Electrolyte. *ACS Omega*, 7, 8, 7190–7198. Copyright 2022, American Chemical Society.

storage kinetics, and microstructure. The symmetric two-electrode device based on f-Mo$_2$Ti$_2$C$_3$ demonstrated high capacitance (152 F g^{-1}), extraordinary specific energy (188 Wh kg^{-1}), and power density (22 kW kg^{-1}). The Ragone plot (Fig. 9.5 [23]) depicts the specific energy and power density of the f-Mo$_2$Ti$_2$C$_3$ cell computed from the specific capacitance using GCD curves.

The MXene-based symmetric supercapacitors are summarized in Table 9.1. A research team led by A.S. Etman [24] has developed a simple one-step process for producing composite electrodes based on Mo$_{1.33}$CT$_z$-cellulose, with high MXene loading. For 45 wt% cellulose content, they achieved up to 440 F g^{-1} with a thickness of 5.9 mm. Similarly for 5 wt%, they were able to attain a capacitance of up to 1178 F cm^{-3} (with a thickness of 4.8 mm). These values were observed to be higher than that of pristine films with a similar MXene loading (i.e., 272 F g^{-1}). Additionally, these electrodes outperformed previous Mo$_{1.33}$CT$_z$-based electrodes by demonstrating ~95% capacitance retention (after 30,000 cycles). Furthermore, a high areal capacitance of around 1.4 F cm^{-2} was also observed due to the presence of cellulose inside a thick composite electrode, which facilitates the opening of structure during cycling. By using 25% L-MoxC electrodes, a symmetric device was developed in a 1M H$_2$SO$_4$ solution. Figure 9.6a [24] displays the CV curves of the device at different scan rates. The variation in the capacitance of a device is illustrated in Fig. 9.6b, and it was approximately 95, 50, 30, and 23 F g^{-1} for 2, 10, 50, and 100 mV s^{-1}, respectively. Furthermore, the variation in capacitance with regard to the current densities is also presented (Fig. 9.6c). Figure 9.6d shows the capacitance retention after 35,000 cycles.

TABLE 9.1

MXene-Based Symmetric Supercapacitors

Electrode Material	Capacitance (F g^{-1})/ (F cm^{-2})/ (F cm^{-3})	Cyclic Stability	Electrolyte	Ref.
V_2CT_x and Ti_2CT_x	420 F g^{-1} 1315 F cm^{-3}	77%/1 million cycles	H_2SO_4, Na_2SO_4, K_2SO_4, and LiCl	[25]
UN-Ti_3C_2	927 F g^{-1} 2836 F cm^{-3}	81.7%/20,000 cycles	3 M H_2SO_4	[26]
MN-Ti_3C_2	786 F g^{-1} 2643 F cm^{-3}	100%/20,000 cycles	3 M H_2SO_4	[26]
f-$Mo_2Ti_2C_3$	152 F g^{-1}	86%/5000 cycles	1 M EMIMTFSI/ACN	[23]
N, O Co-doped C@ Ti_3C_2	250.6 F g^{-1}	94%/5000 cycles	6 M KOH	[27]
Alkalinized and annealed $Ti_3C_2T_x$ film	1805 F cm^{-3}	98%/8000 cycles	2 M KOH	[28]
$Mo_{1.33}CT_z$–cellulose composite	440 F g^{-1} 1178 F cm^{-3} 1.4 F cm^{-2}	~95%/30,000 cycles	1 M H_2SO_4	[24]
1T-MoS_2/Ti_3C_2	347 mF cm^{-2}	91.1%/20,000 cycles	PVA- H_2SO_4	[29]
CN-MX	616 F cm^{-3}	88.4%/5000 cycles	–	[30]
BiOCl-$Ti_3C_2T_x$ (TCBOC)	396.5 F cm^{-3} at 1 A g^{-1} and 228 F cm^{-3} at 15 A g^{-1}	85%/5000 cycles	1 M KOH	[31]
MXene/ rGO-5 wt%	1040 F cm^{-3}	61%/20,000 cycles	3 M H_2SO_4	[32]
$Ti_3C_2T_x$/CNTs 3D-PMCF	375 F g^{-1}	95.9%/10,000 cycles	3 M H_2SO_4	[33]

9.3.2 ASYMMETRIC SUPERCAPACITORS

Symmetric supercapacitors developed using MXenes as negative electrode materials in aqueous solutions have a narrow voltage window, which is a limitation for achieving significant energy density. This can be overcome and high energy density can be achieved by developing asymmetric supercapacitors. rGO oxide can be used as double layer capacitive electrode materials, which can match $Ti_3C_2T_x$. However, this electrode material does not exhibit high value of capacitance. Therefore, researchers have been investigating both battery-type and pseudocapacitive electrode materials based on MXenes to overcome this limitation [9]. A summary of these MXene-based asymmetric supercapacitors is presented in Table 9.2. For example, the electrochemical behavior of V_2NT_x MXenes in 3.5 M KOH was evaluated for asymmetric supercapacitors by using CV, GCD, and electrochemical impedance spectroscopy (EIS) measurements [34]. This electrode material displayed high specific capacitance at 1.85 mA cm^{-2} (i.e., 112.8 F g^{-1}). Similarly, high power and energy densities (3748.4 W kg^{-1} and 15.66 W hkg^{-1}, respectively) were observed for the asymmetric supercapacitor, along with 96% capacitance retention even after 10,000 cycles of charging

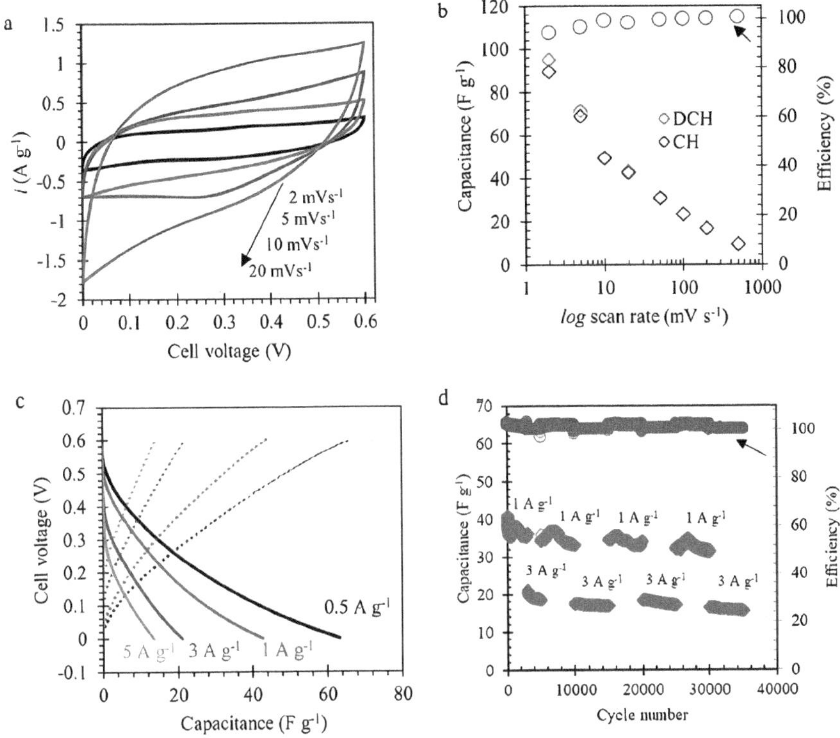

FIGURE 9.6 Electrochemical performance of the symmetric device of $Mo_{1.33}CT_z$–cellulose composite: (a) CV curves of the device at different scan rates. (b) Variation in the charge–discharge capacitances with scan rates. (c) Potential capacitance profiles for the symmetric device. (d) Variation in discharge capacitance with cycle number at applied current densities.

Source: [24] A. S. Etman, J. Halim and J. Rosen. 2021. Fabrication of $Mo_{1.33}CT_z$ (MXene)–cellulose free-standing electrodes for supercapacitor applications. *Mater. Adv.*, 2, 743–753. Reproduced with permission from the Royal Society of Chemistry.

and discharging. In KOH electrolyte, this asymmetric supercapacitor achieved a cell voltage of 1.8 V, where positive and negative electrodes are developed using Mn_3O_4 NWs and V_2NT_x, respectively.

Because of water dissociation, a cell voltage greater than >1.2 V cannot be achieved in a symmetric device. Therefore, in order to expand the voltage window, an asymmetric device (ASC) was developed by using a safe aqueous electrolyte rather than flammable organic electrolytes, which resulted in producing high energy density. SEM images showed that there is a uniform growth of Mn_3O_4 NWs on porous CF. At different scan rates, the CV curves of Mn_3O_4 NWs in 3.5 M KOH were also presented. According to the results, these NWs exhibited Faradaic charge storage mechanism with certain oxidation/reduction peaks, which resulted in high values of capacitances and energy densities. The asymmetric device exhibited pseudocapacitive behavior, as both electrodes display the same mechanism of storing charges. The

TABLE 9.2

MXene-Based Asymmetric Supercapacitors

Electrode Material	Capacitance (F g^{-1})/ (F cm^{-2})/ (F cm^{-3})	Cyclic Stability	Electrolyte	Ref.
V2NTx	112.8 F g^{-1}	96%/10,000 cycles	3.5 M KOH	[34]
MD-Ti$_3$C$_2$	3123 F cm^{-3}	~100%/10,000 cycles	3 M H$_2$SO$_4$	[36]
Ti$_3$C$_2$-MXenes/1D NiCo$_2$S$_4$	1927 F g^{-1}	67.57%/4000 cycles	3 M KOH	[37]
Mo$_{1.33}$CT$_z$ and Ti$_3$C$_2$T$_z$	460 F g^{-1} 1380 F cm^{-3}	96%/17,000 cycles	1 M H$_2$SO$_4$	[38]
Nb$_2$CT$_x$/CNT	462 mF cm^{-2}	73.3%/2,000 cycles	1 M H$_2$SO$_4$	[39]
RuO$_2$//Ti$_3$C$_2$T$_x$	60 mF cm^{-2}	86%/20,000 cycles	1 M H$_2$SO$_4$	[40]
Mo$_{1.33}$C//Mn$_x$O$_n$	815 F cm^{-3}	92%/10,000 cycles	5 M LiCl	[35]
V$_2$CT$_x$/NiV-LDH-10	1658.19 F g^{-1}	80.95%/10,000 cycles	–	[41]
Co-MXene//AC	1061 F g^{-1}	93%/1,000 cycles	–	[42]
LS modified-MXene (Ti$_3$C$_2$T$_x$)-rGO	386 F g^{-1} 1967 mF cm^{-2}	96.3%/10,000 cycles	3 M H$_2$SO$_4$	[43]
Ti$_3$C$_2$T$_x$/Ag NPs	332.2 mF cm^{-2}	87%/10,000 cycles	1 M Na$_2$SO$_4$	[44]
Ti$_3$C$_2$T$_x$/MWCNT	48 F g^{-1} 78 F cm^{-3}	97%/1,000 cycles	1 M H$_2$SO$_4$ 1 M Et$_4$NBF$_4$/ ACN	[45]
Ti$_3$C$_2$/CuS	169.5 C g^{-1} at 1 A g^{-1}	82.4%/5,000 cycles	1 M KOH	[46]
MXene/CNT	219 mF cm^{-2} 35.5 F cm^{-3}	~105%/1,000 cycles	PVA/LiCl	[47]
Co-MOF/Ti$_3$C$_2$Tx@Ni	3741 F g^{-1}	92.1%/3,000 cycles	3 M KOH	[48]
Ti$_3$C$_2$/Ni-Co-Al-LDH	748.2 F g^{-1}	97.8%/10,000 cycles	PVA-KOH	[49]
MoS$_2$/MXene	342 F g^{-1}	99%/10,000 cycles	1 M H$_2$SO$_4$	[50]
Ti$_3$C$_2$ aerogel	1012.5 mF cm^{-2}	82%/2,500 cycles	1 M KOH	[51]
MXene/CoS$_2$	1320 F g^{-1}	78.4%/3,000 cycles	2 M KOH	[52]

nonlinear GCD profiles further demonstrated that the V$_2$NT$_x$ MXene and Mn$_3$O$_4$ NWs electrodes exhibit Faradaic charge storage [34].

The energy storage applications of 2D Mo$_{1.33}$C have been investigated in an acidic electrolyte, that is, H$_2$SO$_4$. However, this electrolyte limits the potential window of both symmetric and asymmetric devices to 0.9 and 1.3 V, respectively. Therefore, in recent studies of this MXene, researchers utilized a neutral salt, that is, LiCl electrolyte, that exhibit minimal toxicity and good solubility compared to the acidic electrolyte. As a result, a high value of capacitance, that is, 815 F cm^{-3} at 2 mV s^{-1}, was observed with a wide range of potential window (−1.2 to +0.3V vs. Ag/AgCl). Figure 9.7a [35] illustrates the device design of the asymmetric supercapacitor. The electrochemical performance of the as-developed ASC device is presented in Figs. 9.7b–f, with the help of CV curves and GCD profiles at variable scan rates and current densities, respectively. Furthermore, the asymmetric Mo$_{1.33}$C//M$_n$xO$_n$ device, with a potential window of 2V was developed by using 5M LiCl electrolyte. Therefore, this asymmetric device exhibited high power and energy densities (31 W cm^{-3} and 58 mWh cm^{-3}, respectively), along with 92% of capacitance retention after 10,000 cycles of charging and discharging (as illustrated in Figs. 9.8a and b [35], respectively) [35].

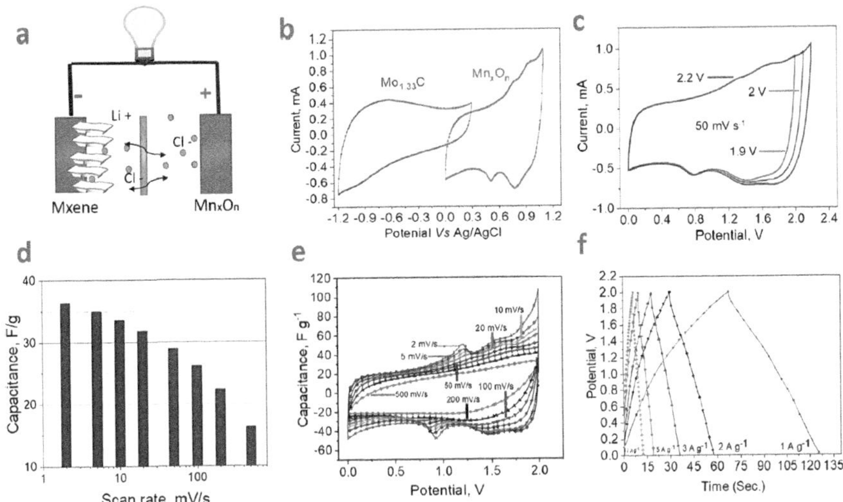

FIGURE 9.7 Electrochemical performance of MXene//Mn$_x$O$_n$ ASC. (a) Schematic illustration of the device design; (b) CV curves of the positive and negative electrodes at 10 mV s^{-1}; (c) CV curves at the scan rate of 50 mV s^{-1} in different voltage windows; (d) gravimetric capacitance at different scan rates; (e) CV curves at different scan rates; and (f) GCD profiles at different current densities.

Source: Reprinted from [35] *Energy Storage Materials*, 41, Ahmed El Ghazaly, Wei Zheng, Joseph Halim, et al., Enhanced supercapacitive performance of Mo$_{1.33}$C MXene based asymmetric supercapacitors in lithium chloride electrolyte, 203–208, Copyright 2021, with permission from Elsevier.

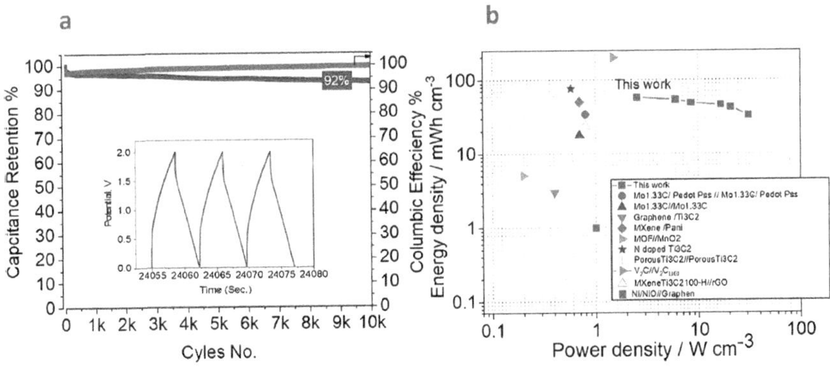

FIGURE 9.8 (a) Capacitance retention and columbic efficiency of the ASC. b) The Ragone plot compares the energy and power densities of the ASC device with previously reported MXene-based devices.

Source: Reprinted from [35] *Energy Storage Materials*, 41, Ahmed El Ghazaly, Wei Zheng, Joseph Halim, et al., Enhanced supercapacitive performance of Mo$_{1.33}$C MXene based asymmetric supercapacitors in lithium chloride electrolyte, 203–208, Copyright 2021, with permission from Elsevier.

TABLE 9.3

MXene-Based Hybrid Supercapacitors

Electrode Material	Capacitance (F g^{-1})/ (F cm^{-2})/ (F cm^{-3})	Cycling Stability	Electrolyte	Ref.
PDA-MXene	124.4 F g^{-1} at 0.2 A g^{-1}	85%/10,000 cycles	2 M ZnSO$_4$	[56]
CTAB–Sn(IV) @Ti$_3$C$_2$	268, 220, 181, 173, and 132 F g^{-1}	71.1%/4,000 cycles	1 M LiPF$_6$	[57]
Mo$_2$N	250.8 F g^{-1} 384.12 F cm^{-3}	~89.93%/4,500 cycles	1 M Na$_2$SO$_4$	[58]
N-Ti3C2Tx	325 mAh g^{-1} at 0.5 C	75%/3,500 cycles	1 M NaClO$_4$	[59]
Ti$_3$C$_2$Cl$_2$	536 mAh g^{-1} at 0.2 A g^{-1}	>99.5%/1,000 cycles	–	[60]
K–V$_2$C	195 mAh g^{-1} at 50 mA g^{-1}	95%/200 cycles	2 M KOH	[61]
NiCoAl-LDH/ V$_4$C$_3$T$_x$	300 C g^{-1} at 20 A g^{-1}	98%/10,000 cycles	1 M KOH	[62]
Ti$_3$C$_2$T$_x$/CoS@CC	120 C g^{-1} at 1 A g^{-1}	93.41%/10,000 cycles	–	[63]
V2C	100 F g^{-1}	70%/300 cycles	1 M NaPF$_6$	[54]
Mo$_{1.33}$CT$_z$ eTi$_3$C$_2$T$_z$	114 mAh g^{-1} at 0.5 mA g^{-1}	90%/8,000 cycles	1 M LiCl	[55]
Ti$_3$C$_2$-D.	93.1 F g^{-1}	–	1 M H$_2$SO$_4$	[64]
Bi-stacked Ti$_3$C$_2$T$_x$ electrode	62 F g^{-1}	96.6%/4,000 cycles	1 M NaClO$_4$	[65]

9.3.3 Hybrid Supercapacitors

Interlayer separators such as rGO, CNTs, TMOs, BP, and polymers can be introduced between the MXene NSs to develop hybrid electrodes, which results in improving their electrochemical performance [53]. Some of the hybrid supercapacitors are presented in Table 9.3. Ion capacitors store energy faster than batteries by intercalating cations into an electrode and within a greater potential window. These devices exhibit higher energy densities than EDLCs. Lithium-ion capacitors are now commercially available; however, the development of sodium-ion capacitors face certain limitations regarding the availability of intercalants for quick sodium-ion adsorption. To overcome this issue, the electrochemical behavior of 2D V$_2$C was studied. Researchers used XRD to explore the sodium intercalation and obtained ~100 F g^{-1} (at 0.2 mV s^{-1}) [54]. Figure 9.9a [54] presents the CV curves at various scan rates, while in Fig. 9.9b, we can observe the change in capacitance with scan rate. At slow scanning rates, V$_2$CT$_x$ exhibits excellent power performance for Na intercalation, producing 170 F cm^{-3}. Moreover, at low scan speeds, the CV shows two distinct zones, representing two separate electrochemical processes.

In another study, a zinc-ion hybrid capacitor was developed by A. S. Etman et al. using mixed MXene sheets of Mo$_{1.33}$CT$_z$-Ti$_3$C$_2$T$_z$ [55]. The resultant MXene exhibited capacities of 159 and 59 mAh g^{-1}, at scan rates of 0.5 and 100 mV s^{-1}, respectively. These capacity values were discovered to be greater than those previously reported for pristine MXenes and electrodes of hybrid capacitors. Furthermore, the mixed MXene displayed high power (0.143 and 10.6 kW kg^{-1}) and energy densities (103 and 38 Wh kg^{-1}), along with 90% capacity retention even after 8,000 cycles. Figures 9.10a

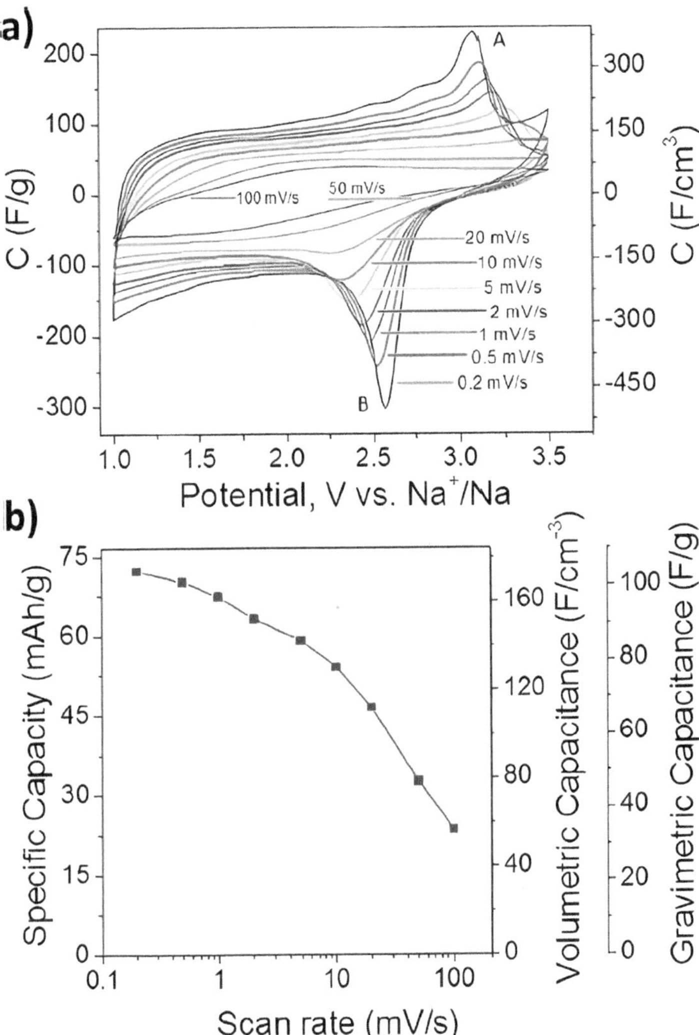

FIGURE 9.9 (a) CV of V_2CT_x at different scan rates and (b) summary of rate performance.

Source: Reprinted with permission from [54] Yohan Dall'Agnese, Pierre-Louis Taberna, Yury Gogotsi, et al. 2015. Two-Dimensional Vanadium Carbide (MXene) as Positive Electrode for Sodium-Ion Capacitors. *J. Phys. Chem. Lett.*, 6, 2305–2309. Copyright 2015, American Chemical Society.

and b [55] show the schematic illustration of mixed and individual MXene films. Figures 9.10c, e, and f demonstrate the electrochemical behavior of mixed and pure MXene films. At low scan rates, mixed MXene displayed CV curves similar to those of $Mo_{1.33}CT_z$. However, due to limitations of diffusion and slow electron transfer, CV curves of $Mo_{1.33}CT_z$ showed severe distortion at high scan rates (10 mV s⁻¹). Furthermore, the CV curves of mixed MXene also became distorted, as the scan

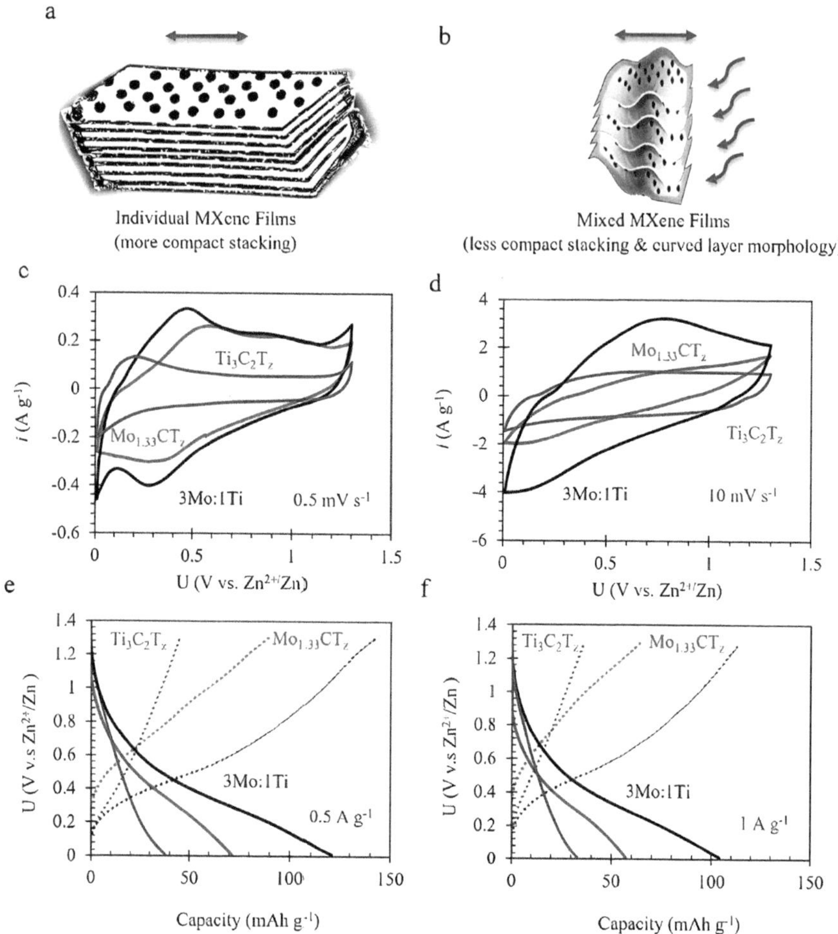

FIGURE 9.10 (a, b) Schematic illustration of the individual MXene films and mixed MXene films. (c, d) CVs at scan rates of 0.5 and 10 mV s^{-1}. (e, f) Charge (dotted lines) and discharge (solid lines) profiles at current densities of 0.5 and 1 A g^{-1}.

Source: Reprinted from [55] *Materials Today Energy*, 22, Ahmed S. Etman, Joseph Halim, Johanna Rosen, $Mo_{1.33}CT_z$–$Ti_3C_2T_z$ mixed MXene freestanding films for zinc-ion hybrid supercapacitors, 100878, Copyright 2021, with permission from Elsevier.

rates increased to 100 mV s^{-1}. A similar behavior was observed in the GCD experiments. The $Mo_{1.33}CT_z$ (red lines in Figs. 9.10e and f) film showed greater IR decline than the mixed MXene (black lines) and $Ti_3C_2T_z$ (blue lines).

9.3.4 MICRO-SUPERCAPACITORS

MXene-based micro-supercapacitors (M-MSCs) have helped to reduce the size of supercapacitors while providing high conductivity, flexibility, and capacitance. Most investigations on M-MSCs have used a planar interdigital architecture. This topology has clear advantages, such as a thinner device, fast ion transport, easy integration with on-chip electronics, and a good match between the 2D structure and planar geometry. Screen printing, vacuum filtering, stamping, and spray printing are the commonly used techniques for producing M-MSCs. These techniques are applicable to different substrates, ranging from silicon wafers to polymer films and flexible filter papers. The most important advantage of M-MSCs is their exceptional flexibility, which allows them to be easily molded into various shapes, making them ideal for various applications. Additionally, M-MSCs exhibit superior areal capacitance, which further enhances their utility [6].

Polymer-based gel electrolytes are considered the most suitable option for creating MSCs due to their added safety feature. Researchers have successfully developed planar M-MSCs, which are highly flexible due to the malleable planar MXene-based microelectrodes enclosed in a hydrogel electrolyte. However, planar M-MSCs have a significant drawback in terms of poor volumetric energy density. To address this, asymmetric M-MSCs were designed by using complementary electrodes such as MnO_2. Designing microelectrodes that strike a balance between volumetric and areal capacitance is a challenge. While simple, thicker microelectrodes may reduce volumetric capacitance, porous microelectrodes may reduce volumetric capacitance while increasing areal capacitance. So, there is a need to find a trade-off between the two for optimal microelectrode performance. However, high-throughput techniques such as screen printing make it easy to produce M-MSCs in various form factors. Moreover, the ability to integrate M-MSCs with microelectronics, such as sensors, in a single step is promising and opens up possibilities for the future of planar M-MSCs [6, 9].

The electrochemical properties of MXene-based MSCs are summarized in Table 9.4. H. Huang et al. [66] developed $Ti_3C_2T_x$-based on-chip MSCs using a treat–cutting–coating method. Interface adhesion of MXene-silicon can be implemented by hydrophilic treatment, leading to better MXene film integrity and uniformity. High volumetric and areal capacitances (21.4 F cm^{-3} and 472 F cm^{-2}, respectively) were reported for the resultant MSC, along with an 87.6% capacitance retention (after 10,000 cycles). Figure 9.11a [66] illustrates $Ti_3C_2T_x$-based on-chip MSCs with in-series, single, and parallel devices. On-chip MSCs can be easily coupled with other components into all-in-one microelectronic devices, thereby increasing integration density while reducing the size of the device (Fig. 9.11b). Figure 9.11c presents the detailed manufacturing steps.

For ejection printing at room temperature, highly concentrated MXene inks were observed to exhibit viscoelastic properties, which enable them to be employed in developing MSCs with diverse structures and properties [67]. This printing method facilitates the production of MSCs on paper or polymer substrates. As a result, the devices exhibit high values of areal capacitance (i.e., ~1035 mF cm^{-2}). MSCs printed on polymer substrates were bent and twisted at various angles and directions but did not shatter or detach from the substrate after the testing (Fig. 9.12a [67]). The electrochemical performances of the printed devices were examined and labeled as MSCF-n (n represents the number of electrode layers). CV curves of MSCF-1 at different scan rates are illustrated in Fig. 9.12b, whereas Fig. 9.12c compares the CV curves of printed devices for n = 1, 2, 5, and 10 at 5 mV s^{-1}.

TABLE 9.4

MXene-Based Micro-supercapacitors

Electrode Material	Capacitance $(F\ g^{-1})/\ (F\ cm^{-2})/\ (F\ cm^{-3})$	Cycling Stability	Energy Density (Wh kg^{-1}) and Power Density (W kg^{-1})	Ref.
$Ti_3C_2T_x$ MXene-based on-chip MSCs	472 $\mu F\ cm^{-2}$ 21.4 $F\ cm^{-3}$	87.6%/10,000 cycles	1.1 mWh cm^{-3} 189 mW cm^{-3}	[66]
$Ti_3C_2T_x$ based- MSCF-n	1035 $mF\ cm^{-22}$	–	1.7 $\mu Wh\ cm^{-2}$ 5.7 mW cm^{-2}	[67]
$Ti_3C_2T_x$ organic inks- based MSCs	562 $F\ cm^{-3}$	~97% and ~100%/16,000 cycles	0.32 $\mu W\ h\ cm^{-2}$ 11.4 $\mu W\ cm^{-2}$	[68]
$Ti_3C_2T_x$/PE	276 $mF\ cm^{-2}$	95%/1,000 cycles	28 $\mu Wh\ cm^{-2}$ and 12 $\mu Wh\ cm^{-2}$ 8 mW cm^{-2} and 1 mW cm^{-2}	[69]
L-s-$Ti_3C_2T_x$ MSC	27 $mF\ cm^{-2}$ 357 $F\ cm^{-3}$	100%/10,000 cycles	11–18 mWh cm^{-3} 0.7–15 W cm^{-3}	[70]
MXene-based symmetric MSCs	27.29 $mF\ cm^{-2}$	70%/5,000 cycles	5.48–6.10 mWh cm^{-3}	[71]
rGO//$Ti_3C_2T_x$	2.4 $mF\ cm^{-2}$ 80 $F\ cm^{-3}$	97%/10,000 cycles	8.6 mWh cm^{-3} 0.2 W cm^{-3}	[72]
SA-MXene	108.1 $mF\ cm^{-2}$ 720.7 $F\ cm^{-3}$	94.7%/4,000 cycles	100.2 mWh cm^{-3} 1.9 W cm^{-3}	[73]
$Ti_3C_2T_x$ based flexible solid-state MSCs	340 $mF\ cm^{-2}$ 183 $F\ cm^{-3}$	82.5%/5,000 cycles	12.4 mWh cm^{-3}	[74]
$Ti_3C_2T_x$-rGO -based MSC	34.6 $mF\ cm^{-2}$ At 1 $mV\ s^{-1}$	91%/15,000 cycles	1.33 $\mu Wh\ cm^{-2}$ 180 $\mu W\ cm^{-2}$	[75]
$Ti_3C_2T_x$-based MSC	7.8 $mF\ cm^{-2}$ 36.5 $F\ cm^{-3}$	91.4%/10,000 cycles	3.5 mWh cm^{-3} 100 W cm^{-3}	[76]
MXene/BC (bacterial cellulose) composite	111.5 $mF\ cm^{-2}$ At 2 $mA\ cm^{-2}$	100%/5,000 cycles	0.00552 mWh cm^{-2}	[77]
Laser-$Ti_3C_2T_x$-based MSC	322 $F\ g^{-1}$ 15.03 $mF\ cm^{-2}$ at 10 $mV\ s^{-1}$	105%/10,000 cycles	0.25 $\mu Wh\ cm^{-2}$ 2.94 mW cm^{-2}	[78]
Double-side MSC	52 $mF\ cm^{-2}$	86%/5,000 cycles	2.62 $\mu Wh\ cm^{-2}$	[79]

At different scan rates, fitting of specific capacitances revealed that improving the elevation of the electrodes can increase the areal capacitance of the devices (Fig. 9.12d, inset). The effect of bending on the performance of MSCF devices was explored to test their flexibility. The CV curves (at 10 mV s^{-1}) and rate capabilities of MSCFs are presented in Figs. 9.12e and f. The results demonstrated that with a change in bending angle, the rate capability, specific capacitance, and time constant do not change considerably, demonstrating the flexibility of the printed devices. Furthermore, the comparison in GCD curves of one, two, and three cell devices is given in Fig. 9.12g. The operational voltage window expands as expected for devices connected in series.

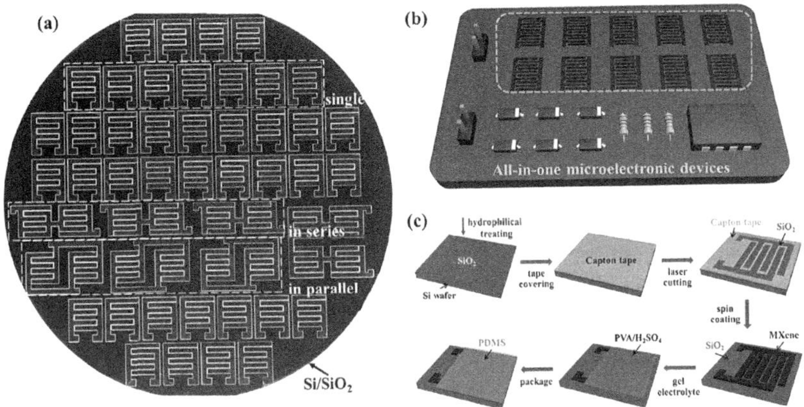

FIGURE 9.11 (a) Digital illustration of the scalable MXene-based on-chip MSCs. (b) Preview of the MSCs integrated with other components. (c) Detailed manufacturing process of on-chip MSCs.

Source: Reprinted from [66] *Nano Energy*, 69, H. Huang, J. He, Z. Wang, et al., Scalable, and low-cost treating-cutting-coating manufacture platform for MXene-based on-chip micro-supercapacitors, 104431, Copyright 2020, with permission from Elsevier.

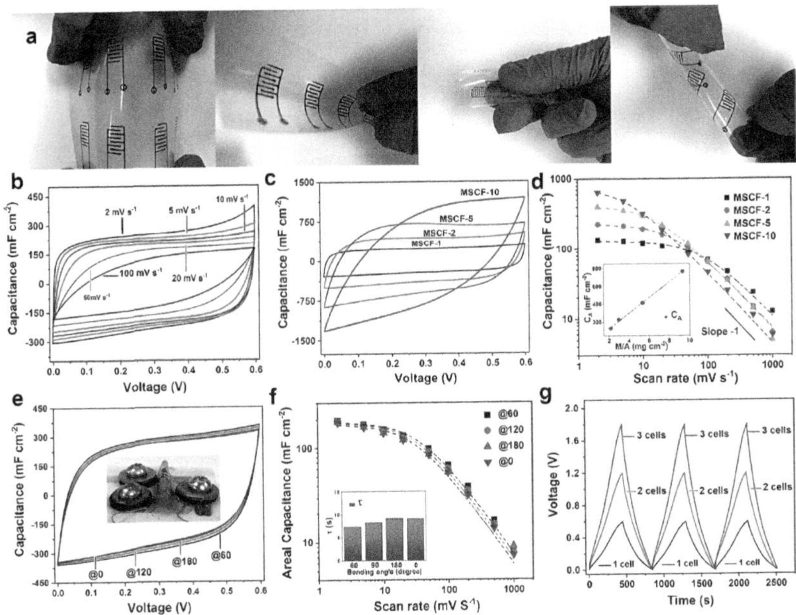

FIGURE 9.12 (a) Excellent adhesion of the printed devices to the substrate. (b) CV curves of MSCF-1 at various scan rates. (c) CV curves at 5 mV s⁻¹ for different MSCFs. (d) Areal capacitance vs. scan rate. (e) Electrochemical performance of MSCF-1 under various bending angles. (f) Areal capacitance vs. scan rate for MSCF-1. (g) Voltage profile of MSCs for different numbers of cells in series.

Source: Reprinted with permission from [67] Jafar Orangi, Fatima Hamade, Virginia A. Davis, et al. 2020. 3D Printing of Additive-Free 2D Ti$_3$C$_2$T$_x$ (MXene) Ink for Fabrication of Micro-Supercapacitors with Ultra-High Energy Densities. *ACS Nano*, 14, 640–650. Copyright 2020, American Chemical Society.

9.4 MXENE/CARBON-BASED ELECTRODES

MXenes are versatile materials that can be combined with a range of other materials, including rGO, CNTs, and others, that do not offer pseudocapacitance. When MXenes are combined with an rGO film, for instance, this film can penetrate the MXene layers, thereby preventing the restacking of layers and allowing the electrode material to retain a large surface area. Moreover, rGO films possess high conductivity, ensuring rapid charge transfer. Because both $Ti_3C_2T_x$ and rGO films are flexible, their composite flakes can be even more flexible. Technologies have been developed to optimize the structure of MXene/rGO composites, enhancing their electrochemical performance. MXene/CNT composites not only behave similarly to MXene/rGO composites, but they also establish a 3D electric network that connects both close and far MXene layers [80]. Table 9.5 summarizes the electrochemical behavior of MXene/carbon-based electrodes. These composites display better electrochemical behavior in supercapacitors with organic electrolytes. In particular, MXene/CNT composites generated through alternate filtration have demonstrated excellent cycling performance in $MgSO_4$ electrolyte, with increased gap between structures of the composite after cycling, thereby boosting capacitance [80].

Another work comprehensively characterized $Ti_3C_2T_x$ and a 3D $Ti_3C_2T_x$@NC nanostructure, demonstrating that NC was equitably decorated between the layers and on the surface of MXene sheets [81]. This nanostructure was observed to exhibit a large surface area (due to the NC layer), better pseudocapacitance, and high

TABLE 9.5

MXene/carbon-Based Electrodes

Electrode Material	Capacitance (F g^{-1})/ (F cm^{-2})/ (F cm^{-3})	Cycling Stability	Ref.
$Ti_3C_2T_x$@NC	442.2 F g^{-1}	91.9%/5,000 cycles	[81]
Ti_2CT_x//OLC (5%)	106 to 147 F g^{-1}	100%/10,000 cycles	[82]
$Ti_3C_2T_x$/CNT	22.7 F cm^{-3} at 0.1 A cm^{-3}	99%/1,500 cycles	[83]
CNT-$Ti_3C_2T_x$	85 F g^{-1} 245 F cm^{-3}	90%/1,000 cycles	[84]
MXene/CNT	300 F g^{-1} at 1 A g^{-1}	92%/10,000 cycles	[85]
d-Ti_3C_2	393 F cm^{-3}	80%/10,000 cycles	[86]
$Ti_3C_2T_x$/CNT	515.3 F g^{-1} 694 F cm^{-3} at 2 mV s^{-1}	95.3%/5,000 cycles	[87]
MXene/SWCNT	345 F cm^{-3} at 5 A g^{-1}	No degradation after 10,000 cycles	[88]
MXene//CNT-HQ	176 F g^{-1} at 2 mV s^{-1}	100%/5,000 cycles	[89]
BMX yarn-shaped supercapacitors (YSCs)	523 F g^{-1} 118 F cm^{-1} 1083 F cm^{-3} 3188 F cm^{-2}	97.4%/10,000 cycles	[90]

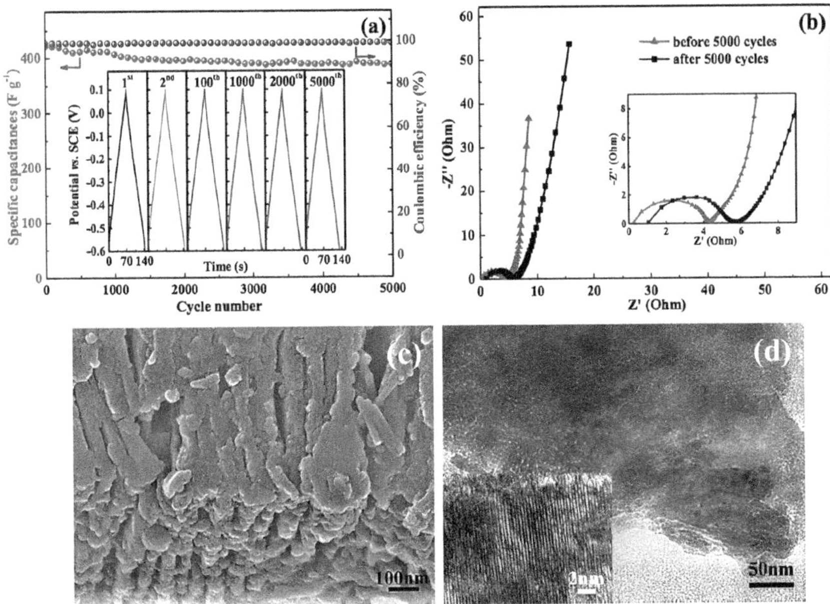

FIGURE 9.13 (a) Cycling performances of the $Ti_3C_2T_x$@NC-2 electrode at 4 A g^{-1} after 5,000 charge–discharge cycles; (b) Nyquist plots of the $Ti_3C_2T_x$@NC-2 electrode before and after cycles; (c) SEM image and (d) TEM image of $Ti_3C_2T_x$@NC-2 after cycles.

Source: Reprinted from [81] *Electrochimica Acta*, 254, T. Zhao, J. Zhang, et al., Dopamine-derived N-doped carbon decorated titanium carbide composite for enhanced supercapacitive performance, 308–319, Copyright 2017, with permission from Elsevier.

conductivity. As a result, the $Ti_3C_2T_x$@NC-2 had a high capacitance value, that is, 442.2 F g^{-1} at 1 A g^{-1}, which is better than that of $Ti_3C_2T_x$. Figure 9.13a [81] presents the GCD curves, demonstrating that the composite electrode has excellent cycling reversibility of 387.6 F g^{-1}, along with 91.9% initial capacitance retention that outperformed earlier MXene-based electrodes. Nyquist charts of $Ti_3C_2T_x$@NC-2 before and after 5,000 cycles are shown in Fig. 9.13b. After 5,000 cycles, diffusive impendence and an increase in charge transfer resistance were observed. The cycling stability after 5,000 cycles is confirmed by the SEM and TEM images of the composite electrode, as shown in Figs. 9.13c and d, respectively [81].

I. Habib et al. [82] investigated the effect of onion-like carbons (OLCs) on the supercapacitive characteristics of Ti_2C. The addition of OLCs was observed to increase the cycling stability of layered MXenes, from ~80% to more than 95%. It was reported that for a range of current densities (i.e., 0.5–5 A/g), the 5% OLC-doped sample outperformed the pristine sample in terms of energy and power density. The capacitance retention after 10,000 cycles is illustrated in Fig. 9.14 [82]. This shows that even after 10,000 cycles, OLC-doped samples retain their capacitance in comparison to the pristine material, which lost 20% of its initial capacitance.

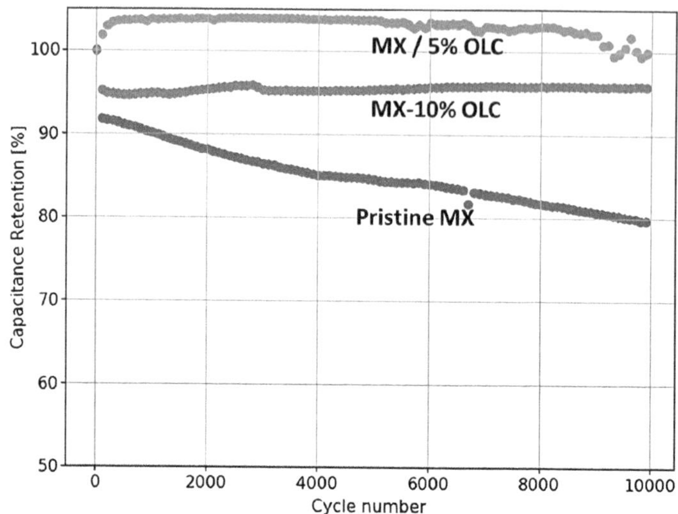

FIGURE 9.14 Plots of the capacitance retention for each sample: pristine and the 5% and 10% OLC-doped MXene samples.

Source: Reprinted with permission from [82] Habib I., Ferrer P. et al. 2019. Interrogating the impact of onion-like carbons on the supercapacitive properties of MXene (Ti$_2$CT$_x$). *J. Appl. Phys.* 126, 134301. Copyright 2019, American Institute of Physics.

9.5 MXENE/METAL OXIDE-BASED ELECTRODES

Metal oxides and conductive polymers are important components of pseudocapacitors due to their large capacity. As a result, they are often used in supercapacitor electrode materials. The high voltage in a MXene/metal oxide composite can lead to oxidation–reduction reactions, resulting in increased capacity in both MXenes and the metal oxides. The oxides form layers with a "sandwich structure" similar to that of Ti$_3$C$_2$T$_x$ layers. This layering structure results in a sequence of micro-capabilities. Moreover, the oxide layers act as septa, preventing restacking of the Ti$_3$C$_2$T$_x$ layers and increasing layer surfing. Furthermore, because the MXene layers are "cemented" in the oxide structures, they are hard to break down. Therefore, due to the better electrochemical properties and structural impacts of MXenes and oxides, the electrodes made up of their composite material can exhibit good cycling stability and high values of capacitance [80]. The capacitance and cycling stability of MXene/metal oxide-based electrodes are summarized in Table 9.6.

S. A. Zahra et al. [91] outlined a ZrO$_2$-MXene composite containing ZrO$_2$ nanoparticles produced on V$_2$CT$_x$ MXene sheets. The linked design increases conductivity while decreasing the possibility of restacking. In 3 M H$_2$SO$_4$, the ZrO$_2$-V$_2$CT$_x$ composite demonstrated an exceptional capacitance of 1200 F g^{-1} (at 5 mV s^{-1}), which is more than two times that of pure MXene. The composite was tested for cycling stability, and it retained 97% of its capacitance, after 10,000 cycles. M. Fatima et al. [92] performed a thorough experimental and theoretical investigation

TABLE 9.6
MXene/metal Oxide-Based Electrodes

Electrode Material	Capacitance (F g⁻¹)/ (F cm⁻²)/(F cm⁻³)	Cycling Stability	Electrolyte	Ref.
ZrO_2-V_2CT_x	$\approx 1{,}200$ F g⁻¹ at 5 mV s⁻¹	97%/10,000 cycles	3 M H_2SO_4	[91]
MnO_2-V_2C	551.8 F g⁻¹	96.5%/5,000 cycles	1 M KOH	[92]
$Mo_2C@MoO_3$	322 F g⁻¹ at 0.5 A g⁻¹	–	3 M Glycerol/ KOH	[93]
T-Nb_2O_5	330 C g⁻¹ 660 mF cm⁻²	–	1 M $LiClO_4$	[94]
$NiMoO_4$/$Ti_3C_2T_x$	483.8 C g⁻¹ at 1 A g⁻¹	72.6%/10,000 cycles	3 M KOH	[95]
Ti_3C_2/MnO_2	254 F g⁻¹	95.5%/5,000 cycles	6 M KOH	[96]
$Ti_3C_2T_x$/MnO_2 NWs	205 mF cm⁻² 1025 F cm⁻³	98.38%/10,000 cycles	PVA/LiCl	[97]
MnO_2/Ti_3C_2Tx	130.5 F g⁻¹	100%/1,000 cycles	PVA-H_2SO_4 gel	[98]
Co_3O_4 nanoparticles-MXene	1081 F g⁻¹ at 0.5 A g⁻¹	83%/8,000 cycles	6 M KOH	[99]
MoO_3/d-Ti_3C_2 films	631 F cm⁻³ at 1 A g⁻¹ 474 F cm⁻³ at 10 A g⁻¹	96.3%/20,000 cycles	5 M LiCl	[100]
$Co(OH)_2$/$Ti_3C_2T_x$	153.0 F g⁻¹	99%/1,000 cycles	1 M KOH	[101]
TiO_2-Ti_3C_2	143 F g⁻¹ at 5 mV s⁻¹	92%/6,000 cycles	6 M KOH	[102]
MnO_2-Ti_3C_2	377 mF cm⁻² at 5 mV s⁻¹	95%/5,000 cycles	6 M KOH	[103]
$Ti_3C_2T_x$/ACF	246.9 F g⁻¹ 197.5 mF cm⁻² at 4 mA cm⁻²	96.7%/5,000 cycles	PVA-KOH	[104]
(rGO)/MXene hybrids	345 F cm⁻³ 195 F g⁻¹ at 0.1 A g⁻¹	124.8%/7,500 cycles	1 M H_2SO_4	[105]
Ti_3C_2/FeOOH QDs	485 mF cm⁻²	94.8%/5,000 cycles	1 M Li_2SO_4	[106]
RuO_2@MXene	864.2 F cm⁻³ at 1 mV s⁻¹	90%/10,000 cycles	PVA-KOH	[107]
M/MoO_3 ($Ti_3C_2T_x$/MoO_3)	1817 F cm⁻³ 545 F g⁻¹	90%/5,000 cycles	1 M H_2SO_4	[108]
MXene/ Fe_3O_4/MXene	46.4 mF cm⁻² 0.5 mA cm⁻²	91.7%/2,000 cycles	1 M Li_2SO_4	[109]
FeOOH-$Ti_3C_2T_x$	217 F g⁻¹ at 1 A g⁻¹	81%/3,000 cycles	1 M Li_2SO_4	[110]
Co-Fe oxide/$Ti_3C_2T_x$	2467.6 F cm⁻³	88.2%/10,000 cycles	1 M LiCl	[111]
MXene/rGO	607.9 mF cm⁻² 463.0 F cm⁻³ at 2 μA	80%/1,000 cycles	PVA-H_3PO_4	[112]
Ti_3C_2Tx/ α-Fe_2O_3	405.4 F g⁻¹ at 2 A g⁻¹ 197.6 F g⁻¹ at 20 A g⁻¹	97.7%/2,000 cycles	5 M LiCl	[113]
$Ti_3C_2T_x$/RGO	140 F g⁻¹ 490 F cm⁻³	Under various strain conditions/10,000 cycles	PVA-H_2SO_4 gel	[114]
Ti_3C_2/ZnO	120 F g⁻¹ at 2 mV s⁻¹	85%/10,000 cycles	1 M KOH	[115]

of V_2CT_x and MnO_2-V_2C nanocomposites for supercapacitors. The MnO_2-V_2C nanocomposite attained a high value of capacitance, that is, 551.8 F g⁻¹ in 1 M KOH, which was observed to be double that for pristine MXene (i.e., 196.5 F g⁻¹), as shown in Fig. 9.15a [92]. Furthermore, 96.5% capacitive retention was reported for the nanocomposite, after 5,000 cycles for cycling stability (Fig. 9.15d). The comparison of the

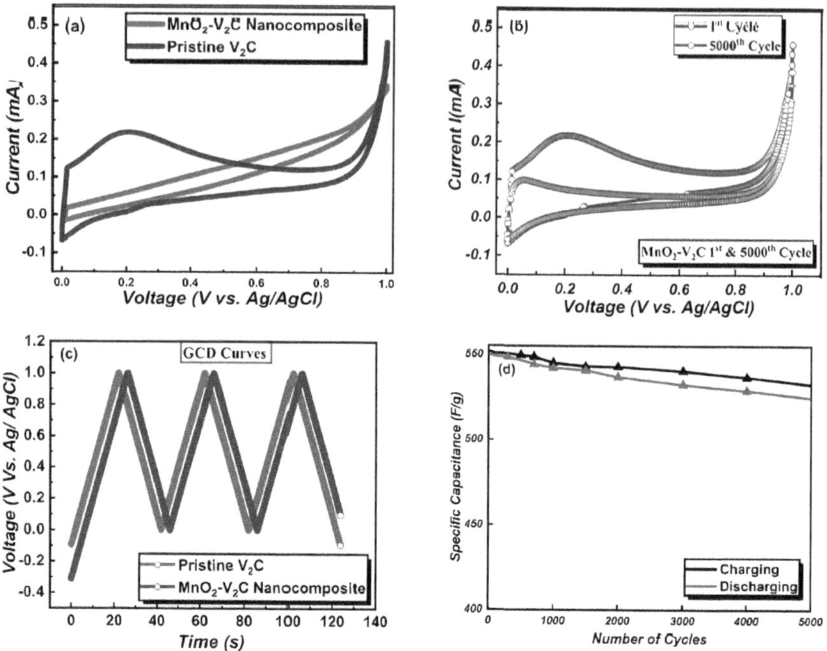

FIGURE 9.15 (a) Current vs. voltage of pristine MXene and MnO_2-V_2C nanocomposite. (b) Comparison of the first and the 5,000th cycle of the nanocomposite. (c) GCD curves of V_2C and MnO_2-V_2C. (d) Specific capacitance vs. number of cycles.

Source: Reprinted from [92] M. Fatima, S. A. Zahra, S. A. Khan, et al.: Experimental and Computational Analysis of $MnO_2@V_2C$-MXene for Enhanced Energy Storage. *Nanomaterials* 2021, 11(7), 1707. Open access.

first and the 5,000th cycle is presented in Fig. 9.15b. The GCD curves of pure and composite electrodes reveal a better performance of the composite electrode, which exhibits high capacitance even after 5,000 cycles [92].

In another study, electrodes based on nitrogen-doped $Mo_2C@MoO_3$ nanocomposites were developed, by considering carbon composite as a host and inserting these nanocomposites (N-MoCX) at different weight percentages (i.e., X = 1, 5, 10, or 15) [93]. Of all used percentages, N-MoC10 demonstrated better electrochemical properties by displaying a broader potential window (0–2 V) and a high capacitance of 322 F g^{-1} in 3 M KOH/glycerol electrolyte. Furthermore, a symmetrical device was also developed by using this nanocomposite, which showed a high energy and power density (37.5 Wh kg^{-1} and 2.495 W kg^{-1}, respectively). C. Zhang et al. [94] reported a simple method for producing a T-Nb_2O_5/carbon/Nb_2CT_x composite by oxidizing Nb_2CT_x in CO_2 in a single step. This composite was utilized as an electrode material (50-μm thick) of a hybrid device in a nonaqueous electrolyte, which demonstrated high areal and specific capacitances with good cycling performance.

The size of the Nb_2O_5 crystallites gradually increases with increasing oxidation temperature, which, combined with Li$^+$ intercalation/de-intercalation peaks in the

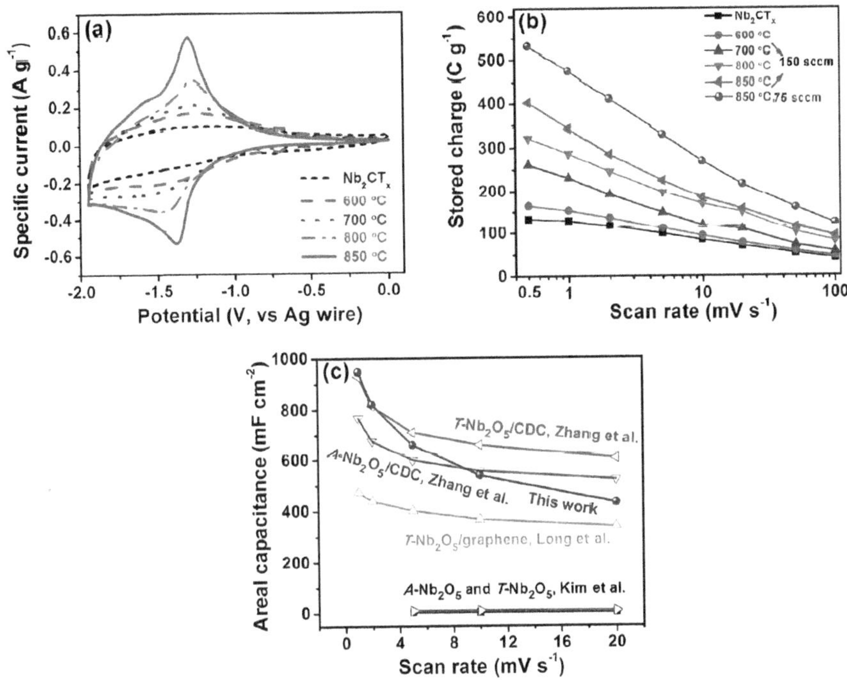

FIGURE 9.16 (a) CV curves at 1 mV s⁻¹ and (b) stored charge comparison of Nb_2CT_x and different oxidized Nb_2CT_x materials; (c) areal capacitance of oxidized Nb_2CT_x compared to other Nb_2O_5-based electrode systems.

Source: Reprinted with permission from [94] Chuanfang Zhang, Majid Beidaghi, Michael Naguib, et al. 2016. Synthesis and Charge Storage Properties of Hierarchical Niobium Pentoxide/Carbon/Niobium Carbide (MXene) Hybrid Materials. *Chem. Mater.*, 28, 11, 3937–3943. Copyright 2016, American Chemical Society.

CV curves, indicates that the values of capacitance are dependent on the formation of Nb_2O_5 (as shown in Fig. 9.16a [94]). It can be generalized that for improved charge storage, more active sites are produced during oxidation at 850°C (Fig. 9.16b). Extensive comparison of oxidized Nb_2CT_x with other materials in Fig. 9.16c provides a fuller perspective on the hybrid material's electrochemical properties. These results show that thin-film electrodes (or electrodes with low mass loading) typically possess more stored charge or high capacitance than thick-film electrodes.

9.6   MXENE/CONDUCTIVE POLYMER-BASED ELECTRODES

MXene/conductive polymer composites exhibit excellent electrical conductivity, which facilitates the charge transport on polymer networks, along with fast charging or discharging. Repeated cycles of this process may alter the bulk of the polymer in polymer-based electrodes. However, in composite materials, such as $Ti_3C_2T_x$/PPy, the PPy hydrogen bonds can firmly hold the surface groups on the MXene surface. Consequently, PPy and $Ti_3C_2T_x$ films have greater stability. These two materials can maintain their structures due to their integral bonding, resulting in a supercapacitor with excellent cycle performance. Furthermore, the polymer can form bonds with distant $Ti_3C_2T_x$ flakes to reduce electrical resistance and increase the charge transfer. As a result, they exhibit good performance and a stable volume capacity, and it may be employed as an all-solid-state supercapacitor with exceptional cycling stability [80]. MXene/conductive polymer-based electrodes are summarized in Table 9.7.

The surface of MXenes has a large number of electrochemically active sites, which facilitate aniline polymerization. This process would prevent the MXene layers from stacking, thereby increasing electrochemical activity. X. Wang et al. [116] reported the use of $PANI/V_2C$ composites as an electrode material for supercapacitors. As a result, a capacitance of 337.5 F g^{-1} (at 1 A g^{-1}) and high values of energy and power densities (11.25 Wh kg^{-1} and 415.38 W kg^{-1}, respectively) were observed by using AC as the cathode and this composite as the anode. Figure 9.17a [116] depicts the CV curves of as-prepared samples from −0.6 to 0.2 V at a scan rate of 10 mV s^{-1}. Figure 9.17b illustrates the GCD curves of a PANI/MXene electrode at various

TABLE 9.7
MXene/conductive Polymer-Based Electrodes

Electrode Material	Capacitance (F g^{-1})/(F cm^{-2})/(F cm^{-3})	Cycling Stability	Energy Density (Wh kg^{-1}) and Power Density (W kg^{-1})	Ref.
CT-S@ Ti$_3$C$_2$_450//AC	231.5, 218, 196.5, 176.5, 167, and 138.5 F g^{-1}	73.3%/10,000 cycles	263.2 Wh kg^{-1} 8240 W kg^{-1}	[118]
N-doped Ti$_2$CT$_x$	327 F g^{-1} at 1 A g^{-1}	96.2%/5,000 cycles	19.9 Wh kg^{-1} 10.6 kW kg^{-1}	[119]
Mo$_{1.33}$C MXene/PEDOT:PSS	1310 F cm^{-3}	84%/10,000 cycles	33.2 mWh cm^{-3} 19,470 mW cm^{-3}	[120]
CTAB–Sn(IV) @Ti$_3$C$_2$	268, 220, 181, 173, and 132 F g^{-1}	71.1%/4,000 cycles	239.50 Wh kg^{-1} 10.8 kW kg^{-1}	[57]
PANI/MXene	337.5 F g^{-1} at 1 A g^{-1}	97.6%/10,000 cycles	11.25 Wh kg^{-1} 415.38 W kg^{-1}	[116]
Sn@Mo$_2$TiC$_2$	670 F g^{-1}	99%/10,000 cycles	–	[117]
EDA-Ti$_3$C$_2$T$_x$	249.4 F g^{-1} 673.4 F cm^{-3}	94.3%/10,000 cycles at 1 A/g 89.7%/10,000 cycles at 10 A/g	–	[121]
Ti$_3$C$_2$T$_x$//rGO/CNT/ PANI	1277 F cm^{-3}	89.3%/10,000 cycles	70 Wh L^{-1} 111 kW L^{-1}	[122]
Ti$_3$C$_2$/PANI-NTs-1	596.6 F g^{-1} at 0.1 A g^{-1}	94.7%/5,000 cycles	25.6 Wh kg^{-1} 1610.8 W kg^{-1}	[123]
PANI@TiO$_2$/Ti$_3$C$_2$T$_x$	188.3 F g^{-1} at 10 mV s^{-1}	94%/8,000 cycles	–	[124]
PPy/l-Ti$_3$C$_2$	35 mF cm^{-2}	~100%/20,000 cycles	10 mWh/cm^{-3} 500 mW/cm^{-3}	[125]
PEDOT/Ti$_3$C$_2$T$_x$	455 µF cm^{-2} at 10 mV s^{-1}, 120 µF cm^{-2} at 1,000 mV s^{-1}	90%/10,000 cycles	8.7 mWh/cm^{-3} 4.5 W cm^{-3}	[126]

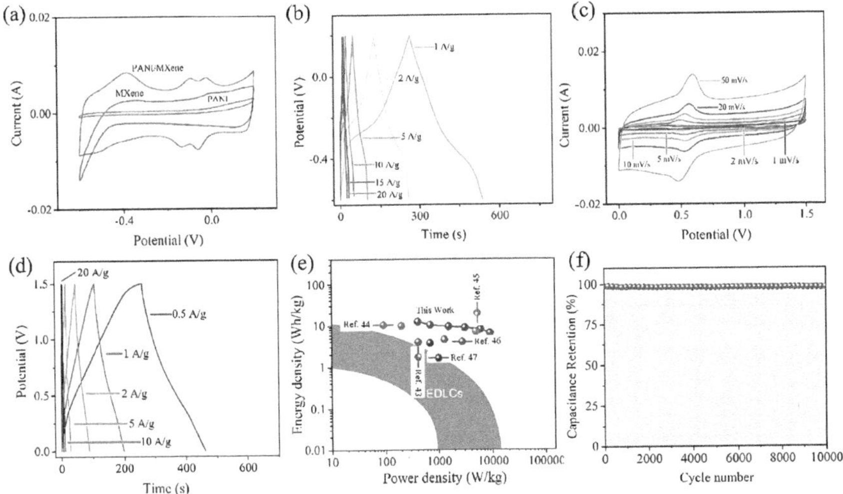

FIGURE 9.17 (a) CV curves of electrode materials at 5 mV s^{-1}. (b) GCD curves of the PANI/ MXene electrode. (c) CV curves of the button supercapacitor at 5 mV s^{-1}. (d) CV curves of the button supercapacitor at different scan rates. (e) Ragone plots of the devices compared with previously reported results. (f) Cycling performances of the device at 5 A g^{-1}.

Source: Reprinted from [116] *Nano Energy*, 88, X. Wang et al., In situ polymerized polyaniline/MXene (V$_2$C) as building blocks of supercapacitor and ammonia sensor self-powered by electromagnetic-triboelectric hybrid generator, 106242, Copyright 2021, with permission from Elsevier.

current densities; the nonlinear nature of the curves suggests typical pseudocapacitive behavior. Because of its great coulomb efficiency, the curve exhibits extraordinarily high symmetry.

The CV curves of the devices exhibit identical shapes from 1 mV s^{-1} to 50 mV s^{-1} (as presented in Fig. 9.17c), indicating high-rate performance. Figure 9.17d shows the GCD curves with varying current densities. As demonstrated in Fig. 9.17e, the devices outperform others in terms of electrochemical performance, with high energy and power densities. Figure 9.17f depicts the cycling stability of the device with 97.6% retention of its initial capacitance after 10,000 cycles [116].

The Sn^{2+}-intercalated Mo$_2$TiC$_2$ (or Sn@Mo$_2$TiC$_2$) exhibits lower bandgap (1.3 eV), larger interlayer spacing (1.47 nm), higher surface area (30 m^2 g^{-1}), and high specific capacitance (670 F g^{-1} at 2 mVs^{-1}), compared to the pristine MXene [117]. The composite electrode also demonstrated coulombic efficiency, excellent capacity retention, and better cycling performance over 10,000 cycles. Figure 9.18a [117] shows that at all scan rates, the peak electrode currents for the Sn@Mo$_2$TiC$_2$ are higher than that of the pristine MXene, which indicates the ability of the hybrid material to store charge and ion intercalation/de-intercalation due to its good electrochemical activity. A comparison with previous hybrid electrodes in different electrolytes is presented in Fig. 9.18b. Furthermore, Fig. 9.18c illustrates the cycling performance of a hybrid electrode at 10 A g^{-1}.

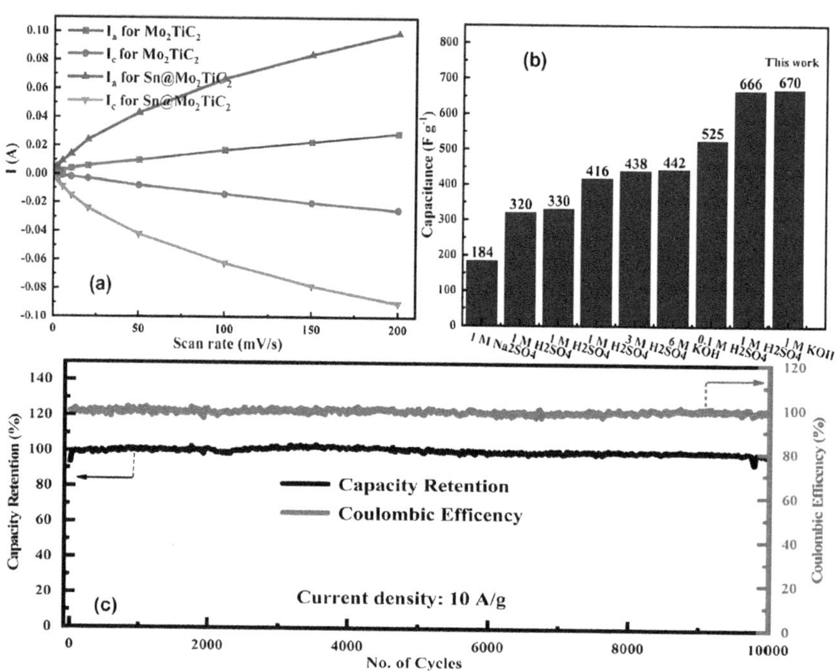

FIGURE 9.18 (a) Anodic and cathodic peak currents vs. the scan rates calculated for both Mo$_2$TiC$_2$ and Sn@Mo$_2$TiC$_2$ electrodes. (b) Comparison of hybrid electrode capacitance with the reported data in the literature. (c) Long-term cycling performance of a nanohybrid at 10 A g^{-1}.

Source: [117] Irfan Ali, Zulqarnain Haider and Syed Rizwan. 2022. Enhanced pseudocapacitive energy storage and thermal stability of Sn^{2+} ion-intercalated molybdenum titanium carbide (Mo$_2$TiC$_2$) MXene. *RSC Adv.*, 12, 31923–31934. Reproduced with permission from the Royal Society of Chemistry.

REFERENCES

[1] Pang, J., et al., *Applications of 2D MXenes in energy conversion and storage systems.* Chemical Society Reviews, 2019. **48**(1): p. 72–133.

[2] Ibrahim, Y., et al., *The recent advances in the mechanical properties of self-standing two-dimensional MXene-based nanostructures: deep insights into the supercapacitor.* Nanomaterials, 2020. **10**(10): p. 1916.

[3] Dey, A., et al., *Doped MXenes—a new paradigm in 2D systems: synthesis, properties and applications.* Progress in Materials Science, 2023: p. 101166.

[4] Zhang, C., et al., *Two-dimensional transition metal carbides and nitrides (MXenes): synthesis, properties, and electrochemical energy storage applications.* Energy & Environmental Materials, 2020. **3**(1): p. 29–55.

[5] Lokhande, P., et al., *Prospects of MXenes in energy storage applications.* Chemosphere, 2022. **297**: p. 134225.

[6] Das, P. and Z.-S. Wu, *MXene for energy storage: present status and future perspectives.* Journal of Physics: Energy, 2020. **2**(3): p. 032004.

[7] Meng, W., et al., *Advances and challenges in 2D MXenes: from structures to energy storage and conversions.* Nano Today, 2021. **40**: p. 101273.

[8] Forouzandeh, P. and S.C. Pillai, *MXenes-based nanocomposites for supercapacitor applications*. Current Opinion in Chemical Engineering, 2021. **33**: p. 100710.

[9] Hu, M., et al., *Emerging 2D MXenes for supercapacitors: status, challenges and prospects*. Chemical Society Reviews, 2020. **49**(18): p. 6666–6693.

[10] Sun, S., et al., *Two-dimensional MXenes for energy storage*. Chemical Engineering Journal, 2018. **338**: p. 27–45.

[11] Shao, H., et al., *Unraveling the charge storage mechanism of $Ti_3C_2T_x$ MXene electrode in acidic electrolyte*. ACS Energy Letters, 2020. **5**(9): p. 2873–2880.

[12] Wang, H. and X. Wu, *High capacitance of dipicolinic acid-intercalated MXene in neutral water-based electrolyte*. Chemical Engineering Journal, 2020. **399**: p. 125850.

[13] Wang, X., et al., *Influences from solvents on charge storage in titanium carbide MXenes*. Nature Energy, 2019. **4**(3): p. 241–248.

[14] Meshkian, R., et al., *Theoretical analysis, synthesis, and characterization of 2D $W_{1.33}C$ (MXene) with ordered vacancies*. ACS Applied Nano Materials, 2019. **2**(10): p. 6209–6219.

[15] Guan, Y., et al., *A hydrofluoric acid-free synthesis of 2D vanadium carbide (V_2C) MXene for supercapacitor electrodes*. 2D Materials, 2020. **7**(2): p. 025010.

[16] Ai, W., et al., *Synthesis of high-quality two-dimensional V_2C MXene for supercapacitor application*. Energies, 2022. **15**(10): p. 3696.

[17] Shan, Q., et al., *Two-dimensional vanadium carbide (V_2C) MXene as electrode for supercapacitors with aqueous electrolytes*. Electrochemistry Communications, 2018. **96**: p. 103–107.

[18] Syamsai, R. and A.N. Grace, *Synthesis, properties and performance evaluation of vanadium carbide MXene as supercapacitor electrodes*. Ceramics International, 2020. **46**(4): p. 5323–5330.

[19] Xin, Y. and Y.-X. Yu, *Possibility of bare and functionalized niobium carbide MXenes for electrode materials of supercapacitors and field emitters*. Materials & Design, 2017. **130**: p. 512–520.

[20] Zhao, S., et al., *Flexible $Nb_4C_3T_x$ film with large interlayer spacing for high-performance supercapacitors*. Advanced Functional Materials, 2020. **30**(47): p. 2000815.

[21] Syamsai, R. and A.N. Grace, *Ta_4C_3 MXene as supercapacitor electrodes*. Journal of Alloys and Compounds, 2019. **792**: p. 1230–1238.

[22] Ma, W., et al., *A new $Ti_2V_{0.9}Cr_{0.1}C_2T_x$ MXene with ultrahigh gravimetric capacitance*. Nano Energy, 2022. **96**: p. 107129.

[23] Gandla, D., F. Zhang, and D.Q. Tan, *Advantage of larger interlayer spacing of a $Mo_2Ti_2C_3$ MXene free-standing film electrode toward an excellent performance supercapacitor in a binary ionic liquid–organic electrolyte*. ACS Omega, 2022. **7**(8): p. 7190–7198.

[24] Etman, A.S., J. Halim, and J. Rosen, *Fabrication of $Mo_{1.33}CT_z$ (MXene)–cellulose free-standing electrodes for supercapacitor applications*. Materials Advances, 2021. **2**(2): p. 743–753.

[25] VahidMohammadi, A., et al., *Assembling 2D MXenes into highly stable pseudocapacitive electrodes with high power and energy densities*. Advanced Materials, 2019. **31**(8): p. 1806931.

[26] Yang, C., et al., *Flexible nitrogen-doped 2D titanium carbides (MXene) films constructed by an ex situ solvothermal method with extraordinary volumetric capacitance*. Advanced Energy Materials, 2018. **8**(31): p. 1802087.

[27] Pan, Z. and X. Ji, *Facile synthesis of nitrogen and oxygen co-doped C@ Ti_3C_2 MXene for high performance symmetric supercapacitors*. Journal of Power Sources, 2019. **439**: p. 227068.

[28] Zhang, X., et al., *Surface modified MXene film as flexible electrode with ultrahigh volumetric capacitance*. Electrochimica Acta, 2019. **294**: p. 233–239.

[29] Wang, X., et al., *2D/2D 1T-MoS₂/Ti₃C₂ MXene heterostructure with excellent supercapacitor performance*. Advanced Functional Materials, 2020. **30**(15): p. 0190302.

[30] Fan, Z., et al., *A compact MXene film with folded structure for advanced supercapacitor electrode material*. ACS Applied Energy Materials, 2020. **3**(2): p. 1811–1820.

[31] Xia, Q.X., et al., *Bismuth oxychloride/MXene symmetric supercapacitor with high volumetric energy density*. Electrochimica Acta, 2018. **271**: p. 351–360.

[32] Yan, J., et al., *Flexible MXene/graphene films for ultrafast supercapacitors with outstanding volumetric capacitance*. Advanced Functional Materials, 2017. **27**(30): p. 1701264.

[33] Zhang, P., et al., *In situ ice template approach to fabricate 3D flexible MXene film-based electrode for high performance supercapacitors*. Advanced Functional Materials, 2020. **30**(47): p. 2000922.

[34] Venkateshalu, S., et al., *New method for the synthesis of 2D vanadium nitride (MXene) and its application as a supercapacitor electrode*. ACS Omega, 2020. **5**(29): p. 17983–17992.

[35] El Ghazaly, A., et al., *Enhanced supercapacitive performance of Mo₁.₃₃C MXene-based asymmetric supercapacitors in lithium chloride electrolyte*. Energy Storage Materials, 2021. **41**: p. 203–208.

[36] Yang, C., et al., *Methanol and diethanolamine assisted synthesis of flexible nitrogen-doped Ti₃C₂ (MXene) film for ultrahigh volumetric performance supercapacitor electrodes*. ACS Applied Energy Materials, 2019. **3**(1): p. 586–596.

[37] Wu, W., et al., *Hierarchical materials constructed by 1D hollow nickel–cobalt sulfide nanotubes supported on 2D ultrathin MXenes nanosheets for high-performance supercapacitor*. Ceramics International, 2020. **46**(8): p. 12200–12208.

[38] Etman, A.S., J. Halim, and J. Rosen, *Mixed MXenes: Mo₁.₃₃CTz and Ti₃C₂Tz freestanding composite films for energy storage*. Nano Energy, 2021. **88**: p. 106271.

[39] Xiao, J., et al., *A safe etching route to synthesize highly crystalline Nb₂CTₓ MXene for high performance asymmetric supercapacitor applications*. Electrochimica Acta, 2020. **337**: p. 135803.

[40] Jiang, Q., et al., *All pseudocapacitive MXene-RuO₂ asymmetric supercapacitors*. Advanced Energy Materials, 2018. **8**(13): p. 1703043.

[41] Pan, J., et al., *Designed formation of 2D/2D hierarchical V₂CTₓ MXene/NiV layered double hydroxide heterostructure with boosted electrochemical performance for asymmetric supercapacitors*. Journal of Energy Storage, 2022. **55**: p. 105415.

[42] Shen, B., et al., *Synthesis of Nb₂C MXene-based 2D layered structure electrode material for high-performance battery-type supercapacitors*. Electrochimica Acta, 2022. **413**: p. 140144.

[43] Ma, L., et al., *A dual utilization strategy of lignosulfonate for MXene asymmetric supercapacitor with high area energy density*. Chemical Engineering Journal, 2021. **405**: p. 126694.

[44] Li, L., et al., *Ag-nanoparticle-decorated 2D titanium carbide (MXene) with superior electrochemical performance for supercapacitors*. ACS Sustainable Chemistry & Engineering, 2018. **6**(6): p. 7442–7450.

[45] Navarro-Suárez, A.M., et al., *Development of asymmetric supercapacitors with titanium carbide-reduced graphene oxide couples as electrodes*. Electrochimica Acta, 2018. **259**: p. 752–761.

[46] Pan, Z., et al., *A facile method for synthesizing CuS decorated Ti₃C₂ MXene with enhanced performance for asymmetric supercapacitors*. Journal of Materials Chemistry A, 2019. **7**(15): p. 8984–8992.

[47] Park, J.W., et al., *Highly loaded MXene/carbon nanotube yarn electrodes for improved asymmetric supercapacitor performance*. MRS Communications, 2019. **9**(1): p. 114–121.

[48] Ramachandran, R., et al., *Influence of Ti$_3$C$_2$T$_x$ (MXene) intercalation pseudocapacitance on electrochemical performance of Co-MOF binder-free electrode*. Ceramics International, 2018. **44**(12): p. 14425–14431.

[49] Zhao, R., et al., *Molecular-level heterostructures assembled from titanium carbide MXene and Ni–Co–Al layered double-hydroxide nanosheets for all-solid-state flexible asymmetric high-energy supercapacitors*. ACS Energy Letters, 2017. **3**(1): p. 132–140.

[50] Chandran, M., et al., *MoS$_2$ confined MXene heterostructures as electrode material for energy storage application*. Journal of Energy Storage, 2020. **30**: p. 101446.

[51] Li, L., et al., *New Ti$_3$C$_2$ aerogel as promising negative electrode materials for asymmetric supercapacitors*. Journal of Power Sources, 2017. **364**: p. 234–241.

[52] Liu, H., et al., *One-step synthesis of nanostructured CoS$_2$ grown on titanium carbide MXene for high-performance asymmetrical supercapacitors*. Advanced Materials Interfaces, 2020. **7**(6): p. 1901659.

[53] Xu, X., et al., *MXenes with applications in supercapacitors and secondary batteries: a comprehensive review*. Materials Reports: Energy, 2022. **2**(1): p. 100080.

[54] Dall'Agnese, Y., et al., *Two-dimensional vanadium carbide (MXene) as positive electrode for sodium-ion capacitors*. The Journal of Physical Chemistry Letters, 2015. **6**(12): p. 2305–2309.

[55] Etman, A.S., J. Halim, and J. Rosen, *Mo$_{1.33}$CTz–Ti$_3$C$_2$T$_z$ mixed MXene freestanding films for zinc-ion hybrid supercapacitors*. Materials Today Energy, 2021. **22**: p. 100878.

[56] Peng, M., et al., *Manipulating the interlayer spacing of 3D MXenes with improved stability and zinc-ion storage capability*. Advanced Functional Materials, 2022. **32**(7): p. 2109524.

[57] Luo, J., et al., *Pillared structure design of MXene with ultralarge interlayer spacing for high-performance lithium-ion capacitors*. ACS Nano, 2017. **11**(3): p. 2459–2469.

[58] Ranjan, B., G.K. Sharma, and D. Kaur, *Rationally synthesized Mo2N nanopyramids for high-performance flexible supercapacitive electrodes with deep insight into the Na-ion storage mechanism*. Applied Surface Science, 2022. **588**: p. 152925.

[59] Fan, Z., et al., *3D printing of porous nitrogen-doped Ti$_3$C$_2$ MXene scaffolds for high-performance sodium-ion hybrid capacitors*. ACS Nano, 2020. **14**(1): p. 867–876.

[60] Zhu, L., et al., *Pseudocapacitance of Cl-terminated MXene nanosheets for efficient chloride-ion hybrid capacitors*. Energy & Fuels, 2023. **37**(7): p. 5607–5612.

[61] Ming, F., et al., *Porous MXenes enable high performance potassium ion capacitors*. Nano Energy, 2019. **62**: p. 853–860.

[62] Wang, X., et al., *Heterostructures of Ni–Co–Al layered double hydroxide assembled on V$_4$C$_3$ MXene for high-energy hybrid supercapacitors*. Journal of Materials Chemistry A, 2019. **7**(5): p. 2291–2300.

[63] Singh, M.K., S. Krishnan, and D.K. Rai, *Rational design of Ti$_3$C$_2$T$_x$ MXene coupled with hierarchical CoS for a flexible supercapattery*. Electrochimica Acta, 2023. **441**: p. 141825.

[64] Vyskočil, J., et al., *2D stacks of MXene Ti$_3$C$_2$ and 1T-phase WS2 with enhanced capacitive behavior*. ChemElectroChem, 2019. **6**(15): p. 3982–3986.

[65] Kurra, N., et al., *Bistacked titanium carbide (MXene) anodes for hybrid sodium-ion capacitors*. ACS Energy Letters, 2018. **3**(9): p. 2094–2100.

[66] Huang, H., et al., *Scalable, and low-cost treating-cutting-coating manufacture platform for MXene-based on-chip micro-supercapacitors*. Nano Energy, 2020. **69**: p. 104431.

[67] Orangi, J., et al., *3D printing of additive-free 2D Ti$_3$C$_2$T$_x$ (MXene) ink for fabrication of micro-supercapacitors with ultra-high energy densities*. ACS Nano, 2019. **14**(1): p. 640–650.

[68] Zhang, C., et al., *Additive-free MXene inks and direct printing of micro-supercapacitors*. Nature Communications, 2019. **10**(1): p. 1795.

[69] Xu, S., et al. *A MXene based all-solid-state microsupercapacitor with 3D interdigital electrode*, in *2017 19th International Conference on Solid-State Sensors, Actuators and Microsystems (Transducers)*. 2017, IEEE.

[70] Peng, Y.-Y., et al., *All-MXene (2D titanium carbide) solid-state microsupercapacitors for on-chip energy storage*. Energy & Environmental Science, 2016. **9**(9): p. 2847–2854.

[71] Hu, H. and T. Hua, *An easily manipulated protocol for patterning of MXenes on paper for planar micro-supercapacitors*. Journal of Materials Chemistry A, 2017. **5**(37): p. 19639–19648.

[72] Couly, C., et al., *Asymmetric flexible MXene-reduced graphene oxide micro-supercapacitor*. Advanced Electronic Materials, 2018. **4**(1): p. 1700339.

[73] Wu, C.-W., et al., *Excellent oxidation resistive MXene aqueous ink for micro-supercapacitor application*. Energy Storage Materials, 2020. **25**: p. 563–571.

[74] Huang, H., et al., *Extraordinary areal and volumetric performance of flexible solid-state micro-supercapacitors based on highly conductive freestanding $Ti_3C_2T_x$ films*. Advanced Electronic Materials, 2018. **4**(8): p. 1800179.

[75] Yue, Y., et al., *Highly self-healable 3D microsupercapacitor with MXene—graphene composite aerogel*. ACS Nano, 2018. **12**(5): p. 4224–4232.

[76] Xie, Y., et al., *High-voltage asymmetric MXene-based on-chip micro-supercapacitors*. Nano Energy, 2020. **74**: p. 104928.

[77] Jiao, S., et al., *Kirigami patterning of MXene/bacterial cellulose composite paper for all-solid-state stretchable micro-supercapacitor arrays*. Advanced Science, 2019. **6**(12): p. 1900529.

[78] Tang, J., et al., *Laser writing of the restacked titanium carbide MXene for high performance supercapacitors*. Energy Storage Materials, 2020. **32**: p. 418–424.

[79] Wang, N., et al., *Laser-cutting fabrication of MXene-based flexible micro-supercapacitors with high areal capacitance*. ChemNanoMat, 2019. **5**(5): p. 658–665.

[80] Yang, J., et al., *MXene-based composites: synthesis and applications in rechargeable batteries and supercapacitors*. Advanced Materials Interfaces, 2019. **6**(8): p. 1802004.

[81] Zhao, T., et al., *Dopamine-derived N-doped carbon decorated titanium carbide composite for enhanced supercapacitive performance*. Electrochimica Acta, 2017. **254**: p. 308–319.

[82] Habib, I., et al., *Interrogating the impact of onion-like carbons on the supercapacitive properties of MXene (Ti_2CT_x)*. Journal of Applied Physics, 2019. **126**(13).

[83] Yu, C., et al., *A solid-state fibriform supercapacitor boosted by host—guest hybridization between the carbon nanotube scaffold and MXene nanosheets*. Small, 2018. **14**(29): p. 1801203.

[84] Dall'Agnese, Y., et al., *Capacitance of two-dimensional titanium carbide (MXene) and MXene/carbon nanotube composites in organic electrolytes*. Journal of Power Sources, 2016. **306**: p. 510–515.

[85] Chen, H., et al., *Carbon nanotubes enhance flexible MXene films for high-rate supercapacitors*. Journal of Materials Science, 2020. **55**: p. 1148–1156.

[86] Yan, P., et al., *Enhanced supercapacitive performance of delaminated two-dimensional titanium carbide/carbon nanotube composites in alkaline electrolyte*. Journal of Power Sources, 2015. **284**: p. 38–43.

[87] Li, L., et al., *The facile synthesis of layered Ti_2C MXene/carbon nanotube composite paper with enhanced electrochemical properties*. Dalton Transactions, 2017. **46**(43): p. 14880–14887.

[88] Zhao, M.Q., et al., *Flexible MXene/carbon nanotube composite paper with high volumetric capacitance*. Advanced Materials, 2015. **27**(2): p. 339–345.

[89] Hu, M., et al., *High-energy-density hydrogen-ion-rocking-chair hybrid supercapacitors based on $Ti_3C_2T_x$ MXene and carbon nanotubes mediated by redox active molecule*. ACS Nano, 2019. **13**(6): p. 6899–6905.

[90] Wang, Z., et al., *High-performance biscrolled MXene/carbon nanotube yarn supercapacitors.* Small, 2018. **14**(37): p. 1802225.

[91] Zahra, S.A., E. Ceesay, and S. Rizwan, *Zirconia-decorated V_2CT_x MXene electrodes for supercapacitors.* Journal of Energy Storage, 2022. **55**: p. 105721.

[92] Fatima, M., et al., *Experimental and computational analysis of Mno_2@ V_2C-Mxene for enhanced energy storage.* Nanomaterials, 2021. **11**(7): p. 1707.

[93] Cevik, E., et al., *Synthesis of hierarchical multilayer N-doped Mo_2C@ MoO_3 nanostructure for high-performance supercapacitor application.* Journal of Energy Storage, 2022. **46**: p. 103824.

[94] Zhang, C., et al., *Synthesis and charge storage properties of hierarchical niobium pentoxide/carbon/niobium carbide (MXene) hybrid materials.* Chemistry of Materials, 2016. **28**(11): p. 3937–3943.

[95] Wang, Y., et al., *2D/2D heterostructures of nickel molybdate and MXene with strong coupled synergistic effect towards enhanced supercapacitor performance.* Journal of Power Sources, 2019. **414**: p. 540–546.

[96] Yuan, W., et al., *2D-Ti_3C_2 as hard, conductive substrates to enhance the electrochemical performance of MnO_2 for supercapacitor applications.* Ceramics International, 2018. **44**(14): p. 17539–17543.

[97] Zhou, J., et al., *A conductive and highly deformable all-pseudocapacitive composite paper as supercapacitor electrode with improved areal and volumetric capacitance.* Small, 2018. **14**(51): p. 1803786.

[98] Jiang, H., et al., *A novel $MnO_2/Ti_3C_2T_x$ MXene nanocomposite as high performance electrode materials for flexible supercapacitors.* Electrochimica Acta, 2018. **290**: p. 695–703.

[99] Zhang, Y., et al., *Assembling Co_3O_4 nanoparticles into MXene with enhanced electrochemical performance for advanced asymmetric supercapacitors.* Journal of Colloid and Interface Science, 2021. **599**: p. 109–118.

[100] Zheng, W., et al., *Boosting the volumetric capacitance of MoO_3-x free-standing films with Ti_3C_2 MXene.* Electrochimica Acta, 2021. **370**: p. 137665.

[101] Jiang, H., et al., *$Co(OH)_2$/MXene composites for tunable pseudo-capacitance energy storage.* Electrochimica Acta, 2020. **353**: p. 136607.

[102] Zhu, J., et al., *Composites of TiO_2 nanoparticles deposited on Ti_3C_2 MXene nanosheets with enhanced electrochemical performance.* Journal of the Electrochemical Society, 2016. **163**(5): p. A785.

[103] Tang, Y., et al., *Enhanced supercapacitive performance of manganese oxides doped two-dimensional titanium carbide nanocomposite in alkaline electrolyte.* Journal of Alloys and Compounds, 2016. **685**: p. 194–201.

[104] Patil, A.M., et al., *Fabrication of a high-energy flexible all-solid-state supercapacitor using pseudocapacitive 2D-$Ti_3C_2T_x$-MXene and battery-type reduced graphene oxide/nickel—cobalt bimetal oxide electrode materials.* ACS Applied Materials & Interfaces, 2020. **12**(47): p. 52749–52762.

[105] Wang, Z., et al., *Facile fabrication of flexible rGO/MXene hybrid fiber-like electrode with high volumetric capacitance.* Journal of Power Sources, 2020. **448**: p. 227398.

[106] Zhao, K., et al., *Free-standing MXene film modified by amorphous FeOOH quantum dots for high-performance asymmetric supercapacitor.* Electrochimica Acta, 2019. **308**: p. 1–8.

[107] Li, H., et al., *Hydrous RuO_2-decorated MXene coordinating with silver nanowire inks enabling fully printed micro-supercapacitors with extraordinary volumetric performance.* Advanced Energy Materials, 2019. **9**(15): p. 1803987.

[108] Wang, Y., et al., *Intercalating ultrathin MoO_3 nanobelts into MXene film with ultrahigh volumetric capacitance and excellent deformation for high-energy-density devices.* Nano-Micro Letters, 2020. **12**: p. 1–14.

[109] Li, H., et al., *Laser crystallized sandwich-like MXene/Fe$_3$O$_4$/MXene thin film electrodes for flexible supercapacitors.* Journal of Power Sources, 2021. **497**: p. 229882.

[110] Zhang, X., et al., *Low-temperature synthesized nanocomposites with amorphous FeOOH on Ti$_3$C$_2$T$_x$ for supercapacitors.* Journal of Alloys and Compounds, 2018. **744**: p. 507–515.

[111] Xie, W., et al., *MOF-derived CoFe$_2$O$_4$ nanorods anchored in MXene nanosheets for all pseudocapacitive flexible supercapacitors with superior energy storage.* Applied Surface Science, 2020. **534**: p. 147584.

[112] Yang, Q., et al., *MXene/graphene hybrid fibers for high performance flexible supercapacitors.* Journal of Materials Chemistry A, 2017. **5**(42): p. 22113–22119.

[113] Zou, R., et al., *Self-assembled MXene (Ti$_3$C$_2$T$_x$)/α-Fe$_2$O$_3$ nanocomposite as negative electrode material for supercapacitors.* Electrochimica Acta, 2018. **292**: p. 31–38.

[114] Zhou, Y., et al., *Ti$_3$C$_2$T$_x$ MXene-reduced graphene oxide composite electrodes for stretchable supercapacitors.* ACS Nano, 2020. **14**(3): p. 3576–3586.

[115] Wang, F., et al., *ZnO nanoparticle-decorated two-dimensional titanium carbide with enhanced supercapacitive performance.* RSC Advances, 2016. **6**(92): p. 88934–88942.

[116] Wang, X., et al., *In situ polymerized polyaniline/MXene (V$_2$C) as building blocks of supercapacitor and ammonia sensor self-powered by electromagnetic-triboelectric hybrid generator.* Nano Energy, 2021. **88**: p. 106242.

[117] Ali, I., Z. Haider, and S. Rizwan, *Enhanced pseudocapacitive energy storage and thermal stability of Sn^{2+} ion-intercalated molybdenum titanium carbide (Mo$_2$TiC$_2$) MXene.* RSC Advances, 2022. **12**(49): p. 31923–31934.

[118] Luo, J., et al., *Atomic sulfur covalently engineered interlayers of Ti$_3$C$_2$ MXene for ultrafast sodium-ion storage by enhanced pseudocapacitance.* Advanced Functional Materials, 2019. **29**(10): p. 1808107.

[119] Yoon, Y., et al., *A strategy for synthesis of carbon nitride induced chemically doped 2D MXene for high-performance supercapacitor electrodes.* Advanced Energy Materials, 2018. **8**(15): p. 1703173.

[120] Qin, L., et al., *High-performance ultrathin flexible solid-state supercapacitors based on solution processable Mo$_{1.33}$C MXene and PEDOT: PSS.* Advanced Functional Materials, 2018. **28**(2): p. 1703808.

[121] Xu, P., et al., *A MXene-based EDA-Ti$_3$C$_2$T$_x$ intercalation compound with expanded interlayer spacing as high performance supercapacitor electrode material.* Carbon, 2021. **173**: p. 135–144.

[122] Li, K., et al., *All-pseudocapacitive asymmetric MXene-carbon-conducting polymer supercapacitors.* Nano Energy, 2020. **75**: p. 104971.

[123] Wu, W., et al., *Facile strategy of hollow polyaniline nanotubes supported on Ti$_3$C$_2$-MXene nanosheets for high-performance symmetric supercapacitors.* Journal of Colloid and Interface Science, 2020. **580**: p. 601–613.

[124] Lu, X., et al., *Hierarchical architecture of PANI@ TiO$_2$/Ti$_3$C$_2$T$_x$ ternary composite electrode for enhanced electrochemical performance.* Electrochimica Acta, 2017. **228**: p. 282–289.

[125] Zhu, M., et al., *Highly flexible, freestanding supercapacitor electrode with enhanced performance obtained by hybridizing polypyrrole chains with MXene.* Advanced Energy Materials, 2016. **6**(21): p. 1600969.

[126] Li, J., et al., *MXene-conducting polymer electrochromic microsupercapacitors.* Energy Storage Materials, 2019. **20**: p. 455–461.

10 MXenes for Batteries

10.1 MXENE-BASED COMPOSITES FOR BATTERIES

Over the past few years, scientists have successfully created many compositions of MXene materials. These materials possess a distinctive 2D layered structure, rapid ion diffusion, and high electronic conductivity, along with other properties. Due to these features, MXenes are recognized as efficient materials to be used in batteries. Therefore, review of the advancements in the synthesis and novel applications of MXenes is crucial. These materials exhibit the potential to be used as an electrode material for both lithium-ion and non-lithium secondary batteries, including magnesium, sodium, calcium, potassium, and aluminum ions. As lithium-ion batteries (LiBs) become more widely used, it is becoming clear that lithium resources are scarce [1–5].

The research on non-LIBs has gained increasing interest due to their vast reserves and reduced costs. Ti_3C_2 has theoretical capacities of 351.8, 191.8, and 319.8 mAh g^{-1} for Na-, K-, and Mg-ion batteries, respectively [1–5]. MXene materials in their bare state exhibit a lower specific gravimetric capacity compared to typical electrode materials. For instance, Ti_2CT_x MXene has a gravimetric capacity of only 160 mA h g^{-1} when utilized as an anode for LIBs. The low specific capacity of MXenes is primarily due to its large unit molecular mass. Furthermore, the assembly of 2D sheets into an electrode presents a significant challenge, resulting in restacking issues. Therefore, MXene composites for electrode applications have been developed using various strategies, such as increasing interlayer spacing and blending with other nanostructures through heteroatom or molecule insertion. Such approaches have been shown to significantly enhance the electrochemical properties of MXenes [1].

10.2 HIGH AREAL AND VOLUMETRIC CAPACITY

Batteries are widely used in energy storage and as power supplies for electric vehicles (EVs) due to their significantly higher energy densities than supercapacitors. In practice, the focus should be on achieving optimal areal and volumetric capacity, rather than merely on gravimetric capacity. This emphasizes the significance of improving electrode nanostructures, particularly through the incorporation of tap density and increasing the mass loading of electrode and high-theoretical-capacity materials. Commercialized devices typically have graphite anodes with ~2 mAh cm^{-2} and <600 mAh cm^{-3}, that is, values of areal and volumetric capacities, respectively. However, these values fall far short of meeting the ever-increasing demands of modern technological devices. On the other hand, MXenes, with their excellent metallic conductivity, large theoretical storage capacity, and effective Li$^+$ diffusion barrier, are identified as promising anode materials for next-generation LiB development [5].

DOI: 10.1201/9781003465768-10

Areal capacity is represented by C/A (mAh cm^{-2}), which has been observed to be a combination of specific capacity (C/M) and electrode mass loading (M/A). High C/A electrodes are critical for optimal battery performance and can be achieved by improving C/M, increasing M/A, or a combination of the two. Improving C/M is a simple matter of incorporating highly redox-active elements into the composite electrode. However, increasing M/A without influencing the structure of the electrode is challenging due to crack formation when thickness exceeds the critical cracking thickness (CCT), but increasing the CCT improves electrode performance. However, as the electrode becomes thicker, the diffusion of Li$^+$ ions is affected, which limits the specific capacity, C/M, and the ability to handle high rates. This phenomenon is best illustrated by the fact that low-mass-loading electrodes often achieve large gravimetric capacities due to improved ion diffusion kinetics. However, when the electrode thickness increases, the specific capacities decline rapidly [5].

The demand for thinner LIBs is growing rapidly to meet the needs of flexible electronics industries. This has led to the production of high volumetric LIBs, which can be achieved through the use of MXenes. These materials are known for their low Li$^+$ diffusion barriers, exceptional electrical conductivity, and superior tap density. The design and manufacture of electrodes are crucial in increasing the volumetric capacity of these batteries. By incorporating more active locations for redox reactions, like increasing interlayer spacing and using high-theoretical-capacity materials in electrodes, it is possible to increase electrode-specific capacities. Furthermore, in order to achieve increased volumetric capacity at the device level, high tap density electrodes must be manufactured while ion diffusion channels are preserved [5].

10.3 LITHIUM-ION BATTERIES

LIBs hold great promise as an efficient energy storage device. However, due to their low energy density and the cost of lithium resources, their usage is currently limited. Therefore, it is crucial to develop high-energy and cost-effective electrode materials [1]. One promising option is conductive MXene nanosheets (NSs), which have a large and tunable interlayer spacing similar to layered graphite anodes, making them ideal for lithium-ion storage. The surface functionalization of NSs offer Li-storage sites, making pure MXene a capable anode with a potential range of 0.5–1.5 V compared to Li/Li$^+$. Nonetheless, the specific surface functional groups of MXene electrodes have a significant influence on their effectiveness in this regard, chemical components, porous architectures, and doped atoms [6]. The drawbacks of electrochemical performance pure $M_3X_2T_x$ include quick oxidation, below-maximum electrical conductivity, and a tendency to restack. MXenes can enhance the electrochemical performance of other electro-active materials when combined with them to create active electrode composites. It has been observed that pristine MXenes exhibit high metallic conductivity, while functionalized MXenes display semimetallic conductivity due to the presence of surface groups [1].

The $M_{n+1}X_n$ thin film exhibits a delocalization of electrons across the metal surface, with the presence of functionalized groups influencing this delocalization. MXenes' hierarchical structure may exhibit increased disorderliness due to the presence of $-T_x$ groups, which are also thought to react with specific molecules, resulting in a delamination effect. Delaminated MXene films, which have a lower thickness, have better electrochemical properties than non-delaminated MXene flakes. The ultrathin layer structure is widely acknowledged to improve the performance of transporting and storing Li^+ ions. Exploring the MXene surface and controlling the layer restacking of MXene are important aspects of improving the electrical properties of LIBs that use MXene as an electrode material. Regulating layer restacking allows the electrolyte and MXene to maintain intimate contact, allowing access to extra electrochemically active sites. Li^+ ions saturate the surface of the anode during charging, especially when LiPF6 is used as the electrolyte, and the porous structure of MXene-based composites facilitates the intercalation of these ions [1]. The electrochemical behavior of MXene-based composites for LIBs is summarized in Table 10.1.

H. Dong et al. [7] described the Lewis acid etching of Nb_2CT_x MXene using molten salts. The synthesized MXene showed a high lithium storage capacity value, that is, up to 330 mAh g^{-1} (at 0.05 A g^{-1}) and high-rate performance and was observed to outperform the highest capacity of $Ti_3C_2T_x$ (205 mAh g^{-1}), synthesized by the same

TABLE 10.1
MXene-Based LIBs

Electrode Material	Initial Charge/Discharge Capacities (mAh g^{-1})	Cycling Stability (mAh g^{-1})	Potential Window (V)	Ref.
$V_2C@Co$	1006.5/1117.3 at 0.1 A g^{-1}	248.8 at 8 A g^{-1} after 15,000 cycles	0.01–3.5	[12]
$V_2C@Sn$	1115.6/1284.6 at 0.1 A g^{-1}	1,262.9 at 0.1 A g^{-1} after 90 cycles	0.01–3.5	[13]
P-V_2C/ NiCo-LDH	≈735/≈756 at 200 mA g^{-1}	1,077 at 500 mA g^{-1} after 700 cycles	0.01–3.0	[14]
V_2C MXene	260 at 370 mA g^{-1}	243 at 500 mA g^{-1} after 500 cycles	0.01–3.0	[15]
$(V_{1-x}Ti_x)_2C$ (x = 0.1, 0.2, 0.3)	290/492 at 372 mA g^{-1}	238 at 372 mA g^{-1} after 140 cycles	0.01–3.0	[16]
$(V_{0.5},Ti_{0.5})_2C$	286.6/445.9 at 1 A g^{-1}	204.9 at 1 A g^{-1} after 1,000 cycles	0.01–3.0	[17]
$Mo_{1.33}CT_x$	603.7 at 0.2 A g^{-1}	525.7 at 1 A g^{-1} after 1,300 cycles	0.01–3.0	[18]
$V_4C_3T_x$-BM-HF	600.8/376.2 at 0.1 A g^{-1}	225 at 0.1 A g^{-1} after 300 cycles	0.01–3.0	[11]
V_4C_3-MXene/ MoS_2/C	916.4/1156.9 at 0.1 A g^{-1}	622.6 at 1 A g^{-1} after 450 cycles	0.01–3.0	[19]
Nb_2CT_x	113 and 80 at 50 C and 100 C	330 at 1 A g^{-1} after 1,000 cycles	0.01–3.0	[7]
N-doped Nb_2CT_x	315 at 0.5 C	288 at 0.5 C after 1,500 cycles	0.01–3.0	[8]
$Nb_4C_3T_x$	333/546 at 100 mA g^{-1}	320 at 1 A g^{-1} after 1,000 cycles	0.01–3.0	[10]
$Ti_xTa_{(4-x)}C_3$	1411 at 0.05 C	459 at 0.5 C after 200 cycles	0.01–3.0	[20]
Fe_3O_4@MXene/ CNFs	985.92 at 2 A g^{-1}	806 at 2 A g^{-1} after 500 cycles	0.01–3.0	[21]
N–$Ti_3C_2T_x$/P	666.5/1160.4 at 100 mA g^{-1}	801 at 500 mA g^{-1} after 1,040 cycles	0.01–3.0	[22]
MoS_2/Mo_2TiC_2Tx-500	554/646 at 100 mA g^{-1}	509 at 100 mA g^{-1} after 100 cycles	0.01–3.0	[23]
p-MXene-71	455.5 at 50 mA g^{-1}	220 at 1 A g^{-1} after 3,500 cycles	0.01–3.0	[24]

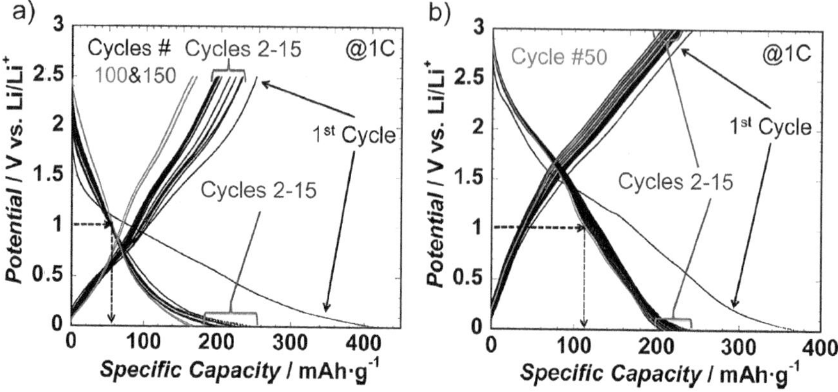

FIGURE 10.1 (a) Nb_2CT_x voltage profile among 0 and 2.5 V vs. Li/Li⁺. (b) V_2CT_x voltage profile between 0 and 3 V vs. Li/Li⁺.

Source: Reprinted with permission from [9] Michael Naguib, Joseph Halim, Jun Lu, et al. 2013. New Two-Dimensional Niobium and Vanadium Carbides as Promising Materials for Li-Ion Batteries. *J. Am. Chem. Soc.*, 135, 15966–15969. Copyright 2013, American Chemical Society.

method. In another work, the N-doped Nb_2CT_x had a substantially higher value of reversible capacity, that is, 360 mAh g⁻¹ (at 0.2 C), than the pure material, which displayed 190 mAh g⁻¹. Furthermore, the doped MXene exhibited remarkable cycling stability and capacity retention after 1,500 cycles [8]. Similarly, M. Naguib et al. [9] investigated the use of Nb_2CT_x and V_2CT_x as LIB electrode materials. At lower lithiation voltages, Nb_2CT_x demonstrated 170 mAh g⁻¹ at 1 C. At higher lithiation voltage, V_2CT_x demonstrated increased capacity (210 mAh g⁻¹ at 1 C), as illustrated in Fig. 10.1 [9]. At the cycling rate of 1 C, the voltage profile for Nb_2CT_x produces a capacity of ~422 mAh g⁻¹ after the first cycle, around 250 mAh g⁻¹ after the second cycle, and 170 mAh g⁻¹ after 100 cycles [9].

In another study, researchers reported Li insertion into a 2D $Nb_4C_3T_x$, and it was discovered that the charge–discharge capacity of the as-synthesized anode material can be improved by cycling [10]. For instance, after 100 cycles, the value of specific capacity increased from 310 to 380 mAh g⁻¹ at 0.1 A g⁻¹ and from 116 to 320 mAh g⁻¹ at 1 A g⁻¹. J. Zhou et al. [11] demonstrated that by utilizing ball-milling treatment to develop V_4AlC_3, the interlayer spacing and specific surface area of the corresponding MXene can be increased, which is advantageous for Li-ion storage. As a result, this substance demonstrated significant capacities, that is, 225 and 125 mAh g⁻¹ at 1 A g⁻¹, and impressive rate capability along with cycling stability after 300 cycles. The coulombic efficiency and cycling endurance of the MAX phase and the associated MXene electrodes are depicted in Fig. 10.2a [11]. The V_4AlC_3 electrode's low reversible capacity suggests that it is unsuitable for use as an anode in LIBs.

Following treatment with an HF solution, $V_4C_3T_x$-HF and $V_4C_3T_x$-BM-HF underwent functionalization with distinct surface groups. Energy storage sites were identified within the layers and at the surface–layer interface. Consequently, the electrodes derived from these MXene variants exhibited a relatively elevated reversible

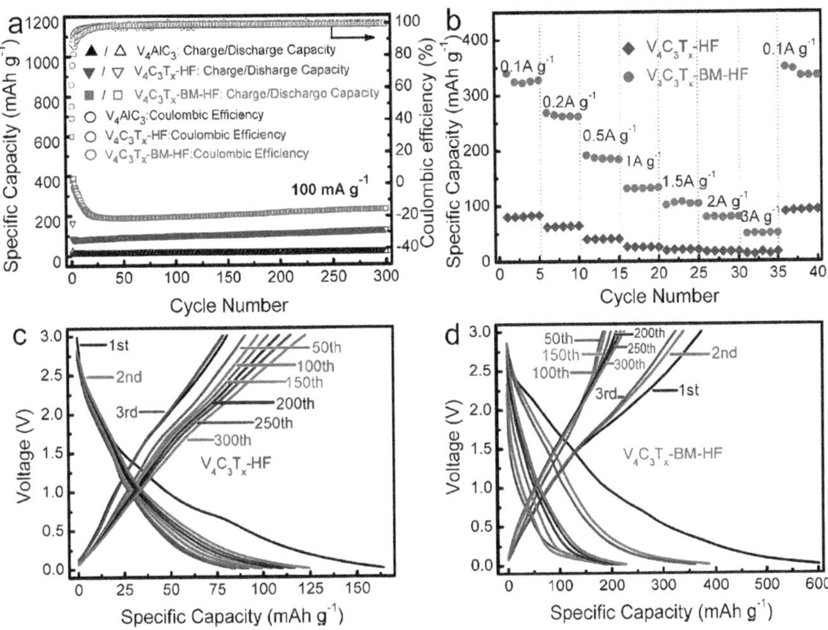

FIGURE 10.2 (a) Cycling performance and coulombic efficiency for V_4AlC_3, $V_4C_3T_x$-HF, and V_4C_3T-BM-HF at 0.1 A g^{-1}. (b) Rate capacity of $V_4C_3T_x$-HF and $V_4C_3T_x$-BM-HF. GCD curves for selected cycles of (c) $V_4C_3T_x$-HF and (d) $V_4C_3T_x$-BM-HF.

Source: Reprinted from [11] *Chemical Engineering Journal*, 373, J. Zhou, et al., Synthesis and lithium ion storage performance of two-dimensional V_4C_3 MXene, 203–212, Copyright 2019, with permission from Elsevier.

capacity, accompanied by an observed capacity increase as the number of cycles increased. Figure 10.2b depicts the rate capabilities of these two electrodes under varying current densities. For these electrodes, the representative charge–discharge potential curves at 0.1 A g^{-1} are shown in Figs. 10.2c and d. The outcomes show that the reversible lithiation capacity is less than 1 V in more than two-thirds of cases, which is a crucial factor when selecting an anode material for LIBs.

10.4 LITHIUM–SULFUR BATTERIES

Secondary lithium–sulfur (Li-S) batteries have been observed to exhibit a high theoretical capacity of 1,675 mA h g^{-1}, along with a high energy density of up to 2,500 W h kg^{-1}. MXenes with distinct structural diversity, metallic conductivity, an effective catalytic influence promoting rapid kinetics, the ability to induce uniform Li growth, and the ability to chemically adsorb polysulfides position themselves as potential contenders for Li-S batteries. MXenes can serve as a sulfur host, a lithium anode, and an interlayer for modifying the separator. The field of battery technology is confronted with several challenges that require attention, including low-rate capacity caused by the insulating properties of active sulfur and discharge products, electrode

TABLE 10.2

MXene-Based LIBs

Electrode Material	Initial Charge/Discharge Capacities (mAh g^{-1})	Cycling Stability (mAh g^{-1})	Potential Window (V)	Ref.
N-Ti$_3$C$_2$T$_x$/S	1,144 at 335 mA g^{-1}	610 at 2C after 1,000 cycles	1.7–2.8	[30]
N-MX-CoS$_2$	1031 at 1 C	651 at 1 C after 700 cycles	1.7–2.8	[31]
MCoNPCNSs/S-M-PP	1,340.2/1,188.4 at 0.2 C	914.7 at 1 C after 1,000 cycles	1.7–2.8	[32]
S/P-NTC	993 at 0.2 C	820 at 2 C after 1,200 cycles	1.7–2.8	[26]
N-Ti$_3$C$_2$@CNTs/S	1,124. 6 at 0.2 C	788.6 at 4 C after 650 cycles	1.7–2.8	[27]
Ti$_3$C$_2$T$_x$/S	1,196 at 0.1 C	970 at 1 C after 1,500 cycles	1.7–2.8	[28]
S@TiO$_2$/Ti$_2$C	1,408.6 at 0.2 C	464.0 and 227.3 at 2 C and 5 C after 200 cycles	1.5–3.0	[33]
TCD-TCS/S	1,609 at 0.05 C	815 at 2 C after 400 cycles	1.5–3.0	[34]
VO$_2$(p)-V$_2$C/S	1,250 at 0.2 C	69.1% retention at 2 C after 500 cycles	1.7–2.8	[29]
TiO$_2$ QDs@ MXene/S	1,158, 1,037, 925, 812, and 663 at C/5, C/2, 1 C, 2 C, and 5 C	680 at 2 C after 500 cycles	1.5–3.0	[35]
MXene/SiO$_2$	1,007 at 0.1 C	623 at 0.5 C after 200 cycles	−0.4–0.4	[36]

instability caused by significant volume changes (up to 80%) at the cathode during cycling, and decreased coulombic efficiency and capacity deterioration caused by the shuttle effect. To address these issues, researchers are focusing their efforts on the development of novel sulfur host and separator materials. The goal is to improve cycling stability while increasing cathode conductivity and sulfur utilization efficiency, and suppress the this effect [1, 2, 4, 25]. Some of the MXene-based Li-S batteries are presented in Table 10.2.

Song, Y. et al. [26] demonstrated the scalable sacrificial templating approach for the fabrication of a cathode for Li-S batteries, that is, P-NTC (porous N-doped Ti$_3$C$_2$). At over 1,200 cycles, the cathode exhibited a capacity retention of only 0.033% (at 2.0 C). Another study successfully synthesized N-Ti$_3$C$_2$@CNT microspheres using a simple spray-drying process. Doped MXene and carbon nanotubes were discovered to effectively chemically immobilize polysulfides, trapping them within porous microspheres. The cathode made of N-Ti$_3$C$_2$@CNT microspheres/S displayed high-capacity retention after 1,000 cycles (with 775 mAh g^{-1}), with a fading rate

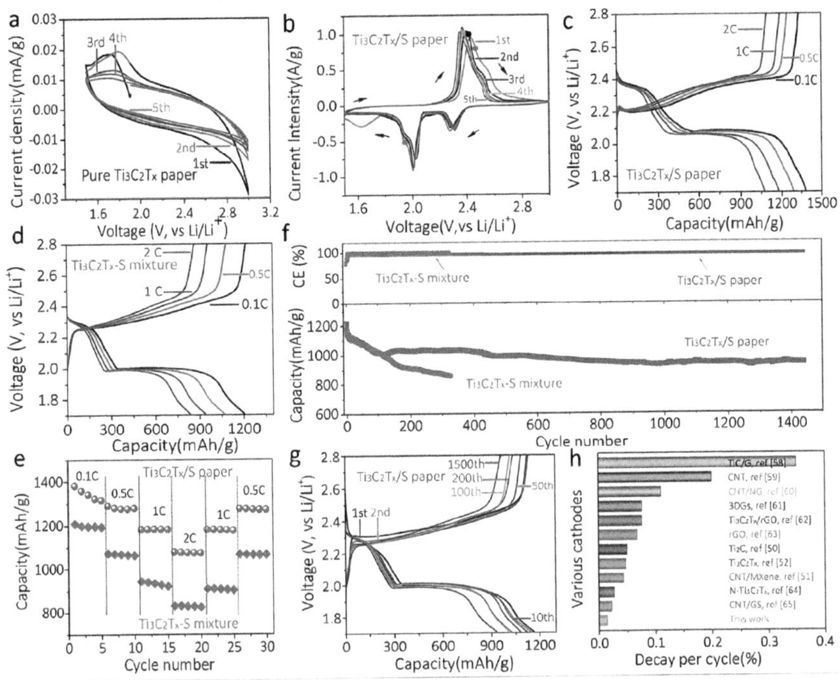

FIGURE 10.3 (a) CV curves for pure $Ti_3C_2T_x$ paper and (b) CV curves for $Ti_3C_2T_x/S$ paper were generated, displaying different cycles at 0.2 mV s^{-1}. At various C rates, GCD profiles were recorded for (c) $Ti_3C_2T_x/S$ paper and (d) $Ti_3C_2T_x$-S mixture. (e) The rate capability of $Ti_3C_2T_x/S$ paper was compared to that of the $Ti_3C_2T_x$-S mixture. (f) At 1 C, Coulombic efficiency and long-term cycling performance of $Ti_3C_2T_x/S$ paper was also investigated. (g) $Ti_3C_2T_x/S$ paper charge–discharge curves were investigated. h) The rate of capacity decay was compared for various conditions.

Source: [28] Tang H., Li W., Pan L., et al.: A Robust, Freestanding MXene-Sulfur Conductive Paper for Long-Lifetime Li–S Batteries. *Adv. Funct. Mater.* 2019, 1901907. Copyright WILEY-VCH Verlag GmbH & Co. KGaA, Weinheim. Reproduced with permission.

(FR) of only 0.016% per cycle [27]. Furthermore, binder-free $Ti_3C_2T_x/S$ paper was also observed to exhibit high capacities (i.e., 1,383 and 1,075 mAh g^{-1} at 0.1 and 2 C, respectively) and a capacity decay rate of 0.014% (after 1,500 cycles). These efficient electrochemical properties can be attributed to the physicochemical behavior of MXene NSs. These properties not only facilitate rapid kinetics of electron and ion transport but also deal with the mechanical stress resulting from the expansion of sulfur volume [28].

Figure 10.3a [28] illustrates the quasi-rectangular CV curves and quasi-linear GCD profiles for the Ti_3C_2Tx paper. The comparatively slow lithium diffusion kinetics within the dense NS network are responsible for the limited Li storage capacity of 9.5 mAh g^{-1}. The formation of long- and short-chain polysulfides (Li$_2$Sn) was revealed by the cathodic peaks (at 0.3 and 2 V) of CV curves (Fig. 10.3b). The appearance of elemental sulfur is accompanied by a prominent anodic shoulder peak at 2.4 V. Figure 10.3c shows the GCD curves of $Ti_3C_2T_x$/S paper at different C rates. The obtained discharge curves were observed to be aligned with cathodic peaks evident in the CV curves. Correspondingly, the charge curve (at 2.4 V) corresponds to the position of the anodic peak. Consequently, the redox process kinetics in the $Ti_3C_2T_x$-S mixture are more constrained, as indicated by the higher voltage polarization (430 mV) compared to that of the $Ti_3C_2T_x$/S paper (290 mV) (as shown in Fig. 10.3d).

Under varying C rates, the $Ti_3C_2T_x$/S paper outperforms the $Ti_3C_2T_x$-S combination in terms of capacity. For example, 77.7% capacity retention was observed as the current density increased from 0.1 to 2 C, which shows a decrease in capacity from 1,383 to 1,075 mAh g^{-1} (as illustrated Fig. 10.3e). The initial capacity and first cycle coulombic efficiency in the $Ti_3C_2T_x$/S paper are shown in Fig. 10.3f. The capacity of the $Ti_3C_2T_x$/S paper stabilizes after an initial relatively rapid decay in the first 100 cycles, and after 1,500 cycles at a charge–discharge rate of 1 C, it maintains a capacity of 970 mAh g^{-1}. The reduced performance of the $Ti_3C_2T_x$-S combination emphasizes the critical importance of uniform distribution of sulfur on a well-connected conductive network. Figure 10.3g depicts the $Ti_3C_2T_x$/S paper's galvanostatic charge–discharge (GCD) curves across different cycles, resulting in 0.014% capacity decay rate per cycle. This rate was observed to be significantly lower than the values reported for other sulfur-containing cathodes, as shown in Fig. 10.3h [28].

Wang, Z. et al. [29] reported a high-performance VO_2(p)-V_2C/S catalytic host for sulfur cathodes, and it was observed to exhibit a high sulfur-loading cycling capacity, reversible discharge capacity of 1,250 mAh g^{-1} (at 0.2 C), and 69.1% capacity retention after 500 cycles (at 2 C). The study utilized different concentrations of Li_2Sn/DME solutions to analyze the interaction of a hybrid cathode with lithium polysulfides (LiPSs). As seen in Figs. 10.4a and b [29], VO_2(p)-V_2C demonstrated exceptional adsorption ability, causing the solution's color to gradually shift to transparent. The adsorption impact of both pure and hybrid structures on LiPSs was assessed using a UV-Vis spectrophotometer and a Li_2S_4/DME solution as a representative, as shown in Fig. 10.4b. Asymmetric batteries with a Li_2S_6/DOL/DME electrolyte were also developed to further investigate the catalytic conversion capabilities of VO_2(p) nanorods (Fig. 10.4c).

10.5 LITHIUM–METAL BATTERY ELECTRODES

Metallic lithium exhibits remarkable potential as a material for LIBs due to its potential (3.04 V) and impressive theoretical capacity of 3,860 mA h g^{-1}. However, it faces challenges such as a limited life cycle and notable drawbacks like expansion in volume and the formation of dendritic Li. Dendritic Li poses a risk by piercing the separator membrane, leading to potential short circuits and damage to the cell. Consequently, research on pure Li anodes has been restricted. Despite these challenges, advancements in nanotechnology have spurred exploration into lithium metals as electrode materials, particularly in the form of nanosized 2D materials with substantial specific areas. This research has become a crucial aspect of rechargeable lithium–metal cells, as well as lithium–sulfur and lithium–oxygen batteries. Researchers have developed three key solutions to address these challenges.

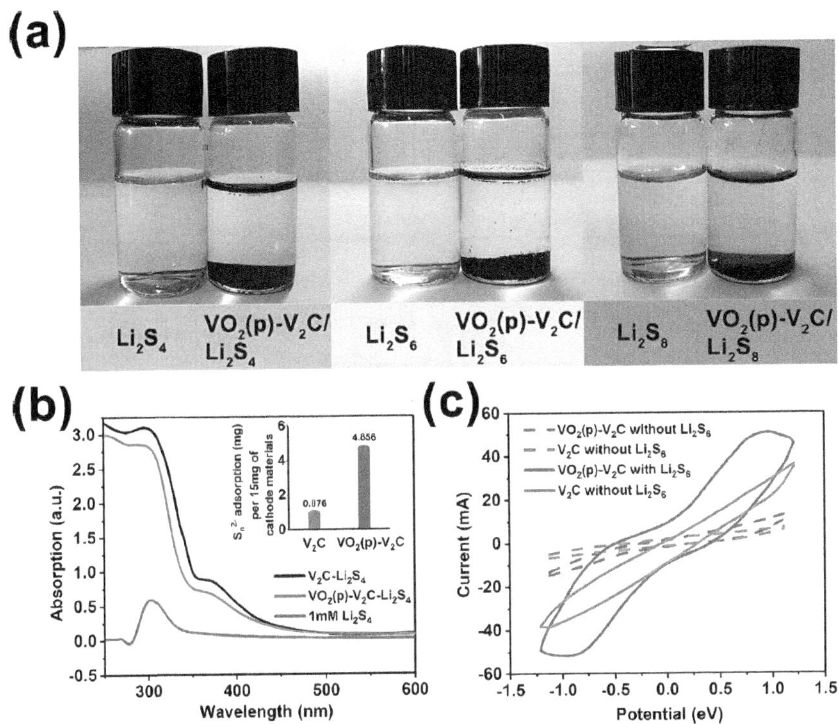

FIGURE 10.4 (a) Adsorption performance of V_2C and the composite. (b) After 8 hours, UV-Vis spectra of 1 mM Li_2S_4, V_2C-Li_2S_4, and $VO_2(p)$-V_2C-Li_2S_4 were examined. (c) CV curve for symmetric cells at 50 mV s^{-1}.

Source: Reprinted with permission from [29] Zhenguo Wang, Ke Yu, Yu Feng, et al. 2019. $VO_2(p)$-V_2C(MXene) Grid Structure as a Lithium Polysulfide Catalytic Host for High-Performance Li–S Battery. *ACS Appl. Mater. Interfaces*, 11, 44282–44292. Copyright 2019, American Chemical Society.

First, enhancing the strength of separators can resist the penetration of Li dendrites. Second, modifying the surfaces of Li metal can generate a stable solid electrolyte interface (SEI) coating to prevent passivation and ensure better performance. Third, the set-up of Li metal in hosting materials can help limit volume fluctuations and reduce the formation of vertical dendrites. This involves incorporating Li metal into conductive porous materials with large specific areas. However, this topic has not been widely studied [4, 25].

K. Shen et al. [37] developed a dendrite-free anode exhibiting high areal capacities, minimal overpotential (~10 mV), and an extended cycling life. MXene arrays on 3D-printed anodes enhance lithium nucleation and structure development, resulting in high-rate capacity and capacity retention after over 300 cycles when combined with the LiFePO$_4$ cathode. The 3DP-MXene arrays-Li electrodes exhibited favorable electrochemical characteristics. The corresponding anode cell demonstrated significant capacities of 149.4 mAh g^{-1} and 127.6 mAh g^{-1} in tests conducted at high current rates (from 0.2 C to 30 C). Even at a high current rate of 30 C, the cell displayed a

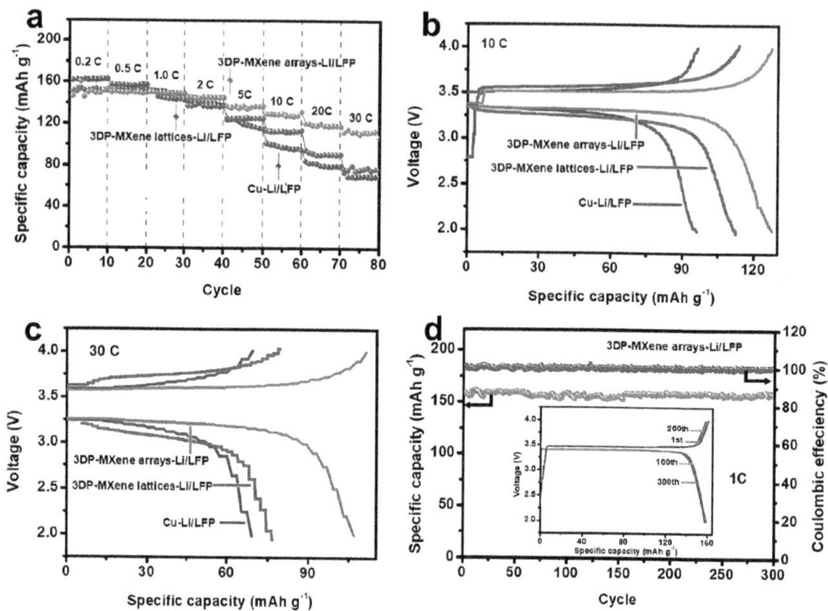

FIGURE 10.5 (a) Evaluating the rate capabilities of 3D-printed MXene arrays-Li/LFP, MXene lattices-Li/LFP, and Cu-Li/LFP over a temperature range of 0.2–30°C. Examining selected charge–discharge profiles at (b) 10°C and (c) 30°C. (d) Testing the cycling stability of the electrode, with typical voltage profiles depicted in the inset figure.

Source: Reprinted from [37] *Energy Storage Materials*, 24, Kai Shen, Bin Li, Shubin Yang, 3D printing dendrite-free lithium anodes based on the nucleated MXene arrays, 670–675, Copyright 2020, with permission from Elsevier.

minimal voltage hysteresis of only 0.4 V, significantly lower than that of cells with 3DP-MXene lattices-Li and Cu-Li. Furthermore, the stability of the complete cell was evident up to 300 cycles, demonstrating exceptional capacity retention at 99.4% under 1 C conditions (Fig. 10.5d [37]).

B. Li et al. [38] investigated an effective technique for creating a flexible Ti_3C_2-based lithium anode. It was discovered that by modifying the properties of atomic layers, growth, and nucleation, lithium dendrites can be regulated and lithium plating can also be restricted in the nanoscaled gaps of these NSs. As a result, the anode material displayed a low overpotential of 32 mV (at 1.0 mA cm^{-2}), cycling stability after 200 cycles, and high-rate performances. In another study, ultrathin polypropylene (PP) intercalated (CTAB)/CNT/$Ti_3C_2T_x$ hybrid separators were used to create very stable Li/Na-Se batteries. In situ permeation experiments were conducted to visually observe polyselenide properties, reveal the shuttle effect, and protect the Li anode from corrosion using specially designed separators. Consequently, the Li-Se batteries showed remarkable cycling stability for 500 cycles (at 1 C), with a 0.05% capacity decline per cycle being the very least [39].

10.6 SODIUM-ION BATTERIES

Sodium-ion batteries (SIBs) are becoming a better option for large-scale energy storage due to their abundance and cost-effectiveness, compared to LIBs. However, the large size of Na^+ makes diffusion in electrode materials slow. To address this issue, researchers have turned to creating nanomaterials as cathode materials for these batteries. By reducing the paths for Na^+ to move in and out of the cathode, the time for Na^+ diffusion is significantly shortened. Additionally, increasing electronic conductivity can moderately increase Na^+ diffusion. Therefore, MXenes and MXene-based composites have been identified as excellent electrode materials, as they can intercalate distinct ions of varying sizes. MXene exhibits a useful property that makes it suitable for the development of SIBs. Additionally, the double metal ion layer is expected to increase the capacity twofold. Surface and structural tuneability of MXenes make them suitable to be used as electrode materials. Researchers have developed the methods for the insertion of sodium ions in various MXene materials at the atomic level. The computed findings showed that functionalized materials exhibit a very low barrier for Na^+ and Al^{3+} diffusion [25, 40–42].

A number of studies on MXene-based SIBs are summarized in Table 10.3. Y. Wu et al. [43] reported on the development of a heterolayered MX/SnS_2 electrode. This

TABLE 10.3

MXene-Based SIBs

Electrode Material	Rate Capability (mAh g^{-1})	Cycling Stability (mAh g^{-1})	Potential Window (V)	Ref.
Sulfur-decorated Ti$_3$C$_2$ MXenes	136.6 at 5 A g^{-1}	135 at 2 A g^{-1} after 2,000 cycles	0.01–3.0	[47]
S-doped Ti$_3$C$_2$T$_x$	121.3 at 2 A g^{-1} and 113.9 at 4 A g^{-1}	138.2 at 0.5 A g^{-1} after 2000 cycles	0.01–3.0	[48]
FePS$_3$ nanosheets@ MXene	676.1 at 100 mA g^{-1}	527.7 at 0.5 A g^{-1} after 90 cycles	0.01–3.0	[49]
MXene/SnS$_2$	901, 377, 286, 184, 134, and 78 at 50, 100, 200, 500, 1,000, and 2,000 mA g^{-1}	155 at 100 mA g^{-1} after 250 cycles	0.01–2.5	[43]
VO$_2$/MXene	206 at 1.6 A g^{-1}	141% capacity retention at 0.1 A g^{-1} after 200 cycles	0.01–3.0	[44]
CoNiO$_2$/MXene	188 at 300 mA g^{-1}	223 at 100 mA g^{-1} after 140 cycles	0.01–3.0	[45]
CoS/MXene	272 at 5 A g^{-1}	267 at 2 A g^{-1} after 1,700 cycles	0.01–2.5	[46]
f-Ti$_3$C$_2$T$_x$_DMSO	267 at 0.1 A g^{-1}	196 at 100 mA g^{-1} after 500 cycles	0.01–2.5	[50]
Ti$_3$C$_2$ MXene/ CNT-SA	422 and 124 at 50 mA g^{-1}	164 at 50 mA g^{-1} after 100 cycles	0.01–4.0	[51]
Sb$_2$O$_3$/Ti$_3$C$_2$T$_x$	295 at 2 A g^{-1}	472 at 100 mA g^{-1} after 100 cycles	0.01–2.5	[52]
SnS /Ti$_3$C$_2$T$_x$	255.9 at 1,000 mA g^{-1}	410 at 100 mA g^{-1} after 50 cycles	0.01–3.0	[53]

electrode has shown exceptional performance for SIBs. The MX/SnS$_2$ electrode with a ratio of 1:5 has demonstrated a high value of reversible capacity after 200 cycles (i.e., 322 mAh g^{-1}). Even after temperature changes from 0°C to 20°C, long-term cyclability and an outstanding rate capacity were observed. The GCD curves for the first, second, fifth, and tenth cycles of the composite electrode are shown in Fig. 10.6a [43]. The initial charge–discharge values of capacity were observed to be 882 and 460 mAh g^{-1}, respectively. During the first cycle, layers of SEI were formed, which resulted in a long sloping curve with the value of the potential ranging from 0.5 to 1.25 V. Furthermore, the MX/SnS$_2$ 10:1 electrode (Fig. 10.6b) has a lower capacity and no discernible SnS$_2$ discharge plateau.

On the other hand, multilevel discharge plateaus of SnS$_2$ are seen in the MX/SnS$_2$ 2:1 electrode (Fig. 10.6c), which causes a rapid capacity decline after 10 cycles. The MX/SnS$_2$ 5:1 electrode exhibits a notably greater reversible capacity than other MXenes and their composites even after 200 cycles (Fig. 10.6d). At high current density, the electrode was observed to exhibit a consistent discharge–charge

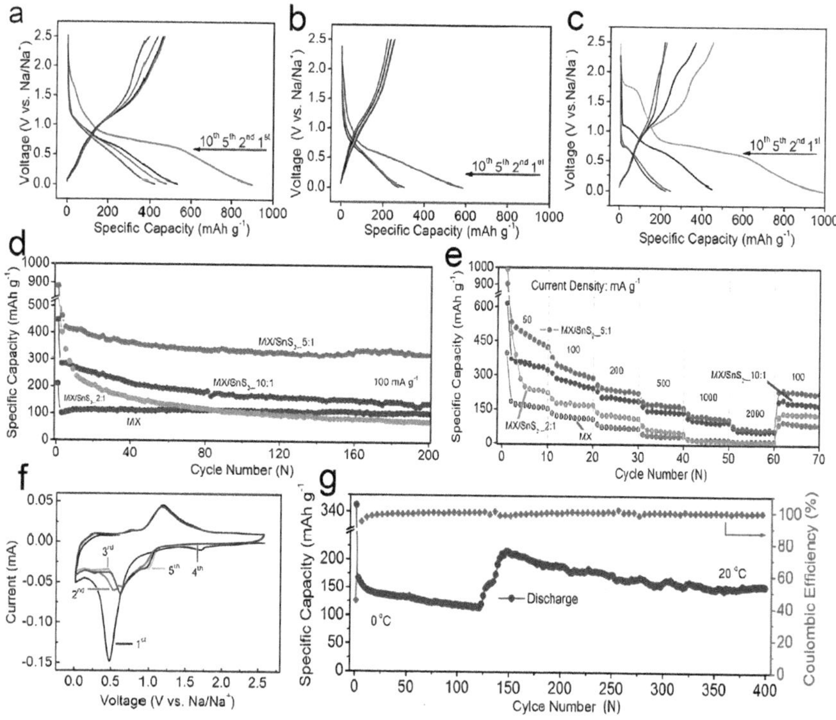

FIGURE 10.6 GCD profiles for MX/SnS$_2$ (a) 10:1, (b) 5:1, and (c) 2:1. (d) Cycling performance and (e) rate performance assessment for MX/SnS$_2$ 10:1, MX/SnS$_2$ 5:1, and MX/SnS$_2$ 2:1. (f) MX/SnS$_2$ 5:1 electrode CV curves at 0.1 mV s^{-1}. (g) Cycling performance of the MX/SnS$_2$ 5:1 electrode.

Source: Reprinted from [43] *Chemical Engineering Journal*, 334, Yuting Wu, Ping Nie, et al., 2D MXene/SnS$_2$ composites as high-performance anodes for sodium ion batteries, 932–938, Copyright 2018, with permission from Elsevier.

capacity. This electrode provided a range of capacities, that is, 901, 377, 286, 184, 134, and 78mAh g^{-1} at 50, 100, 500, 1,000, and 2,000 mA g^{-1}, as shown in Fig. 10.6e. Furthermore, at 100 mA g^{-1}, the discharge capacity returns to 230 mAh g^{-1}, indicating that the MX/SnS$_2$ 5:1 electrode has superior structural stability. Furthermore, the CVs between 0.01 and 2.5 V (at 0.1 mV s^{-1}) are displayed in Fig. 10.6f [43].

F. Wu et al. [44] demonstrated a hydrothermal synthesis method for a VO$_2$/MXene hybrid which demonstrated superior sodium storage behavior due to the combination of MXene and the oxide material. The results showed improved cycling performance and a significant reversible capacity of 280.9 mA h g^{-1} (at 0.1 A g^{-1}). When used as the anode in SIBs, another layered structure, the CoNiO$_2$/MXene composite, demonstrated enhanced rate capacity (188 mA h g^{-1}) and a significant discharge capacity (223 mA h g^{-1}) [45]. Additionally, a hetero-layered CoS/Ti$_3$C$_2$ composite showed excellent rate performance after 1,700 cycles, a high value of reversible capacity of 267 mA h g^{-1}, and a capacity decay rate per cycle of 0.0072% [46]. Figure 10.7a

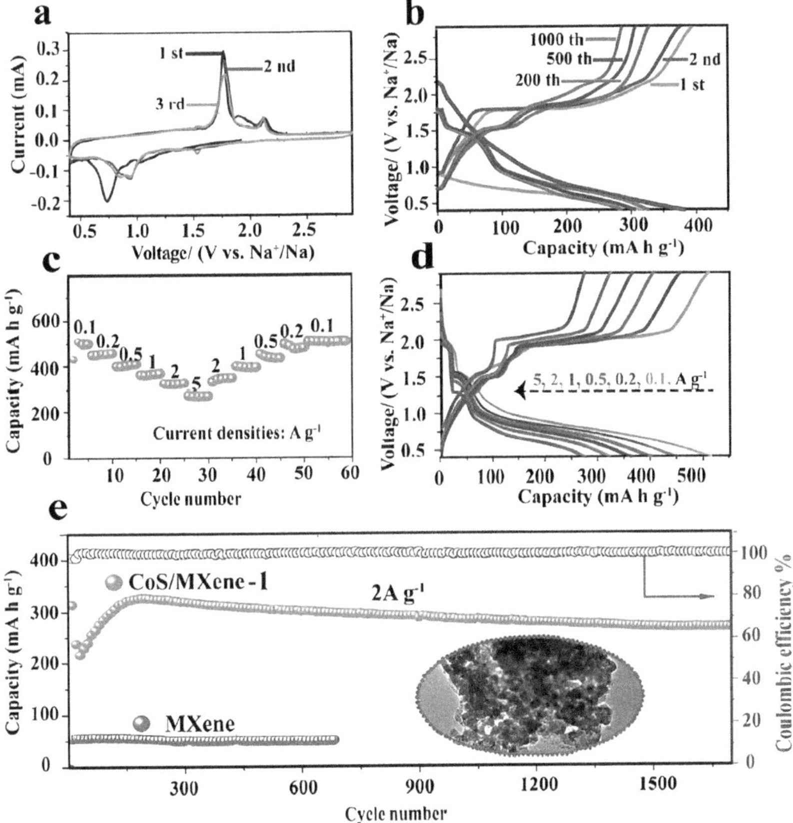

FIGURE 10.7 (a) CV curves recorded a 0.1-mV s^{-1} scan rate. At 2 A g^{-1}, charge–discharge profiles were obtained. (c) Rate performance assessment and (d) examination of charge–discharge curves at various current densities. (e) Evaluation of cycling performances at 2 A g^{-1}.

Source: Reprinted from [46] *Chemical Engineering Journal*, 357, Y. Zhang, R. Zhan, et al., Circuit board-like CoS/MXene composite with superior performance for sodium Storage, 220–225, Copyright 2019, with permission from Elsevier.

[46] displays CV curves showcasing the performance of composite material during the first three cycles, spanning 0.4–2.9 V at a 0.1 mV s^{-1} rate. Similarly, Fig. 10.7b displays the GCD curves of the electrode at 2 A g^{-1}. The discharge capacity in the first cycle, 315 mA h g^{-1}, is lower than that in the second cycle due to the lower initial voltage. Figures 10.7c and d show the composite electrode's charge–discharge profiles and rate capabilities, demonstrating different values of discharge capacity. Notably, the CoS/MXene electrode exhibits sustained reversible capabilities at 0.1 A g^{-1}, registering a discharge capacity of 505 mA hg^{-1}. [46].

10.7 ALUMINUM-ION BATTERIES

The most prevalent metal in the upper crust of the Earth is aluminum. In theory, an aluminum-based redox pair has the ability to transfer three electrons, outperforming LIBs. When compared to LIBs, aluminum-ion batteries (AIBs) are not only safer but also more cost-effective. Nonetheless, the significant dimensions of Al^{3+} ions when solvated in electrolytes and their strong bonding with host materials are noteworthy and pose significant challenges to AIB development. Previous research has indicated that both bare MXenes and O-terminated MXenes have significant capacities for storing aluminum ions. Unfortunately, only a few MXenes have demonstrated a reasonable Al-ion storage capacity so far. In summary, theoretical predictions suggested that MXenes have high capacities for storing various multivalent ions such as Mg^{2+}, Zn^{2+}, and Al^{3+}. However, experimental progress in this area has been limited. Existing literature suggests that modifying the surface and manipulating the structure of MXenes are effective ways to increase their metal ion storage ability. Therefore, controlling MXene surface chemistry is critical for future MXene electrode development for multivalent metal ion storage [4, 41].

H. Liu et al. [54] investigated the electrochemical characteristics of bare Mo-based double transition metal Mo$_2$MC$_2$ MXenes as AIB anodes using the SCAN-rVV10 density functional theory (DFT) calculations. In a multilayered adsorption configuration, the average energy for each layer was calculated, and these MXenes were discovered to be capable of adsorbing three layers of Al atoms, resulting in capacities ranging from 888.98 to 1,170.33 mAh g^{-1} for Mo$_2$TaC$_2$ and Mo$_2$ScC$_2$, respectively. Figure 10.8a [54] displays the Al-saturated Mo$_2$MC$_2$ MXene. They discovered that Mo$_2$MC$_2$ MXenes can hold three aluminum atom layers on both upper and lower surfaces, resulting in a saturated structure with six Al atomic layers. Figure 10.8b depicts the mean Al adsorption energy. Furthermore, as the molar mass of Mo$_2$MC$_2$ increases, their corresponding capacities are observed to decrease consistently. This is due to the constant maximum absorption of Al atoms on the monolayer for all Mo$_2$MC$_2$ MXenes under study [54].

A. V. Mohammadi et al. [55] described how V$_2$CT$_x$ was used as a cathode material in the development of a rechargeable AIB. The mechanism of charge storage involved the reversible intercalation of Al^{3+} between layers of the MXene. Optimizing electrochemical performance was discovered to entail reducing the multilayered structure to sheets with fewer layers. Under high discharge rates, V$_2$CT$_x$ electrodes demonstrated impressive specific capacities exceeding 300 mAh g^{-1}, ranking first among all documented cathodes for these batteries. The synthesized MXene powder was

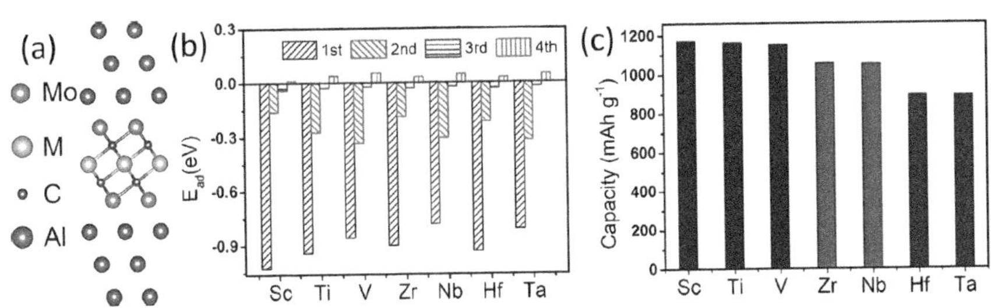

FIGURE 10.8 (a) Saturation of Al adsorption on Mo_2MC_2 MXenes. (b) Determination of average adsorption energy for individual Al layers on the surfaces of various Mo_2MC_2 configurations. (c) Estimation of the theoretical capacity of these materials as AIB anodes.

Source: Reprinted with permission from [54] Haoliang Liu, Hangyu Wang, Ziang Jing, et al. 2020. Bare Mo-Based Ordered Double-Transition Metal MXenes as High-Performance Anode Materials for Aluminum-Ion Batteries. *J. Phys. Chem. C*, 124, 25769–25774. Copyright 2020, American Chemical Society.

used as the cathode in the construction of an AIB cell by casting it onto a porous carbon paper substrate. The anode was made of high-purity aluminum foil, and the electrolyte was an ionic liquid consisting of AlCl/[EMIm]Cl. The components and mechanisms of the produced aluminum battery are shown in Fig. 10.9 [55].

10.8 ZINC-ION BATTERIES

Zinc is a low-cost and easily available metal that has a sufficiently high redox potential of 0.76 V versus SHE. This makes it possible to create a broad voltage range in an aqueous electrolyte. An aqueous zinc-ion battery (ZIB) that is generally composed of a cathode, a zinc foil anode, and an electrolyte, such as 2 M $ZnSO_4$ is safe and environmentally friendly. MXene NSs have hydrophilic surfaces and conductive cores that make them attractive for ZIB electrode designs because they are inherently wettable with aqueous electrolytes. Although ZIBs do not have the same energy densities as LIBs, they are ideal for situations where safety is paramount, as they eliminate the risk of combustion associated with organic electrolyte-based metal-ion batteries (MIBs). Vanadium (V)- and manganese (Mn)-based materials and their hybrids with MXene NSs or nanocarbon are typical ZIB cathodes [4, 6].

The electrochemical performance of MXene-based ZIBs is presented in Table 10.4. For example, in a study, the designed Zn/MS-S-V_2CT_x battery demonstrated a high reversible capacity of 411.3 mAh g^{-1} (at 0.5 A g^{-1}), 80% capacity retention after 3,000 cycles, and high energy density in a 2 M $ZnSO_4$ electrolyte. According to DFT calculations, the ion diffusion energy barrier for Zn^{2+} ions can significantly be reduced by the heteroatom doping of S atoms, along with improvement in the stability of V_2O_5 [56]. Y. Liu et al. [57] reported the production of ZIBs with high-rate capability and high energy density by using V_2C MXene. The high valency of outer VOx and electronic conductivity of inner MXene resulted into a high-rate performance with 358 mAh g^{-1}, as well as high power and energy densities (i.e., 318 Wh kg^{-1} at 22.5 kW kg^{-1}).

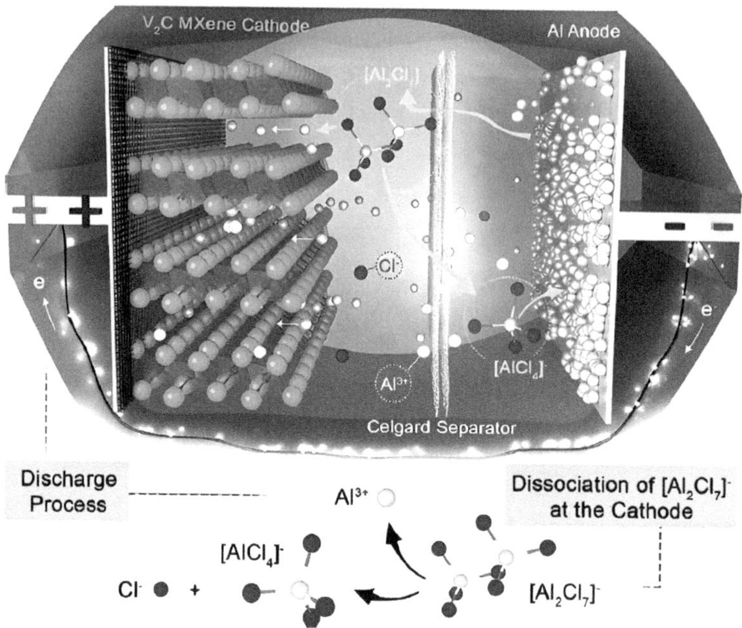

FIGURE 10.9 Diagram illustrating the proposed mechanism for an aluminum battery with V$_2$CT$_x$ MXene as the cathode, observed during discharge in a 1.3:1 AlCl$_3$/[EMIm]Cl ionic liquid electrolyte.

Source: Reprinted with permission from [55] Vahid Mohammadi, Ali Hadjikhani, et al. 2017. Two-Dimensional Vanadium Carbide (MXene) as a High-Capacity Cathode Material for Rechargeable Aluminum Batteries. *ACS Nano*, 11, 11135–11144. Copyright 2017, American Chemical Society.

TABLE 10.4
MXene-Based ZIBs

Electrode Material	Specific Capacities (mAh g^{-1})	Cycling Stability (mAh g^{-1})	Potential Window (V)	Ref.
Zn/MS-S-V$_2$CT$_x$	411.3 at 0.5 A g^{-1}	80% capacitance retention at 10 A g^{-1} after 3,000 cycles	0.01–2.0	[56]
V$_2$CT$_x$	409.7 at 0.5 A g^{-1}	97.5 at 64 A g^{-1} after 4,000 cycles	0.01–2.0	[60]
V$_2$CT$_x$	358 at 30 A g^{-1}	283.7 at 30 A g^{-1} after 2,000 cycles	0.01–1.6	[57]
K–V$_2$C@MnO$_2$	408.1 at 0.3 A g^{-1}	119.2 at 10 A g^{-1} after 10,000 cycles	0.7–1.8	[58]
MnO$_2$– CNTs cathode and MXene anode	115.1 F g^{-1} at 1 mV s^{-1}	~83.6% capacitive retention after 15,000 cycles	0.01–0.8	[59]
Zn@Ti$_3$C$_2$	132 F g^{-1} at 0.5 A g^{-1}	82.5% of capacitive retention after 1,000 cycles	0.01–2.8	[61]
MXene-integrated Zn anode	145.7 at 3 A g^{-1}	205.5 at 1 A g^{-1} after 500 cycles	1.0–1.8	[62]
H$_2$V$_3$O$_8$/MXene	365 at 0.2 A g^{-1}	~84% capacitive retention at 5 A g^{-1} after 5,600 cycles	0.01–1.6	[63]

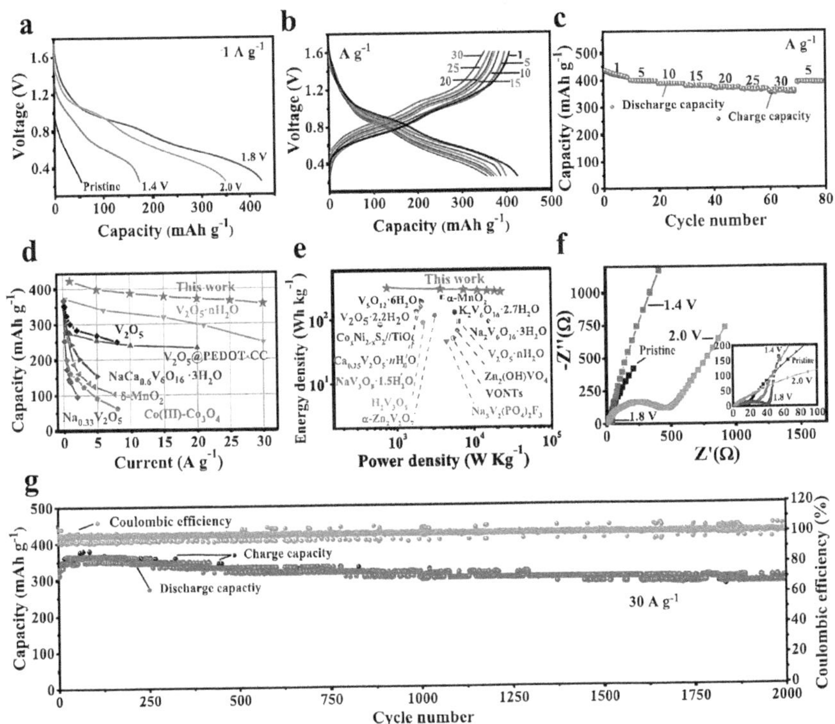

FIGURE 10.10 (a) Discharge profiles at 1 A g^{-1} for V$_2$CT$_x$, 1.4-V$_2$CT$_x$, 1.8-V$_2$CT$_x$, and 2.0-V$_2$CT$_x$. (b) Charge–discharge patterns specific to the 1.8-V$_2$CT$_x$. (c) Rate capability evaluation for 1.8-V$_2$CT$_x$. (d) Comparison of rate performance for 1.8-V$_2$CT$_x$. (e) Ragone plot comparison for 1.8-V$_2$CT$_x$ in the context of ZIBs. (f) EIS spectra for V$_2$CT$_x$, 1.4-V$_2$CT$_x$, 1.8-V$_2$CT$_x$, and 2.0-V$_2$CT$_x$, with an enlarged view of the spectra in the inset. (g) Evaluation of 1.8-V$_2$CT$_x$ cycling performance at 30 A g^{-1}.

Source: [57] Liu Y., Jiang Y., Hu Z. et al.: In-Situ Electrochemically Activated Surface Vanadium Valence in V$_2$C MXene to Achieve High Capacity and Superior Rate Performance for Zn-Ion Batteries. *Adv. Funct. Mater.* 2020, 2008033. Copyright WILEY-VCH Verlag GmbH & Co. KGaA, Weinheim. Reproduced with permission.

Without first charging activation, pure V$_2$CT$_x$ has a relatively low capacity, as shown in Fig. 10.10a [57], to which CNTs contribute negligibly. The supplied capacity only slightly decreases to 398.1 mAh g^{-1} between 1 and 5 A g^{-1}, with approximately 94% of the capacity still present at 1 A g^{-1}. It is possible to achieve remarkably high discharge capacities of 386, 371, 378, 365, and 358 mAh g^{-1} even at 10, 15, 20, 25, and 30 A g^{-1}, respectively (Figs. 10.10b and c). The rate capability of 1.8-V$_2$CT$_x$ is much superior to that of several documented high-performance cathodes, as expected (see Fig. 10.10d for a thorough comparison). The Ragone figure clearly shows the exceptionally high power and energy densities of 1.8-V$_2$CT$_x$, as demonstrated in Fig. 10.10e. It stabilizes the reversible capacity with 100% Coulombic efficiency after 2,000 cycles (Fig. 10.10g), which is superior to 1.4-V$_2$CT$_x$ and 2.0-V$_2$CT$_x$. [57].

X. Zhu et al. [58] reported the synthesis and properties of the K-V$_2$C@MnO$_2$ cathode for aqueous ZIBs. This hybrid structure exhibited abundant active sites, high

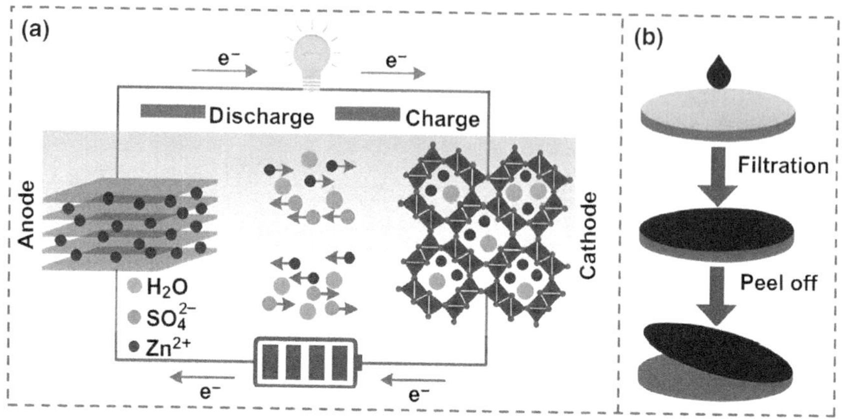

FIGURE 10.11 (a) Diagram illustrating the ZIC configuration, in which MnO_2-CNTs serve as the cathode, MXene serves as the anode, and the electrolyte consists of 2 M $ZnSO_4$ and 0.1 M MXene. (b) Fabrication procedure of electrodes.

Source: Reprinted with permission from [59] Springer Nature. *Nano-Micro Letters*. Siliang Wang et al. 2019. A New Free-Standing Aqueous Zinc-Ion Capacitor Based on MnO_2–CNTs Cathode and MXene Anode. 11:70, copyright 2019.

conductivity, and effect of Mn^{2+} electrodeposition which prevented structural degradation of oxides. As a result, this cathode material demonstrated high specific capacities of 408.1 and 119.2 mAh g^{-1} at 0.3 and 10 A g^{-1}, respectively, over 10,000 cycles. S. Wang et al. [59] developed a new type of capacitor called the zinc-ion capacitor (ZIC). To create the ZIC, they combined a battery-type cathode made of freestanding MnO_2-CNTs with a capacitor-type anode made of $Ti_3C_2T_x$. The resultant ZIC demonstrated outstanding electrochemical performance in an aqueous electrolyte. It displayed high specific capacitance, a high energy density of 77.5 W kg^{-1}, a high-power density of 29.7 Wh kg^{-1}, and a capacitance retention of 83.6% after 15,000 cycles. In Fig. 10.11a [59], we can see the schematic diagram of the ZIC. It consists of a pre-activated MnO_2-CNTs as a cathode, a MXene as an anode, and electrolytes (0.1 M $MnSO_4$ and 2 M $ZnSO_4$). When charging, zinc ions exit the MnO_2 tunnels and move into the electrolyte, followed by intercalation into the $Ti_3C_2T_x$ interlayer. The cycling life of the hybrid capacitor was observed to be improved by the addition of $MnSO_4$ at an adequate concentration, which blocks the MnO_2 dissolution. The fabrication technique for the freestanding cathode and anode is presented in Fig. 10.11b [59].

10.9 POTASSIUM-ION BATTERIES

In recent years, potassium-ion batteries (PIBs) have been gaining a lot of attention due to their various advantages over Li, Na, Mg, and Al batteries. These advantages include the abundance of raw resources, lower cost, and a lower standard potential (−2.922 V vs. SHE). In theory, PIBs have a higher discharge potential and energy density than LIBs, making them an excellent option. Additionally, the use of commercial graphite anodes in PIBs paves the way for future PIB commercialization.

However, during the insertion or removal process, the internal structure of the active material is greatly influenced by the large sizes of K^+, which results in a poor rate performance and short cycling life. As a result, formation of electrode materials with extensive tolerance of volume strain and high cycling stability is critical to the effective deployment of PIBs. Due to abundance and cost-effectiveness, carbon-based materials have been observed to exhibit applications in PIBs [41, 64].

MXene is a material that possess hydrophilicity, tunable interlayer spacing, electronic conductivity, and a high specific surface area, which distinguishes it from other materials being used for PIBs. Despite having outstanding rate performance and cycling stabilities, MXenes exhibit a very low theoretical specific capacity, just 191.8 mAh g^{-1}, for PIBs. Due to stacking of MXene layers and irreversible agglomeration, Ti_3C_2 shows a substantially lower actual specific capacity than the theoretical value. These problems can be solved by focusing on certain factors, including the formation of a 2D heterostructure, optimization of the interlayer structure, reducing the self-stacking effect, ensuring the K^+ fast transmission channel, and building additional potassium storage sites, that can effectively improve the electrochemical performance of PIBs [41, 64]. A summary of MXene-based PIBs is presented in Table 10.5.

TABLE 10.5

MXene-Based PIBs

Electrode Material	Initial Specific Capacities (mAh g^{-1})	Cycling Stability (mAh g^{-1})	Potential Window (V)	Ref.
Alkalized Ti3C2	502 and 136 at 20 mA g^{-1}	~42 at 200 mA g^{-1} after 500 cycles	0.01–3.0	[65]
K@DN-MXene/CNT	638 at 0.5 C	230 at 0.5 C after 500 cycles	0.5–3.5	[66]
MoSe$_2$/ MXene@C)	355 at 200 mA g^{-1}	183 at 10 A g^{-1} after 100 cycles	0.01–3.0	[67]
Sb/Na-Ti$_3$C$_2$T$_x$	573.3 at 0.1 A g^{-1}	392.2 at 0.1 A g^{-1} after 450 cycles	0.01–3.0	[68]
2D/1D MXene@ NCRib	287.2 at 0.1 A g^{-1}	201.5 at 1 A g^{-1} after 1,000 cycles	0.01–3.0	[69]
MXene-derived TiO$_2$/ rGO composite	104 at 1 A g^{-1}	~85% cycling retention after 1,000 cycles	0.01–3.0	[70]
o-P-CoTe$_2$/MXene	168.2 at 20 A g^{-1}	0.011% capacity decay per cycle over 2,000 cycles at 2 A g^{-1}	0.01–3.0	[71]
MXene-derived TiO$_x$N$_y$/C	765, 127, 102, 84, and 72 at 100, 200, 400, 800, and 1,600 mA g^{-1}	150 at 200 mA g^{-1} after 1,250 cycles	0.01–3.0	[72]

Titanium carbonitride MXene, Ti_3CNT_z, performed well as an anode for PIBs. The first and second cycle capacities were observed to be 710 and 275 mAh g^{-1}, respectively (at 20 mA g^{-1}). A reversible capacity of 75 mAh g^{-1} was retained after 100 discharge–charge cycles [73]. Similarly, H. Huang et al. [67] demonstrated that a hybrid MoSe$_2$/MXene@C can effectively serve as an anode for PIBs, achieving an excellent rate performance of 183 mA h g^{-1} and high reversible capacity after 100 cycles. The first three CV curves of this anode material are shown in Fig. 10.12a [67], which were reported at a scan rate of 0.1 mV s^{-1} with the normal potential range

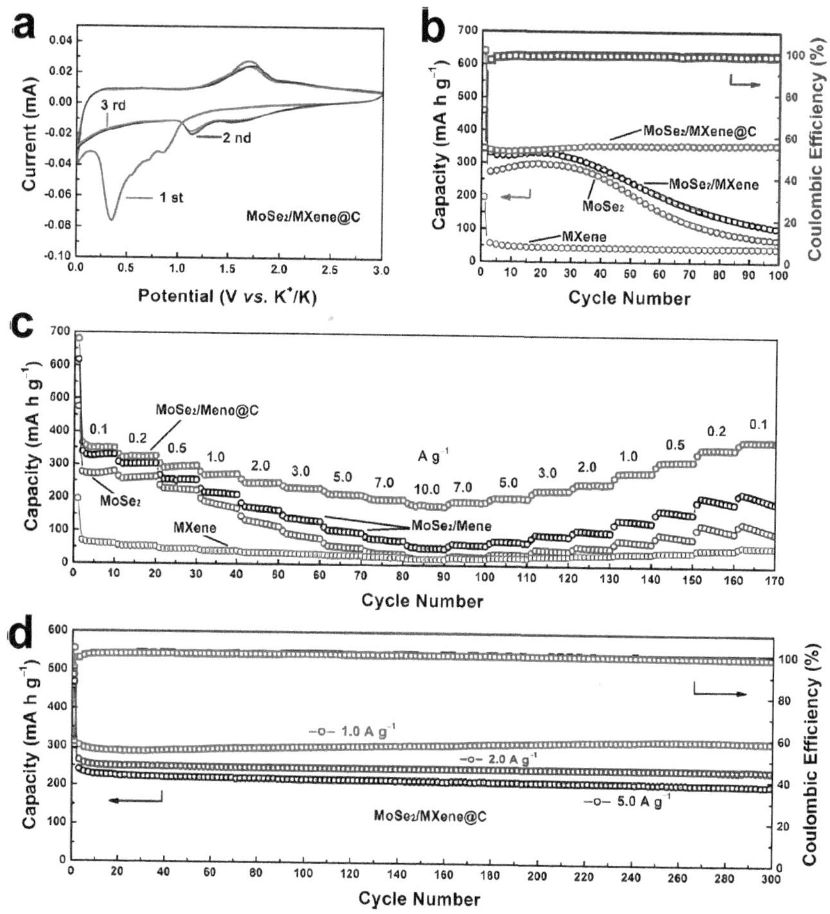

FIGURE 10.12 (a) CV curves for MoSe$_2$/MXene@C were obtained at a scanning rate of 0.1 mV s^{-1}. (b) Cycling performance at 200 mA g^{-1} of MoSe$_2$/MXene@C, MoSe$_2$/MXene, MoSe$_2$, and MXene was compared, along with the Coulombic efficiency of MoSe$_2$/MXene@C. (c) Comparison of rate performance. (d) Cycling stability and discharge capacities of MoSe$_2$/MXene@C.

Source: Reprinted with permission from [67] Huawen Huang, Jie Cui, Guoxue Liu, et al. 2019. Carbon-Coated MoSe$_2$/MXene Hybrid Nanosheets for Superior Potassium Storage. *ACS Nano*, 13, 3448–3456. Copyright 2019, American Chemical Society.

(0.01–3.00 V). Three distinct reduction peaks can be detected in the first cathodic sweep at 0.86, 0.72, and 0.35 V. The initial peak (0.86 V) might be attributed to the development of KxMoSe$_2$.

Due to the reduction of KxMoSe$_2$ to K$_5$Se$_3$ and metallic Mo, the next two peaks were observed, along with the formation of an SEI layer on the surface. Figure 10.12b depicts the cycling performance of this hybrid structure at 200 mA g^{-1}. After 100 cycles, the MoSe$_2$/MXene@C reserved 99.7% capacity retention and a reversible capacity of 355 mA h g^{-1}, indicating a highly stable cycling feature. Except for the initial cycles, around 99% CE were found in subsequent cycles as well. As seen in Fig. 10.12c, the rate performance of each of the as-prepared samples was evaluated at different current densities. At higher current densities, these results outperformed MoSe$_2$, MXene, and MoSe$_2$/MXene composites. However, a significant decrease in the current density was observed to cause the reversible capacity to start again. The cycling performances at 1.0, 2.0, and 5.0 A g^{-1} are shown in Fig. 10.12d [67].

10.10 HYBRID-ION BATTERIES

MXene-based materials have the ability to incorporate potassium ions, magnesium ions, aluminum ions, lithium ions, and sodium ions into their layers. This property makes them suitable for use in K$^+$, Mg^{2+}, and Al^{3+} cells. For instance, in a hybrid battery that uses Mg^{2+} and Li$^+$ ions, Ti$_3$C$_2$T$_x$/CNT composite was utilized as an electrode material. Since Mg^{2+} ions have a larger radius, their movement is slower than that of Li$^+$ ions. Consequently, Li$^+$ ions were the primary contributors to the capacitance. MXenes also work well in combination with high-temperature materials, as demonstrated by the polybenzimidazole/Ti$_3$C$_2$T$_x$ composite used in fuel cells. Additionally, MXene-based compounds exhibited better performance in PIBs. Ti$_3$CNT$_x$, for example, has been displayed to exhibit 20 mA g^{-1} after 100 cycles [1]. R. Meng et al. [74] developed composite anodes for LIBs and SIBs using black phosphorus quantum dots (BPQDs) and titanium carbide nanosheets (TNSs), incorporating a dual-model energy storage mechanism. The incorporation of BPQDs onto TNS, in particular, increased conductivity and reduced stress during cycling, resulting in a composite electrode with excellent battery performance. The lithium storage exhibited noteworthy features such as complete capacity retention (even after 2,400 cycles), remarkable high-rate capability, and a high capacity of 910 mAh g^{-1} at 100 mA g^{-1}.

Similarly, Q. Wang et al. [75] designed a zinc-ion hybrid SC (ZHSC) with a zinc foil anode and porous 3D Ti$_3$C$_2$T$_x$-rGO aerogel cathode. The ZHSC demonstrated exceptional electrochemical performance with a high specific capacitance (128.6 F g^{-1}), a high energy density, and 95% retention after 75,000 cycles. Figure 10.13a [75] presents the CV curves of rGO//ZnSO$_4$//Zn, MXene-rGO1//ZnSO$_4$//Zn, MXene-rGO2//ZnSO$_4$//Zn, and MXene-rGO3//ZnSO$_4$//Zn ZHSCs, at 2 mV s^{-1}. Figure 10.13b presents/illustrates the GCD curves of various ZHSCs at 0.4 A g^{-1}. The MXenerGO2//ZnSO$_4$//Zn ZHSC was observed to exhibit the longest discharge duration, and therefore, it showed the superior electrochemical performance. The CV curves of this ZHSC are depicted in Figs. 10.13c and d, at 2–100 mV s^{-1} that are close to rectangles, demonstrating excellent capacitive behavior. Furthermore, at varied current densities, all of the GCD curves are symmetrical triangular, as illustrated in Fig. 10.13e,

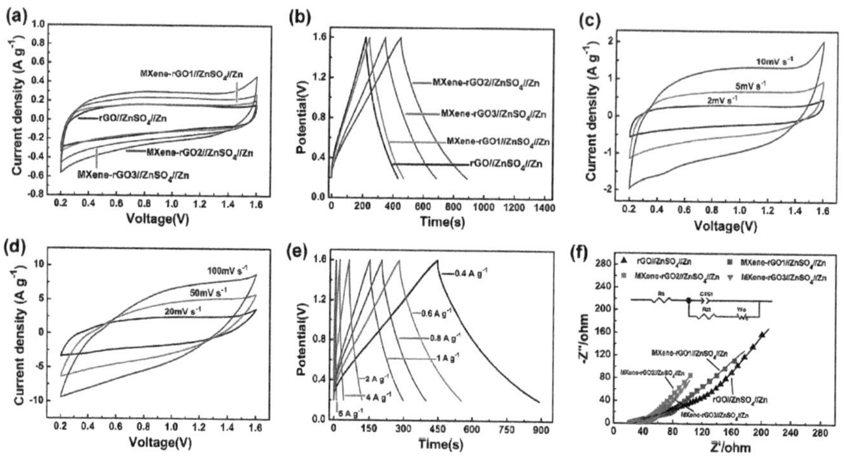

FIGURE 10.13 (a) CV plots were generated (at 2 mV s^{-1}) for the rGO//ZnSO$_4$//Zn and MXene-rGO//ZnSO$_4$//Zn ZHSCs. (b) At 0.4 A g^{-1}, GCD curves were also recorded. (c, d) CV curves were obtained at 2–100 mV s^{-1}, (e) GCD curves were obtained at 0.4–6 A g^{-1}, and (f) Nyquist plots for the MXene-rGO$_2$//ZnSO$_4$//Zn ZHSCs were constructed.

Source: [75] Wang Q., Wang S., Guo X. et al: MXene-Reduced Graphene Oxide Aerogel for Aqueous Zinc-Ion Hybrid Supercapacitor with Ultralong Cycle Life. *Adv. Electron. Mater.* 2019, 1900537. Copyright WILEY-VCH Verlag GmbH & Co. KGaA, Weinheim. Reproduced with permission.

indicating that the ZHSC is reversible. The electrochemical behaviors of ZHSCs are studied using electrochemical impedance spectroscopy (EIS). Figure 10.13f shows two EIS curves: a semicircle and a straight line [75].

The researchers demonstrated another unique FeS$_2$@MXene nanohybrid, created by sulfurizing a self-assembled iron hydroxide@MXene precursor in a single step. The nanohybrid demonstrated high-rate performance with LIBs and SIBs, achieving a reversible specific capacity of 762 mAh g^{-1} for Li-ion storage after 1,000 cycles. Likewise, for Na-ion storage, after 100 cycles, a reversible specific capacity of 563 mAh g^{-1} was reported [76]. In another study [77], researchers reported the use of Ti$_3$C$_2$T$_x$ as a cathode in a hybrid Mg^{2+}/Li$^+$ device, in conjunction with a Mg metal as an anode. The Ti$_3$C$_2$T$_x$/CNT composite paper demonstrated high capacity, delivering 100 and 50 mAh g^{-1} at different temperatures, and maintaining its capacity of 80 mAh g^{-1} after 500 cycles.

10.11   OTHER METAL-ION BATTERIES

A typical rechargeable magnesium-ion battery (MgIB) is made up of an abundant and environmentally friendly Mg anode, a Mg(AlCl$_2$BuEt)^2-based electrolyte, and a cathode. Usually, the reported cathodes are chalcogenides, polyanions, and Chevrel phases, and the majority of them have limited cycling life with low-rate capability. These limitations are due to their structures, weak conductivities, and inactive kinetics. In contrast to these materials, MXenes provide an active surface for Mg^{2+} storage, large interlayer spacing, and high electronic conductivity. Therefore, several studies were carried out by researchers to examine the MgIBs. For example, $Mg_{0.21}Ti_3C_2T_x$ (i.e., Mg^{2+} pre-intercalated porous films) demonstrated a high value capacity. However, after 100 cycles, the value of capacity reduced to 50 mAh g^{-1}, which shows that the design of electrode structure and overall optimization of the electrolyte with an electrode material are required [4, 6, 78].

M. Xu et al. [79] used a cationic surfactant cetyltrimethylammonium bromide to pre-intercalate into Ti_3C_2 MXene, resulting in an easy approach for high magnesium storage capability. According to the DFT calculations, intercalated cations considerably improve the insertion–desertion of Mg^{2+} ions between layers of the MXene, by reducing the diffusion barrier of these ions on the surface of material. As a result, these electrodes exhibited a high-rate performance at 50 mA g^{-1}. Y. Wang et al. [80] synthesized S-functionalized Ti_2C and investigated its features as anode materials for rechargeable MgIBs using first-principle calculations. Ti_2CS_2 monolayer's metallic character, great structural stability, and low diffusion barriers made it a viable anode material. Most importantly, the multilayered adsorption of Na and Mg ensured outstanding capacity of Ti_2CS_2 monolayer in SIBs and MgIBs, with 935.57 mA h g^{-1} and 1871.13 mA h g^{-1}, respectively. Similarly, as a cathode material, the MoS_2 NSs/MXene composite demonstrated exceptional rate-performance increased capacities of 165 mAh/g for MgIBs [81].

The nanolayered Ti_2NT_x was reported to exhibit high areal capacitances and reversible redox peaks (in 1 M $MgSO_4$ electrolyte), indicating their pseudocapacitive mechanism to store charge [82]. Figure 10.14 shows how capacitance retention varies with cycle number, indicating the electrochemical stability and offering more information about the pseudocapacitive process in the nanolayered Ti_2NT_x MXene. In each electrolyte, a capacitance measurement is made every 100 cycles, with 1,000 cycles of MXene at a scan rate of 50 mV s^{-1}. Most electrolytic solutions have a capacitance retention of more than 94% after 1,000 cycles, demonstrating that the Ti_2NT_x MXene is stable in aqueous conditions. The H_2SO_4 electrolyte exhibits a low capacitance retention of only 60% after 600 cycles (Fig. 10.14a [82]), indicating its use in strongly acidic environments. This could be because the nanolayered Ti_2NT_x oxidizes slowly in acidic conditions. Using GCD analysis, the study was conducted to investigate these nanolayered Ti_2NT_x MXenes in a Mg-ion electrolyte. Figure 10.14b displays the Ti_2NT_x MXene galvanostatic cycling in Mg-ion electrolyte at 5 A g^{-1} [82].

Calcium has a lower potential of –2.868 V vs. SHE, and it shows a lesser polarizing characteristic than magnesium. This indicates a tremendous potential for calcium-ion batteries (CIBs). However, reported metal oxide- and chalcogenide-based cathodes, like those used in MgIBs, could not sustain fast Ca^{2+} insertion and de-insertion without structural fatigue. To solve this issue, scientists have been exploring 2D MXene NSs, which are a developing material star. These NSs have huge interlayer spacing and strong conductivity, which opens up new possibilities for CIBs. In a pioneer work, first-principle simulations suggested that MXene NSs might improve calcium storage. The simulations showed that O-functionalized NSs tend to easily adsorb Ca^{2+}. Based on the data, it has been found that Ca can transmit 0.79 e per atom to Ti_2CO_2 surfaces. Ti_2CO_2 has a remarkable Ca^{2+} storage of 487 mAh g^{-1}, which is higher than the values for SIBs and PIBs. Although the theoretical results show that

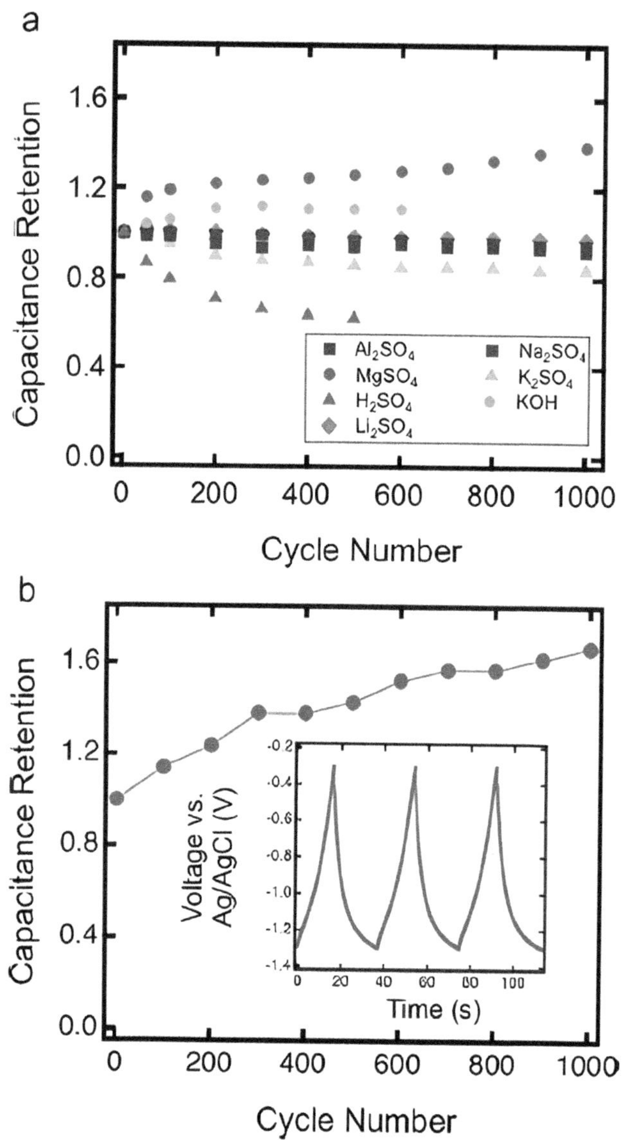

FIGURE 10.14 (a) The electrochemical behavior of nanolayered Ti_2NT_x electrodes were investigated in 1 M electrolytes. (b) Correlation between capacitance retention and cycle number for Ti_2NT_x.

Source: Reprinted with permission from [82] Abdoulaye Djire, Andre Bos, Jun Liu, et al. 2019. Pseudo-capacitive Storage in Nanolayered Ti_2NT_x MXene Using Mg-Ion Electrolyte. *ACS Appl. Nano Mater.*, 2, 2785–2795. Copyright 2019, American Chemical Society.

MXenes have the potential to be used in Ca^{2+} storage, more experimental studies and results are required to assess their true feasibility [6].

10.12 METAL–AIR BATTERIES

Z. Zeng et al. [83] reported a porous and highly conductive $N\text{-}CoSe_2/3D\ Ti_3C_2T_x$ composite with a large surface area. The intrinsic activity of $CoSe_2$, doping, and 3D structure of the MXene contribute to the excellent performance of the composite structure. Furthermore, when compared to combined Pt/C and RuO_2 batteries, Zn-air batteries with the developed material as the air cathode demonstrated a longer cycling life (over 500 cycles) and higher power and energy densities. In another study, O-functionalized Nb_2C as a cathode for lithium–oxygen batteries (LOBs) was demonstrated. DFT computations found that Nb_2CO_2 had a higher catalytic activity than other surface groups and bare surfaces. The cathodes exhibited cycling stability over 130 cycles at 3 A g^{-1} [84]. In Fig. 10.15a [84], the CV curves of a Nb_2C-based

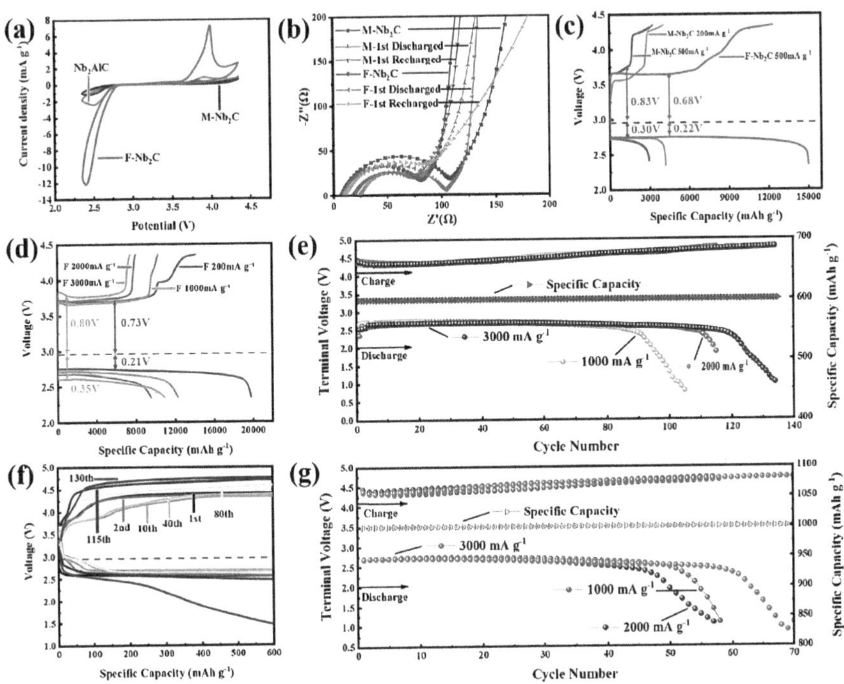

FIGURE 10.15 (a) CV curves were recorded for various samples at 0.1 mV s^{-1}. (b) EIS spectra for $M\text{-}Nb_2C$ and $F\text{-}Nb_2C$ electrodes. (c) Comparison of the initial GCD profiles of LOBs for M, $F\text{-}Nb_2C$ MXene at 500 mA g^{-1} and $M\text{-}Nb_2C$ MXene at 200 mA g^{-1}. (d) Initial GCD curves for the $F\text{-}Nb_2C$ MXene cathode. (f) Discharge–charge profiles for various cycles at 3 A g^{-1}. Terminal voltage of cycles for (e) 600 mAh g^{-1} and (g) 1,000 mAh g^{-1}.

Source: [84] Li, G., et al., *Highly efficient Nb₂C MXene cathode catalyst with uniform O-terminated surface for lithium—oxygen batteries*. Advanced Energy Materials, 2020: p. 2002721. Copyright WILEY-VCH Verlag GmbH & Co. KGaA, Weinheim. Reproduced with permission.

cathode are presented, recorded in a 1.0 M $LiNO_3$/dimethyl sulfoxide (DMSO) electrolyte. The voltage cutoff has a scanning rate of 0.1 mV s^{-1} and ranges from 2.35 V to 4.35 V. The use of DMSO in the electrolyte aims to preserve the layer structure of the MXene. Figure 10.15b depicts the EIS spectra obtained both before and after the initial cycle, encompassing a frequency range spanning from 105 to 0.1 Hz. Furthermore, Fig. 10.15c displays the GCD curves of cathodes based on F-Nb_2C and M-Nb_2C, at 500 mA g^{-1}.

For the oxygen reduction reaction/oxygen evolution reaction process, the F-Nb_2C MXene cathode demonstrated a high initial discharge–charge capacity and minimal overpotentials (0.22/0.68 V). As a result, the discharge to charge voltage ratio yielded a round-trip efficiency of 75.3% and a coulombic efficiency of 82.0%. A high-rate performance of the F-Nb_2C cathode is depicted in Fig. 10.15d across various current densities within a voltage range of 2.35–4.435 V. The observed decrease in charge polarization and increase in overpotentials at higher current density were suggested due to the internal charge transfer resistance of cell. The study investigated the cycling stability of a Nb_2C MXene cathode under various current densities with fixed capacity. The results are presented in Figs. 10.15e–g, along with the consistent discharge–charge curves in Fig. 10.15f [84]. Similarly, a recent study explored the potential of MXene-based cathodes for rechargeable LOBs. The study explained trends in the thermodynamic nucleation overpotential, which was calculated using the free energy of the adsorbed intermediate LiO_2*. The investigation revealed that $Ni_4N_3(OH)_2$ had the lowest nucleation overpotential of all the analyzed MXenes, indicating that mixed-transition metal MXenes are a promising exploration direction [85].

REFERENCES

[1] Yang, J., et al., *MXene-based composites: synthesis and applications in rechargeable batteries and supercapacitors.* Advanced Materials Interfaces, 2019. **6**(8): p. 1802004.

[2] Xu, X., et al., *MXenes with applications in supercapacitors and secondary batteries: a comprehensive review.* Materials Reports: Energy, 2022. **2**(1): p. 100080.

[3] Dey, A., et al., *Doped MXenes—a new paradigm in 2D systems: synthesis, properties and applications.* Progress in Materials Science, 2023: p. 101166.

[4] Ming, F., et al., *MXenes for rechargeable batteries beyond the lithium-ion.* Advanced Materials, 2021. **33**(1): p. 2004039.

[5] Zhang, C., et al., *Two-dimensional transition metal carbides and nitrides (MXenes): synthesis, properties, and electrochemical energy storage applications.* Energy & Environmental Materials, 2020. **3**(1): p. 29–55.

[6] Dong, Y., H. Shi, and Z.S. Wu, *Recent advances and promise of MXene-based nanostructures for high-performance metal ion batteries.* Advanced Functional Materials, 2020. **30**(47): p. 2000706.

[7] Dong, H., et al., *Molten salt derived Nb_2CT_x MXene anode for Li-ion batteries.* ChemElectroChem, 2021. **8**(5): p. 957–962.

[8] Liu, R., et al., *Nitrogen-doped Nb_2CT_x MXene as anode materials for lithium ion batteries.* Journal of Alloys and Compounds, 2019. **793**: p. 505–511.

[9] Naguib, M., et al., *New two-dimensional niobium and vanadium carbides as promising materials for Li-ion batteries.* Journal of the American Chemical Society, 2013. **135**(43): p. 15966–15969.

[10] Zhao, S., et al., *Li-ion uptake and increase in interlayer spacing of Nb_4C_3 MXene.* Energy Storage Materials, 2017. **8**: p. 42–48.

[11] Zhou, J., et al., *Synthesis and lithium ion storage performance of two-dimensional V$_4$C$_3$ MXene.* Chemical Engineering Journal, 2019. **373**: p. 203–212.

[12] Wang, C., et al., *Atomic cobalt covalently engineered interlayers for superior lithium-ion storage.* Advanced Materials, 2018. **30**(32): p. 1802525.

[13] Wang, C., et al., *Atomic Sn^{4+} decorated into vanadium carbide MXene interlayers for superior lithium storage.* Advanced Energy Materials, 2019. **9**(4): p. 1802977.

[14] Guo, X., et al., *Constructing P-doped self-assembled V$_2$C MXene/NiCo-layered double hydroxide hybrids toward advanced lithium storage.* Materials Advances, 2023. **4**(6): p. 1523–1533.

[15] Liu, F., et al., *Preparation of high-purity V$_2$C MXene and electrochemical properties as Li-ion batteries.* Journal of the Electrochemical Society, 2017. **164**(4): p. A709.

[16] Zhou, J., et al., *Ti-enhanced exfoliation of V$_2$AlC into V$_2$C MXene for lithium-ion battery anodes.* Ceramics International, 2017. **43**(14): p. 11450–11454.

[17] Wang, Y., et al., *Preparation of (Vx, Ti$_{1-x}$)$_2$C MXenes and their performance as anode materials for LIBs.* Journal of Materials Science, 2019. **54**: p. 11991–11999.

[18] Guo, X., et al., *Vacancy manipulating of molybdenum carbide MXenes to enhance Faraday reaction for high performance lithium-ion batteries.* Nano Research Energy, 2022. **1**(3): p. e9120026.

[19] Bai, J., et al., *Construction of hierarchical V$_4$C$_3$-MXene/MoS$_2$/C nanohybrids for high rate lithium-ion batteries.* Nanoscale, 2020. **12**(2): p. 1144–1154.

[20] Syamsai, R., et al., *Double transition metal MXene (Ti$_x$Ta$_{4-x}$C$_3$) 2D materials as anodes for Li-ion batteries.* Scientific Reports, 2021. **11**(1): p. 688.

[21] Guo, Y., et al., *MXene-encapsulated hollow Fe$_3$O$_4$ nanochains embedded in N-doped carbon nanofibers with dual electronic pathways as flexible anodes for high-performance Li-ion batteries.* Nanoscale, 2021. **13**(8): p. 4624–4633.

[22] Zhang, S., et al., *Vapor deposition red phosphorus to prepare nitrogen-doped Ti$_3$C$_2$T$_x$ MXenes composites for lithium-ion batteries.* The Journal of Physical Chemistry Letters, 2019. **10**(21): p. 6446–6454.

[23] Chen, C., et al., *MoS$_2$-on-MXene heterostructures as highly reversible anode materials for lithium-ion batteries.* Angewandte Chemie International Edition, 2018. **57**(7): p. 1846–1850.

[24] Zhao, Q., et al., *Flexible 3D porous MXene foam for high-performance lithium-ion batteries.* Small, 2019. **15**(51): p. 1904293.

[25] Pang, J., et al., *Applications of 2D MXenes in energy conversion and storage systems.* Chemical Society Reviews, 2019. **48**(1): p. 72–133.

[26] Song, Y., et al., *Rational design of porous nitrogen-doped Ti$_3$C$_2$ MXene as a multifunctional electrocatalyst for Li–S chemistry.* Nano Energy, 2020. **70**: p. 104555.

[27] Wang, J., et al., *Rational design of porous N-Ti$_3$C$_2$ MXene@ CNT microspheres for high cycling stability in Li-S battery.* Nano-Micro Letters, 2020. **12**: p. 1–14.

[28] Tang, H., et al., *A robust, freestanding MXene-sulfur conductive paper for long-lifetime Li-S batteries.* Advanced Functional Materials, 2019. **29**(30): p. 1901907.

[29] Wang, Z., et al., *VO$_2$(p)-V$_2$C (MXene) grid structure as a lithium polysulfide catalytic host for high-performance Li-S battery.* ACS Applied Materials & Interfaces, 2019. **11**(47): p. 44282–44292.

[30] Bao, W., et al., *Facile synthesis of crumpled nitrogen-doped MXene nanosheets as a new sulfur host for lithium–sulfur batteries.* Advanced Energy Materials, 2018. **8**(13): p. 1702485.

[31] Yang, C., et al., *In situ N-doped CoS$_2$ anchored on MXene toward an efficient bifunctional catalyst for enhanced lithium-sulfur batteries.* Chemical Engineering Journal, 2022. **427**: p. 131792.

[32] Wang, J., et al., *MXene-based co, N-codoped porous carbon nanosheets regulating polysulfides for high-performance lithium–sulfur batteries.* ACS Applied Materials & Interfaces, 2019. **11**(42): p. 38654–38662.

[33] Du, C., et al., *Embedding S@TiO₂ nanospheres into MXene layers as high rate cyclability cathodes for lithium-sulfur batteries.* Electrochimica Acta, 2019. **295**: p. 1067–1074.

[34] Xiao, Z., et al., *Ultrafine Ti₃C₂ MXene nanodots-interspersed nanosheet for high-energy-density lithium–sulfur batteries.* ACS Nano, 2019. **13**(3): p. 3608–3617.

[35] Gao, X.T., et al., *Ultrathin MXene nanosheets decorated with TiO₂ quantum dots as an efficient sulfur host toward fast and stable Li-S batteries.* Small, 2018. **14**(41): p. 1802443.

[36] Zhou, J., et al., *Hierarchical porous and three-dimensional MXene/SiO₂ hybrid aerogel through a sol-gel approach for lithium–sulfur batteries.* Molecules, 2022. **27**(20): p. 7073.

[37] Shen, K., B. Li, and S. Yang, *3D printing dendrite-free lithium anodes based on the nucleated MXene arrays.* Energy Storage Materials, 2020. **24**: p. 670–675.

[38] Li, B., et al., *Flexible Ti₃C₂ MXene-lithium film with lamellar structure for ultrastable metallic lithium anodes.* Nano Energy, 2017. **39**: p. 654–661.

[39] Zhang, F., et al., *Interface engineering of MXene composite separator for high-performance Li–Se and Na–Se batteries.* Advanced Energy Materials, 2020. **10**(20): p. 2000446.

[40] Bhat, A., et al., *Prospects challenges and stability of 2D MXenes for clean energy conversion and storage applications.* npj 2D Materials and Applications, 2021. **5**(1): p. 61.

[41] Aslam, M.K., Y. Niu, and M. Xu, *MXenes for non-lithium-ion (Na, K, Ca, Mg, and Al) batteries and supercapacitors.* Advanced Energy Materials, 2021. **11**(2): p. 2000681.

[42] Aslam, M.K. and M. Xu, *A mini-review: MXene composites for sodium/potassium-ion batteries.* Nanoscale, 2020. **12**(30): p. 15993–16007.

[43] Wu, Y., et al., *2D MXene/SnS₂ composites as high-performance anodes for sodium ion batteries.* Chemical Engineering Journal, 2018. **334**: p. 932–938.

[44] Wu, F., et al., *A 3D flower-like VO₂/MXene hybrid architecture with superior anode performance for sodium ion batteries.* Journal of Materials Chemistry A, 2019. **7**(3): p. 1315–1322.

[45] Tao, M., et al., *A chemically bonded CoNiO₂ nanoparticles/MXene composite as anode for sodium-ion batteries.* Materials Letters, 2018. **230**: p. 173–176.

[46] Zhang, Y., et al., *Circuit board-like CoS/MXene composite with superior performance for sodium storage.* Chemical Engineering Journal, 2019. **357**: p. 220–225.

[47] Sun, S., et al., *Hybrid energy storage mechanisms for sulfur-decorated Ti₃C₂ MXene anode material for high-rate and long-life sodium-ion batteries.* Chemical Engineering Journal, 2019. **366**: p. 460–467.

[48] Li, J., et al., *Improved sodium-ion storage performance of Ti₃C₂Tx MXenes by sulfur doping.* Journal of Materials Chemistry A, 2018. **6**(3): p. 1234–1243.

[49] Ding, Y., et al., *Facile synthesis of FePS₃ nanosheets@ MXene composite as a high-performance anode material for sodium storage.* Nano-Micro Letters, 2020. **12**: p. 1–12.

[50] Wu, Y., et al., *Few-layer MXenes delaminated via high-energy mechanical milling for enhanced sodium-ion batteries performance.* ACS Applied Materials & Interfaces, 2017. **9**(45): p. 39610–39617.

[51] Xie, X., et al., *Porous heterostructured MXene/carbon nanotube composite paper with high volumetric capacity for sodium-based energy storage devices.* Nano Energy, 2016. **26**: p. 513–523.

[52] Guo, X., et al., *Sb₂O₃/MXene (Ti₃C₂Tₓ) hybrid anode materials with enhanced performance for sodium-ion batteries.* Journal of Materials Chemistry A, 2017. **5**(24): p. 12445–12452.

[53] Zhang, Y., et al., *Synthesis of SnS nanoparticle-modified MXene (Ti₃C₂Tₓ) composites for enhanced sodium storage.* Journal of Alloys and Compounds, 2018. **732**: p. 448–453.

[54] Liu, H., et al., *Bare Mo-based ordered double-transition metal MXenes as high-performance anode materials for aluminum-ion batteries.* The Journal of Physical Chemistry C, 2020. **124**(47): p. 25769–25774.

[55] VahidMohammadi, A., et al., *Two-dimensional vanadium carbide (MXene) as a high-capacity cathode material for rechargeable aluminum batteries.* ACS Nano, 2017. **11**(11): p. 11135–11144.

[56] Jiang, W., et al., *Molten salt thermal treatment synthesis of S-doped V_2CT_x and its performance as a cathode in aqueous Zn-ion batteries.* ACS Applied Materials & Interfaces, 2022. **14**(12): p. 14482–14491.

[57] Liu, Y., et al., *In-situ electrochemically activated surface vanadium valence in V_2C MXene to achieve high capacity and superior rate performance for Zn-ion batteries.* Advanced Functional Materials, 2021. **31**(8): p. 2008033.

[58] Zhu, X., et al., *Superior-performance aqueous zinc-ion batteries based on the in situ growth of MnO_2 nanosheets on V_2CT_x MXene.* ACS Nano, 2021. **15**(2): p. 2971–2983.

[59] Wang, S., et al., *A new free-standing aqueous zinc-ion capacitor based on MnO_2—CNTs cathode and MXene anode.* Nano-Micro Letters, 2019. **11**: p. 1–12.

[60] Li, X., et al., *In situ electrochemical synthesis of MXenes without acid/alkali usage in/for an aqueous zinc ion battery.* Advanced Energy Materials, 2020. **10**(36): p. 2001791.

[61] Yang, Q., et al., *A wholly degradable, rechargeable $Zn–Ti_3C_2$ MXene capacitor with superior anti-self-discharge function.* ACS Nano, 2019. **13**(7): p. 8275–8283.

[62] Zhang, N., et al., *Direct self-assembly of MXene on Zn anodes for dendrite-free aqueous zinc-ion batteries.* Angewandte Chemie International Edition, 2021. **60**(6): p. 2861–2865.

[63] Liu, C., et al., *Highly stable $H_2V_3O_8$/Mxene cathode for Zn-ion batteries with superior rate performance and long lifespan.* Chemical Engineering Journal, 2021. **405**: p. 126737.

[64] Wu, Y., et al., *MXenes: advanced materials in potassium ion batteries.* Chemical Engineering Journal, 2021. **404**: p. 126565.

[65] Lian, P., et al., *Alkalized Ti_3C_2 MXene nanoribbons with expanded interlayer spacing for high-capacity sodium and potassium ion batteries.* Nano Energy, 2017. **40**: p. 1–8.

[66] Tang, X., et al., *MXene-based dendrite-free potassium metal batteries.* Advanced Materials, 2020. **32**(4): p. 1906739.

[67] Huang, H., et al., *Carbon-coated $MoSe_2$/MXene hybrid nanosheets for superior potassium storage.* ACS Nano, 2019. **13**(3): p. 3448–3456.

[68] Zhao, R., et al., *Encapsulating ultrafine Sb nanoparticles in Na+ pre-intercalated 3D porous $Ti_3C_2T_x$ MXene nanostructures for enhanced potassium storage performance.* ACS Nano, 2020. **14**(10): p. 13938–13951.

[69] Cao, J., et al., *Microbe-assisted assembly of $Ti_3C_2T_x$ MXene on fungi-derived nanoribbon heterostructures for ultrastable sodium and potassium ion storage.* ACS Nano, 2021. **15**(2): p. 3423–3433.

[70] Fang, Y., et al., *MXene-derived TiO_2/reduced graphene oxide composite with an enhanced capacitive capacity for Li-ion and K-ion batteries.* Journal of Materials Chemistry A, 2019. **7**(10): p. 5363–5372.

[71] Xu, X., et al., *Orthorhombic cobalt ditelluride with Te vacancy defects anchoring on elastic MXene enables efficient potassium-ion storage.* Advanced Materials, 2021. **33**(31): p. 2100272.

[72] Tao, M., et al., *TiOxNy nanoparticles/c composites derived from MXene as anode material for potassium-ion batteries.* Chemical Engineering Journal, 2019. **369**: p. 828–833.

[73] Naguib, M., et al., *Electrochemical performance of MXenes as K-ion battery anodes.* Chemical Communications, 2017. **53**(51): p. 6883–6886.

[74] Meng, R., et al., *Black phosphorus quantum dot/Ti_3C_2 MXene nanosheet composites for efficient electrochemical lithium/sodium-ion storage.* Advanced Energy Materials, 2018. **8**(26): p. 1801514.

[75] Wang, Q., et al., *MXene-reduced graphene oxide aerogel for aqueous zinc-ion hybrid supercapacitor with ultralong cycle life.* Advanced Electronic Materials, 2019. **5**(12): p. 1900537.

[76] Du, C.-F., et al., *Porous MXene frameworks support pyrite nanodots toward high-rate pseudocapacitive Li/Na-ion storage.* ACS Applied Materials & Interfaces, 2018. **10**(40): p. 33779–33784.

[77] Byeon, A., et al., *Two-dimensional titanium carbide MXene as a cathode material for hybrid magnesium/lithium-ion batteries.* ACS Applied Materials & Interfaces, 2017. **9**(5): p. 4296–4300.

[78] Garg, R., A. Agarwal, and M. Agarwal, *A review on MXene for energy storage application: effect of interlayer distance.* Materials Research Express, 2020. **7**(2): p. 022001.

[79] Xu, M., et al., *Opening magnesium storage capability of two-dimensional MXene by intercalation of cationic surfactant.* ACS Nano, 2018. **12**(4): p. 3733–3740.

[80] Wang, Y., et al., *Achieving superior high-capacity batteries with the lightest Ti_2C MXene anode by first-principles calculations: overarching role of S-functionate (Ti_2CS_2) and multivalent cations carrier.* Journal of Power Sources, 2020. **451**: p. 227791.

[81] Xu, M., et al., *Synthesis of MXene-supported layered MoS_2 with enhanced electrochemical performance for Mg batteries.* Chinese Chemical Letters, 2018. **29**(8): p. 1313–1316.

[82] Djire, A., et al., *Pseudocapacitive storage in nanolayered Ti_2NT_x MXene using Mg-ion electrolyte.* ACS Applied Nano Materials, 2019. **2**(5): p. 2785–2795.

[83] Zeng, Z., et al., *Bifunctional N-$CoSe_2$/3D-MXene as highly efficient and durable cathode for rechargeable Zn–air battery.* ACS Materials Letters, 2019. **1**(4): p. 432–439.

[84] Li, G., et al., *Highly efficient Nb_2C MXene cathode catalyst with uniform O-terminated surface for lithium—oxygen batteries.* Advanced Energy Materials, 2021. **11**(1): p. 2002721.

[85] Lee, A., D. Krishnamurthy, and V. Viswanathan, *Exploring MXenes as cathodes for non-aqueous lithium—oxygen batteries: design rules for selectively nucleating Li_2O_2.* ChemSusChem, 2018. **11**(12): p. 1911–1918.

11 Future Perspectives

MXenes possess unique properties, which makes them a promising material for several applications. Their scalable synthesis has also contributed to their demand for research and development, as can be observed by the significant number of research papers and patents published in recent years. Additionally, MXenes have undergone extensive study through first-principle calculations, which have played an important role in understanding and predicting their properties. The composition of MXenes can be varied through changes in M, X, n, Tx, and intercalants. MXenes are becoming increasingly popular with countless possible compositions due to ordered structures and solid solutions on the M, X, and T sites. Depending upon the types of compositions, it can be expected that many new stable structures will be developed in the future. Although theoretical studies showed that nitride MXenes exhibit better electronic stability and high values of capacitance, there are very few experimental studies on these materials compared to carbides. There has been minimal success in synthesizing MXenes with $X = B$ (MBenes), while other 2D borides that have been predicted theoretically are still waiting to be synthesized. So, future researchers should focus on synthesizing these new compositions, to expand the family of 2D MXenes.

MXenes can be made hydrophilic by adding hydroxyl, oxygen, or fluorine terminals during the synthesis process. This makes them compatible with water-based colloids, which do not require surfactants for stabilization due to the hydrophilicity and high surface charge of the materials. Different synthesis techniques can be used to optimize the surface termination alterations and shape of MXenes, making them versatile for different applications such as supercapacitors, alkaline ion batteries, and more. MXenes' layered structure, distinctive surface chemistry, high carrier density, and photoelectric properties contribute to their performance in various applications, and their conductivity is determined by the etching and stacking procedures used, as well as post-synthesis processing. MXenes also show high sensitivity to external stimuli, making them a promising material for chemical, light, and pressure sensors. The work function and flexible bandgap allow these materials to be utilized in photocatalysis. Similarly, the inter-band transitions of MXene-based quantum dots result in biocompatibility, making them suitable for white laser production and high-resolution cellular imaging [1, 2].

It is undeniable that more MXenes will be discovered in the future, but it is crucial to focus on important issues that this community has not addressed yet. A team of renowned MXene researchers created a list of questions related to these materials that need to be answered. The list was reviewed and ranked by hundreds of representatives during the third International Conference on MXenes in October 2020. The topmost priorities for promoting the expansion of this family and moving toward practical applications were the following: environmentally friendly, scalable, and safe synthesis methods; control of surface chemistry; improvement in chemical

DOI: 10.1201/9781003465768-11

stability; and understanding of their physical properties. Various types of MXenes have been synthesized till date. However, except for nitrides and carbonitrides, most of the MXenes are carbides. Only a few MXenes, such as $Ti_3C_2T_x$ and Nb_2CT_x, have been studied for their oxidation performance. Despite the theoretical calculations based on binding energy, there is still inadequate understanding of the stability differences between different MXenes. Therefore, efficient studies on the synthesis of MXenes with different values of M, multi-M, and X are essential for advancing our knowledge of their intrinsic stability [2, 3].

Several important characteristics of MXenes remain unexplored. One of these is the nature of their surface termination, which is currently a topic of debate. It is imperative to develop a uniform strategy for terminating the same groups (such as hydrogen, oxygen, hydroxyl, or fluorine) for different purposes. Additionally, approaches for size selection should be introduced to create MXene flakes with adjustable thickness and size. Despite the documentation of various synthetic processes like chemical vapor deposition, some manufacturing processes still use harmful etching agents such as F-containing acidic solutions. The preparation procedure for MXene synthesis is associated with substantial safety dangers and environmental contamination issues, making it difficult to implement in large-scale industrial production. Therefore, researchers should focus on developing safe and environmentally friendly etching agents for this purpose. Most of the unique electrical (insulator or semiconductor) and magnetic properties (e.g., half-metallicity) have been predicted for pristine MXenes. But the synthesis of pristine or uniformly terminated MXenes is found to be challenging. To address this challenge, it is particularly necessary to develop new etching approaches or post-processing methods that offer improved surface functionalization control [1, 3].

Despite the fact that diverse chemistry of MXenes makes their properties adjustable for a wide range of applications, there are several obstacles and opportunities for future study in case of doped MXenes. While the literature shows that the active sites are the dopant atoms, we must not overlook MXenes' complex surface chemistry, which provides a range of adsorption pathways. As a result, several reactions will take place due to the synergistic effect of elements or components being doped on the surface of MXenes, other than the desired reaction. Moreover, the studies on doped MXenes largely depend on theoretical research, particularly with latest catalyst designs (such as bifunctional or trifunctional catalysts). These studies simulate the dynamic behavior of MXenes during catalysis and predict the mechanisms of reaction along with the intermediate products. The results of these findings will help the researchers develop novel doped MXenes with better electrocatalytic activity. Additionally, due to the significant influence of the synthesis procedures, it is essential to develop new approaches to produce high-quality MXenes with uniform surface terminations [4].

Understanding the role of chemistry and structure in properties like transport, magnetic behavior, and electrochemistry is crucial for designing materials with specific desired qualities. However, most studies focus on pristine or uniformly functionalized MXenes, which is not entirely representative of the experimental conditions. Therefore, it is essential to model the experimental settings accurately while considering heterogeneous terminations in calculations to better represent the real situation and retrieve the experimental studies. Furthermore, while most theoretical investigations use density functional theory (DFT) computations, only a few ab initio molecular dynamics simulations have been conducted. In order to more accurately model chemical etching, it is important to develop or utilize more advanced tools. Additionally, it is necessary to investigate transport parameters such as thermal transport, thermoelectric power factor, and electronic bandgap. Despite predictions from DFT, the intrinsic MXene semiconductor has yet to be experimentally achieved. By overcoming these challenges and improving MXene synthesis and simulations, performance of many applications can be improved [2, 3].

In recent years, there has been an increasing interest in the studies of the magnetic properties of MXenes. These semimetallic crystals have ferromagnetic properties and possess spin-polarized electrons at the Fermi level, making them useful for electronics applications. Recent computational research has shown that Janus MXenes can breach symmetry by surface functionalization, leading to increased orderliness in ferromagnetic and antiferromagnetic materials. The magnetic characteristics of MXenes can be altered by modifying their top and bottom layers. However, the discovery of RE-based i-MAX phases with varying magnetic ordering has opened up new possibilities. By etching them into corresponding 2D MXenes, the intrinsic magnetic properties of these 2D crystals can be expanded and diversified. Additionally, it is important to define thoroughly the optical and thermal properties of all MXenes, along with the stability of van der Waals (vdW) heterostructures. To effectively anticipate both electrical and elastic properties, simulations should include vdW interactions in this context [1, 3].

The choice of transition metals and surface groups has a direct influence on the stability of MXenes. MXenes experience structural changes at high temperatures due to the molecular hydrogen trapped between the nanosheets, even when no external oxidant is supplied. The oxidation process can start at the most sensitive surface group sites, such as $-OH$. To improve their thermal stability, MXenes can undergo annealing in an inert gas or vacuum to remove surface groups. Heat treatment can lead to changes in the crystal structure of materials, such as reducing the interlayer spacing, healing defects, and transforming pairing. Additionally, MXenes that are terminated with $-Cl$ have greater stability compared to those terminated with $-F$ and $=O$. There may be novel termination systems yet to be discovered that could create MXenes with even better performance. Surface terminations on MXenes (e.g., $T = Cl, Br, Se, S,$ or Te) can be adjusted by substituting or eliminating components in various molten inorganic salt compositions [2, 5]. It is essential to address the holdup of oxidation before developing MXenes and their derivatives. The oxidation rate is significantly higher in their aqueous dispersions than in their freestanding film form. However, the mechanism by which MXenes oxidize in this state remains unknown. An initial study of $Ti_3C_2T_x$ revealed that oxygen and water initiate the oxidation, which was later lessened to water as the main reason for material hydrolysis. It is inspiring to note that some of the issues raised less than a year ago are already being resolved. However, the long-term effectiveness of techniques to avoid oxidation is still uncertain, which needs to be addressed in the future [2, 6].

In order to develop a particular crystal structure with specific properties and fewer defects, it is necessary to have a detailed understanding of optimized synthesis techniques to achieve maximum stability. In addition, to have a comprehensive knowledge of the hydration layer creation on MXenes, it is important to investigate the role of water molecules and their interactions within the bulk and surface, as well as the interaction of surface termination. At present, studies on thermal stability are limited to a few members. It is expected that using nonstoichiometric transition metal carbide phase diagrams to model and predict specific phase stabilities will help estimate the thermal transitions of MXenes. The use of antioxidation techniques in stabilizing MXene materials heavily relies on the capping of their surfaces. To better understand the molecular interactions between antioxidants and MXenes, it is necessary to conduct more systematic studies on the stereospecific orientation of antioxidants over the surface of MXenes. Additionally, the concentration of the antioxidants and their removal are also a cause for concern. Hence, there is a need to focus research efforts on discovering more antioxidant molecules that can efficiently stabilize MXenes at lower concentrations, with ease of removal or without the need for removal. Over the last decade, research on MXene materials has grown exponentially, resulting in advancements in nanotechnology and materials engineering [6].

MXene has become a popular material in catalysis due to its exceptional properties and performance. However, restacking can hinder ion and electron transport in the electrode, which leads to poor charge transfer behavior and limits its applicability. To enhance the performance of MXene catalysts, several strategies have been proposed, including functional group control, reducing the defects, 2D material confinement effects, and heterostructure fabrication. For example, different etchants can induce different groups to terminate the surface of MXene, changing its intrinsic properties and thus improving catalytic efficacy. As a result, more research is required to gain a complete understanding of catalytic mechanism of MXenes in order to rationally select the modification approach, perform comparison, combination, and identify the optimal answer in continuous tests. MXenes' photocatalytic and electrocatalytic performance for fuel production is largely theoretical. Theoretical simulations show that hydrogen generation, CO_2 hydrogenation, and ammonia fuel production are all energetically favorable reaction pathways for catalytic fuel production [7, 8].

Several investigations have shown that the electrochemical performance of MXenes in H_2SO_4 solution can be significantly influenced by their surface chemical composition and state. However, currently available methods for distinguishing among hydrogen, hydroxyl, oxygen, or fluorine is not reliable. To achieve a qualitative increase in capacitance, a controlled strategy is being developed to achieve consistent terminations with lower electronegativity. MXenes without terminations have not yet been synthesized, but they may exhibit fascinating electrochemical properties. It is also important to understand why pseudocapacitance is exclusively observed in H_2SO_4 solutions. In addition to H_2SO_4 solution, we should explore more environmentally friendly electrolytes. MXenes have been extensively studied for their electrochemical performance in aqueous solutions, but their performance in organic solutions or ionic liquids has received less attention, particularly for supercapacitor applications. Nonaqueous electrolytes offer a wider voltage window, resulting in a higher energy density. Therefore, it is important to shift the focus of MXene research in supercapacitor applications toward organic or ionic liquid systems. Additionally, further examination is required to determine the energy storage mechanism and ion intercalation dynamics in these systems [9].

MXenes have potential applications in energy storage, especially for Li-ion and Na-ion intercalation. It is important to minimize irreversibility during the first cycle of intercalation. Understanding how to intercalate multivalent ions such as Al^{3+}, Mg^{2+}, and Zn^{2+} and large organic ions is critical for developing electrodes for next-generation batteries. Theoretical predictions will expand our knowledge of MXene attributes and lead to the growth of this family through diverse compositions. To enhance the capabilities of MXenes as anodes for lithium-ion batteries, heterostructures of MXenes with different elements and compounds can be used. MXene composites play a crucial role in the performance enhancement of Li-S batteries by binding lithium polysulfides (LiPSs). By modifying the structure of MXene-based composites, the LiPSs and sulfur content can be varied. To achieve significant sulfur loading, hollow porous MXene-based composites were produced via the template approach [10]. To conclude, more MXene-based composites should be developed to enhance the electrochemical performance of different metal-ion and metal–air batteries. Currently, new hybrid structures are being developed, named as supercabattery, which combine the electrochemical properties of batteries and supercapacitors. These devices deliver high energy and power densities. Therefore, further studies are needed to develop more of these devices in the future.

REFERENCES

[1] Meng, W., et al., *Advances and challenges in 2D MXenes: from structures to energy storage and conversions.* Nano Today, 2021. **40**: p. 101273.

[2] Naguib, M., M.W. Barsoum, and Y. Gogotsi, *Ten years of progress in the synthesis and development of MXenes.* Advanced Materials, 2021. **33**(39): p. 2103393.

[3] Champagne, A. and J.-C. Charlier, *Physical properties of 2D MXenes: from a theoretical perspective.* Journal of Physics: Materials, 2020. **3**(3): p. 032006.

[4] Dey, A., et al., *Doped MXenes—a new paradigm in 2D systems: synthesis, properties and applications.* Progress in Materials Science, 2023: p. 101166.

[5] Cao, F., et al., *Recent advances in oxidation stable chemistry of 2D MXenes.* Advanced Materials, 2022. **34**(13): p. 2107554.

[6] Razium, A.S. and F. Baomin, *Progression in the Oxidation Stability of MXenes.* 2023, Springer.

[7] Pang, J., et al., *Applications of 2D MXenes in energy conversion and storage systems.* Chemical Society Reviews, 2019. **48**(1): p. 72–133.

[8] Ahmad Junaidi, N.H., et al., *A comprehensive review of MXenes as catalyst supports for the oxygen reduction reaction in fuel cells.* International Journal of Energy Research, 2021. **45**(11): p. 15760–15782.

[9] Hu, M., et al., *Emerging 2D MXenes for supercapacitors: status, challenges and prospects.* Chemical Society Reviews, 2020. **49**(18): p. 6666–6693.

[10] Yang, J., et al., *MXene-based composites: synthesis and applications in rechargeable batteries and supercapacitors.* Advanced Materials Interfaces, 2019. **6**(8): p. 1802004.

Index

For Product Safety Concerns and Information please contact our
EU representative GPSR@taylorandfrancis.com Taylor & Francis
Verlag GmbH, Kaufingerstraße 24, 80331 München, Germany